PWS PUBLISHERS

Prindle, Weber & Schmidt · ✿ · Willard Grant Press · **wG** · Duxbury Press · ♠
Statler Office Building · 20 Providence Street · Boston, Massachusetts 02116

PWS Publishers is a division of Wadsworth, Inc.

Library of Congress Cataloging in Publication Data

Zill, Dennis G.
    A first course in differential equations with applications.

    Includes index.
    1. Differential equations.   I.   Title.
QA372.Z54   1982   515.3'5   81-15790
ISBN 0-87150-319-0   AACR2
ISBN 0-534-98011-2 (International Edition)

ISBN 0-87150-319-0

ISBN 0-534-98011-2   (International Edition)

Printed in the United States of America.
    82 83 84 85 — 10 9 8 7 6 5 4 3 2

This book was filmset by Interactive Composition
Corporation, Pleasant Hill, California. The text and
cover were designed by John Servideo. Art was drawn
by Phil Carver & Friends, Inc. Printing and binding
were done by The Alpine Press, Inc., Stoughton,
Massachusetts.

Second Edition

A First Course in

Differential Equatio

with Applications

Dennis G. Zill
Loyola Marymount Universi

Prindle, Weber & ?

Boston

Preface

This text is intended for a first course in the methods of solution, applications, and theory of ordinary differential equations. Since it has been my experience that the majority of students enrolled in a course in differential equations represents the major areas of science and engineering, the emphasis of the text is on both how to solve differential equations and how to interpret these equations in a physical setting. However, an attempt has been made to strike a balance between methodology, applications, and the theoretical foundations of the subject. Students are encouraged to look beyond merely finding an answer and to think in terms of possible applications as well as problems concerning existence and uniqueness of solutions. Detailed proofs of the more difficult theorems are omitted in favor of additional examples or a plausible argument. At times intuition is favored over rigor. The style of the text is straightforward, no nonsense, readable, and I hope, understandable.

In doing a revision of a text an author must incorporate improvements as suggested by those teachers and students who have used the first edition, while simultaneously trying to maintain the original intent of the text. Instructors who considered using the first edition have also supplied ideas that must be included in the revision. With this in mind the following modifications have been made:

- All of the exercise sets have been scrutinized with an eye to giving them a better balance as to the difficulty of the problems. Most of the exercise sets have been expanded; over 640 problems are new.

- Many new examples and figures have been added.

- The use of the metric system of units has been increased.

- Commonly used systems of units are reviewed in Chapter 1.

- The concepts of the complete solution and the fundamental set of solutions of a linear equation have been introduced.

- Chapter 2 now concludes with a brief discussion of Picard's method of successive approximations.

- The discussion of the Method of Undetermined Coefficients has been completely rewritten. The notion of the annihilator differential operator has been introduced as the most reasonable alternative to the usually unmotivated rules for forming the particular solution of a nonhomogeneous linear differential equation. It is felt that the student can now see why, under certain circumstances, the structure of a particular solution is prescribed.

- A chapter appendix on the Gamma Function now follows Chapter 6.

- In Chapter 7, a discussion of the Laplace transform of periodic functions has been added, along with problems involving functions such as the meander and staircase functions. Also, Chapter 7 ends with a new section on the Dirac delta function.

- A substantial portion of Chapter 8 on systems of linear differential equations has been revised. The discussion and utilization of matrices has been greatly expanded. The eigenvalue problem and its role in solving linear systems with constant coefficients is now given the prominence it deserves. Furthermore, the cases of complex and repeated eigenvalues are treated fully. The dual discussion on variation of parameters for linear systems has been reduced to a single discussion employing fundamental matrices. Chapter 8 ends with an introduction to the matrix exponential.

- The Review Exercises that were keyed to each section of the text have been reorganized into Chapter Tests.

This second edition, like the first, is designed for a one-semester, or one-quarter, introductory course. The format of the first edition has been retained: all examples are set off by three-sided boxes and important formulas are highlighted to delineate them clearly from the textual discussion and for easy reference. Answers to odd-numbered problems are given at the end of the book. Even-numbered problems marked with a star are worked out in detail in an accompanying Student Solutions Manual. Sections marked with [O] are optional and may be skipped without loss of continuity. Finally, some of the longer sections (such as the method of Frobenius and the operational properties of the Laplace transform), although unified by content, are partitioned into subsections, with the exercises appropriately marked, in order to accommodate those instructors who wish to cover that material in more than one lecture.

In conclusion, I would like to express my gratitude to:

The many users of the first edition who, unsolicited, have taken the time to send a word of support or to offer a constructive criticism.

The reviewers of the second edition: Philip Bacon (University of

Florida), Paul W. Davis (Worcester Polytechnic Institute), Raymond Fabec (Louisiana State University), Paul J. Gormley (Villanova University), Anthony J. John (Southeastern Massachusetts University), C. J. Neugebauer (Purdue University), Nancy J. Poxon (California State University-Sacramento), and Robert Pruitt (San Jose State University).

Mr. Andy Castanon for checking many of the answers.

The production staff of Prindle, Weber & Schmidt; former editors John Kimmel, Barbara Schott, and current editor Mary Lu Walsh. Their help, encouragement, and cooperation left little to be desired.

Dennis G. Zill
Los Angeles

Contents

CHAPTER 1

An Introduction to Differential Equations

1.1 Basic Definitions and Terminology

In calculus the reader learned that given a function $y = f(x)$, the derivative

$$\frac{dy}{dx} = f'(x),$$

is itself a function of x and is found by some appropriate rule. For example, if $y = e^{x^2}$ then

$$\frac{dy}{dx} = 2xe^{x^2}$$

or
$$\frac{dy}{dx} = 2xy. \tag{1}$$

The problem that we face in this course is not: given a function $y = f(x)$, find its derivative; but rather our problem is: if we are given an equation such as $dy/dx = 2xy$, to somehow find a function $y = f(x)$ which satisfies the equation. In a word, we wish to *solve* differential equations.

> **DEFINITION 1.1** An equation containing the derivatives or differentials of one or more dependent variables, with respect to one or more independent variables, is said to be a **differential equation.** ☐

Differential equations are classified according to the following three properties.

Classification by type

If an equation contains only ordinary derivatives of one or more dependent variables, with respect to a single independent variable, it is then said to be

1

an **ordinary differential equation.** For example,

$$\frac{dy}{dx} - 5y = 1$$

$$(x + y)\, dx - 4y\, dy = 0$$

$$\frac{du}{dx} - \frac{dv}{dx} = x$$

$$\frac{d^2y}{dx^2} - 2\frac{dy}{dx} + 6y = 0$$

are ordinary differential equations. An equation involving the partial derivatives of one or more dependent variables of two or more independent variables is called a **partial differential equation.** For example,

$$\frac{\partial u}{\partial y} = -\frac{\partial v}{\partial x}$$

$$x\frac{\partial u}{\partial x} + y\frac{\partial u}{\partial y} = u$$

$$\frac{\partial^2 u}{\partial x\, \partial y} = x + y$$

$$a^2\frac{\partial^2 u}{\partial x^2} = \frac{\partial^2 u}{\partial t^2} - 2k\frac{\partial u}{\partial t}$$

are partial differential equations.

Classification by order

The order of the highest derivative in a differential equation is called the **order of the equation.** For example,

$$\frac{d^2y}{dx^2} + 5\left(\frac{dy}{dx}\right)^2 - 4y = x$$

is a second-order ordinary differential equation. Since the differential equation

$$x^2\, dy + y\, dx = 0$$

can be put into the form

$$x^2\frac{dy}{dx} + y = 0$$

by dividing by the differential dx, it is an example of a first-order ordinary differential equation. The equation

$$c^2\frac{\partial^4 u}{\partial x^4} + \frac{\partial^2 u}{\partial t^2} = 0$$

is a fourth-order partial differential equation.

Although partial differential equations are very important, their study demands a good foundation in the theory of ordinary differential equations. Consequently, in the discussion that follows (as well as in the next eight chapters of this text) we shall confine our attention to ordinary differential equations.

A general nth-order, ordinary differential equation is often represented by the symbolism

$$F\left(x, y, \frac{dy}{dx}, \ldots, \frac{d^n y}{dx^n}\right) = 0 \tag{2}$$

The following is a special case of (2).

Classification as linear or nonlinear

A differential equation is said to be **linear** if it has the form

$$a_n(x)\frac{d^n y}{dx^n} + a_{n-1}(x)\frac{d^{n-1} y}{dx^{n-1}} + \cdots + a_1(x)\frac{dy}{dx} + a_0(x)y = g(x).$$

It should be observed that linear differential equations are characterized by two properties: (a) the dependent variable y and all its derivatives are of the first degree, that is, the power of each term involving y is 1; and (b) each coefficient depends only on the independent variable x. An equation that is not linear is said to be **nonlinear.**

The equations
$$x\,dy + y\,dx = 0$$
$$y'' - 2y' + y = 0$$

and
$$x^3\frac{d^3 y}{dx^3} - x^2\frac{d^2 y}{dx^2} + 3x\frac{dy}{dx} + 5y = e^x$$

are linear first-, second-, and third-order ordinary differential equations, respectively. On the other hand,

$$\frac{dy}{dx} = xy^{1/2}$$

$$yy'' - 2y' = x + 1$$

and
$$\frac{d^3 y}{dx^3} + y^2 = 0$$

are nonlinear first-, second-, and third-order ordinary differential equations, respectively.

Solutions

As mentioned before our goal in this course is to solve, or find solutions of, differential equations.

> **DEFINITION 1.2**   Any function f defined on some interval I^* which when substituted into a differential equation reduces the equation to an identity, is said to be **solution** of the equation on the interval.   ☐

---

*The nature of I is purposely left vague. Depending on the context I could represent (1) $a < x < b$, (2) $a \le x \le b$, (3) $0 < x < \infty$, (4) $-\infty < x < \infty$, and so on.

In other words, a solution of a differential equation (2) is a function $y = f(x)$ possessing at least n derivatives such that

$$F(x, f(x), f'(x), \ldots, f^{(n)}(x)) = 0$$

for every x in I.

EXAMPLE

The function $y = x^4/16$ is a solution of the nonlinear equation

$$\frac{dy}{dx} - xy^{1/2} = 0$$

on $-\infty < x < \infty$. Since

$$\frac{dy}{dx} = 4 \cdot \frac{x^3}{16} = \frac{x^3}{4}$$

we see

$$\frac{dy}{dx} - xy^{1/2} = \frac{x^3}{4} - x\left(\frac{x^4}{16}\right)^{1/2} = \frac{x^3}{4} - \frac{x^3}{4} = 0$$

for every real number.

Many differential equations possess an important but seemingly trivial solution. Notice in the last example that $y \equiv 0*$ is a solution of $dy/dx - xy^{1/2} = 0$ along with $y = x^4/16$.

EXAMPLE

The first-order differential equations

$$\left(\frac{dy}{dx}\right)^2 + 1 = 0 \quad \text{and} \quad (y')^2 + y^2 + 4 = 0$$

possess no real solutions. Why?

Explicit and implicit solutions

Solutions of differential equations are further distinguished as either **explicit** or **implicit solutions**. For example, we have already seen that $y = e^{x^2}$ is an explicit solution of equation (1). A relation $G(x, y) = 0$ is said to define a solution of (2) implicitly on an interval I provided it defines one or more explicit solutions on I.

EXAMPLE

$y = xe^x$ is an explicit solution of $y'' - 2y' + y = 0$ on $-\infty < x < \infty$. To see this, we compute

$$y' = xe^x + e^x$$

and

$$y'' = xe^x + 2e^x.$$

---

*That is, $y = 0$ for every real number x.

Observe $y'' - 2y' + y = (xe^x + 2e^x) - 2(xe^x + e^x) + xe^x$

$$= 0$$

for every real number.

EXAMPLE   For $-2 < x < 2$ the relation $x^2 + y^2 - 4 = 0$ is an implicit solution of the differential equation

$$\frac{dy}{dx} = -\frac{x}{y}.$$

By implicit differentiation it follows that

$$\frac{d}{dx}(x^2) + \frac{d}{dx}(y^2) = 0$$

$$2x + 2y\frac{dy}{dx} = 0$$

or $$\frac{dy}{dx} = -\frac{x}{y}.$$

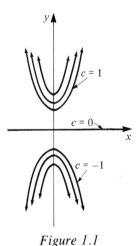

Figure 1.1

The relation $x^2 + y^2 - 4 = 0$ in the preceding example defines two functions: $y = \sqrt{4 - x^2}$ and $y = -\sqrt{4 - x^2}$ on the interval $-2 < x < 2$. Also note that any relation of the form $x^2 + y^2 - c = 0$ will *formally* satisfy $dy/dx = -x/y$ for any constant c. However, it is naturally understood that the relation should always make sense in the real number system; thus we cannot say that $x^2 + y^2 + 1 = 0$ determines a solution of the differential equation.

Since the distinction between an explicit and an implicit solution should be intuitively clear, we shall not belabor the phraseology "here is an explicit (implicit) solution."

The student should become accustomed to the fact that a given differential equation will usually possess an infinite number of solutions. By direct substitution, we can prove that any curve, that is, function, in the one-parameter family $y = ce^{x^2}$, where c is any arbitrary constant, also satisfies equation (1). As indicated in Figure 1.1, $y \equiv 0$, obtained by setting $c = 0$, is also a solution of the equation. In the third example we saw that $y = xe^x$ is a solution of $y'' - 2y' + y = 0$; tracing back through the work reveals $y = cxe^x$ represents a family of solutions.

EXAMPLE   For any value of c the function $y = c/x + 1$ is a solution of the first-order differential equation

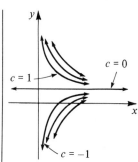

Figure 1.2

$$x\frac{dy}{dx} + y = 1$$

on the interval $0 < x < \infty$.

We have
$$\frac{dy}{dx} = c\frac{d}{dx}(x^{-1}) + \frac{d}{dx}(1) = -cx^{-2} = -\frac{c}{x^2}$$

so that
$$x\frac{dy}{dx} + y = x\left(-\frac{c}{x^2}\right) + \left(\frac{c}{x} + 1\right) = 1.$$

By choosing c to be any real number we can generate an infinite number of solutions; in particular, for $c = 0$ we obtain a constant solution $y \equiv 1$. See Figure 1.2.

In the preceding example, $y = c/x + 1$ is a solution of the differential equation on any interval not containing the origin. The function is not differentiable at $x = 0$.

EXAMPLES

(a) The functions $y = c_1 \cos 4x$ and $y = c_2 \sin 4x$, where c_1 and c_2 are arbitrary constants, are solutions of the differential equation
$$y'' + 16y = 0.$$
For $y = c_1 \cos 4x$ the first and second derivatives are
$$y' = -4c_1 \sin 4x$$
$$y'' = -16c_1 \cos 4x$$
and so $y'' + 16y = -16c_1 \cos 4x + 16(c_1 \cos 4x) = 0.$
Similarly, for $y = c_2 \sin 4x$
$$y'' + 16y = -16c_2 \sin 4x + 16(c_2 \sin 4x) = 0.$$
(b) The function
$$y = c_1 \cos 4x + c_2 \sin 4x$$
can also be shown to be a solution of the given equation.

EXAMPLE

The reader should be able to show that
$$y = e^x$$
$$y = e^{-x}$$
$$y = c_1 e^x$$
$$y = c_2 e^{-x}$$
and $$y = c_1 e^x + c_2 e^{-x}$$

are all solutions of the linear second-order differential equation

$$y'' - y = 0.$$

Note that $y = c_1 e^x$ is a solution for any choice of c_1, but $y = e^x + c_1, c_1 \neq 0$, does *not* satisfy the equation since for this latter family of functions we would get $y'' - y = -c_1$.

A solution of a differential equation can be defined in a piecewise manner.

EXAMPLE

Any function in the one-parameter family $y = cx^4$ is a solution of the differential equation

$$xy' - 4y = 0.$$

We have $xy' - 4y = x(4cx^3) - 4cx^4 = 0$. The piecewise-defined function

$$y = \begin{cases} -x^4, & x < 0 \\ x^4, & x \geq 0 \end{cases}$$

is also a solution. Observe, however, that this function is not a member of the given family. That is, this function cannot be obtained from $y = cx^4$ by a single selection of the parameter c. See Figure 1.3(b).

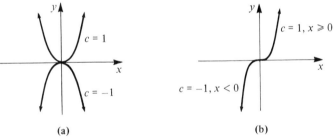

(a)

(b)

Figure 1.3

Further terminology

The study of differential equations is similar to integral calculus.* When evaluating an antiderivative or indefinite integral we utilize a single constant of integration. In like manner, when solving a first-order differential equation $F(x, y, y') = 0$ we shall usually obtain a family of curves or functions $G(x, y, c) = 0$ containing one arbitrary parameter such that each member of the family is a solution of the differential equation. In fact, when solving an nth-order equation $F(x, y, y', \ldots, y^{(n)}) = 0$, where $y^{(n)}$ means $d^n y/dx^n$, we expect an **n-parameter family of solutions** $G(x, y, c_1, \ldots, c_n) = 0$.

---

*A solution, either explicit, implict, or otherwise, (see Problems 43–46) is sometimes referred to as an **integral curve** or simply as an **integral** of the equation.

A solution of a differential equation that is free of arbitrary parameters is called a **particular solution.** One way of obtaining a particular solution is to choose specific values of the parameter(s) in a family of solutions. For example, it is readily seen that $y = ce^x$ is a one-parameter family of solutions of the simple first-order equation $y' = y$. For $c = 0, -2$, and 5, we get the particular solutions $y \equiv 0$, $y = -2e^x$, and $y = 5e^x$, respectively.

Sometimes a differential equation possesses a solution that cannot be obtained by specializing the parameters in a family of solutions. Such a solution is called a **singular solution.**

EXAMPLE

In Section 2.1 we shall prove that a one-parameter family of solutions of $y' - xy^{1/2} = 0$ is given by $y = (x^2/4 + c)^2$. When $c = 0$ the resulting particular solution is $y = x^4/16$. In this case the zero function, $y \equiv 0$, is a singular solution of the equation since it cannot be obtained from the family for any choice of the parameter c.

If *every* solution of $F(x, y, y', \ldots, y^{(n)}) = 0$ on an interval I can be obtained from $G(x, y, c_1, \ldots, c_n) = 0$ by appropriate choices of the c_i, $i = 1, 2, \ldots, n$, we then say that the n-parameter family is the **general** or **complete** solution of the differential equation.

Remark: There are two schools of thought concerning the concept of a "general solution." An alternative viewpoint holds that a general solution of an nth-order differential equation is a family of solutions containing n essential* parameters. Period! In other words, the family is not required to contain all solutions of the differential equation on some interval. The difference in these opinions is really a distinction between the solutions to linear and nonlinear equations. In solving linear differential equations we shall impose relatively simple restrictions on the coefficients; with these restrictions one can always be assured that not only does a solution exist on an interval but that a family of solutions will indeed yield all possible solutions.

Another fact deserves mention at this time. Nonlinear equations, with the exception of some first-order equations, are usually difficult or *impossible* to solve in terms of the standard elementary functions.† Furthermore, if we happen to have a family of solutions for a nonlinear equation it is not obvious when this family constitutes a general solution. On a practical level then, the designation "general solution" is applied only to linear differential equations.

---

*We won't try to define this concept. But roughly it means don't play games with the constants. Certainly $y = x + c_1 + c_2$ represents a family of solutions of $y' = 1$. By renaming $c_1 + c_2$ as c, the family has *essentially* one constant: $y = x + c$. The reader should verify that $y = c_1 + \ln c_2 x$ is a solution of $x^2y'' + xy' = 0$ on the interval $0 < x < \infty$ for any choice of c_1 and $c_2 > 0$. Are c_1 and c_2 essential parameters?

†For example, algebraic functions, exponential and logarithmic functions, trigonometric and inverse trigonometric functions.

EXERCISES 1.1   *Answers to odd-numbered problems begin on page A-1.*

In Problems 1–10 state whether the given differential equations are linear or nonlinear. Give the order of each equation.

1. $(1 - x)y'' - 4xy' + 5y = \cos x$ **2.** $x\dfrac{d^3y}{dx^3} - 2\left(\dfrac{dy}{dx}\right)^4 + y = 0$

3. $yy' + 2y = 1 + x^2$ **★4.** $x^2\, dy + (y - xy - xe^x)\, dx = 0$

5. $x^3y^{(4)} - x^2y'' + 4xy' - 3y = 0$ **6.** $\dfrac{d^2y}{dx^2} + 9y = \sin y$

7. $\dfrac{dy}{dx} = \sqrt{1 + \left(\dfrac{d^2y}{dx^2}\right)^2}$ **8.** $\dfrac{d^2r}{dt^2} = -\dfrac{k}{r^2}$

9. $(\sin x)y''' - (\cos x)y' = 2$ **10.** $(1 - y^2)\, dx + x\, dy = 0$

In Problems 11–40 verify that the indicated function is a solution of the given differential equation. Where appropriate, c_1 and c_2 denote constants.

11. $2y' + y = 0; \quad y = e^{-x/2}$

12. $y' + 4y = 32; \quad y \equiv 8$

13. $\dfrac{dy}{dx} - 2y = e^{3x}; \quad y = e^{3x} + 10e^{2x}$

14. $\dfrac{dy}{dt} + 20y = 24; \quad y = \tfrac{6}{5} - \tfrac{6}{5}e^{-20t}$

15. $y' = 25 + y^2; \quad y = 5\tan 5x$

16. $\dfrac{dy}{dx} = \sqrt{\dfrac{y}{x}}; \quad y = (\sqrt{x} + c_1)^2, \quad x > 0$

17. $y' + y = \sin x; \quad y = \tfrac{1}{2}\sin x - \tfrac{1}{2}\cos x + 10e^{-x}$

★18. $2xy\, dx + (x^2 + 2y)\, dy = 0; \quad x^2y + y^2 = c_1$

19. $x^2\, dy + 2xy\, dx = 0; \quad y = -\dfrac{1}{x^2}$

20. $(y')^3 + xy' = y; \quad y = x + 1$

21. $y = 2xy' + y(y')^2; \quad y^2 = c_1(x + \tfrac{1}{4}c_1)$

★22. $y' = 2\sqrt{|y|}; \quad y = x|x|$

23. $y' - \dfrac{1}{x}y = 1; \quad y = x\ln x, \quad x > 0$

24. $\dfrac{dP}{dt} = P(a - bP); \quad P = \dfrac{ac_1e^{at}}{1 + bc_1e^{at}}$

25. $\dfrac{dX}{dt} = (2 - X)(1 - X); \quad \ln\dfrac{2 - X}{1 - X} = t$

26. $y' + 2xy = 1;$ $y = e^{-x^2} \displaystyle\int_0^x e^{t^2}\, dt + c_1 e^{-x^2}$

27. $(x^2 + y^2)\, dx + (x^2 - xy)\, dy = 0;$ $c_1(x + y)^2 = xe^{y/x}$

28. $y'' + y' - 12y = 0;$ $y = c_1 e^{3x} + c_2 e^{-4x}$

29. $y'' - 6y' + 13y = 0;$ $y = e^{3x} \cos 2x$

30. $\dfrac{d^2 y}{dx^2} - 4\dfrac{dy}{dx} + 4y = 0;$ $y = e^{2x} + xe^{2x}$

31. $y'' = y;$ $y = \cosh x + \sinh x$*

32. $y'' + 25y = 0;$ $y = c_1 \cos 5x$

33. $y'' + (y')^2 = 0;$ $y = \ln|x + c_1| + c_2$

34. $y'' + y = \tan x;$ $y = -\cos x\, \ln(\sec x + \tan x)$

35. $x\dfrac{d^2 y}{dx^2} + 2\dfrac{dy}{dx} = 0;$ $y = c_1 + c_2 x^{-1}$

★**36.** $x^2 y'' - xy' + 2y = 0;$ $y = x \cos(\ln x),$ $x > 0$

37. $x^2 y'' - 3xy' + 4y = 0;$ $y = x^2 + x^2 \ln x,$ $x > 0$

38. $y''' - y'' + 9y' - 9y = 0;$ $y = c_1 \sin 3x + c_2 \cos 3x + 4e^x$

39. $y''' - 3y'' + 3y' - y = 0;$ $y = x^2 e^x$

40. $x^3 \dfrac{d^3 y}{dx^3} + 2x^2 \dfrac{d^2 y}{dx^2} - x\dfrac{dy}{dx} + y = 12x^2;$ $y = c_1 + c_2 x \ln x + 4x^2,$ $x > 0$

In Problems 41 and 42 verify that the indicated piecewise defined function is a solution of the given differential equation.

41. $xy' - 2y = 0;$ $y = \begin{cases} -x^2, & x < 0 \\ x^2, & x \geq 0 \end{cases}$

42. $(y')^2 = 9xy;$ $y = \begin{cases} 0, & x < 0 \\ x^3, & x \geq 0 \end{cases}$

In Problems 43–46 proceed formally to show that the indicated parametric equations form a solution of the given differential equation.

EXAMPLE

$4\left(\dfrac{dy}{dx}\right)^2 = y + 2,$ $x = 4t + 1,$ $y = t^2 - 2.$

---

*Recall, the hyperbolic cosine and hyperbolic sine are defined by

$$\cosh x = (e^x + e^{-x})/2 \quad \text{and} \quad \sinh x = (e^x - e^{-x})/2.$$

Solution: Recall from calculus that

$$\frac{dy}{dx} = \frac{dy/dt}{dx/dt}$$

$$= \frac{2t}{4}$$

$$= \frac{1}{2}t$$

and so
$$4\left(\frac{dy}{dx}\right)^2 = 4\left(\frac{1}{2}t\right)^2$$

$$= t^2$$

$$= t^2 - 2 + 2$$

$$= y + 2.$$

43. $\left(\dfrac{dy}{dx}\right)^3 + 2x\dfrac{dy}{dx} = 2y + 1; \quad x = -\dfrac{3}{2}t^2, \quad y = -t^3 - \dfrac{1}{2}$

44. $y = xy' + (y')^2; \quad x = -2t, \quad y = -t^2$

45. $y = xy' + (y')^2 - \ln y'; \quad x = -2t + \dfrac{1}{t}, \quad y = -t^2 - \ln t + 1$

46. $y[1 + (y')^2] = c; \quad x = \dfrac{c}{2}(2\theta - \sin 2\theta), \quad y = \dfrac{c}{2}(1 - \cos 2\theta)$

47. Verify that a one-parameter family of solutions for

$$y = xy' + (y')^2 \quad \text{is} \quad y = cx + c^2.$$

   Determine a value of k such that $y = kx^2$ is a singular solution of the differential equation.

48. Verify that a one-parameter family of solutions for

$$y = xy' + \sqrt{1 + (y')^2} \quad \text{is} \quad y = cx + \sqrt{1 + c^2}.$$

   Show that the relation $x^2 + y^2 = 1$ defines a singular solution of the equation on the interval $-1 < x < 1$.

49. A one-parameter family of solutions for

$$y' = y^2 - 1 \quad \text{is} \quad y = \frac{1 + ce^{2x}}{1 - ce^{2x}}.$$

   By inspection, determine a singular solution of the differential equation.

★**50.** On page 5 we have seen that $y = \sqrt{4 - x^2}$ and $y = -\sqrt{4 - x^2}$ are solutions of $\dfrac{dy}{dx} = -\dfrac{x}{y}$ on the interval $-2 < x < 2$. Explain why

$$y = \begin{cases} \sqrt{4 - x^2}, & -2 < x < 0 \\ -\sqrt{4 - x^2}, & 0 \le x < 2 \end{cases}$$

is not a solution of the differential equation on the interval.

Miscellaneous problems

In Problems 51 and 52 find values of m so that $y = e^{mx}$ is a solution of each differential equation.

51. $y'' - 5y' + 6y = 0$ **52.** $y'' + 10y' + 25y = 0$

In Problems 53 and 54 find values of m so that $y = x^m$ is a solution of each differential equation.

53. $x^2 y'' - y = 0$ ★**54.** $x^2 y'' + 6xy' + 4y = 0$

55. Show that $y_1 = x^2$ and $y_2 = x^3$ are both solutions of

$$x^2 y'' - 4xy' + 6y = 0.$$

Are the constant multiples $c_1 y_1$ and $c_2 y_2$, with c_1 and c_2 arbitrary, also solutions? Is the sum $y_1 + y_2$ a solution?

56. Show that $y_1 = 2x + 2$ and $y_2 = -x^2/2$ are both solutions of

$$y = xy' + (y')^2/2$$

Are the constant multiples $c_1 y_1$ and $c_2 y_2$, with c_1 and c_2 arbitrary, also solutions? Is the sum $y_1 + y_2$ a solution?

57. By inspection* determine, if possible, a real solution of the given differential equation.

(a) $\left| \dfrac{dy}{dx} \right| + |y| = 0$ **(b)** $\left| \dfrac{dy}{dx} \right| + |y| + 1 = 0$

(c) $\left| \dfrac{dy}{dx} \right| + |y| - 1 = 0$

1.2 Origins of Differential Equations

It would be a shame for a student to pass through a course such as this (as some do) and not have a modicum of appreciation for some of the origins of

---

*Translated, this means take a good guess and see if it works.

the subject matter. In the discussion that follows we shall see how specific differential equations arise not only out of consideration of families of geometric curves, but also how differential equations result from an attempt to describe, in mathematical terms, physical problems in the sciences and engineering. It would not be overly presumptive to state that differential equations form the backbone of subjects such as physics and electrical engineering, and even provide an important working tool in such diverse areas as biology and economics. Several of the examples and problems in this section will serve as previews of coming attractions for the material in Chapters 3 and 5.

1.2.1   The Differential Equation of a Family of Curves

At the end of the preceding section we expressed the expectation that an nth-order ordinary differential equation will yield an n-parameter family of solutions. However, the reader should not get the impression that we can *always* find an n-parameter family of solutions for every conceivable nth-order differential equation. On the other hand, suppose we turn the problem around: starting with an n-parameter family of curves, can we then find an associated nth-order differential equation which is entirely free of arbitrary parameters which represents the given family? In most cases the answer is yes.*

In the discussion of particular solutions we saw that each function in the one-parameter family $y = ce^x$ satisfies the same first-order differential equation $y' = y$. Suppose that we now seek to find the differential equation of the two-parameter family

$$y = c_1 e^x + c_2.$$

The first two derivatives are

$$\frac{dy}{dx} = c_1 e^x$$

$$\frac{d^2 y}{dx^2} = c_1 e^x.$$

Thus

$$\frac{d^2 y}{dx^2} = \frac{dy}{dx}$$

or

$$\frac{d^2 y}{dx^2} - \frac{dy}{dx} = 0.$$

---

*See Problem 30 of this section. You should develop a suspicion that exceptions might exist to "general" discussions unless the points under consideration are summarized by means of a theorem. The hypothesis of the theorem sets the conditions under which the conclusion must always follow.

EXAMPLE   By taking two derivatives we find that the differential equation of the two-parameter family of straight lines

$$y = c_1 x + c_2$$

is simply

$$\frac{d^2y}{dx^2} = 0.$$

EXAMPLE   Find the differential equation of the family

$$y = cx^3$$

indicated in Figure 1.4.

Solution: We expect a first-order differential equation since the family contains only one parameter. It follows that

$$\frac{dy}{dx} = 3cx^2,$$

but $c = y/x^3$ so that

$$\frac{dy}{dx} = 3\left(\frac{y}{x^3}\right)x^2$$

$$= 3\frac{y}{x}.$$

Thus we obtain the linear first-order equation

$$x\frac{dy}{dx} - 3y = 0.$$

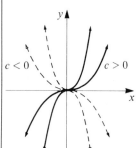

Figure 1.4

A first-order differential equation is often given in differential form. The differential equation in the preceding example can be written as

$$x\,dy - 3y\,dx = 0.$$

EXAMPLE   Find the differential equation of the two-parameter family

$$y = c_1 e^{2x} + c_2 e^{-2x}. \tag{1}$$

Solution: Taking two derivatives we obtain

$$\frac{dy}{dx} = 2c_1 e^{2x} - 2c_2 e^{-2x}$$

$$\frac{d^2y}{dx^2} = 4c_1 e^{2x} + 4c_2 e^{-2x}$$

$$= 4[c_1 e^{2x} + c_2 e^{-2x}].$$

Using (1) we find

$$\frac{d^2y}{dx^2} = 4y \qquad \text{or} \qquad y'' - 4y = 0.$$

EXAMPLE   Find the differential equation of the family of circles centered at the origin.

Solution: Concentric circles with center at the origin are described by the one-parameter equation

$$x^2 + y^2 = c^2, \qquad c > 0.$$

By implicit differentiation we obtain

$$2x + 2y\frac{dy}{dx} = 0$$

$$\frac{dy}{dx} = -\frac{x}{y} \qquad \text{or} \qquad x\,dx + y\,dy = 0.$$

EXAMPLE   Find the differential equation of the family of circles passing through the origin with center on the y-axis.

Solution: This family of circles is characterized by the one-parameter equation

$$x^2 + y^2 = cy.$$

Thus
$$2x + 2y\frac{dy}{dx} = c\frac{dy}{dx}.$$

Substituting $c = (x^2 + y^2)/y$ into the last equation and simplifying gives

$$(x^2 - y^2)\frac{dy}{dx} = 2xy \qquad \text{or} \qquad (x^2 - y^2)\,dy - 2xy\,dx = 0.$$

EXAMPLE   Find the differential equation of the family of parabolas
$$y = (x + c)^2.$$

Solution: The first derivative is

$$\frac{dy}{dx} = 2(x + c).$$

From the original equation we have $x + c = \pm y^{1/2}$ so that the differential equation representing the family is

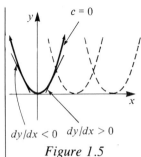

$$\frac{dy}{dx} = \pm 2y^{1/2} \quad \text{or} \quad \left(\frac{dy}{dx}\right)^2 = 4y.$$

It should be observed that

$$\frac{dy}{dx} = 2y^{1/2} \tag{2}$$

dy/dx < 0   dy/dx > 0

Figure 1.5

does not describe the complete family since by convention $y^{1/2} \geq 0$. Equation (2) would give the slope only of a right-hand branch ($x > -c$) of any particular parabola. Figure 1.5 illustrates the case when $c = 0$. In this case we could say that $y = x^2$ is a solution of (2) on the interval $x > 0$.

EXAMPLE   Find the differential equation of the family

$$y = \frac{2ce^{2x}}{1 + ce^{2x}}.$$

Solution: By the quotient rule and algebra we find

$$\frac{dy}{dx} = \frac{4ce^{2x}}{(1 + ce^{2x})^2}$$

$$= \frac{y^2 e^{-2x}}{c}.$$

Solving the given equation for c gives $c = e^{-2x}y/(2 - y)$ and thus we obtain

$$\frac{dy}{dx} = y^2 e^{-2x}\frac{1}{\dfrac{e^{-2x}y}{2-y}} \quad \text{or} \quad \frac{dy}{dx} = y(2 - y). \tag{3}$$

In the preceding two examples the differential equation actually gives a bit more than we bargained for. By inspection we can see that the trivial function $y \equiv 0$ is a solution of (2) and the constant function $y \equiv 2$ is a solution of (3). In neither case is this particular solution a member of the given family.

A two-parameter family of curves can sometimes lead to a rather complicated differential equation.

EXAMPLE   Find the differential equation that describes the family of circles passing through the origin.

Solution: As Figure 1.6 indicates, the general form of the equation of these circles is

$$(x - h)^2 + (y - k)^2 = (\sqrt{h^2 + k^2})^2$$

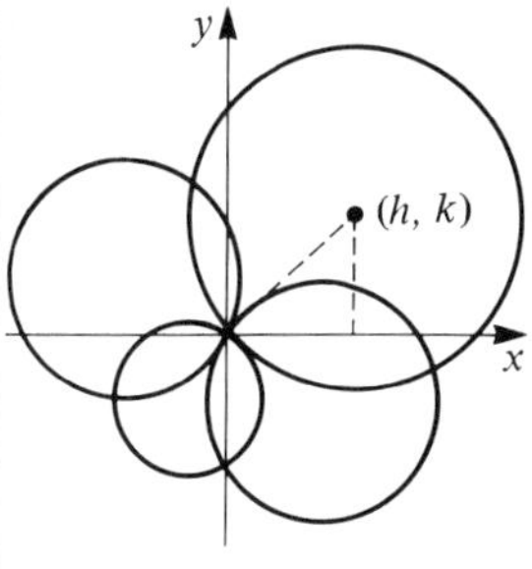

or $$x^2 - 2xh + y^2 - 2ky = 0. \tag{4}$$

Using implicit differentiation twice we find

$$x - h + yy' - ky' = 0 \tag{5}$$

and $$1 + yy'' + (y')^2 - ky'' = 0 \tag{6}$$

We then use the original equation of the family (4) to solve for h:

$$h = \frac{x^2 + y^2 - 2ky}{2x}$$

Figure 1.6

and substitute in (5),

$$x - \frac{x^2 + y^2 - 2ky}{2x} + yy' - ky' = 0. \tag{7}$$

Now solving (7) for k gives

$$k = \frac{x^2 - y^2 + 2xyy'}{2(xy' - y)}. \tag{8}$$

Substituting this latter value in (6) and simplifying yields the nonlinear equation

$$1 + yy'' + (y')^2 - \frac{x^2 - y^2 + 2xyy'}{2(xy' - y)} y'' = 0$$

or $$(x^2 + y^2)y'' + 2[(y')^2 + 1](y - xy') = 0.$$

Alternatively, we can obtain the last result directly by differentiating (8) by the quotient rule.

1.2.2 Some Physical Origins of Differential Equations

EXAMPLE

It is well known that free-falling objects close to the surface of the earth accelerate at a constant rate g. Acceleration is the derivative of velocity, and this in turn, is the derivative of distance s. Thus if we assume that the upward direction is positive, the statement

$$\frac{d^2s}{dt^2} = -g$$

Figure 1.7

is the differential equation governing the vertical distance that the falling body travels. The minus sign is used since the weight of the body is a force directed opposite to the positive direction.

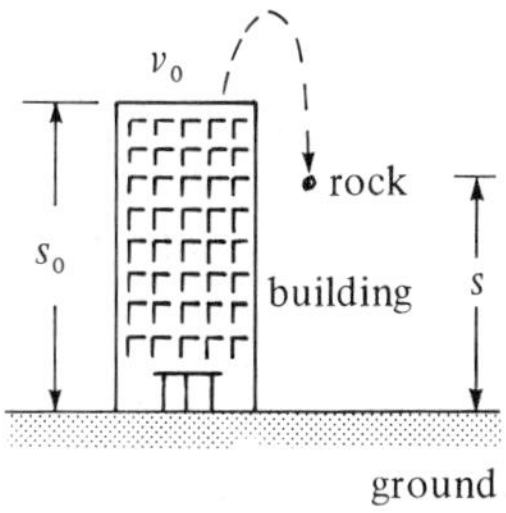

If we suppose further that a rock is tossed off the roof of a building of height s_0 (see Figure 1.7) with an initial upward velocity of, say v_0, then we must solve

$$\frac{d^2s}{dt^2} = -g, \qquad 0 < t < t_1,$$

subject to the side conditions

$$s(0) = s_0, \qquad s'(0) = v_0.$$

Here $t = 0$ is taken to be the initial time when the rock leaves the roof of the building and t_1 is the time required to hit the ground. Since the rock is thrown upward it would naturally be assumed that $v_0 > 0$. This formulation of the problem ignores other forces such as air resistance acting on the body.

EXAMPLE

To find the vertical displacement $x(t)$ of a mass attached to a spring we use two different empirical laws: Newton's second law of motion and Hooke's law. The former law states that the net force acting on the system in motion is $F = ma$ where m is the mass and a is acceleration. Hooke's law states that the restoring force of a stretched spring is proportional to the elongation $s + x$. That is, the restoring force is $k(s + x)$ where $k > 0$ is a constant. As shown in Figure 1.8(b), s is the elongation of the spring after the mass has been attached and the system hangs at rest in the *equilibrium position*. When the system is in motion, the variable x represents a directed distance of the mass beyond the equilibrium position. In Chapter 5 we shall prove that when the system is in motion the *net force* acting on the mass is simply $F = -kx$.

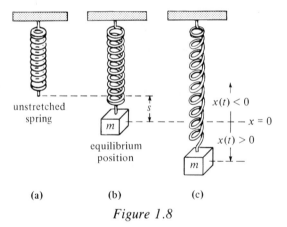

(a)   (b)   (c)

Figure 1.8

Thus in the absence of damping and other external forces which might be impressed on the system, the differential equation of the vertical motion through the center of gravity of the mass can be obtained by equating:

$$m\frac{d^2x}{dt^2} = -kx.$$

Here the minus sign means that the restoring force of the spring acts opposite to the direction of motion, that is, toward the equilibrium position. In practice this second-order differential equation is often written as

$$\frac{d^2x}{dt^2} + \omega^2 x = 0 \tag{9}$$

where $\omega^2 = k/m$.

Units

A word is in order regarding the system of units that are used in describing dynamic problems such as illustrated in the last two examples. Three commonly used systems of units are summarized in the following table. In each system the basic unit of time is the second.

Quantity	Engineering System*	mks	cgs
Force	pound (lb)	newton (nt)	dyne
Mass	slug	kilogram (kg)	gram (g)
Distance	foot (ft)	meter (m)	centimeter (cm)
Acceleration of gravity g (approximate)	32 ft/sec²	9.8 m/sec²	980 cm/sec²

The gravitational *force* exerted by the earth on a body of mass m is called its *weight W*. In the absence of air resistance the only force acting on a freely falling body is its weight. Hence from Newton's second law of motion it follows that mass m and weight W are related by

$$W = mg.$$

For example, in the engineering system a mass of 1/4 slug corresponds to an 8-lb weight. Since $m = W/g$, a 64-lb weight corresponds to a mass of $64/32 = 2$ slugs. In the cgs system a weight of 2450 dynes has a mass of $2450/980 = 2.5$ grams. In the mks system a weight of 50 newtons has a mass of $50/9.8 = 5.1$ kilograms. We note that

$$1 \text{ newton} = 10^5 \text{ dynes} = 0.2247 \text{ pounds.}$$

In the next example we derive the differential equation that describes the motion of a *simple pendulum*.

EXAMPLE

A mass m having weight W is suspended from the end of a rod of constant length l. For motion in a vertical plane, we would like to determine the displacement angle θ, measured from the vertical, as a function of time t (we consider $\theta > 0$ to the right of OP and $\theta < 0$ to the left of OP). Recall, an arc s of a circle of radius l is related to the central angle θ through the formula

$$s = l\theta.$$

---

*Also known as the English gravitational system or British engineering system.

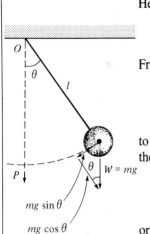

Figure 1.9

Hence the angular acceleration is

$$a = \frac{d^2s}{dt^2} = l\frac{d^2\theta}{dt^2}.$$

From Newton's second law we then have

$$F = ma = ml\frac{d^2\theta}{dt^2}.$$

From Figure 1.9 we see that the tangential component of the force due to the weight W is $mg \sin \theta$. When the mass of the rod is ignored we equate the two different formulations of the tangential force to obtain

$$ml\frac{d^2\theta}{dt^2} = -mg \sin \theta$$

or

$$\frac{d^2\theta}{dt^2} + \frac{g}{l} \sin \theta = 0. \tag{10}$$

Unfortunately the nonlinear equation (10) of the preceding example cannot be solved in terms of the familiar elementary functions (see Exercise 3.3), so usually a further simplifying assumption is made. If the angular displacements θ are not too large we can use the approximation $\sin \theta \approx \theta$* so that (10) can be replaced with the linear second-order differential equation

$$\frac{d^2\theta}{dt^2} + \frac{g}{l} \theta = 0. \tag{11}$$

If we set $\omega^2 = g/l$, observe that (11) has the exact same structure as the differential equation governing the free vibrations of a weight on a spring (equation (9)). The fact that one basic differential equation can describe many diverse physical or even economic phenomena is a common occurrence in the study of applicable mathematics.

EXAMPLE

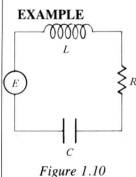

Figure 1.10

Consider the single loop series circuit containing an inductor, resistor, and capacitor, shown in Figure 1.10. Kirchoff's second law states that the *sum* of the voltage drops across each part of the circuit is the same as the impressed voltage $E(t)$. If $q(t)$ denotes the charge of the capacitor at any time, then the current $i(t)$ is given by $i = dq/dt$. Now it is known that the voltage drops across an

$$\text{inductor} = L\frac{di}{dt} = L\frac{d^2q}{dt^2}$$

$$\text{capacitor} = \frac{1}{C}q$$

---

*The reader is encouraged to get out his or her calculus text and inspect its table of trigonometric functions and compare the numerical values of $\sin \theta$ with the value of θ in radians. It is also a good idea to look up the Maclaurin expansion for the sine function.

$$\text{resistor} = iR = R\frac{dq}{dt}$$

where L, C, and R are constants called the inductance, capacitance, and resistance, respectively. Therefore, to determine $q(t)$ we must solve the second-order differential equation

$$L\frac{d^2q}{dt^2} + R\frac{dq}{dt} + \frac{1}{C}q = E(t). \tag{12}$$

In the previous example, the side conditions $q(0)$ and $q'(0)$ represent the charge on the capacitor and the current in the circuit, respectively, at $t = 0$. Also, the impressed voltage $E(t)$ is said to be an **electromotive force** or **emf.** In other words, the emf causes the current to flow. The following table shows the basic units of measurement used in circuit analysis.

Quantity	Unit
Impressed voltage or emf	volt (v)
Inductance L	henry (h)
Capacitance C	farad (f)
Resistance R	ohm (Ω)
Charge q	coulomb
Current i	ampere (amp)

EXAMPLE   It seems plausible to expect that the rate at which a population P expands is proportional to the population that is present at any time. Roughly put, the more people there are, the more there are going to be. Thus one model for population growth is given by the differential equation

$$\frac{dP}{dt} = kP \tag{13}$$

where k is a constant of proportionality. Since we also expect the population to expand we must have $dP/dt > 0$ and thus $k > 0$.

EXAMPLE   In the spread of a contagious disease, for example a flu virus, it is reasonable to assume that the rate, dx/dt, at which the disease spreads is proportional not only to the number of people, $x(t)$, who have contracted the disease, but also to the number of people, $y(t)$, who have not yet been exposed. That is,

$$\frac{dx}{dt} = kxy \tag{14}$$

where k is the usual constant of proportionality. If one infected person is introduced into a fixed population of n people, then x and y are related by

$$x + y = n + 1. \tag{15}$$

Using (15) to eliminate y in (14) then gives

$$\frac{dx}{dt} = kx(n + 1 - x). \tag{16}$$

The obvious side condition accompanying equation (16) is $x(0) = 1$.

The logistic equation   The nonlinear first-order equation (16) is a special case of a more general equation

$$\frac{dP}{dt} = P(a - bP), \qquad a \text{ and } b \text{ constants,} \tag{17}$$

known as the **logistic equation** (see Section 3.3). The solution of this equation is very important in ecological, sociological, and even managerial sciences.

EXAMPLE   Newton's law of cooling states that the time rate at which a body cools is proportional to the difference between the temperature of the body and the temperature of the surrounding medium. If $T(t)$ denotes the temperature of the body at any time t, and T_0 is the constant temperature of the outside medium, it follows that

$$\frac{dT}{dt} = k(T - T_0) \tag{18}$$

where k is the constant of proportionality. Note that when $T_0 = 0$, equation (18) reduces to (13). However, in this case $T(t)$ is decreasing so we want $k < 0$.

Equation (13) also appears in a different context in the next example.

EXAMPLE   When interest is compounded **continuously** the rate at which an amount of money S grows is proportional to the amount of money present at any time. That is,

$$\frac{dS}{dt} = rS \tag{19}$$

where r is the annual rate of interest.* This is analogous to the population growth of an earlier example. The rate of growth is large when the amount

---

*Both dS/dt and r are rates. A ratio such as $(dS/dt)/S$ is often called the *growth rate*, *specific growth rate*, *relative growth rate*, or *average growth rate*.

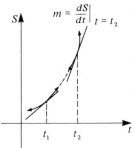

$$m = \frac{dS}{dt}\bigg|_{t\,=\,t_2}$$

Figure 1.11

of money present in the account is also large. Translated geometrically, this means the tangent line is steep when S is large (see Figure 1.11).

The definition of a derivative provides an interesting derivation of equation (19). Suppose $S(t)$ is the amount accrued in a savings account after t years when the annual rate of interest r is compounded continuously. If h denotes an increment in t, then the interest obtained in the time span $(t + h) - t$ is the difference in amounts accrued:

$$S(t + h) - S(t). \tag{20}$$

Since interest is given by

$$(\text{rate}) \times (\text{time}) \times (\text{principal}) \tag{21}$$

we can approximate the interest earned in this same time period by either

$$rhS(t) \tag{22}$$

or

$$rhS(t + h). \tag{23}$$

Intuitively (22) and (23) are lower and upper bounds, respectively, for the actual interest (20), that is,

$$rhS(t) \le S(t + h) - S(t) \le rhS(t + h)$$

or

$$rS(t) \le \frac{S(t + h) - S(t)}{h} \le rS(t + h). \tag{24}$$

Taking the limit of (24) as $h \to 0$ gives

$$rS(t) \le \lim_{h \to 0} \frac{S(t + h) - S(t)}{h} \le rS(t),$$

and so it must follow that

$$\lim_{h \to 0} \frac{S(t + h) - S(t)}{h} = rS(t) \qquad \text{or} \qquad \frac{dS}{dt} = rS.$$

EXAMPLE

Suppose a suspended wire hangs under its own weight. As Figure 1.12(a) shows, this could be a long telephone wire between two posts. Our goal here is to determine the differential equation governing the shape that the hanging wire assumes.

Let us examine only a portion of the wire between the lowest point P_1 and any arbitrary point P_2. See Figure 1.12(b). Three forces are acting on the wire: the weight of the segment $P_1 P_2$, and the tensions $\mathbf{T}_1$ and $\mathbf{T}_2$ in the wire at P_1 and P_2, respectively. If w is the linear density (measured, say, in lb/ft) and s is the length of the segment $P_1 P_2$, its weight is necessarily ws.

Now the tension $\mathbf{T}_2$ resolves into horizontal and vertical components (scalar quantities) $T_2 \cos \theta$ and $T_2 \sin \theta$. Because of equilibrium we can write

$$|\mathbf{T}_1| = T_1 = T_2 \cos \theta,$$

and

$$ws = T_2 \sin \theta.$$

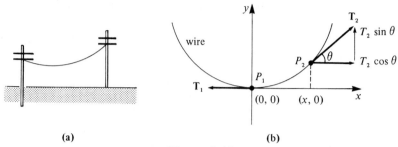

(a)   **(b)**

Figure 1.12

Dividing the last two equations we find

$$\tan \theta = \frac{ws}{T_1}$$

or
$$\frac{dy}{dx} = \frac{ws}{T_1}. \tag{25}$$

Now since the length of the arc between points P_1 and P_2 is

$$s = \int_0^x \sqrt{1 + \left(\frac{dy}{dx}\right)^2}\, dx$$

it follows from one form of the fundamental theorem of calculus that

$$\frac{ds}{dx} = \sqrt{1 + \left(\frac{dy}{dx}\right)^2}. \tag{26}$$

Differentiating (25) with respect to x and using (26) leads to

$$\frac{d^2y}{dx^2} = \frac{w}{T_1}\frac{ds}{dx}$$

or
$$\frac{d^2y}{dx^2} = \frac{w}{T_1}\sqrt{1 + \left(\frac{dy}{dx}\right)^2}. \tag{27}$$

One might conclude from Figure 1.12 that the shape which the hanging wire assumes is parabolic. However, this is not the case; a wire or heavy rope hanging only under its own weight takes on the shape of a hyperbolic cosine (see Problem 12, Exercises 3.3). Recall that the graph of the hyperbolic cosine is called a **catenary** which stems from the Latin word *catena* meaning "chain." The Romans used the catena as a dog leash. Probably the most graphic example of the shape of a catenary is the 630-ft-high Gateway arch in St. Louis, Missouri.

A system of differential equations

It may take more than one differential equation to describe a physical situation. The following mechanical system is said to have *two degrees of freedom*.

EXAMPLE   Two masses m_1 and m_2 are connected to two springs A and B having spring constants k_1 and k_2, respectively. In turn, the two springs are attached as shown in Figure 1.13. Let $x_1(t)$ and $x_2(t)$ denote the vertical displacements of the masses from their equilibrium positions. When the system is in motion spring B is subject to both an elongation and a compression, hence its net elongation is $x_2 - x_1$. Therefore it follows from Hooke's law that springs A and B exert forces

$$- k_1 x_1 \quad \text{and} \quad k_2(x_2 - x_1),$$

respectively, on m_1. The net force acting on m_1 is then

$$-k_1 x_1 + k_2(x_2 - x_1).$$

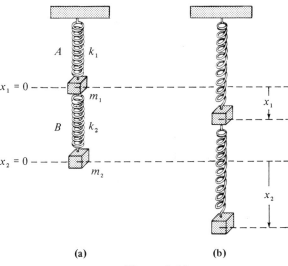

(a)   (b)

Figure 1.13

By Newton's second law we can write

$$m_1 \frac{d^2 x_1}{dt^2} = -k_1 x_1 + k_2(x_2 - x_1). \tag{28}$$

Similarly, the net force exerted on mass m_2 is due solely to the net elongation of B, that is,

$$-k_2(x_2 - x_1).$$

Thus it follows that $\quad m_2 \dfrac{d^2 x_2}{dt^2} = -k_2(x_2 - x_1). \tag{29}$

In other words, the motion of the coupled system is represented by the *simultaneous* second-order differential equations

$$m_1 x_1'' = -k_1 x_1 + k_2(x_2 - x_1)$$
$$m_2 x_2'' = -k_2(x_2 - x_1). \tag{30}$$

In the derivation of the coupled differential equations given in (30) we have made the usual simplifying assumptions: we assume that no external force is impressed on the system (for example, no one is shaking the system), that there are no damping forces, and that the masses of the two springs are neglibible.

EXERCISES 1.2   *Answers to odd-numbered problems begin on page A-2.*

[1.2.1]   In Problems 1–22 find the differential equation of the given family of curves.

1. $y = c_1 x + 2$

2. $c_1 y + 2x = 3$

3. $y = c_1 e^{-x}$

4. $y = e^x + c_1 e^{-x}$

5. $y^2 = c_1(x + 1)$

6. $c_1(y + 1)^2 = x$

7. $c_1 y^2 + 4y = 2x^2$

★ **8.** $c_1 x^2 - y^2 = 1$

9. $y = c_1 + c_2 e^x$

10. $y = c_1 e^{3x} + c_2 e^{-4x}$

11. $y = c_1 \sin \omega t + c_2 \cos \omega t$,   where ω is a constant not to be eliminated.

12. $y = c_1 \sin (\omega t + c_2)$,   where ω is a constant not to be eliminated.

13. $y = c_1 \sinh kt + c_2 \cosh kt$,   where k is a constant not to be eliminated.

14. $y = c_1 e^{kt} + c_2 e^{-kt}$,   where k is a constant not to be eliminated.

15. $y = c_1 e^{4x} + c_2 x e^{4x}$

★**16.** $y = c_1 e^x \cos x + c_2 e^x \sin x$

17. $y = c_1 + c_2 \ln x$

18. $y = c_1 x + c_2 x^2$

19. $y = c_1 e^x + c_2 e^{2x} + c_3 e^{3x}$

20. $y = c_1 + c_2 e^x + c_3 x e^x$

21. $r = c_1(1 + \cos \theta)$

22. $r = c_1(\sec \theta + \tan \theta)$

23. Find the differential equation of the family of straight lines passing through the origin.

24. Find the differential equation of the family of circles with centers on the y-axis.

25. Find the differential equation of the family of circles passing through the origin with centers on the x-axis.

26. Find the differential equation of the family of circles passing through $(0, -3)$ and $(0, 3)$, whose centers are on the x-axis.

27. Find the differential equation of the family of parabolas whose vertex is at the origin but whose focus is on the x-axis.

★**28.** Find the differential equation of the family of tangent lines to the parabola $y^2 = 2x$.

29. Find the differential equation of the family of parabolas whose vertex and focus are on the x-axis.

★30. Show that the two-parameter family

$$y^2 = 2c_1 x^2 y + c_2 x^4$$

yields the first-order differential equation

$$x\frac{dy}{dx} = 2y.$$

[1.2.2]   In Problems 31–44, derive the appropriate differential equation(s) describing the given physical situation.

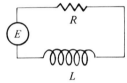

Figure 1.14

31. A series circuit contains a resistor and an inductor as shown in Figure 1.14. Determine the differential equation for the current $i(t)$ if the resistance is R, the inductance is L, and the impressed voltage is $E(t)$.

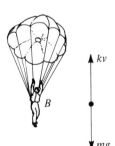

Figure 1.15

32. A series circuit contains a resistor and a capacitor as shown in Figure 1.15. Determine the differential equation for the charge $q(t)$ on the capacitor if the resistance is R, the capacitance is C, and the impressed voltage is $E(t)$.

EXAMPLE

Figure 1.16

Under some circumstances a falling body B of mass m (such as a man hanging from a parachute) encounters air resistance proportional to its instantaneous velocity, $v(t)$. Use Newton's second law to find the differential equation for the velocity of the body at any time.

Solution: Assuming that the downward direction is positive, the sum of the forces acting on the body is

$$mg - kv \qquad (31)$$

where k is a constant of proportionality, and the minus sign indicates that the resistance acts in a direction opposite to the motion. See Figure 1.16. Newton's second law can be written as

$$ma = m\frac{dv}{dt} \qquad (32)$$

where a represents acceleration. Equating (31) and (32) then gives

$$m\frac{dv}{dt} = mg - kv \quad \text{or} \quad \frac{dv}{dt} + \frac{k}{m}v = g.$$

33. What is the differential equation for the velocity v of a body of mass m falling vertically downward through a medium offering a resistance proportional to the square of the instantaneous velocity.

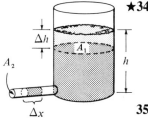

★**34.** Determine the differential equation governing the height h, at any time, of water flowing through an orifice at the bottom of a cylindrical tank. See Figure 1.17. Use the fact that the decrease in the volume of the water $-A_1 \, \Delta h \, (\Delta h < 0)$ is the same as the volume of the element of length Δx in a time Δt. Also use the fact that an object falling from rest acquires a velocity $\sqrt{2gh}$ ft/sec, $g = 32$, in h feet. (Where did this come from?)

Figure 1.17

35. A man M, starting at the origin moves in the direction of the positive x-axis pulling a weight along the curve C (called a **tractrix**) indicated in Figure 1.18. The weight, initially located on the y-axis at $(0, s)$, is pulled by a rope of constant length s which is kept taut throughout the motion.

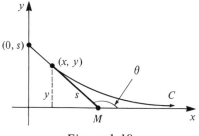

Figure 1.18

Find the differential equation of the path of motion. [*Hint:* The rope is always tangent to C; consider the angle of inclination θ as shown in the figure.]

★**36.** A projectile shot from a gun has weight $w = mg$ and velocity $\mathbf{v}$ tangent to its path of motion. Ignoring air resistance and all other forces except its weight, find the system of differential equations which describes the motion. [*Hint:* Use Newton's second law in the x and y direction. See Figure 1.19.]

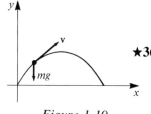

Figure 1.19

37. Determine the equations of motion if the projectile in Problem 36 encounters a retarding force $\mathbf{k}$ (of magnitude k) acting tangent to the path but opposite to the motion. [*Hint:* $\mathbf{k}$ is a multiple of the velocity, say $c\mathbf{v}$. See Figure 1.20.]

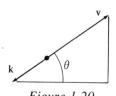

Figure 1.20

38. The rate at which a radioactive substance decays is proportional to the amount of the substance remaining at any time. Determine the differential equation for the amount $A(t)$.

39. A drug is infused into a patient's bloodstream at a constant rate r grams/sec. Simultaneously, the drug is removed at a rate proportional to the amount $x(t)$ of the drug present at any time. Determine the differential equation governing the amount $x(t)$.

40. Two chemicals A and B react to form a new chemical C. Assuming that the concentrations of both A and B decrease by the amount of C formed, find the differential equation governing the concentration $x(t)$ of the chemical C if the rate at which the chemical reaction takes place is proportional to the product of the remaining concentrations of A and B.

41. Light strikes a plane curve C in such a manner that all beams L parallel to the y-axis are reflected to a single point 0. Determine the differential equation for the function $y = f(x)$ describing the shape of the curve. (The fact that the angle of incidence is equal to the angle of reflection is a principle of optics.) [*Hint:* Inspection of Figure 1.21 shows that the inclination of the tangent line from the horizontal at $P(x, y)$ is $\pi/2 - \theta$ and that we can write $\phi = 2\theta$. (Why?) Also, don't be afraid to use a trigonometric identity.]

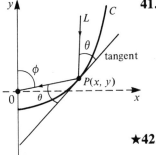

Figure 1.21

★**42.** A cylindrical barrel s feet in diameter of weight w lb is floating in water. After an initial depression the barrel exhibits an up and down bobbing motion along a vertical line. Using Figure 1.22(b), determine the differential equation for the vertical displacements $y(t)$, if the origin is taken to be on the vertical axis at the surface of the water when the barrel is at rest. Use Archimedes' principle that the buoyancy, or upward force of the water on the barrel, is equal to the weight of the water displaced, and the fact that the density of water is 62.4 lb/ft³. Assume that the downward direction is positive.

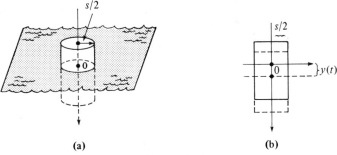

(a) (b)

Figure 1.22

43. A rocket is shot vertically upward from the surface of the earth. After all its fuel has been expended the mass of the rocket is a constant m. Use Newton's second law of motion, and the fact that the force of gravity varies inversely as the square of the distance, to find the differential equation for the distance y from the earth's center to the rocket at any time after burnout. State appropriate conditions at $t = 0$ associated with this differential equation.

44. Newton's second law $F = ma$ can be written $F = d/dt \, (mv)$. When the mass of an object is variable this latter formulation is used. The mass $m(t)$ of a rocket launched upward changes as its fuel is consumed.* If $v(t)$ denotes its velocity at any time it can be shown that

$$-mg = m\frac{dv}{dt} - V\frac{dm}{dt} \tag{33}$$

where V is the constant velocity of the exhaust gases relative to the rocket. Use (33) to find the differential equation for v if it is known that $m(t) = m_0 - at$ and $V = -b$ where m_0, a and b are constants.

CHAPTER 1 SUMMARY

We classify a differential equation by its type: **ordinary** or **partial,** by its **order,** and whether it is **linear** or **nonlinear.**

A **solution** of a differential equation is any function, having a sufficient number of derivatives, which satisfies the equation identically on some interval.

When solving an nth-order ordinary differential equation we expect to find an n-parameter family of solutions. A **particular solution** is any solution free of arbitrary parameters that satisfies the differential equation. A **singular solution** is any solution that cannot be obtained from an n-parameter family of solutions by assigning values to the parameters. When an n-parameter family of solutions gives every solution of a differential equation on some interval it is then called a **general** or a **complete** solution.

Starting with an n-parameter family of curves in the plane, we can find, in *most* cases, an nth-order differential equation representing the family. In the analysis of physical problems, many differential equations can be obtained by equating two different empirical formulations of the same situation. For example, a differential equation of motion can sometimes be obtained by simply equating Newton's second law of motion with the net forces acting on a body.

---

*It is assumed that the *total* mass:

mass of vehicle + mass of fuel + mass of exhaust gases

is constant. In this case $m(t)$ = mass of vehicle + mass of fuel.

CHAPTER 1 TEST

Answers to odd-numbered problems begin on page A-2.

1. Classify the following differential equations as to type and order. Classify the ordinary differential equations as to linearity.

   (a) $(2xy - y^2) \, dx + e^x dy = 0$ _____

   (b) $(\sin xy) \, y''' + 4xy' = 0$ _____

   (c) $\dfrac{\partial^2 u}{\partial x^2} + \dfrac{\partial^2 u}{\partial y^2} = u$ _____

   (d) $x^2 \dfrac{d^2 y}{dx^2} - 3x \dfrac{dy}{dx} + y = x^2$ _____

2. Verify that the indicated function is a solution of the given differential equation.

   (a) $y' + 2xy = 2 + x^2 + y^2; \quad y = x + \tan x$

   (b) $x^2 y'' + xy' + y = 0; \quad y = c_1 \cos (\ln x) + c_2 \sin (\ln x), \; x > 0$

   (c) $y''' - 2y'' - y' + 2y = 6; \quad y = c_1 e^x + c_2 e^{-x} + c_3 e^{2x} + 3$

3. By inspection determine at least one solution for each of the following differential equations.

   (a) $y' = 2x$ _____

   (b) $y'' = 1$ _____

   (c) $y'' = y'$ _____

   (d) $\dfrac{dy}{dx} = 5y$ _____

   (e) $y' = y^3 - 8$ _____

   (f) $2y \dfrac{dy}{dx} = 1$ _____

4. Explain why the differential equation

   $$\left(\frac{dy}{dx} \right)^2 = \frac{4 - y^2}{4 - x^2}$$

   possesses no real solutions for $|x| < 2, |y| > 2$. Are there other regions in the xy-plane for which the equation has no solutions?

5. Determine an interval on which $y^2 - 2y = x^2 - x - 1$ defines a solution of $2(y - 1) \, dy + (1 - 2x) \, dx = 0$.

6. **(a)** Show that the differential equation of the family of circles $(x - c)^2 + y^2 = 25$ is $(y')^2 = (25 - y^2)/y^2$.

   (b) Find two singular solutions of the differential equation in part (a).

   continued

**CHAPTER 1
TEST**

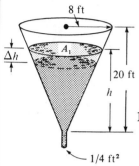

Figure 1.23

7. Find the differential equation which represents the family of straight lines passing through the point (2, 1).

8. Find the differential equation representing the family of circles passing through the origin, with centers on the line $y = x$.

9. The conical tank shown in Figure 1.23 loses water out of an orifice at its bottom. If the cross-sectional area of the orifice is $(1/4)\text{ft}^2$, find the differential equation representing the height of the water h at any time. (See Problem 34, Exercises 1.2.)

10. A weight of 96 lb slides down an incline making a 30° angle with the horizontal. If the coefficient of sliding friction is μ, determine the differential equation for the velocity $v(t)$ of the weight at any time. Use the fact that the force of friction opposing the motion is μN, where N is the normal component of the weight. See Figure 1.24.

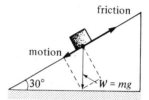

Figure 1.24

First-Order Differential Equations

2.1 Preliminary Theory

Initial-value problem

We are often interested in solving a first-order differential equation

$$\frac{dy}{dx} = f(x, y)^*$$ (1)

subject to a side condition

$$y(x_0) = y_0$$ (2)

where x_0 is a point in I and y_0 is an arbitrary real number. The problem

$$Solve: \quad \frac{dy}{dx} = f(x, y)$$

$$Subject\ to: \quad y(x_0) = y_0$$ (3)

is called an **initial-value problem.** The side condition (2) is known as an **initial condition.**

EXAMPLE

We have seen that $y = ce^x$ is a one-parameter family of solutions for $y' = y$ on the interval $-\infty < x < \infty$. If we specify, say, $y(0) = 3$, then substituting

---

*In this text we shall assume that a differential equation $F(x, y, y', \ldots, y^{(n)}) = 0$ can be solved for the highest order derivative: $y^{(n)} = f(x, y, y', \ldots, y^{(n-1)})$. There are exceptions.

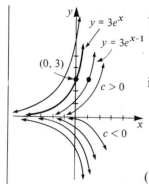

Figure 2.1

$x = 0$, $y = 3$ in the family yields $3 = ce^0 = c$. Thus, as shown in Figure 2.1,

$$y = 3e^x$$

is a solution of the initial-value problem

$$y' = y$$

$$y(0) = 3.$$

Had we demanded that a solution of $y' = y$ pass through the point $(1, 3)$ rather than $(0, 3)$ then $y(1) = 3$ would yield $c = 3e^{-1}$ and so $y = 3e^{x-1}$. The graph of this function is also indicated in Figure 2.1.

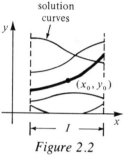

Figure 2.2

Two fundamental questions arise in the consideration of an initial-value problem such as (3):

Does a solution of the problem *exist*?

If a solution exists, is it *unique*, or the only solution of the problem?

Geometrically we are asking in the second question: Of all the solutions of a differential equation (1) that exist on an interval I, is there only one whose graph passes through (x_0, y_0)? See Figure 2.2.

As the next example shows the answer to the second question is sometimes no.

EXAMPLE

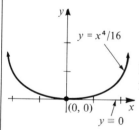

Figure 2.3

Figure 2.3 shows that the initial-value problem

$$\frac{dy}{dx} - xy^{1/2} = 0$$

$$y(0) = 0$$

has at least *two* solutions on the interval $-\infty < x < \infty$. The graphs of the functions

$$y \equiv 0 \quad \text{and} \quad y = \frac{x^4}{16}$$

both pass through $(0, 0)$.

It is often desirable to know in advance of tackling an initial-value problem whether a solution exists, and when it does, whether it is the only solution of the problem. The following theorem due to Picard[*] gives *sufficient* conditions for the existence of a unique solution of (3).

---

[*]Emile Picard, a French mathematician (1856–1941).

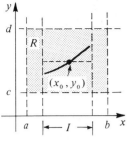

Figure 2.4

THEOREM 2.1 Let R be a rectangular region in the xy-plane defined by $a \leq x \leq b$, $c \leq y \leq d$ that contains the point (x_0, y_0) in its interior. If $f(x, y)$ and $\partial f/\partial y$ are continuous on R, then there exists an interval I centered at x_0 and a unique function $y(x)$ defined on I satisfying the initial-value problem (3). $\square$

The foregoing is one of the most popular existence and uniqueness theorems for first-order differential equations because the criteria of continuity of $f(x, y)$ and $\partial f/\partial y$ are relatively easy to check. In general, it is not always possible to find a specific interval I on which a solution is defined without actually solving the differential equation (see Problem 16). The geometry of Theorem 2.1 is illustrated in Figure 2.4.

EXAMPLE

We have already seen in the last example that the differential equation

$$\frac{dy}{dx} - xy^{1/2} = 0$$

possesses at least two solutions whose graphs pass through $(0, 0)$. By writing the equation in the form $dy/dx = xy^{1/2}$ we can identify

$$f(x, y) = xy^{1/2} \qquad \text{and} \qquad \frac{\partial f}{\partial y} = \frac{x}{2y^{1/2}}.$$

We note that both functions are continuous in the upper half plane defined by $y > 0$. We conclude from Theorem 2.1 that through any point (x_0, y_0), $y_0 > 0$ (say, for example, $(0, 1)$) there is some interval around x_0 on which the given differential equation has a unique solution.

EXAMPLE

Theorem 2.1 guarantees that there exists an interval about $x = 0$ on which $y = 3e^x$ is the only solution of the problem

$$y' = y$$

$$y(0) = 3.$$

This follows from the fact that $f(x, y) = y$ and $\partial f/\partial y = 1$ are continuous throughout the entire xy-plane. It can be further shown that this interval is $-\infty < x < \infty$.

EXAMPLE

For

$$\frac{dy}{dx} = x^2 + y^2$$

we observe that $f(x, y) = x^2 + y^2$ and $\partial f/\partial y = 2y$ are continuous throughout the entire xy-plane. Therefore through any given point (x_0, y_0) there passes one and only one solution of the differential equation.

The reader should be aware of the distinction between a solution existing and exhibiting a solution. Clearly if we find a solution by exhibiting it we certainly can say that it exists, but on the other hand a solution can exist but we may not be able to display it. From the last example we know that a solution of the problem $dy/dx = x^2 + y^2$, $y(0) = 1$ exists on some interval around $x = 0$ and is unique. However, the best we can do in this case is to approximate the solution (see Chapter 9).

Remark: The conditions stated in Theorem 2.1 are *sufficient* but not *necessary*. When $f(x, y)$ and $\partial f/\partial y$ are continuous on a rectangular region R it must *always* follow that there exists a unique solution of (3) when (x_0, y_0) is a point interior to R. However, if the conditions stated in the hypothesis of the theorem do not hold, then the initial-value problem (3) may still have either (a) no solution, (b) more than one solution, or (c) a unique solution. Furthermore, the continuity condition on $\partial f/\partial y$ can be relaxed somewhat without changing the conclusion of the theorem. This results in a stronger theorem but is, unfortunately, not as easy to apply as Theorem 2.1. Indeed if we are not interested in uniqueness, then a famous theorem due to Peano[*] states that the continuity of $f(x, y)$ on R is sufficient to guarantee the existence of at least one solution of $dy/dx = f(x, y)$ through a point (x_0, y_0) interior to R.

EXERCISES 2.1   *Answers to odd-numbered problems begin on page A-3.*

In Problems 1–10 determine a region of the xy-plane for which the given differential equation would have a unique solution through a point (x_0, y_0) in the region.

1. $\dfrac{dy}{dx} = y^{2/3}$ 2. $\dfrac{dy}{dx} = \sqrt{xy}$

3. $x\dfrac{dy}{dx} = y$ 4. $\dfrac{dy}{dx} - y = x$

5. $(4 - y^2)y' = x^2$ 6. $(1 + y^3)y' = x^2$

7. $(x^2 + y^2)y' = y^2$ ★ 8. $(y - x)y' = y + x$

9. $\dfrac{dy}{dx} = x^3 \cos y$ 10. $\dfrac{dy}{dx} = (x - 1)e^{y/(x-1)}$

11. By inspection find at least two solutions of the initial-value problem

$$y' = 3y^{2/3}$$

$$y(0) = 0.$$

---

[*]Guiseppe Peano, an Italian mathematician (1858–1932).

12. By inspection find at least two solutions of the initial-value problem

$$x\frac{dy}{dx} = 2y$$

$$y(0) = 0.$$

13. By inspection determine a solution of the nonlinear differential equation $y' = y^3$ satisfying $y(0) = 0$. Is the solution unique?

★**14.** By inspection find a solution of the initial-value problem

$$y' = |y - 1|$$

$$y(0) = 1.$$

State why the conditions of Theorem 2.1 do not hold for this differential equation. Although we shall not prove it, the solution to this initial-value problem is unique.

15. Verify that $y = cx$ is a solution of the differential equation $xy' = y$ for every value of the parameter c. Find at least two solutions of the initial-value problem

$$xy' = y$$

$$y(0) = 0.$$

Observe that the piecewise-defined function

$$y = \begin{cases} 0, & x < 0, \\ x, & x \geq 0 \end{cases}$$

satisfies the condition $y(0) = 0$. Is it a solution of the initial-value problem?

★**16. (a)** Consider the differential equation

$$\frac{dy}{dx} = 1 + y^2.$$

Determine a region of the xy-plane for which the equation has a unique solution through a point (x_0, y_0) in the region.

(b) Formally show that $y = \tan x$ satisfies the differential equation and the condition $y(0) = 0$.

(c) Explain why $y = \tan x$ is not a solution of the initial-value problem

$$\frac{dy}{dx} = 1 + y^2$$

$$y(0) = 0$$

on the interval $-2 < x < 2$.

(d) Explain why $y = \tan x$ is a solution of the initial-value problem in part (c) on the interval $-1 < x < 1$.

In Problems 17–20 determine whether Theorem 2.1 guarantees that the differential equation $y' = \sqrt{y^2 - 9}$ possesses a unique solution through the given point.

17. $(1, 4)$ **18.** $(5, 3)$

19. $(2, -3)$ **20.** $(-1, 1)$

2.2   Separable Variables

We begin our study of the methodology of solving a first-order equation with the simplest of all differential equations.

If $g(x)$ is a given continuous function then the first-order equation

$$\frac{dy}{dx} = g(x) \tag{1}$$

can be solved by integration. The solution of (1) is

$$y = \int g(x)\, dx + c.$$

EXAMPLES   Solve **(a)** $\dfrac{dy}{dx} = 1 + e^{2x}$,   **(b)** $\dfrac{dy}{dx} = \sin x$.

Solution: **(a)** $y = \int (1 + e^{2x})\, dx = x + \dfrac{1}{2}e^{2x} + c$

(b) $y = \int \sin x\, dx = -\cos x + c.$

Equation (1), as well as its method of solution, is just a special case of the following.

DEFINITION 2.1   A differential equation of the form

$$\frac{dy}{dx} = \frac{g(x)}{h(y)}$$

is said to be **separable** or to have **separable variables.**   ☐

Observe that a separable equation can be written as

$$h(y)\frac{dy}{dx} = g(x). \tag{2}$$

It is seen immediately that (2) reduces to (1) when $h(y) \equiv 1$.

Now if $y = f(x)$ denotes a solution of (2) we must have

$$h(f(x))\, f'(x) = g(x)$$

and therefore $$\int h(f(x))\, f'(x)\, dx = \int g(x)\, dx + c. \tag{3}$$

But $dy = f'(x)\,dx$ so (3) is the same as

$$\int h(y)\,dy = \int g(x)\,dx + c \qquad (4)$$

The method of solution

Equation (4) indicates the procedure for solving separable differential equations. A one-parameter family of solutions, usually given implicitly, is obtained by integrating both sides of $h(y)\,dy = g(x)\,dx$.

Note: There is no need to use two constants in the integration of a separable equation since

$$\int h(y)\,dy + c_1 = \int g(x)\,dx + c_2$$

$$\int h(y)\,dy = \int g(x)\,dx + c_2 - c_1$$

$$= \int g(x)\,dx + c$$

where c is completely arbitrary. In many instances throughout the following chapters, we shall not hesitate to relabel constants in a manner which may prove convenient for a given equation. For example, multiples of constants or combinations of constants can sometimes be replaced by one constant.

EXAMPLE

Solve $(1 + x)\,dy - y\,dx = 0$.

Solution: Dividing by $(1 + x)y$ we can write $dy/y = dx/(1 + x)$ from which it follows

$$\int \frac{dy}{y} = \int \frac{dx}{1 + x}.$$

$$\ln|y| = \ln|1 + x| + c_1$$

$$y = e^{\ln|1+x|+c_1}$$

$$= e^{\ln|1+x|} \cdot e^{c_1}$$

$$= (1 + x)e^{c_1}.$$

Relabeling e^{c_1} as c then gives $y = c(1 + x)$.*

Alternative solution: Since each integral results in a logarithm, a judicious choice for the constant of integration is $\ln|c|$ rather than c:

---

*One might object that since $e^{c_1} > 0$ then necessarily $c > 0$. Because $\ln|y| - \ln|1 + x| = c_1$ implies $|y/(1 + x)| = e^{c_1}$ and $y = \pm e^{c_1}(1 + x)$ we are, strictly speaking, replacing $\pm e^{c_1}$ by c. In other words, $y = c(1 + x)$ is a solution of the differential equation on the interval $-\infty < x < \infty$ *for any real choice of* c. The reader is encouraged to worry about such things. Albeit unfortunate, to save time in solving differential equations many manipulations are carried out without much thought to mathematical rigor. This is what is known in the trade as "proceeding formally."

$$\ln |y| = \ln |1 + x| + \ln |c|$$

or

$$\ln |y| = \ln |c(1 + x)|$$

so that

$$y = c(1 + x).$$

Even if not *all* the indefinite integrals are logarithms, it may still be advantageous to use $\ln |c|$. However, no firm rule can be given.

EXAMPLE   Solve

$$xy^4 \, dx + (y^2 + 2)e^{-3x} \, dy = 0. \tag{5}$$

Solution: By multiplying the given equation by e^{3x} and dividing by y^4 we obtain

$$xe^{3x} \, dx + \frac{y^2 + 2}{y^4} \, dy = 0 \qquad \text{or} \qquad xe^{3x} \, dx + (y^{-2} + 2y^{-4}) \, dy = 0. \tag{6}$$

Using integration by parts on the first term yields

$$\tfrac{1}{3}xe^{3x} - \tfrac{1}{9}e^{3x} - y^{-1} - \tfrac{2}{3}y^{-3} = c_1.$$

The one-parameter family of solutions can also be written as

$$e^{3x}(3x - 1) = \frac{9}{y} + \frac{6}{y^3} + c \tag{7}$$

where the constant $9c_1$ is rewritten as c.

Two points are worth mentioning at this time. Firstly, unless it is important or convenient there is no need to try to solve an expression representing a family of solutions for y explicitly in terms of x. Equation (7) shows this task may present more problems than just the drudgery of symbol pushing. As a consequence it is often the case that the interval over which a solution is valid is not apparent. Secondly, some care should be exercised when separating variables to make certain that divisors are not zero. A constant solution may sometimes get lost in the shuffle of solving the problem.

EXAMPLE   An equivalent form of equation (5) in the last example is

$$(y^2 + 2)\frac{dy}{dx} = -xe^{3x}y^4 \tag{8}$$

whereas (6) and (7) are restricted by $y \neq 0$. Observe that $y \equiv 0$ is a perfectly good solution of equation (8) but is not a member of the set of solutions defined by (7).

EXAMPLE

Solve
$$x \sin x \, e^{-y} \, dx - y \, dy = 0.$$

Solution: After dividing by e^{-y} the equation becomes
$$x \sin x \, dx = ye^{y} \, dy.$$

Using integration by parts on both sides of the equality gives
$$-x \cos x + \sin x = ye^{y} - e^{y} + c.$$

EXAMPLE

Solve
$$\frac{dy}{dx} = \frac{x - 4}{x - 3}$$

subject to the initial condition $y(1) = 2$.

Solution: The right side of the equation can be written as
$$\frac{dy}{dx} = 1 - \frac{1}{x - 3}$$

by long division. Integrating the last equation gives
$$y = x - \ln |x - 3| + c.$$

When $x = 1$, $y = 2$, we have
$$2 = 1 - \ln |-2| + c$$

so
$$c = 1 + \ln 2.$$

Thus
$$y = x - \ln |x - 3| + 1 + \ln 2$$

or
$$y = x + 1 + \ln \frac{2}{|x - 3|}.$$

It can be shown as a consequence of Theorem 2.1 that the above is the only solution of the problem on the interval $-\infty < x < 3$.

EXAMPLE

Solve
$$\frac{dy}{dx} = -\frac{x}{y}$$

subject to $y(4) = 3$.

Solution: We saw in Chapter 1 that the differential equation of the family of concentric circles $x^2 + y^2 = c^2$ was $dy/dx = -x/y$. We now are in a position of being able to work "backwards." Obviously $y \, dy = -x \, dx$ and so
$$\int y \, dy = -\int x \, dx$$

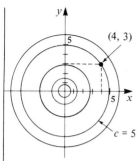

Figure 2.5

$$\frac{y^2}{2} = -\frac{x^2}{2} + c_1$$

or

$$x^2 + y^2 = c^2$$

where the constant $2c_1$ is replaced by c^2.

Now when $x = 4$, $y = 3$ so that $16 + 9 = 25 = c^2$. Thus the initial-value problem determines the solution $x^2 + y^2 = 25$. In view of Theorem 2.1 we can conclude that it is the only circle of the family passing through the point (4, 3). See Figure 2.5.

One should never become too complacent when solving differential equations. Even though the solution of a first-order equation contains an arbitrary constant, we have seen previously that we may not be able to find all solutions by simply assigning different values to this parameter.

EXAMPLE   Solve

$$\frac{dy}{dx} = y^2 - 4$$

subject to $y(0) = -2$.

Solution: We put the equation into the form

$$\frac{dy}{y^2 - 4} = dx \tag{9}$$

and use partial fractions on the left side. We have

$$\left[\frac{-\frac{1}{4}}{y + 2} + \frac{\frac{1}{4}}{y - 2}\right] dy = dx \tag{10}$$

so that

$$-\tfrac{1}{4} \ln |y + 2| + \tfrac{1}{4} \ln |y - 2| = x + c_1. \tag{11}$$

Thus

$$\ln \left|\frac{y - 2}{y + 2}\right| = 4x + c_2 \qquad [c_2 = 4c_1]$$

and

$$\frac{y - 2}{y + 2} = ce^{4x}$$

where we have formally replaced e^{c_2} by c. Finally, we obtain

$$y = 2\frac{1 + ce^{4x}}{1 - ce^{4x}}. \tag{12}$$

Substituting $x = 0$, $y = -2$ leads to the somewhat embarrassing dilemma

$$-2 = 2\frac{1 + c}{1 - c}$$

$$-1 + c = 1 + c \qquad \text{or} \qquad -1 = 1.$$

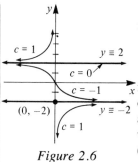

Figure 2.6

Let us consider the differential equation a little more carefully. The fact is, the equation

$$\frac{dy}{dx} = (y + 2)(y - 2)$$

is satisfied by two constant functions, namely, $y \equiv -2$ and $y \equiv 2$. Inspection of equations (9), (10), and (11) clearly indicates we must preclude $y = -2$ and $y = 2$ at those steps in our solution. But it is interesting to observe that we can subsequently recover the solution $y \equiv 2$ by setting $c = 0$ in equation (12). However, there is no finite value of c which will ever yield the solution $y \equiv -2$. This latter constant function is the only solution to the original initial-value problem. See Figure 2.6.

If, in the preceding example, had we used $\ln |c|$ for the constant of integration, then the form of the one-parameter family of solutions would be

$$y = 2\frac{c + e^{4x}}{c - e^{4x}}. \tag{13}$$

Note that (13) reduces to $y \equiv -2$ when $c = 0$, but now there is no finite value of c that will give the constant solution $y \equiv 2$.

If an initial condition leads to a particular solution by finding a specific value of the parameter c in a family of solutions for a first-order differential equation it is a natural inclination of most students (and instructors) to relax and be content. In Secion 2.1 we saw, however, that a solution of an initial-value problem may not be unique. For example, the problem

$$\frac{dy}{dx} - xy^{1/2} = 0$$
$$y(0) = 0 \tag{14}$$

has at least two solutions, namely, $y \equiv 0$ and $y = x^4/16$. We are now in a position to solve the equation. By separation of variables

$$y^{-1/2} \, dy = x \, dx$$
$$2y^{1/2} = \frac{x^2}{2} + c_1$$

or
$$y = \left(\frac{x^2}{4} + c\right)^2.$$

When $x = 0$, $y = 0$ so necessarily $c = 0$. Therefore, $y = x^4/16$. The solution $y \equiv 0$ was lost by dividing by $y^{1/2}$. In addition, the initial-value problem (14) possesses infinitely more solutions, since for any choice of the parameter $a \geq 0$ the piecewise defined function

$$y = \begin{cases} 0, & x < a \\ \dfrac{(x^2 - a^2)^2}{16}, & x \geq a \end{cases}$$

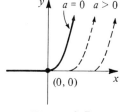

Figure 2.7

satisfies both the differential equation and the initial condition. See Figure 2.7.

EXERCISES 2.2   *Answers to odd-numbered problems begin on page A-3.*

In Problems 1–40 solve the given differential equation by separation of variables.

1. $\dfrac{dy}{dx} = \cos 2x$

2. $\dfrac{dy}{dx} = (x + 1)^2$

3. $dx - x^2\, dy = 0$

4. $dx + e^{3x}\, dy = 0$

5. $(x + 1)\dfrac{dy}{dx} = x$

6. $e^x\dfrac{dy}{dx} = 2x$

7. $xy' = 4y$

8. $\dfrac{dy}{dx} + 2xy = 0$

9. $\dfrac{dy}{dx} = \dfrac{y^3}{x^2}$

10. $\dfrac{dy}{dx} = \dfrac{y + 1}{x}$

11. $\dfrac{dx}{dy} = \dfrac{x^2 y^2}{1 + x}$

12. $\dfrac{dx}{dy} = \dfrac{1 + 2y^2}{y \sin x}$

13. $\dfrac{dy}{dx} = e^{3x + 2y}$

★**14.** $e^x y\, dy - (e^{-y} + e^{-2x-y})\, dx = 0$

15. $(4y + yx^2)\, dy - (2x + xy^2)\, dx = 0$

16. $(1 + x^2 + y^2 + x^2 y^2)\, dy = y^2\, dx$

17. $2y(x + 1)\, dy = x\, dx$

18. $x^2 y^2\, dy = (y + 1)\, dx$

19. $y \ln x \dfrac{dx}{dy} = \left(\dfrac{y + 1}{x}\right)^2$

20. $\dfrac{dy}{dx} = \left(\dfrac{2y + 3}{4x + 5}\right)^2$

21. $\dfrac{dS}{dr} = kS$

22. $\dfrac{dQ}{dt} = k(Q - 70)$

23. $\dfrac{dP}{dt} = P(1 - P)$

★**24.** $(yx^2 - y)\, dy + (y + 1)^2\, dx = 0$

25. $\sec^2 x\, dy + \csc y\, dx = 0$

26. $\sin 3x\, dx + 2y \cos^3 3x\, dy = 0$

27. $e^y \sin 2x\, dx + \cos x(e^{2y} - y)\, dy = 0$

28. $\sec x\, dy = x \cot y\, dx$

29. $(e^y + 1)^2 e^{-y} \, dx + (e^x + 1)^3 e^{-x} \, dy = 0$

30. $\dfrac{y}{x} \dfrac{dy}{dx} = (1 + x^2)^{-1/2}(1 + y^2)^{1/2}$

31. $\dfrac{dN}{dt} + N = Nte^{t+2}$ ★**32.** $2\dfrac{dy}{dx} - \dfrac{1}{y} = \dfrac{2x}{y}$

33. $\dfrac{dy}{dx} = \dfrac{xy + 3x - y - 3}{xy - 2x + 4y - 8}$ **34.** $\dfrac{dy}{dx} = \dfrac{xy + 2y - x - 2}{xy - 3y + x - 3}$

35. $\dfrac{dy}{dx} = \sin x \, (\cos 2y - \cos^2 y)$

36. $\sec y \, \dfrac{dy}{dx} + \sin (x - y) = \sin (x + y)$

37. $x\sqrt{1 - y^2} \, dx = dy$

38. $y(4 - x^2)^{1/2} \, dy = (4 + y^2)^{1/2} \, dx$

39. $(e^x + e^{-x}) \dfrac{dy}{dx} = y^2$ ★**40.** $(x + \sqrt{x})\dfrac{dy}{dx} = y + \sqrt{y}$

In Problems 41–48 solve the given differential equation subject to the indicated initial conditon.

41. $\sin x(e^{-y} + 1) \, dx = (1 + \cos x) \, dy$, $y(0) = 0$

42. $(1 + x^4) \, dy + x(1 + 4y^2) \, dx = 0$, $y(1) = 0$

43. $y \, dy = 4x(y^2 + 1)^{1/2} \, dx$, $y(0) = 1$

44. $\dfrac{dy}{dt} + ty = y$, $y(1) = 3$

45. $\dfrac{dx}{dy} = 4(x^2 + 1)$, $x\left(\dfrac{\pi}{4}\right) = 1$

46. $\dfrac{dy}{dx} = \dfrac{y^2 - 1}{x^2 - 1}$, $y(2) = 2$

47. $x^2y' = y - xy$, $y(-1) = -1$

★**48.** $y' + 2y = 1$, $y(0) = \dfrac{5}{2}$

In Problems 49 and 50 find a solution of the given differential equation that passes through the indicated points.

49. $\dfrac{dy}{dx} - y^2 = -9$ **(a)** $(0, 0)$ **(b)** $(0, 3)$ **(c)** $\left(\dfrac{1}{3}, 1\right)$

★**50.** $x \, dy + (y - y^2) \, dx = 0$

     (a) $(0, 1)$ **(b)** $(0, 0)$ **(c)** $\left(\dfrac{1}{2}, \dfrac{1}{2} \right)$

Miscellaneous problems

A differential equation of the form $dy/dx = f(ax + by + c)$, $b \neq 0$, can always be reduced to an equation with separable variables by means of the substitution $u = ax + by + c$. Use this procedure to solve Problems 51–56.

EXAMPLE

Solve
$$\frac{dy}{dx} = \frac{1}{x + y + 1}.$$

Solution: Let $u = x + y + 1$ so that $du/dx = 1 + dy/dx$. The given equation then becomes

$$\frac{du}{dx} - 1 = \frac{1}{u}$$

or

$$\frac{du}{dx} = \frac{1 + u}{u}$$

$$\frac{u \, du}{1 + u} = dx.$$

Integrating

$$\left(1 - \frac{1}{1 + u} \right) du = dx$$

gives

$$u - \ln |1 + u| = x + c_1$$

$$x + y + 1 - \ln |x + y + 2| = x + c_1$$

$$y + 1 - \ln |x + y + 2| = c_1 \qquad \text{or} \qquad x + y + 2 = ce^y$$

where we have replaced $e^{1 - c_1}$ by c.

---

51. $\dfrac{dy}{dx} = (x + y + 1)^2$

52. $\dfrac{dy}{dx} = \dfrac{1 - x - y}{x + y}$

53. $\dfrac{dy}{dx} = \tan^2 (x + y)$

54. $\dfrac{dy}{dx} = \sin (x + y)$

55. $\dfrac{dy}{dx} = 2 + \sqrt{y - 2x + 3}$

56. $\dfrac{dy}{dx} = 1 + e^{y - x + 5}$

2.3 Homogeneous Equations

If an equation in the differential form

$$M(x, y) \, dx + N(x, y) \, dy = 0 \tag{1}$$

has the property that

$$M(tx, ty) = t^n M(x, y) \quad \text{and} \quad N(tx, ty) = t^n N(x, y)$$

we say it has **homogeneous coefficients,** or is a **homogeneous equation.** The important point in the subsequent discussion is the fact that a homogeneous differential equation *can always be reduced to a separable equation* through an appropriate algebraic substitution. Before pursuing the method of solution for this type of differential equation let us closely examine the nature of homogeneous functions.

DEFINITION 2.2 If $f(tx, ty) = t^n f(x, y)$, for some real number n, then $f(x, y)$ is said to be a homogeneous function of *degree n*. □

EXAMPLES

(a) $f(x, y) = x - 3\sqrt{xy} + 5y$

$$f(tx, ty) = (tx) - 3\sqrt{(tx)(ty)} + 5(ty)$$

$$= tx - 3\sqrt{t^2 xy} + 5ty$$

$$= t[x - 3\sqrt{xy} + 5y] = tf(x, y)$$

The function is homogeneous of degree one.

(b) $f(x, y) = \sqrt{x^3 + y^3}$

$$f(tx, ty) = \sqrt{t^3 x^3 + t^3 y^3}$$

$$= t^{3/2} \sqrt{x^3 + y^3} = t^{3/2} f(x, y)$$

The function is homogeneous of degree 3/2.

(c) $f(x, y) = x^2 + y^2 + 1$

$$f(tx, ty) = t^2 x^2 + t^2 y^2 + 1 \ne t^2 f(x, y)$$

since $t^2 f(x, y) = t^2 x^2 + t^2 y^2 + t^2$. The function is not homogeneous.

(d) $f(x, y) = \dfrac{x}{2y} + 4$

$$f(tx, ty) = \frac{tx}{2ty} + 4$$

$$= \frac{x}{2y} + 4 = t^0 f(x, y)$$

The function is homogeneous of degree zero.

As parts (c) and (d) of the above example show, a constant added to a function destroys homogeneity, *unless* the function is homogeneous of degree zero. Also, in many instances a homogeneous function can be recognized by examining the total degree of each term.

EXAMPLES

(a) $f(x, y) = 6xy^3 - x^2y^2$

degree 1, degree 3 $\}$ degree 4

degree 2, degree 2 $\}$ degree 4

The function is homogeneous of degree 4.

(b) $f(x, y) = x^2 - y$

degree 2

degree 1

The function is not homogeneous.

If $f(x, y)$ is a homogeneous function of degree n, notice that we can write

$$f(x, y) = x^n f\left(1, \frac{y}{x}\right) \quad \text{and} \quad f(x, y) = y^n f\left(\frac{x}{y}, 1\right)$$

where $f(1, y/x)$ and $f(x/y, 1)$ are both of degree zero.

EXAMPLE

We see $f(x, y) = x^2 + 3xy + y^2$ is homogeneous of degree 2. Thus

$$f(x, y) = x^2\left[1 + 3\left(\frac{y}{x}\right) + \left(\frac{y}{x}\right)^2\right] = x^2 f\left(1, \frac{y}{x}\right)$$

$$f(x, y) = y^2\left[\left(\frac{x}{y}\right)^2 + 3\left(\frac{x}{y}\right) + 1\right] = y^2 f\left(\frac{x}{y}, 1\right).$$

The method of solution

An equation of the form $M(x, y)\, dx + N(x, y)\, dy = 0$ where M and N have the same degree of homogeneity can be reduced to separable variables by *either* the substitution $y = ux$ or $x = vy$, where u and v are new dependent variables. In particular if we choose $y = ux$ then $dy = u\, dx + x\, du$. Hence the differential equation (1) becomes

$$M(x, ux)\, dx + N(x, ux)[u\, dx + x\, du] = 0.$$

Now by homogeneity of M and N we can write

$$x^n M(1, u)\, dx + x^n N(1, u)[u\, dx + x\, du] = 0$$

or

$$[M(1, u) + uN(1, u)]\, dx + xN(1, u)\, du = 0$$

which gives

$$\frac{dx}{x} + \frac{N(1, u)\, du}{M(1, u) + uN(1, u)} = 0.$$

We hasten to point out that the preceding formula should not be memorized; rather, *the procedure should be worked through each time*. The proof that the substitution $x = vy$ in (1) also leads to a separable equation is left as an exercise.

EXAMPLE

Solve
$$(x^2 + y^2)\, dx + (x^2 - xy)\, dy = 0.$$

Solution: Both $M(x, y)$ and $N(x, y)$ are homogeneous of degree 2. If we let $y = ux$ it follows that

$$(x^2 + u^2 x^2)\, dx + (x^2 - ux^2)[u\, dx + x\, du] = 0$$

$$x^2(1 + u)\, dx + x^3(1 - u)\, du = 0$$

$$\frac{1 - u}{1 + u}\, du + \frac{dx}{x} = 0$$

$$\left[-1 + \frac{2}{1 + u} \right] du + \frac{dx}{x} = 0$$

$$-u + 2 \ln |1 + u| + \ln |x| + \ln |c| = 0$$

$$-\frac{y}{x} + 2 \ln \left| 1 + \frac{y}{x} \right| + \ln |x| + \ln |c| = 0.$$

Using the properties of logarithms the above solution can be written in the alternative form

$$c(x + y)^2 = x e^{\,y/x}.$$

EXAMPLE

Solve
$$(2\sqrt{xy} - y)\, dx - x\, dy = 0.$$

Solution: The coefficients $M(x, y)$ and $N(x, y)$ are homogeneous of degree 1. If $y = ux$ the differential equation becomes

$$\frac{du}{2u - 2u^{1/2}} + \frac{dx}{x} = 0.$$

From calculus the integral of the first term can be evaluated by the further substitution $t = u^{1/2}$. The result is

$$\frac{dt}{t - 1} + \frac{dx}{x} = 0$$

$$\ln |t - 1| + \ln |x| = \ln |c|$$

$$\ln \left| \sqrt{\frac{y}{x}} - 1 \right| + \ln |x| = \ln |c|$$

$$x \left(\sqrt{\frac{y}{x}} - 1 \right) = c$$

$$\sqrt{xy} - x = c.$$

By now the reader may be asking when should the substitution $x = vy$ be used? Although it can be used for every homogeneous differential equation, in practice we try $x = vy$ whenever the function $N(x, y)$ is simpler in structure than $M(x, y)$. In solving $(x^2 + y^2)\, dx + (x^2 - xy)\, dy = 0$, for example, there is no appreciable difference between M and N so either $y = ux$ or $x = vy$ could be used. Also, it could happen that after using one substitution we may encounter integrals that are difficult or impossible to evaluate in closed form; switching substitutions may result in an easier problem.

EXAMPLE   Solve $\qquad\qquad 2x^3y\, dx + (x^4 + y^4)\, dy = 0.$

Solution: Each coefficient is a homogeneous function of degree four. Since the coefficient of dx is slightly simpler than the coefficient of dy we try $x = vy$. It follows that

$$2v^3y^4[v\, dy + y\, dv] + (v^4y^4 + y^4)\, dy = 0$$

$$2v^3y^5\, dv + y^4(3v^4 + 1)\, dy = 0$$

$$\frac{2v^3\, dv}{3v^4 + 1} + \frac{dy}{y} = 0$$

$$\frac{1}{6} \ln (3v^4 + 1) + \ln |y| = \ln |c_1|$$

or $\qquad\qquad\qquad 3x^4y^2 + y^6 = c.$

Had $y = ux$ been used then

$$\frac{dx}{x} + \frac{u^4 + 1}{u^5 + 3u}\, du = 0.$$

It is worth a minute of the reader's time to reflect on how to evaluate the integral of the second term in the last equation.

A homogeneous differential equation can always be expressed in the alternative form

$$\frac{dy}{dx} = F\!\left(\frac{y}{x}\right).$$

To see this suppose we write the equation $M(x, y)\, dx + N(x, y)\, dy = 0$ as $dy/dx = f(x, y)$ where

$$f(x, y) = -\frac{M(x, y)}{N(x, y)}.$$

The function $f(x, y)$ must necessarily be homogeneous of degree zero when M and N are homogeneous of degree n. Using homogeneity it follows that

$$f(x, y) = -\frac{x^n M\left(1, \frac{y}{x}\right)}{x^n N\left(1, \frac{y}{x}\right)} = -\frac{M\left(1, \frac{y}{x}\right)}{N\left(1, \frac{y}{x}\right)}.$$

The last ratio is recognized as a function of the form $F(y/x)$. We leave it as an exercise to demonstrate that a homogeneous differential equation can also be written as $dy/dx = G(x/y)$. See Problem 49.

EXAMPLE

Solve
$$x\frac{dy}{dx} = y + xe^{y/x}$$

subject to $y(1) = 1$.

Solution: In the form
$$\frac{dy}{dx} = \frac{y + xe^{y/x}}{x}$$

We see that the function to the right of the equality is homogeneous of degree zero. It occasionally pays to be clever with the algebra before substituting. We write the equation as
$$\frac{dy}{dx} = \frac{y}{x} + e^{y/x}$$

and then use $u = y/x$. Now by the product rule the derivative of $y = ux$ is
$$\frac{dy}{dx} = u + x\frac{du}{dx}$$

so that the equation becomes
$$u + x\frac{du}{dx} = u + e^{u}$$

$$e^{-u}\, du = dx/x.$$

Hence
$$-e^{-u} + c = \ln|x|$$

$$-e^{-y/x} + c = \ln|x|.$$

Since $y = 1$ when $x = 1$ we get
$$-e^{-1} + c = 0 \quad \text{or} \quad c = e^{-1}.$$

Therefore the solution to the initial-value problem is
$$e^{-1} - e^{-y/x} = \ln|x|.$$

EXERCISES 2.3   *Answers to odd-numbered problems begin on page A-3.*

In Problems 1–10 determine whether the given function is homogeneous. If so, state the degree of homogeneity.

1. $x^2 + 2xy - y^3/x$

2. $\sqrt{x + y}(4x + 3y)$

3. $\dfrac{x^3y - x^2y^2}{x + 8y}$

4. $\dfrac{x}{y^2 + \sqrt{x^4 + y^4}}$

5. $\cos \dfrac{x^2}{x + y}$

6. $\sin \dfrac{x}{x + y}$

7. $\ln x^2 - 2 \ln y$

★ **8.** $\dfrac{\ln x^3}{\ln y^3}$

9. $(x^{-1} + y^{-1})^2$

10. $(x + y + 1)^2$

In Problems 11–30 solve the given differential equation by using an appropriate substitution.

11. $(x - y)\, dx + x\, dy = 0$

12. $(x + y)\, dx + x\, dy = 0$

13. $x\, dx + (y - 2x)\, dy = 0$

14. $y\, dx = 2(x + y)\, dy$

15. $(y^2 + yx)\, dx - x^2\, dy = 0$

16. $(y^2 + yx)\, dx + x^2\, dy = 0$

17. $\dfrac{dy}{dx} = \dfrac{y - x}{y + x}$

18. $\dfrac{dy}{dx} = \dfrac{x + 3y}{3x + y}$

19. $-y\, dx + (x + \sqrt{xy})\, dy = 0$

★**20.** $x\dfrac{dy}{dx} - y = \sqrt{x^2 + y^2}$

21. $2x^2y\, dx = (3x^3 + y^3)\, dy$

22. $(x^4 + y^4)\, dx - 2x^3y\, dy = 0$

23. $\dfrac{dy}{dx} = \dfrac{y}{x} + \dfrac{x}{y}$

24. $\dfrac{dy}{dx} = \dfrac{y}{x} + \dfrac{x^2}{y^2} + 1$

25. $y\dfrac{dx}{dy} = x + 4ye^{-2x/y}$

★**26.** $(x^2e^{-y/x} + y^2)\, dx = xy\, dy$

27. $\left(y + x \cot \dfrac{y}{x}\right) dx - x\, dy = 0$

28. $\dfrac{dy}{dx} = \dfrac{y}{x} \ln \dfrac{y}{x}$

29. $(x^2 + xy - y^2)\, dx + xy\, dy = 0$

★**30.** $(x^2 + xy + 3y^2)\, dx - (x^2 + 2xy)\, dy = 0$

In Problems 31–44 solve the given differential equation subject to the indicated initial condition.

31. $xy^2\dfrac{dy}{dx} = y^3 - x^3, \quad y(1) = 2$

32. $(x^2 + 2y^2)\, dx = xy\, dy, \quad y(-1) = 1$

33. $2x^2\dfrac{dy}{dx} = 3xy + y^2, \quad y(1) = -2$

34. $xy\, dx - x^2\, dy = y\sqrt{x^2 + y^2}\, dy, \quad y(0) = 1$

35. $(x + ye^{y/x}) \, dx - xe^{y/x} \, dy = 0, \quad y(1) = 0$

★36. $y \, dx + \left(y \cos \dfrac{x}{y} - x\right) dy = 0, \quad y(0) = 2$

37. $(y^2 + 3xy) \, dx = (4x^2 + xy) \, dy, \quad y(1) = 1$

38. $y^3 \, dx = 2x^3 \, dy - 2x^2y \, dx, \quad y(1) = \sqrt{2}$

39. $(x + \sqrt{xy}) \dfrac{dy}{dx} + x - y = x^{-1/2}y^{3/2}, \quad y(1) = 1$

40. $y \, dx + x(\ln x - \ln y - 1) \, dy = 0, \quad y(1) = e$

41. $y^2 \, dx + (x^2 + xy + y^2) \, dy = 0, \quad y(0) = 1$

42. $(\sqrt{x} + \sqrt{y})^2 \, dx = x \, dy, \quad y(1) = 0$

43. $(x + \sqrt{y^2 - xy}) \dfrac{dy}{dx} = y, \quad y(1) = 1$

★44. $\dfrac{dy}{dx} - \dfrac{y}{x} = \cosh \dfrac{y}{x}, \quad y(1) = 0$

Miscellaneous problems

A differential equation of the form

$$\frac{dy}{dx} = f\left(\frac{ax + by + c}{Ax + By + C}\right), \quad aB - bA \neq 0,$$

can always be reduced to a homogeneous equation by means of the substitutions $x = u + h$, $y = v + k$. In Problems 45 and 46 use the indicated substitutions to reduce the equation to one with homogeneous coefficients. Solve.

45. $\dfrac{dy}{dx} = \dfrac{x - y - 3}{x + y - 1}; \quad x = u + 2, \, y = v - 1$

46. $\dfrac{dy}{dx} = \dfrac{x + y - 6}{x - y}; \quad x = u + 3, \, y = v + 3$

47. Suppose $M(x, y) \, dx + N(x, y) \, dy = 0$ is a homogeneous equation. Show that the substitution $x = vy$ reduces the equation to one with separable variables.

48. Suppose $M(x, y) \, dx + N(x, y) \, dy = 0$ is a homogeneous equation. Show that the substitutions $x = r \cos \theta$, $y = r \sin \theta$ reduces the equation to one with separable variables.

49. Suppose $M(x, y) \, dx + N(x, y) \, dy = 0$ is a homogeneous equation. Show that the equation has the alternative form $dy/dx = G(x/y)$.

50. If $f(x, y)$ is a homogeneous function of degree n, show that

$$x\frac{\partial f}{\partial x} + y\frac{\partial f}{\partial y} = nf.$$

2.4   Exact Equations

While the simple equation

$$y \, dx + x \, dy = 0$$

is both separable and homogeneous, we should also recognize that it is also equivalent to the differential of the product of x and y. That is,

$$y \, dx + x \, dy = d(xy) = 0.$$

By integrating we immediately obtain the implicit solution $xy = c$.

From calculus you might remember that if $z = f(x, y)$ is a function having continuous first partial derivatives in a region R of the xy-plane then its *total differential* is

$$dz = \frac{\partial f}{\partial x} \, dx + \frac{\partial f}{\partial y} \, dy. \tag{1}$$

Now if $f(x, y) = c$, it follows from (1) that

$$\frac{\partial f}{\partial x} \, dx + \frac{\partial f}{\partial y} \, dy = 0.$$

In other words, given a family of curves $f(x, y) = c$ we can generate a first-order differential equation by computing the total differential.

EXAMPLES

(a)   If $x^2 - 5xy + y^3 = c$ then

$$(2x - 5y) \, dx + (-5x + 3y^2) \, dy = 0 \qquad \text{or} \qquad \frac{dy}{dx} = \frac{5y - 2x}{-5x + 3y^2}.$$

(b)   If $y - \cos x^2 y = 2$ then

$$2xy \sin x^2 y \, dx + (1 + x^2 \sin x^2 y) \, dy = 0.$$

For our purposes, it is more important to turn the problem around, namely, given an equation such as

$$\frac{dy}{dx} = \frac{5y - 2x}{-5x + 3y^2}, \tag{2}$$

can we identify the equation as being equivalent to the statement

$$d(x^2 - 5xy + y^3) = 0?$$

Notice that equation (2) is neither separable nor homogeneous.

Exact equations   **DEFINITION 2.3**   A differential expression

$$M(x, y) \, dx + N(x, y) \, dy$$

is an **exact differential** in a region R of the xy-plane if it corresponds to the total differential of some function $f(x, y)$. An equation

$$M(x, y) \, dx + N(x, y) \, dy = 0$$

is said to be **exact** if the expression on the left side is an exact differential. ☐

EXAMPLE

The equation $x^2y^3\,dx + x^3y^2\,dy = 0$ is exact since it is recognized that

$$d(\tfrac{1}{3}x^3y^3) = x^2y^3\,dx + x^3y^2\,dy.$$

The following theorem is a test for an exact differential.

THEOREM 2.2 Let $M(x, y)$ and $N(x, y)$ be continuous and have continuous first partial derivatives in a region R of the xy-plane. Then a necessary and sufficient condition that

$$M(x, y)\,dx + N(x, y)\,dy$$

be an exact differential is

$$\frac{\partial M}{\partial y} = \frac{\partial N}{\partial x}. \tag{3}$$

Proof of the necessity: For simplicity let us assume that $M(x, y)$ and $N(x, y)$ have continuous first partial derivatives for all (x, y). Now if the expression $M(x, y)\,dx + N(x, y)\,dy$ is exact there exists some function f for which

$$M(x, y)\,dx + N(x, y)\,dy \equiv \frac{\partial f}{\partial x}\,dx + \frac{\partial f}{\partial y}\,dy$$

and

$$M(x, y) = \frac{\partial f}{\partial x}, \qquad N(x, y) = \frac{\partial f}{\partial y}$$

Hence

$$\frac{\partial M}{\partial y} = \frac{\partial}{\partial y}\left(\frac{\partial f}{\partial x}\right) = \frac{\partial^2 f}{\partial y\,\partial x} = \frac{\partial}{\partial x}\left(\frac{\partial f}{\partial y}\right) = \frac{\partial N}{\partial x}$$

The equality of the mixed partials is a consequence of the continuity of the first partial derivatives of $M(x, y)$ and $N(x, y)$. ☐

The sufficiency part of Theorem 2.2 consists of showing there exists a function f for which $\partial f/\partial y = M(x, y)$ and $\partial f/\partial y = N(x, y)$ whenever condition (3) holds. The construction of the function f actually reflects a basic procedure for solving exact equations.

The method of solution

Given the equation

$$M(x, y)\,dx + N(x, y)\,dy = 0 \tag{4}$$

first show

$$\frac{\partial M}{\partial y} = \frac{\partial N}{\partial x}.$$

Then assume that

$$\frac{\partial f}{\partial x} = M(x, y)$$

so we can find f by integrating $M(x, y)$ with respect to x while holding y constant. We write

$$f(x, y) = \int M(x, y) \, dx + g(y) \tag{5}$$

where the arbitrary function $g(y)$ is the "constant" of integration. Now differentiate (5) with respect to y and assume $\partial f / \partial y = N(x, y)$.

$$\frac{\partial f}{\partial y} = \frac{\partial}{\partial y} \int M(x, y) \, dx + g'(y)$$

$$= N(x, y).$$

This gives

$$g'(y) = N(x, y) - \frac{\partial}{\partial y} \int M(x, y) \, dx. \tag{6}$$

It is important to observe that the expression $N(x, y) - (\partial / \partial y) \int M(x, y) \, dx$ is independent of x since

$$\frac{\partial}{\partial x} \left[N(x, y) - \frac{\partial}{\partial y} \int M(x, y) \, dx \right] = \frac{\partial N}{\partial x} - \frac{\partial}{\partial y} \left(\frac{\partial}{\partial x} \int M(x, y) \, dx \right)$$

$$= \frac{\partial N}{\partial x} - \frac{\partial M}{\partial y} = 0.$$

Finally integrate (6) with respect to y and substitute the result in (5). The solution of the equation is $f(x, y) = c$.

Note: In the foregoing procedure we could just as well start out with the assumption that $\partial f / \partial y = N(x, y)$. The analogues of (5) and (6) would be respectively

$$f(x, y) = \int N(x, y) \, dy + h(x)$$

and

$$h'(x) = M(x, y) - \frac{\partial}{\partial x} \int N(x, y) \, dy.$$

In either case *none of these formulas should be memorized*. Also, when testing an equation for exactness make sure it is of form (4). Often a differential equation is written $G(x, y) \, dx = H(x, y) \, dy$. In this case write the equation as $G(x, y) \, dx - H(x, y) \, dy = 0$ and then identify $M(x, y) = G(x, y)$ and $N(x, y) = -H(x, y)$.

EXAMPLE   Solve $2xy \, dx + (x^2 - 1) \, dy = 0.$

Solution: Although the equation is separable it is also exact since

$$\frac{\partial M}{\partial y} = 2x = \frac{\partial N}{\partial x}.$$

Now by Theorem 2.2 there exists a function $f(x, y)$ such that

$$\frac{\partial f}{\partial x} = 2xy \qquad \text{and} \qquad \frac{\partial f}{\partial y} = x^2 - 1.$$

From the first of these equations we obtain

$$f(x, y) = x^2y + g(y).$$

Taking the partial derivative of the last expression with respect to y and setting the result equal to $N(x, y)$ gives

$$\frac{\partial f}{\partial y} = x^2 + g'(y)$$

$$= x^2 - 1.$$

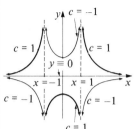

Figure 2.8

It follows that

$$g'(y) = -1 \quad \text{and} \quad g(y) = -y.$$

The constant of integration need not be included in the preceding line since the solution is $f(x, y) = c$. Some of the family of curves $x^2y - y = c$ are given in Figure 2.8.

> *Note:* The solution of the equation is *not* $f(x, y) = x^2y - y$, rather it is $f(x, y) = c$ or $f(x, y) = 0$ if a constant is used in the integration of $g'(y)$.

EXAMPLE

Solve $\qquad (e^{2y} - y \cos xy) \, dx + (2xe^{2y} - x \cos xy + 2y) \, dy = 0.$

Solution: The equation is neither separable nor homogeneous, but is exact since

$$\frac{\partial M}{\partial y} = 2e^{2y} + xy \sin xy - \cos xy = \frac{\partial N}{\partial x}.$$

Thus we know that there exists a function $f(x, y)$ such that

$$M(x, y) = \frac{\partial f}{\partial x} \quad \text{and} \quad N(x, y) = \frac{\partial f}{\partial y}.$$

For variety we shall start with the assumption that $\partial f / \partial y = N(x, y)$.

That is, $\qquad \dfrac{\partial f}{\partial y} = 2xe^{2y} - x \cos xy + 2y$

$$f(x, y) = 2x \int e^{2y} \, dy - x \int \cos xy \, dy + 2 \int y \, dy.$$

Remember, the reason that x can come out in front of the symbol $\int$ is that in the integration with respect to y, x is treated as an ordinary constant. It follows

$$f(x, y) = xe^{2y} - \sin xy + y^2 + h(x)$$

$$\frac{\partial f}{\partial x} = e^{2y} - y \cos xy + h'(x)$$

$$= e^{2y} - y \cos xy,$$

so that $\qquad\qquad h'(x) = 0 \qquad \text{and} \qquad h(x) = c.$

We can then write the solution as

$$xe^{2y} - \sin xy + y^2 + c = 0.$$

EXAMPLE   Solve $(\cos x \sin x - xy^2)\, dx + y(1 - x^2)\, dy = 0$
subject to $y(0) = 2$.

Solution: The equation is exact since

$$\frac{\partial M}{\partial y} = -2xy = \frac{\partial N}{\partial x}.$$

Now $$\frac{\partial f}{\partial y} = y(1 - x^2)$$

$$f(x, y) = \frac{y^2}{2}(1 - x^2) + h(x)$$

$$\frac{\partial f}{\partial x} = -xy^2 + h'(x)$$

$$= \cos x \sin x - xy^2.$$

The last equation implies

$$h'(x) = \cos x \sin x$$

$$h(x) = -\int (\cos x)(-\sin x \, dx)$$

$$= -\frac{1}{2}\cos^2 x.$$

Thus $$\frac{y^2}{2}(1 - x^2) - \frac{1}{2}\cos^2 x = c_1$$

or $$y^2(1 - x^2) - \cos^2 x = c. \qquad (c = 2c_1)$$

The initial condition $y = 2$ when $x = 0$ demands that $4(1) - \cos^2(0) = c$ or
that $c = 3$. The solution is then

$$y^2(1 - x^2) - \cos^2 x = 3.$$

EXAMPLE   Solve $2y(y - 1)\, dx + x(2y - 1)\, dy = 0$ (7)

Solution: The equation is not exact since if

$$M(x, y) = 2y(y - 1), \qquad N(x, y) = x(2y - 1)$$

then $\dfrac{\partial M}{\partial y} = 4y - 2$ $\dfrac{\partial N}{\partial x} = 2y - 1.$

It is interesting to note that if we simply multiply the given equation by x we then obtain

$$2xy(y - 1)\,dx + x^2(2y - 1)\,dy = 0 \qquad (8)$$

so that $M(x, y) = 2xy(y - 1), \qquad N(x, y) = x^2(2y - 1)$

and $\dfrac{\partial M}{\partial y} = 4xy - 2x$ $\dfrac{\partial N}{\partial x} = 4xy - 2x.$

Since the new equation is exact we can use the above method of solution. It is readily shown that $x^2(y^2 - y) = c$ satisfies both (7) and (8).

EXERCISES 2.4   *Answers to odd-numbered problems begin on page A-4.*

In Problems 1–24 determine whether the given equation is exact. If exact, solve.

1. $(2x + 4)\,dx + (3y - 1)\,dy = 0$

2. $(2x + y)\,dx - (x + 6y)\,dy = 0$

3. $(5x + 4y)\,dx + (4x - 8y^3)\,dy = 0$

4. $(\sin y - y \sin x)\,dx + (\cos x + x \cos y - y)\,dy = 0$

5. $(2y^2x - 3)\,dx + (2yx^2 + 4)\,dy = 0$

6. $\left(2y - \dfrac{1}{x} + \cos 3x\right)\dfrac{dy}{dx} + \dfrac{y}{x^2} - 4x^3 + 3y \sin 3x = 0$

7. $(x + y)(x - y)\,dx + x(x - 2y)\,dy = 0$

★ 8. $\left(1 + \ln x + \dfrac{y}{x}\right)dx = (1 - \ln x)\,dy$

9. $(y^3 - y^2 \sin x - x)\,dx + (3xy^2 + 2y \cos x)\,dy = 0$

10. $(x^3 + y^3)\,dx + 3xy^2\,dy = 0$

11. $(y \ln y - e^{-xy})\,dx + \left(\dfrac{1}{y} + x \ln y\right)dy = 0$

12. $\dfrac{2x}{y}\,dx - \dfrac{x^2}{y^2}\,dy = 0$ 13. $x\dfrac{dy}{dx} = 2xe^x - y + 6x^2$

14. $(3x^2y + e^y)\,dx + (x^3 + xe^y - 2y)\,dy = 0$

15. $\left(1 - \dfrac{3}{x} + y\right)dx + \left(1 - \dfrac{3}{y} + x\right)dy = 0$

★16. $(e^y + 2xy \cosh x)y' + xy^2 \sinh x + y^2 \cosh x = 0$

17. $\left(x^2y^3 - \dfrac{1}{1 + 9x^2}\right) dx + x^3y^2 \, dy = 0$

18. $(5y - 2x)y' - 2y = 0$

19. $(\tan x - \sin x \sin y) \, dx + \cos x \cos y \, dy = 0$

20. $(3x \cos 3x + \sin 3x - 3) \, dx + (2y + 5) \, dy = 0$

21. $(1 - 2x^2 - 2y)\dfrac{dy}{dx} = 4x^3 + 4xy$

★22. $(2y \sin x \cos x - y + 2y^2e^{xy^2}) \, dx = (x - \sin^2 x - 4xye^{xy^2}) \, dy$

23. $(4x^3y - 15x^2 - y) \, dx + (x^4 + 3y^2 - x) \, dy = 0$

24. $\left(\dfrac{1}{x} + \dfrac{1}{x^2} - \dfrac{y}{x^2 + y^2}\right) dx + \left(ye^y + \dfrac{x}{x^2 + y^2}\right) dy = 0$

In Problems 25–30 solve the given differential equation subject to the indicated initial conditon.

25. $(x + y)^2 \, dx + (2xy + x^2 - 1) \, dy = 0, \quad y(1) = 1$

26. $(e^x + y) \, dx + (2 + x + ye^y) \, dy = 0, \quad y(0) = 1$

27. $(4y + 2x - 5) \, dx + (6y + 4x - 1) \, dy = 0, \quad y(-1) = 2$

★28. $\left(\dfrac{3y^2 - x^2}{y^5}\right)\dfrac{dy}{dx} + \dfrac{x}{2y^4} = 0, \quad y(1) = 1$

29. $(y^2 \cos x - 3x^2y - 2x) \, dx + (2y \sin x - x^3 + \ln y) \, dy = 0,$
$y(0) = e$

30. $\left(\dfrac{1}{1 + y^2} + \cos x - 2xy\right)\dfrac{dy}{dx} = y(y + \sin x), \quad y(0) = 1$

In Problems 31 and 32 find the value of k so that the given differential equation is exact.

31. $(y^3 + kxy^4 - 2x) \, dx + (3xy^2 + 20x^2y^3) \, dy = 0$

★32. $(6xy^3 + \cos y) \, dx + (kx^2y^2 - x \sin y) \, dy = 0$

33. Determine a function $M(x, y)$ so that the following differential equation is exact.

$$M(x, y) \, dx + \left(xe^{xy} + 2xy + \dfrac{1}{x}\right) dy = 0$$

34. Determine a function $N(x, y)$ so that the following differential equation is exact.

$$\left(y^{1/2}x^{-1/2} + \dfrac{x}{x^2 + y}\right) dx + N(x, y) \, dy = 0$$

Let $M(x, y)\, dx + N(x, y)\, dy = 0$ be an exact differential equation in a region R of the xy-plane. An alternative method for solving the equation is

$$\int_a^x M(u, y)\, du + \int_b^y N(a, v)\, dv = c, \qquad (a, b) \text{ in } R.* \qquad (9)$$

Use (9) to solve the exact equations in Problems 35–38.

EXAMPLE   Solve $(2x + 4y + 1)\, dx + (4x - 3y^2)\, dy = 0$

Solution:

$$\int_a^x (2u + 4y + 1)\, du + \int_b^y (4a - 3v^2)\, dv = c$$

$$\left. (u^2 + 4yu + u) \right|_a^x + \left. (4av - v^3) \right|_b^y = c$$

$$(x^2 + 4xy + x - a^2 - 4ay - a) + (4ay - y^3 - 4ab + b^3) = c.$$

By lumping all the constants under the symbol c_1 we obtain the solution

$$x^2 + 4xy + x - y^3 = c_1.$$

---

35. $(3x + y)\, dx + (x + 8y)\, dy = 0$

36. $(2x + 2y + 5)\, dx + (2x - 6y^2)\, dy = 0$

37. $(2xy^2 + ye^x)\, dx + (2x^2y + e^x - 1)\, dy = 0$

★38. $(2x - y \sin xy - 5y^4)\, dx - (20xy^3 + x \sin xy)\, dy = 0$

Miscellaneous problems

39. We have seen in the last example of this section that the expression $M(x, y)\, dx + N(x, y)\, dy = 0$ may not be exact, however, there could exist a function $\mu(x, y)$ such that $\mu(x, y)M(x, y)\, dx + \mu(x, y)\, N(x, y)\, dy = 0$ is exact. Show that a $\mu M\, dx + \mu N\, dy = 0$ is exact if and only if $\mu(x, y)$ satisfies the partial differential equation

$$N(x, y)\frac{\partial \mu}{\partial x} - M(x, y)\frac{\partial \mu}{\partial y} = \left[\frac{\partial M}{\partial y} - \frac{\partial N}{\partial x}\right]\mu(x, y).$$

The function $\mu(x, y)$ is called an **integrating factor.**

40. The equation $M\, dx + N\, dy = 0$ and $\mu M\, dx + \mu N\, dy = 0$, where $\mu(x, y)$ is an integrating factor, are not necessarily equivalent in the sense that a solution of one is also a solution of the other. Show that $y \equiv 0$ is a solution of the nonexact equation

$$y(2y^2 - 3x^2)\, dx + 2x^3\, dy = 0$$

---

*Roughly, (9) says integrate $M(x, y)$ with respect to x and $N(x, y)$ with respect to y. Try to picture what is going on in the xy-plane.

but is not a solution of the exact equation

$$\left(2 - 3\frac{x^2}{y^2}\right) dx + 2\frac{x^3}{y^3} dy = 0$$

Here the original equation was multiplied by y^{-3}.

In Problems 41–46 solve the equation by verifying that the given $\mu(x, y)$ is an integrating factor. Problem 39 is not of great value in determining $\mu(x, y)$ since we are in no position to solve a partial differential equation.

EXAMPLE

Solve $$(x + y) dx + x \ln x \, dy = 0, \qquad \mu(x, y) = \frac{1}{x}$$

on $0 < x < \infty$.

Solution: Let $M(x, y) = x + y$ and $N(x, y) = x \ln x$ so that $\partial M/\partial y = 1$ and $\partial N/\partial x = 1 + \ln x$. The equation is not exact. However, if we multiply the equation by $\mu(x, y) = 1/x$ we obtain

$$\left(1 + \frac{y}{x}\right) dx + \ln x \, dy = 0.$$

From this latter form we make the identifications:

$$M(x, y) = 1 + y/x, \quad N(x, y) = \ln x, \quad \partial M/\partial y = 1/x = \partial N/\partial x.$$

Therefore the second differential equation is exact. It follows

$$\frac{\partial f}{\partial x} = 1 + \frac{y}{x} = M(x, y)$$

$$f(x, y) = x + y \ln x + g(y)$$

$$\frac{\partial f}{\partial y} = 0 + \ln x + g'(y) = N(x, y)$$

$$= \ln x$$

which implies $$g'(y) = 0 \qquad \text{and} \qquad g(y) = c.$$
Hence $f(x, y) = x + y \ln x + c$ and so the solution of the problem on $0 < x < \infty$ is either

$$x + y \ln x + c = 0 \qquad \text{or} \qquad x + y \ln x = c_1.$$

---

41. $6xy \, dx + (4y + 9x^2) \, dy = 0, \quad \mu(x, y) = y^2$

42. $-y^2 \, dx + (x^2 + xy) \, dy = 0, \quad \mu(x, y) = 1/x^2 y$

43. $(-xy \sin x + 2y \cos x) \, dx + 2x \cos x \, dy = 0, \quad \mu(x, y) = xy$

44. $y(x + y + 1) \, dx + (x + 2y) \, dy = 0, \quad \mu(x, y) = e^x$

45. $(2y^2 + 3x) \, dx + 2xy \, dy = 0, \quad \mu(x, y) = x$

46. $(x^2 + 2xy - y^2) \, dx + (y^2 + 2xy - x^2) \, dy = 0,$
$\mu(x, y) = (x + y)^{-2}$

47. A differential equation can have more than one integrating factor. Show that $\mu_1(x, y) = 1/xy$, $\mu_2(x, y) = 1/y^2$, and $\mu_3(x, y) = 1/(x^2 + y^2)$ are all integrating factors of the equation $y \, dx - x \, dy = 0$. Show that the solutions are formally equivalent.

48. Suppose $M \, dx + N \, dy = 0$ has an integrating factor $\mu(x, y)$ such that $df = \mu M \, dx + \mu N \, dy$ is an exact differential. Show that the equation has an infinite number of integrating factors by demonstrating that the product $\mu G(f)$, where G is an arbitrary function, is also an integrating factor.

EXAMPLE

We have seen in Problem 47 that $\mu(x, y) = 1/y^2$ is an integrating factor of $y \, dx - x \, dy = 0$. The equations

$$\frac{y \, dx - x \, dy}{y^2} = 0 \quad \text{and} \quad d\left(\frac{x}{y}\right) = 0$$

are equivalent, so that we can make the identification $f(x, y) = x/y$. If we now simply make up a function, say $G(u) = u^{-2}$ then by Problem 48 another integrating factor is

$$\mu_1(x, y) = \frac{1}{y^2} G\left(\frac{x}{y}\right)$$

$$= \frac{1}{y^2}\left(\frac{x}{y}\right)^{-2}$$

$$= \frac{1}{x^2}.$$

After multiplying by $-\mu_1(x, y)$ the differential equation becomes

$$\frac{x \, dy - y \, dx}{x^2} = 0 \quad \text{or} \quad d\left(\frac{y}{x}\right) = 0.$$

---

49. Find three more integrating factors of $y \, dx - x \, dx = 0$ and write down the resulting differential equations.

50. Show that $x^{a-1}y^{b-1}$, a and b constants, is an integrating factor of $ay \, dx + bx \, dy = 0$.

51. Show that any separable first-order differential equation is also exact.

2.5   Linear Equations

In Chapter 1 we defined the general form of a *linear* differential equation of order n to be

$$a_n(x)\frac{d^n y}{dx^n} + a_{n-1}(x)\frac{d^{n-1}y}{dx^{n-1}} + \cdots + a_1(x)\frac{dy}{dx} + a_0(x)y = g(x).$$

We remind the reader that linearity means that all coefficients are functions of x only, and that y and all its derivatives are raised to the first power. Now when $n = 1$ we obtain the linear first-order equation

$$a_1(x)\frac{dy}{dx} + a_0(x)y = g(x).$$

Dividing by $a_1(x)$ gives the more useful form

$$\frac{dy}{dx} + P(x)y = f(x). \tag{1}$$

We seek the solution of (1) on an interval I for which $P(x)$ and $f(x)$ are continuous. In the discussion that follows we tacitly assume that (1) has a solution.

An integrating factor

Let us suppose equation (1) is written in the differential form

$$dy + [P(x)y - f(x)]\, dx = 0. \tag{2}$$

Linear equations possess the <u>pleasant property</u> that a function $\mu(x)$ can always be found such that the multiple of (2)

$$\mu(x)\, dy + \mu(x)[P(x)y - f(x)]\, dx = 0 \tag{3}$$

is an exact differential equation. By Theorem 2.2 we know that the left side of equation (3) will be an exact differential if

$$\frac{\partial}{\partial x}\mu(x) = \frac{\partial}{\partial y}\mu(x)[P(x)\,y - f(x)] \tag{4}$$

or

$$\frac{d\mu}{dx} = \mu P(x).$$

This is a separable equation from which we can determine $\mu(x)$. We have

$$\frac{d\mu}{\mu} = P(x)\, dx$$

$$\ln |\mu| = \int P(x)\, dx \tag{5}$$

so that

$$\mu(x) = e^{\int P(x)\, dx}. \tag{6}$$

The function $\mu(x)$ defined in (6) is called an **integrating factor** for the linear equation. Note that we need not use a constant of integration in (5) since (3) is unaffected by a constant multiple. Also, $\mu(x) \neq 0$ for every x in I and is continuous and differentiable.

It is interesting to observe that equation (3) is still an exact differential equation even when $f(x) = 0$. In fact, $f(x)$ plays no part in determining $\mu(x)$ since we see from (4) that $\dfrac{\partial}{\partial y}\mu(x)f(x) = 0$. Thus both

$$e^{\int P(x)\,dx}\,dy + e^{\int P(x)\,dx}[P(x)y - f(x)]\,dx$$

and $\qquad\qquad e^{\int P(x)\,dx}\,dy + e^{\int P(x)\,dx}P(x)y\,dx$

are exact differentials. We now write (3) in the form

$$e^{\int P(x)\,dx}\,dy + e^{\int P(x)\,dx}P(x)y\,dx = e^{\int P(x)\,dx}f(x)\,dx$$

and recognize that we can write the equation as

$$d\left[e^{\int P(x)\,dx}y\right] = e^{\int P(x)\,dx}f(x)\,dx.$$

Integrating the last equation gives

$$e^{\int P(x)\,dx}y = \int e^{\int P(x)\,dx}f(x)\,dx + c$$

or $\qquad\qquad y = e^{-\int P(x)\,dx}\int e^{\int P(x)\,dx}f(x)\,dx + ce^{-\int P(x)\,dx}. \qquad (7)$

In other words, if (1) has a solution it must be of form (7). Conversely, it is a straightforward matter to verify that (7) constitutes a one-parameter family of solutions of equation (1). However, no attempt should be made to memorize formula (7). The procedure should be followed each time, so for convenience we summarize the results.

The method of solution

To solve a linear first-order differential equation first put it into the form (1), that is, make the coefficient of y' unity. Then multiply the entire equation by the integrating factor:

$$e^{\int P(x)\,dx}.$$

The left side of

$$e^{\int P(x)\,dx}\frac{dy}{dx} + P(x)e^{\int P(x)\,dx}y = e^{\int P(x)\,dx}f(x)$$

is the derivative of the product of the integrating factor and the dependent variable: $e^{\int P(x)\,dx}y$. Write the equation in the form

$$\frac{d}{dx}\left[e^{\int P(x)\,dx}y\right] = e^{\int P(x)\,dx}f(x)$$

and finally, integrate both sides.

EXAMPLE

Solve $\qquad\qquad\qquad x\dfrac{dy}{dx} - 4y = x^6 e^x.$

Solution: Write the equation as

$$\frac{dy}{dx} - \frac{4}{x}y = x^5 e^x. \qquad (8)$$

and determine the integrating factor

$$e^{-4\int dx/x} = e^{-4\ln|x|} = e^{\ln x^{-4}} = x^{-4}.$$

Here we have used the basic identity $b^{\log_b N} = N$. Now multiply (8) by this term

$$x^{-4}\frac{dy}{dx} - 4x^{-5}y = x\,e^x \tag{9}$$

and obtain

$$\frac{d}{dx}[x^{-4}y] = x\,e^x.* \tag{10}$$

It follows from integration by parts that

$$x^{-4}y = xe^x - e^x + c$$

or

$$y = x^5 e^x - x^4 e^x + cx^4.$$

EXAMPLE   Solve

$$\frac{dy}{dx} - 3y = 0.$$

Solution: The equation is already in form (1). Hence the integrating factor is

$$e^{\int(-3)\,dx} = e^{-3x}$$

and thus

$$e^{-3x}\frac{dy}{dx} - 3e^{-3x}y = 0$$

$$\frac{d}{dx}[e^{-3x}y] = 0$$

$$e^{-3x}y = c$$

and so

$$y = ce^{3x}.$$

The general solution If it is assumed that $P(x)$ and $f(x)$ are continuous on an interval I and x_0 is any point in the interval, then it follows from Theorem 2.1 that there exists only one solution of the initial-value problem

$$\frac{dy}{dx} + P(x)y = f(x)$$

$$y(x_0) = y_0. \tag{11}$$

But we saw earlier that (1) possesses a family of solutions and that every solution of the equation on the interval I is of form (7). Thus, obtaining the solution of (11) is a simple matter of finding an appropriate value of c in (7).

---

*The reader should perform the indicated differentiations a few times in order to be convinced that all equations, such as (8), (9) and (10) are formally equivalent.

As a consequence we are justified in calling (7) the **general solution** of the differential equation. In retrospect, the reader should recall that in several instances we found singular solutions of nonlinear equations. This cannot happen in the case of a linear equation when proper attention is paid to solving the equation over a common interval on which $P(x)$ and $f(x)$ are continuous.

EXAMPLE

Find the general solution of

$$(x^2 + 9)\frac{dy}{dx} + xy = 0$$

Solution: We write

$$\frac{dy}{dx} + \frac{x}{x^2 + 9}y = 0.$$

The function $P(x) = x/(x^2 + 9)$ is continuous on $-\infty < x < \infty$. Now the integrating factor for the equation is

$$e^{\int x\,dx/(x^2 + 9)} = e^{\frac{1}{2}\int 2x\,dx/(x^2 + 9)} = e^{\frac{1}{2}\ln(x^2 + 9)} = \sqrt{x^2 + 9}$$

so that

$$\sqrt{x^2 + 9}\,\frac{dy}{dx} + \frac{x}{\sqrt{x^2 + 9}}y = 0$$

$$\frac{d}{dx}[\sqrt{x^2 + 9}\,y] = 0$$

$$\sqrt{x^2 + 9}\,y = c.$$

Hence, the general solution on the interval is

$$y = \frac{c}{\sqrt{x^2 + 9}}.$$

EXAMPLE

Solve

$$\frac{dy}{dx} + 2xy = x$$

subject to $y(0) = -3$.

Solution: The functions $P(x) = 2x$ and $f(x) = x$ are continuous on $-\infty < x < \infty$. The integrating factor is

$$e^{2\int x\,dx} = e^{x^2}$$

so that

$$e^{x^2}\frac{dy}{dx} + 2xe^{x^2}y = xe^{x^2}$$

$$\frac{d}{dx}[e^{x^2}y] = xe^{x^2}$$

$$e^{x^2}y = \int xe^{x^2}\,dx$$

$$= \frac{1}{2} \int e^{x^2}(2x\ dx)$$

$$= \frac{1}{2}e^{x^2} + c$$

Thus the general solution of the differential equation is

$$y = \frac{1}{2} + ce^{-x^2}$$

The condition $y(0) = -3$ gives $c = -7/2$ and hence the solution of the initial-value problem on the interval is

$$y = \frac{1}{2} - \frac{7}{2}e^{-x^2}$$

$y \equiv 1/2$

$(0, 3/2)$

$c = 0$ $c = 1$

$c = -1$

$c = -7/2$

$(0, -3)$

Figure 2.9   See Figure 2.9.

EXAMPLE   Solve

$$x\frac{dy}{dx} + y = 2x$$

subject to $y(1) = 0$.

Solution: Write the given equation as

$$\frac{dy}{dx} + \frac{1}{x}y = 2$$

and observe that $P(x) = 1/x$ is continuous on any interval not containing the origin. In view of the initial condition we solve the problem on the interval $0 < x < \infty$.

The integrating factor is

$$e^{\int dx/x} = e^{\ln|x|} = x$$

and so

$$\frac{d}{dx}[xy] = 2x$$

$$xy = x^2 + c.$$

The general solution of the equation is

$$y = x + \frac{c}{x}. \tag{12}$$

But $y(1) = 0$ implies $c = -1$. Hence we obtain

$$y = x - \frac{1}{x}, \qquad 0 < x < \infty. \tag{13}$$

Considered as a one-parameter family of curves the graph of (12) is given in Figure 2.10. The solution (13) of the initial-value problem is indicated by the bold portion of the graph.

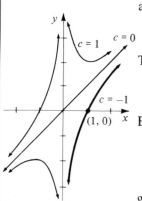

$c = 1$ $c = 0$

$c = -1$

$(1, 0)$

Figure 2.10

EXAMPLE   Solve

$$\frac{dy}{dx} = \frac{1}{x + y^2}$$

subject to $y(-2) = 0$.

Solution: The given differential equation is neither separable, homogeneous, exact, nor linear in the variable y. However if we take the reciprocal then

$$\frac{dx}{dy} = x + y^2 \quad \text{or} \quad \frac{dx}{dy} - x = y^2.$$

This latter equation is *linear in* x so the corresponding integrating factor is $e^{-\int dy} = e^{-y}$. Therefore it follows that for $-\infty < y < \infty$

$$\frac{d}{dy}[e^{-y}x] = y^2 e^{-y}$$

$$e^{-y}x = \int y^2 e^{-y}\, dy$$

$$= -y^2 e^{-y} - 2ye^{-y} - 2e^{-y} + c$$

$$x = -y^2 - 2y - 2 + ce^y$$

When $x = -2$, $y = 0$ we find $c = 0$, and so

$$x = -y^2 - 2y - 2.$$

Remark: Formula (7), representing the general solution of (1), actually consists of the sum of two solutions. We define

$$y = y_c + y_p \qquad\qquad (14)$$

where

$$y_c = ce^{-\int P(x)\, dx} \quad \text{and} \quad y_p = e^{-\int P(x)\, dx} \int e^{\int P(x)\, dx} f(x)\, dx.$$

The function y_c is readily shown to be the general solution of $y' + P(x)y = 0$ whereas y_p is a particular solution of $y' + P(x)y = f(x)$. As we shall see in Chapter 4, the additivity property of solutions (14) to form a general solution is an intrinsic property of linear equations of any order.

EXERCISES 2.5   *Answers to odd-numbered problems begin on page A-5.*

In Problems 1–38 find the general solution of the given differential equation. State an interval on which the general solution is defined.

1. $\dfrac{dy}{dx} = 4y$

2. $\dfrac{dy}{dx} + 2y = 0$

3. $2\dfrac{dy}{dx} + 10y = 1$

4. $x\dfrac{dy}{dx} + 2y = 3$

5. $\dfrac{dy}{dx} + y = e^{3x}$ ★ **6.** $\dfrac{dy}{dx} = y + e^{x}$

7. $y' + 3x^{2}y = x^{2}$ **8.** $y' + 2xy = x^{3}$

9. $x^{2}y' + xy = 1$ **10.** $y' = 2y + x^{2} + 5$

11. $(x + 4y^{2})\,dy + 2y\,dx = 0$ **12.** $\dfrac{dx}{dy} = x + y$

13. $x\,dy = (x\,\sin x - y)\,dx$

★**14.** $(1 + x^{2})\,dy + (xy + x^{3} + x)\,dx = 0$

15. $(1 + e^{x})\dfrac{dy}{dx} + e^{x}y = 0$ **16.** $(1 - x^{3})\dfrac{dy}{dx} = 3x^{2}y$

17. $\cos x\dfrac{dy}{dx} + y\,\sin x = 1$ **18.** $\dfrac{dy}{dx} + y\,\cot x = 2\,\cos x$

19. $x\dfrac{dy}{dx} + 4y = x^{3} - x$ ★**20.** $(1 + x)y' - xy = x + x^{2}$

21. $x^{2}y' + x(x + 2)y = e^{x}$ **22.** $xy' + (1 + x)y = e^{-x}\sin 2x$

23. $\cos^{2} x\,\sin x\,dy + (y\,\cos^{3} x - 1)\,dx = 0$

24. $(1 - \cos x)\,dy + (2y\,\sin x - \tan x)\,dx = 0$

25. $y\,dx + (xy + 2x - ye^{y})\,dy = 0$

26. $(x^{2} + x)\,dy = (x^{5} + 3xy + 3y)\,dx$

27. $x\dfrac{dy}{dx} + (3x + 1)y = e^{-3x}$

28. $(x + 1)\dfrac{dy}{dx} + (x + 2)y = 2xe^{-x}$

29. $y\,dx - 4(x + y^{6})\,dy = 0$ **30.** $xy' + 2y = e^{x} + \ln x$

31. $\dfrac{dy}{dx} + y = \dfrac{1 - e^{-2x}}{e^{x} + e^{-x}}$ **32.** $\dfrac{dy}{dx} - y = \sinh x$

33. $y\,dx + (x + 2xy^{2} - 2y)\,dy = 0$ **34.** $y\,dx = (ye^{y} - 2x)\,dy$

35. $\dfrac{dr}{d\theta} + r\,\sec\theta = \cos\theta$ ★**36.** $\dfrac{dP}{dt} + 2tP = P + 4t - 2$

37. $(x + 2)^{2}\dfrac{dy}{dx} = 5 - 8y - 4xy$ **38.** $(x^{2} - 1)\dfrac{dy}{dx} + 2y = (x + 1)^{2}$

In Problems 39–52 solve the given differential equation subject to the indicated initial condition.

39. $\dfrac{dy}{dx} + 5y = 20, \quad y(0) = 2$

40. $y' = 2y + x(e^{3x} - e^{2x}), \quad y(0) = 2$

41. $L\dfrac{di}{dt} + Ri = E; \quad L, R,$ and E constants; $\quad i(0) = i_0$

42. $y\dfrac{dx}{dy} - x = 2y^2, \quad y(1) = 5$

43. $y' + (\tan x)y = \cos^2 x, \quad y(0) = -1$

44. $\dfrac{dQ}{dx} = 5x^4 Q, \quad Q(0) = -7$

45. $\dfrac{dT}{dt} = k(T - 50), \quad k$ a constant, $\quad T(0) = 200$

★**46.** $x\,dy + (xy + 2y - 2e^{-x})\,dx = 0, \quad y(1) = 0$

47. $(x + 1)\dfrac{dy}{dx} + y = \ln x, \quad y(1) = 10$

48. $xy' + y = e^x, \quad y(1) = 2$

49. $x(x - 2)y' + 2y = 0, \quad y(3) = 6$

50. $\sin x\dfrac{dy}{dx} + (\cos x)y = 0, \quad y\left(-\dfrac{\pi}{2}\right) = 1$

51. $\dfrac{dy}{dx} = \dfrac{y}{y - x}, \quad y(5) = 2$

★**52.** $\cos^2 x\dfrac{dy}{dx} + y = 1, \quad y(0) = -3$

In Problems 53–56 find a continuous solution satisfying each differential equation and the given initial condition.

EXAMPLE

Solve $\quad \dfrac{dy}{dx} + y = f(x) \quad$ where $\quad f(x) = \begin{cases} 1, & 0 \le x \le 1 \\ 0, & x > 1 \end{cases}$

subject to $y(0) = 0$.

Solution: We solve the equation in two parts. For $0 \le x \le 1$ we have

$$\frac{dy}{dx} + y = 1$$

$$\frac{d}{dx}[e^x y] = e^x$$

$$y = 1 + c_1 e^{-x}.$$

Since $y(0) = 0$, we must have $c_1 = -1$, and therefore

$$y = 1 - e^{-x}, \qquad 0 \le x \le 1.$$

For $x > 1$ we then have $\qquad \dfrac{dy}{dx} + y = 0$

which leads to $\qquad\qquad\qquad y = c_2 e^{-x}.$

Hence we can write $\quad y = \begin{cases} 1 - e^{-x}, & 0 \le x \le 1 \\ c_2 e^{-x}, & x > 1. \end{cases}$

Now in order that y be a continuous function we certainly want $\lim_{x \to 1^+} y(x) = y(1)$. This latter requirement is equivalent to $c_2 e^{-1} = 1 - e^{-1}$ or $c_2 = e - 1$. As Figure 2.11 shows, the function

$$y = \begin{cases} 1 - e^{-x}, & 0 \le x \le 1 \\ (e - 1)e^{-x}, & x > 1 \end{cases}$$

Figure 2.11

is continuous but not differentiable at $x = 1$.

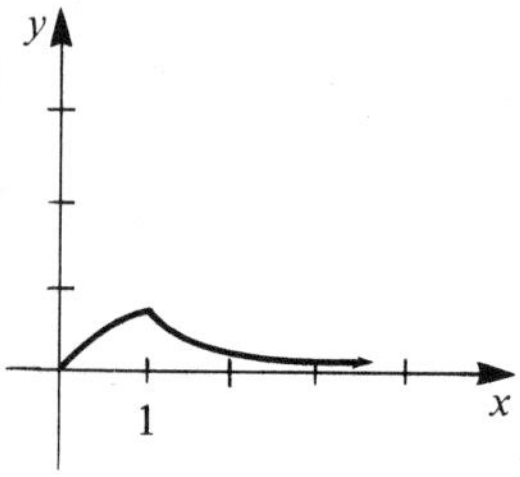

---

53. $\dfrac{dy}{dx} + 2y = f(x), \qquad f(x) = \begin{cases} 1, & 0 \le x \le 3 \\ 0, & x > 3, \end{cases} \qquad y(0) = 0$

54. $\dfrac{dy}{dx} + y = f(x), \qquad f(x) = \begin{cases} 1, & 0 \le x \le 1 \\ -1, & x > 1 \end{cases} \qquad y(0) = 1$

55. $\dfrac{dy}{dx} + 2xy = f(x), \qquad f(x) = \begin{cases} x, & 0 \le x < 1 \\ 0, & x \ge 1 \end{cases} \qquad y(0) = 2$

56. $(1 + x^2)\dfrac{dy}{dx} + 2xy = f(x), \qquad f(x) = \begin{cases} x, & 0 \le x < 1 \\ -x, & x \ge 1 \end{cases}$

$y(0) = 0$

2.6   The Equations of Bernoulli, Ricatti, and Clairaut

In this section we are not going to study any one particular type of differential equation. Rather, we are going to consider three classical equations which, in some instances, can be transformed into equations we have already studied.

EXERCISES 2.6   *Answers to odd-numbered problems begin on page A-6.*

The differential equation

$$\frac{dy}{dx} + P(x)y = f(x)y^n, \tag{1}$$

where n is any real number, is called **Bernoulli's equation** after the Swiss mathematician, James Bernoulli (1654–1705). For $n \ne 0$ and $n \ne 1$ the substitution $w = y^{1-n}$ leads to the linear equation

$$\frac{dw}{dx} + (1 - n)P(x)w = (1 - n)f(x).$$ (2)

In Problems 1–6 solve the given Bernoulli equation.

EXAMPLE   Solve $$\frac{dy}{dx} + \frac{1}{x}y = xy^2.$$

Solution: From (1) we identify $P(x) = 1/x$, $f(x) = x$ and $n = 2$. Thus from (2) we know $w = y^{-1}$ gives

$$\frac{dw}{dx} - \frac{1}{x}w = -x.$$

The integrating factor for this linear equation on, say, $0 < x < \infty$ is

$$e^{-\int dx/x} = e^{-\ln|x|} = e^{\ln|x|^{-1}} = x^{-1}.$$

Hence $$\frac{d}{dx}[x^{-1}w] = -1.$$

Integrating this latter form gives

$$x^{-1}w = -x + c \qquad \text{or} \qquad w = -x^2 + cx.$$

Since $w = y^{-1}$ we obtain $y = 1/w$ and

$$y = \frac{1}{-x^2 + cx}$$

---

1. $x\dfrac{dy}{dx} + y = \dfrac{1}{y^2}$

2. $\dfrac{dy}{dx} - y = e^x y^2$

3. $\dfrac{dy}{dx} = y(xy^3 - 1)$

4. $x\dfrac{dy}{dx} - (1 + x)y = xy^2$

5. $x^2\dfrac{dy}{dx} + y^2 = xy$

★ 6. $3(1 + x^2)\dfrac{dy}{dx} = 2xy(y^3 - 1)$

In Problems 7–10 solve the given differential equation subject to the indicated initial condition.

7. $x^2\dfrac{dy}{dx} - 2xy = 3y^4, \quad y(1) = \dfrac{1}{2}$

★ 8. $y^{1/2}\dfrac{dy}{dx} + y^{3/2} = 1, \quad y(0) = 4$

9. $xy(1 + xy^2)\dfrac{dy}{dx} = 1, \quad y(1) = 0$

10. $2\dfrac{dy}{dx} = \dfrac{y}{x} - \dfrac{x}{y^2}, \quad y(1) = 1$

The **Ricatti equation** is a nonlinear equation

$$\frac{dy}{dx} = P(x) + Q(x)y + R(x)y^2 \tag{3}$$

named after an Italian mathematician-philosopher Count Jacobo Francesco Ricatti (1676–1754). In many cases, depending on $P(x)$, $Q(x)$ and $R(x)$, the solution of (3) cannot be expressed in terms of elementary functions.

11. If y_1 is a known particular solution of the Ricatti equation, show that a family of solutions of (3) is $y = y_1 + u$ where u is the solution of

$$\frac{du}{dx} - (Q + 2y_1R)u = Ru^2. \tag{4}$$

12. Show that (4) can be reduced to the linear equation

$$\frac{dw}{dx} + (Q + 2y_1R)w = -R. \tag{5}$$

by the substitution $w = 1/u$. [*Hint:* (4) is a Bernoulli equation.]

In Problems 13–18 solve the given Ricatti equation.

EXAMPLE   Solve $\dfrac{dy}{dx} = 2 - 2xy + y^2.$

Solution: It is easily verified that a particular solution of this equation is $y_1 = 2x$. From (3) we make the identifications $P(x) = 2$, $Q(x) = -2x$ and $R(x) = 1$ and then solve the linear equation (5),

$$\frac{dw}{dx} + (-2x + 4x)w = -1 \quad \text{or} \quad \frac{dw}{dx} + 2xw = -1.$$

The integrating factor for the last equation is e^{x^2},

$$\frac{d}{dx}[e^{x^2}w] = -e^{x^2}.$$

Now the integral $\int_{x_0}^{x} e^{t^2}\, dt$ cannot be expressed in terms of elementary functions.* Thus, we write

$$e^{x^2}w = -\int_{x_0}^{x} e^{t^2}\, dt + c \quad \text{or} \quad e^{x^2}\left(\frac{1}{u}\right) = -\int_{x_0}^{x} e^{t^2}\, dt + c$$

---

*When an integral $\int f(x)\, dx$ cannot be evaluated in terms of elementary functions it is customary to write $\int_{x_0}^{x} f(t)\, dt$ where x_0 is a constant. When an initial condition is specified it is imperative that this form be used.

so that
$$u = \frac{e^{x^2}}{c - \displaystyle\int_{x_0}^{x} e^{t^2}\, dt}.$$

The solution of the equation is then $y = 2x + u$.

---

13. $\dfrac{dy}{dx} = -2 - y + y^2, \quad y_1 = 2$

14. $\dfrac{dy}{dx} = 1 - x - y + xy^2, \quad y_1 = 1$

15. $\dfrac{dy}{dx} = -\dfrac{4}{x^2} - \dfrac{1}{x}y + y^2, \quad y_1 = \dfrac{2}{x}$

★16. $\dfrac{dy}{dx} = 2x^2 + \dfrac{1}{x}y - 2y^2, \quad y_1 = x$

17. $\dfrac{dy}{dx} = e^{2x} + (1 + 2e^x)y + y^2, \quad y_1 = -e^x$

18. $\dfrac{dy}{dx} = \sec^2 x - (\tan x)y + y^2, \quad y_1 = \tan x$

19. Solve $\dfrac{dy}{dx} = 6 + 5y + y^2$ **★20.** Solve $\dfrac{dy}{dx} = 9 + 6y + y^2$

The differential equation

$$y = xy' + f(y'). \tag{6}$$

is called **Clairaut's equation** after the French mathematician Alexis Claude Clairaut (1713–1765).

21. Show that a solution of equation (6) is the family of straight lines $y = cx + f(c)$ where c is an arbitrary constant.

22. Show that equation (6) may also possess a solution in parametric form:

$$x = -f'(t),$$
$$y = f(t) - tf'(t).$$

This latter solution is called a *singular solution* since, if $f''(t) \neq 0$, it cannot be obtained from the family of solutions given in Problem 21. [*Hint:* Differentiate both sides of the Clairaut equation with respect to x and consider two cases. Use parametric differentiation to show

$$dy/dx = (dy/dt)/(dx/dt) = t, \qquad f''(t) \neq 0.$$

Since the slope of the original solution is constant, it follows that the singular solution cannot be obtained from it.]

In Problems 23–28 solve the given Clairaut equation. Obtain the singular solution.

EXAMPLE

Solve
$$y = xy' + \tfrac{1}{2}(y')^2.$$

Solution: We first make the identification $f(y') = (1/2)(y')^2$ so that $f(t) = (1/2)t^2$. It follows immediately from Problem 21 that a family of solutions is

$$y = cx + \tfrac{1}{2}c^2. \tag{7}$$

The graph of this family is given in Figure 2.12. Since $f'(t) = t$, the singular solution is given by

$$x = -t$$
$$y = \tfrac{1}{2}t^2 - t \cdot t = -\tfrac{1}{2}t^2.$$

After eliminating the parameter we find this latter solution is the same as

$$y = -\tfrac{1}{2}x^2.$$

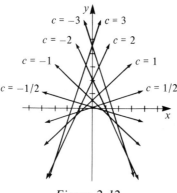

Figure 2.12

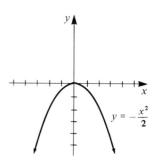

Figure 2.13

One can readily see that this function is not part of the family (7). See Figure 2.13.

---

23. $y = xy' + 1 - \ln y'$

★24. $y = xy' + (y')^{-2}$

25. $y = x\dfrac{dy}{dx} - \left(\dfrac{dy}{dx}\right)^3$

26. $y = (x + 4)y' + (y')^2$

27. $xy' - y = e^{y'}$

28. $y - xy' = \ln y$

Miscellaneous problems

29. When $R(x) = -1$, the Ricatti equation can be written as $y' + y^2 - Q(x)y - P(x) = 0$. Show that the substitution $y = w'/w$ leads to the linear second-order equation $w'' - Q(x)w' - P(x)w = 0$. (When Q and P are also constants there is little difficulty in solving equations of this type.)

30. An alternative definition of Clairaut's equation is any equation of the form $F(y - xy', y') = 0$. (a) Show that a family of solutions of the latter equation is $F(y - cx, c) = 0$. (b) Use the result of part (a) to solve

$$(xy' - y)^3 = (y')^2 + 5.$$

[O] 2.7 Substitutions

In the preceding sections we saw that sometimes a differential equation could be transformed by means of a substitution into a form that could then be solved by one of the standard methods. An equation may look different from any of those that we have just studied, but through a judicious change of variables perhaps an apparently difficult problem may be readily solved. Although we can give no firm rules on what, *if any*, substitution to use, a working axiom might be: Try something! It sometimes pays to be clever.

EXAMPLE

The differential equation

$$y(1 + 2xy)\, dx + x(1 - 2xy)\, dy = 0$$

is neither separable, homogeneous, exact, linear, nor Bernoulli. However, if we stare at the equation long enough we might be prompted to try the substitution

$$u = 2xy \qquad \text{or} \qquad y = \frac{u}{2x}.$$

Since
$$dy = \frac{x\, du - u\, dx}{2x^2},$$

after simplifying the equation becomes

$$2u^2\, dx + (1 - u)x\, du = 0.$$

We recognize the last equation as separable, and so from

$$2\frac{dx}{x} + \frac{1 - u}{u^2}\, du = 0$$

we obtain
$$2 \ln |x| - u^{-1} - \ln |u| = c$$

$$\ln \left| \frac{x}{2y} \right| = c + \frac{1}{2xy}$$

$$\frac{x}{2y} = c_1 e^{1/2xy}$$

$$x = 2c_1 y e^{1/2xy}$$

where e^c was formally replaced by c_1.

Notice again that the solution in the preceding example does not include the solution $y \equiv 0$.

EXAMPLE   Solve

$$x\frac{dy}{dx} - y = \frac{x^3}{y}e^{y/x}.$$

Solution: Let

$$u = \frac{y}{x}$$

so that

$$\frac{du}{dx} = \frac{x\frac{dy}{dx} - y}{x^2} \qquad \text{or} \qquad x\frac{dy}{dx} - y = x^2\frac{du}{dx}.$$

Hence the differential equation becomes

$$\frac{du}{dx} = \frac{e^u}{u} \qquad \text{or} \qquad ue^{-u}\,du = dx.$$

Integration by parts then gives

$$-ue^{-u} - e^{-u} = x + c$$

$$u + 1 = (c_1 - x)e^u \qquad (c_1 = -c)$$

$$\frac{y}{x} + 1 = (c_1 - x)e^{y/x}$$

$$y + x = x(c_1 - x)e^{y/x}.$$

EXAMPLE   Solve

$$2xy\frac{dy}{dx} + 2y^2 = 3x - 6.$$

Solution: The presence of the term $2y\dfrac{dy}{dx}$ prompts us to try $u = y^2$ since

$$\frac{du}{dx} = 2y\frac{dy}{dx}.$$

Now

$$x\frac{du}{dx} + 2u = 3x - 6$$

has the linear form

$$\frac{du}{dx} + \frac{2}{x}u = 3 - \frac{6}{x}$$

so that multiplication by the integrating factor $e^{\int(2/x)\,dx} = e^{\ln x^2} = x^2$ then gives

$$\frac{d}{dx}[x^2u] = 3x^2 - 6x$$

$$x^2u = x^3 - 3x^2 + c \qquad \text{or} \qquad x^2y^2 = x^3 - 3x^2 + c.$$

Some higher-order differential equations can be reduced to first-order equations by a substitution.

EXAMPLE

Solve
$$y'' = 2x(y')^2.$$

Solution: If we let $u = y'$ so that $du/dx = y''$ the equation then reduces to a separable form. We have

$$\frac{du}{dx} = 2xu^2 \qquad \text{or} \qquad \frac{du}{u^2} = 2x \, dx$$

$$\int u^{-2} \, du = \int 2x \, dx$$

$$-u^{-1} = x^2 + c_1^2.$$

The constant of integration is written as c_1^2 for convenience. The reason should be obvious in the next few steps. Since $u^{-1} = 1/y'$ if follows that

$$\frac{dy}{dx} = -\frac{1}{x^2 + c_1^2} \qquad \text{or} \qquad dy = -\frac{dx}{x^2 + c_1^2}$$

$$\int dy = -\int \frac{dx}{x^2 + c_1^2}$$

$$y + c_2 = -\frac{1}{c_1} \tan^{-1} \frac{x}{c_1}.$$

Alternatively, the solution can be written in the implicit form $-c_1 \tan (c_2 + c_1 y) = x$ where the product, $c_1 c_2$, has been replaced by c_2.

It should be observed further that had we written $-u^{-1} = x^2 - c_1^2$, the solution could have been written in terms of the inverse hyperbolic tangent or a logarithm.

EXERCISES 2.7   *Answers to odd-numbered problems begin on page A-6.*

In Problems 1–26 solve the given differential equation by using an appropriate substitution.

1. $xe^{2y}\dfrac{dy}{dx} + e^{2y} = \dfrac{\ln x}{x}$

★ **2.** $y' + y \ln y = ye^x$

3. $y \, dx + (1 + ye^x) \, dy = 0$

4. $(2 + e^{-x/y}) \, dx + 2\left(1 - \dfrac{x}{y}\right) dy = 0$

5. $\dfrac{dy}{dx} - \dfrac{4}{x}y = 2x^5 e^{y/x^4}$

★ **6.** $\dfrac{dy}{dx} + x + y + 1 = (x + y)^2 e^{3x}$

7. $2yy' + x^2 + y^2 + x = 0$ **8.** $y' = y + x(y + 1)^2 + 1$

9. $2x \csc 2y \dfrac{dy}{dx} = 2x - \ln(\tan y)$ **10.** $x^2 \dfrac{dy}{dx} + 2xy = x^4 y^2 + 1$

11. $x^4 y^2 y' + x^3 y^3 = 2x^3 - 3$ **12.** $xe^y y' - 2e^y = x^2$

13. $y' + 1 = e^{-(x+y)} \sin x$

14. $\sin y \sinh x \, dx + \cos y \cosh x \, dy = 0$

15. $y \dfrac{dx}{dy} + 2x \ln x = xe^y$ **16.** $x \sin y \dfrac{dy}{dx} + \cos y = -x^2 e^x$

17. $y'' + (y')^2 + 1 = 0$ **18.** $xy'' = y' + x(y')^2$

19. $xy'' = y' + (y')^3$ **20.** $x^2 y'' + (y')^2 = 0$

21. $y' - xy'' - (y'')^3 = 1$ **★22.** $y'' = 1 + (y')^2$

23. $xy'' - y' = 0$ **24.** $y'' + (\tan x)y' = 0$

25. $y'' + 2y(y')^3 = 0$ **26.** $y^2 y'' = y'$

$$\left[Hint: \text{ Let } u = y' \text{ so that } y'' = \frac{du}{dx} = \frac{du}{dy}\frac{dy}{dx} = \frac{du}{dy}u \right]$$

Miscellaneous problems

27. In calculus the curvature of a curve whose equation is $y = f(x)$ is defined to be the number

$$\kappa = y''/[1 + (y')^2]^{3/2}.$$

Determine a function for which $\kappa = 1$. [*Hint:* For simplicity ignore constants of integration. Also consider a trigonometric substitution.]

[O]   2.8   Picard's Method

The initial-value problem

$$y' = f(x, y)$$
$$y(x_0) = y_0 \tag{1}$$

first considered in Section 2.1 can be written in an alternative manner. Let f be continuous in a region containing the point (x_0, y_0). By integrating both sides of the differential equation with respect to x we get

$$y(x) = c + \int_{x_0}^{x} f(t, y(t)) \, dt.$$

Now

$$y(x_0) = c + \int_{x_0}^{x_0} f(t, y(t)) \, dt = c$$

implies $c = y_0$. Thus

$$y(x) = y_0 + \int_{x_0}^{x} f(t, y(t))\, dt. \tag{2}$$

Conversely, if we start with (2) we can obtain (1). In other words, the integral equation (2) and the initial-value problem (1) are equivalent. We now try to solve (2) by a *method of successive approximations*.

Suppose $y_0(x)$ is an arbitrary continuous function which represents a guess or approximation to the solution of (2). Since $f(x, y_0(x))$ is a known function depending solely on x it can be integrated. With $y(t)$ replaced by $y_0(t)$ the right-hand side of (2) defines another function which we write as

$$y_1(x) = y_0 + \int_{x_0}^{x} f(t, y_0(t))\, dt.$$

It is hoped that this new function is a better approximation to the solution. Repeating the procedure, yet another function is given by

$$y_2(x) = y_0 + \int_{x_0}^{x} f(t, y_1(t))\, dt.$$

In this manner we obtain a sequence of functions $y_1(x)$, $y_2(x)$, $y_3(x)$, . . . , whose nth term is defined by the relation

$$y_n(x) = y_0 + \int_{x_0}^{x} f(t, y_{n-1}(t))\, dt, \qquad n = 1, 2, 3, \ldots . \tag{3}$$

In the application of (3) it is common practice to choose the initial function as $y_0(x) \equiv y_0$. The repetitive use of formula (3) is known as **Picard's method of iteration.**

EXAMPLE   Consider the problem

$$y' = y - 1$$
$$y(0) = 2.$$

Use Picard's method to find the approximations y_1, y_2, y_3, y_4.

Solution: By identifying $x_0 = 0$, $y_0(x) \equiv 2$, and $f(t, y_{n-1}(t)) = y_{n-1}(t) - 1$, equation (3) becomes

$$y_n(x) = 2 + \int_{0}^{x} (y_{n-1}(t) - 1)\, dt, \qquad n = 1, 2, 3, \ldots .$$

Iterating this last expression then gives

$$y_1(x) = 2 + \int_{0}^{x} 1 \cdot dt = 2 + x$$

$$y_2(x) = 2 + \int_{0}^{x} (1 + t)\, dt = 2 + x + \frac{x^2}{2}$$

$$y_3(x) = 2 + \int_0^x (1 + t + \frac{t^2}{2})\, dt$$

$$= 2 + x + \frac{x^2}{2} + \frac{x^3}{2 \cdot 3}$$

$$y_4(x) = 2 + \int_0^x (1 + t + \frac{t^2}{2} + \frac{t^3}{2 \cdot 3})\, dt$$

$$= 2 + x + \frac{x^2}{2} + \frac{x^3}{2 \cdot 3} + \frac{x^4}{2 \cdot 3 \cdot 4}.$$

By induction it can be shown in the last example that the nth term of the sequence of approximations is

$$y_n(x) = 2 + x + \frac{x^2}{2!} + \frac{x^3}{3!} + \cdots + \frac{x^n}{n!}$$

$$= 1 + \sum_{k=0}^{n} \frac{x^k}{k!}.$$

From this latter form we recognize that the *limit* of $y_n(x)$ as $n \to \infty$ is $y(x) = 1 + e^x$. It should come as no surprize to note that this function is an exact solution of the given initial-value problem.

The reader should not be deceived by the relative ease in which the iterants $y_n(x)$ were obtained in the last example. In general, the integration involved in generating each $y_n(x)$ can become complicated very quickly. Nor, for that matter, is it always apparent that the sequence $\{y_n(x)\}$ converges to a nice explicit function. Thus it is fair to ask at this point: Is Picard's method a practical means of solving a first-order equation $y' = f(x, y)$ subject to $y(x_0) = y_0$? The answer is simple: No. One might ask further, in the spirit of a scientist/engineer: What *is* it good for? The answer is not bound to please: Picard's method of iteration is a theoretical tool used in the consideration of existence and uniqueness of solutions of differential equations. Under certain conditions on $f(x, y)$ it can be shown that as $n \to \infty$ the sequence $\{y_n(x)\}$ defined by (3) converges to a function $y(x)$ that satisfies the integral equation (2) and hence the initial-value problem (1). Indeed, it is precisely Picard's method of successive approximations that is used in proving Picard's theorem of Section 2.1. However, the proof of Theorem 2.1 uses concepts from advanced calculus and will not be presented here. Our purpose for introducing this topic was twofold: to gain an appreciation for the potential of the procedure, and to obtain, at least, a nodding acquaintance with an iterative technique. In Chapter 9 we shall consider other methods for approximating solutions of differential equations that also utilize iteration.

EXERCISES 2.8   *Answers to odd-numbered problems begin on page A-6.*

In Problems 1–5 use Picard's method to find y_1, y_2, y_3, y_4. Determine the limit of the sequence $\{y_n(x)\}$ as $n \to \infty$.

1. $y' = -y$, $\quad y(0) = 1$ $\qquad\qquad$ **2.** $y' = x + y$, $\quad y(0) = 1$

3. $y' = 2xy$, $\quad y(0) = 1$ $\qquad\qquad$ ★ **4.** $y' = 2e^x - y$, $\quad y(0) = 1$

5. $y' + y^2 = 0$, $\quad y(0) = 0$

6. In Picard's method the initial choice $y_0(x) \equiv y_0$ is not necessary. Rework Problem 3 with **(a)** $y_0(x) = k$ a constant and $k \neq 1$, **(b)** $y_0(x) = x$.

7. **(a)** Use picard's method to find y_1, y_2, y_3 for the problem

$$y' = 1 + y^2, \quad y(0) = 0.$$

(b) Solve the initial-value problem in part (a) by one of the methods of this chapter.

(c) Compare the results of parts (a) and (b).

CHAPTER 2 SUMMARY

An **initial-value problem** consists of finding a solution of

$$\frac{dy}{dx} = f(x, y)$$

$$y(x_0) = y_0$$

on an interval I containing x_0. If $f(x, y)$ and $\partial f/\partial y$ are continuous in a rectangular region of the xy-plane with (x_0, y_0) in its interior then we are guaranteed there exists an interval around x_0 on which the problem has a unique solution.

The method of solution for a first-order differential equation depends on an appropriate classification of the equation. We summarize five cases.

An equation is **separable** if it can be put into the form $h(y)\, dy = g(x)\, dx$. The solution results from integrating both sides of the equation.

If $M(x, y)$ and $N(x, y)$ are **homogeneous functions** of the same degree, then $M(x, y)\, dx + N(x, y)\, dy = 0$ can be reduced to an equation with separable variables by either the substitution $y = ux$ or $x = vy$. The choice of substitution usually depends on which coefficient is simpler.

The differential equation $M(x, y)\, dx + N(x, y)dy = 0$ is said to be **exact** if the form $M(x, y)\, dx + N(x, y)\, dy$ is an exact differential. When $M(x, y)$ and $N(x, y)$ are continuous and have continuous first partial derivatives, then $\partial M/\partial y = \partial N/\partial x$ is a necessary and sufficient condition that $M(x, y)\, dx + N(x, y)dy$ be exact. This means there exists some function $f(x, y)$ for which $M(x, y) = \partial f/\partial x$ and $N(x, y) = \partial f/\partial y$. The method of solution for an exact equation starts by integrating either of these latter expressions.

If a first-order equation can be put into the form $dy/dx + P(x)y = f(x)$ it is said to be **linear** in the variable y. We solve the equation by first finding the **integrating factor**, $e^{\int P(x)\,dx}$, multiplying both sides of the equation by this factor, and then integrating both sides of

$$\frac{d}{dx}\left[e^{\int P(x)\,dx}y\right] = e^{\int P(x)\,dx}f(x).$$

The **Bernoulli equation** is $dy/dx + P(x)y = f(x)y^n$, where n is any real number. When $n \neq 0$ and $n \neq 1$, Bernoulli's equation can be reduced to a linear equation by the substitution $w = y^{1-n}$.

In certain circumstances a differential equation can be reduced to one of the familiar forms by an appropriate **substitution** or **change of variables**. Of course we already know that this is the procedure when solving a homogeneous or a Bernoulli equation. In the general context, no rule on when to use a substitution can be given.

By converting an initial-value problem to an equivalent integral equation, **Picard's method of iteration** is one way of obtaining an approximation to the solution of the problem.

CHAPTER 2 TEST

Answers to odd-numbered problems begin on page A-7.

Answers Problems 1–4 without referring back to the text. Fill in the blank or answer true/false.

1. The differential equation $y' = 1/(25 - x^2 - y^2)$ will have a unique solution through any point (x_0, y_0) in the region(s) defined by _____.

2. The initial-value problem $xy' = 3y$, $y(0) = 0$ has the solutions $y = x^3$ and _____.

3. The initial-value problem $y' = y^{1/2}$, $y(0) = 0$ has no solution since $\partial f/\partial y$ is discountinuous on the line $y = 0$. _____

4. There exists an interval centered at 2 on which the unique solution of the initial-value problem $y' = (y - 1)^3$, $y(2) = 1$ is $y \equiv 1$. _____

5. Without solving, classify each of the following equations as to: separable, homogeneous, exact, linear, Bernoulli, Ricatti, or Clairaut.

   (a) $\dfrac{dy}{dx} = \dfrac{1}{y - x}$

   (b) $\dfrac{dy}{dx} = \dfrac{x - y}{x}$

   (c) $\left(\dfrac{dy}{dx}\right)^2 + 2y = 2x\dfrac{dy}{dx}$

   (d) $\dfrac{dy}{dx} = \dfrac{1}{x(x - y)}$

   (e) $\dfrac{dy}{dx} = \dfrac{y^2 + y}{x^2 + x}$

   (f) $\dfrac{dy}{dx} = 4 + 5y + y^2$

   (g) $y\,dx = (y - xy^2)\,dy$

   (h) $x\dfrac{dy}{dx} = ye^{x/y} - x$

continued

CHAPTER 2
TEST

(i) $xyy' + y^2 = 2x$

(j) $2xyy' + y^2 = 2x^2$

(k) $y\,dx + x\,dy = 0$

(l) $\left(x^2 + \dfrac{2y}{x}\right)dx = (3 - \ln x^2)\,dy$

(m) $\dfrac{dy}{dx} = \dfrac{x}{y} + \dfrac{y}{x} + 1$ **(n)** $\dfrac{y}{x^2}\dfrac{dy}{dx} + e^{2x^3 + y^2} = 0$

(o) $y = xy' + (y' - 3)^2$ **(p)** $y' + 5y^2 = 3x^4 - 2xy$

6. Solve $(y^2 + 1)\,dx = y\sec^2 x\,dy$.

7. Solve $\dfrac{y}{x}\dfrac{dy}{dx} = \dfrac{e^x}{\ln y}$ subject to $y(1) = 1$.

8. Solve $y(\ln x - \ln y)\,dx = (x\ln x - x\ln y - y)\,dy$.

9. Solve $xyy' = 3y^2 + x^2$ subject to $y(-1) = 2$.

10. Solve $(6x + 1)y^2\dfrac{dy}{dx} + 3x^2 + 2y^3 = 0$.

11. Solve $ye^{xy}\dfrac{dx}{dy} + xe^{xy} = 12y^2$ subject to $y(0) = -1$.

12. Solve $x\,dy + (xy + y - x^2 - 2x)\,dx = 0$.

13. Solve $(x^2 + 4)\dfrac{dy}{dx} = 2x - 8xy$ subject to $y(0) = -1$.

14. Solve $(2x + y)y' = 1$.

15. Solve $x\dfrac{dy}{dx} + 4y = x^4y^2$ subject to $y(1) = 1$.

16. Solve $-xy' + y = (y' + 1)^2$ subject to $y(0) = 0$.

In Problems 17 and 18 solve the given differential equation by means of a substitution.

17. $\dfrac{dy}{dx} + xy^3\sec\dfrac{1}{y^2} = 0$ **18.** $y'' = x - y'$

19. Use the Picard method to find the approximations y_1 and y_2 for $y' = x^2 + y^2$, $y(0) = 1$.

20. Solve $y' + 2y = 4$, $y(0) = 3$ by one of the usual methods. Solve the same problem by Picard's method and compare the results.

ND EDITION A FIRST COURSE IN DIFFERENTIAL EQUATIONS SECOND EDITION A FIRST
SE IN DIFFERENTIAL EQUATIONS SECOND EDITION A FIRST COURSE IN DIFFERENTIAL
TIONS SECOND EDITION A FIRST COURSE IN DIFF~~ERENTIAL EQUATIONS SE~~ND EDI
FIRST COURSE IN DIFFERENTIAL EQUATIONS SEC~~ND EDITION A FIRST COUR~~SE IN D
ENTIAL EQUATIONS SECOND EDITION A FIRST COURSE IN DIFFERENTIAL EQUATIONS
ND EDITION A FIRST COURSE IN DIFFERENTIAL EQUATIONS SECOND EDITION A FIRST
SE IN DIFFERENTIAL EQUATIONS SECOND EDITION A FIRST COURSE IN DIFFERENTIAL
TIONS S~~ECOND EDITION A FI~~RST COURSE IN DIFFERENTIAL EQUATIONS SECOND EDI
FIRST CO~~URSE IN DIFFEREN~~TIAL EQUATIONS SECOND EDITION A FIRST COURSE IN D
ENTIAL E~~QUATIONS SECOND E~~DITION A FIRST COURSE IN DIFFERENTIAL EQUATIONS
ND EDIT~~ION A FIRST COURSE IN~~ DIFFERENTIAL EQUATIONS SECOND EDITION A FIRST
SE IN DI~~FFERENTIAL EQUATIONS SECOND E~~DITION A FIRST COURSE IN DIFFERENTIAL
TIONS S~~ECOND EDITION A FIRST COURSE IN~~ DIFFERENTIAL EQUATIONS SECOND EDI
FIRST COURSE IN DIFFERENTIAL EQUATIONS SECOND EDITION A FIRST COURSE IN D
ENTIAL EQUATIONS SECOND EDITION A FIRST COURSE IN DIFFERENTIAL EQUATIONS

CHAPTER 3

Applications of First-Order Differential Equations

3.1 Orthogonal Trajectories

Recall from your study of analytic geometry that two lines L_1 and L_2, which are not parallel to the coordinate axes, are perpendicular if and only if their respective slopes satisfy the relationship $m_1 m_2 = -1$. For this reason, the graphs of $y = (-1/2)x + 1$ and $y = 2x + 4$ are obviously perpendicular. Indeed, the line $y = (-1/2)x + 1$ is perpendicular to every line in the family of curves $y = 2x + c_1$. See Figure 3.1(a). In fact, as Figure 3.1(b) shows, each curve in the family $y = (-1/2)x + c_2$ is perpendicular to every curve in the family $y = 2x + c_1$.

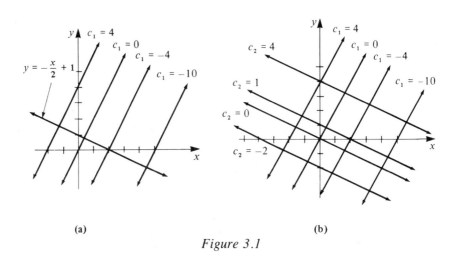

(a)   (b)

Figure 3.1

Orthogonal Curves

In general, two curves $\mathscr{C}_1$ and $\mathscr{C}_2$ are said to be **orthogonal** at a point if and only if their tangents T_1 and T_2 are perpendicular at the point of intersection. See Figure 3.2. Except for the case when T_1 and T_2 are parallel to the

86

coordinate axes, this means the slopes of the tangents are negative reciprocals of one another.

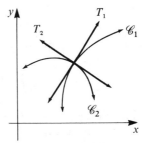

Figure 3.2

EXAMPLE

Show that the curves $y = x^3$ and $x^2 + 3y^2 = 4$ are orthogonal at the point(s) of intersection.

Solution: In Figure 3.3 it is seen that the points of intersection of the graphs are $(1,1)$ and $(-1,-1)$. Now the slope of the tangent line to $y = x^3$ at any point is $dy/dx = 3x^2$

so that
$$\left.\frac{dy}{dx}\right|_{x=1} = \left.\frac{dy}{dx}\right|_{x=-1} = 3.$$

We use implicit differentiation to obtain dy/dx for the second curve:
$$2x + 6y\frac{dy}{dx} = 0$$

$$\frac{dy}{dx} = -\frac{x}{3y}$$

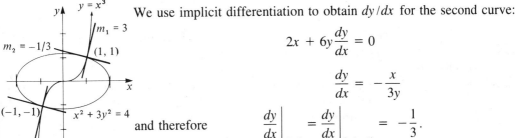

Figure 3.3

and therefore
$$\left.\frac{dy}{dx}\right|_{(1,1)} = \left.\frac{dy}{dx}\right|_{(-1,-1)} = -\frac{1}{3}.$$

Thus, at either $(1,1)$ or $(-1,-1)$ we have
$$\left(\frac{dy}{dx}\right)_{\mathscr{C}_1} \cdot \left(\frac{dy}{dx}\right)_{\mathscr{C}_2} = -1.$$

It is easy to show that any curve $\mathscr{C}_1$ in the family $y = c_1 x^3$; $c_1 \neq 0$, is orthogonal to each curve $\mathscr{C}_2$ in the family $x^2 + 3y^2 = c_2^2$. The differential equation of the first family is:

$$\frac{dy}{dx} = 3c_1 x^2 = 3\left(\frac{y}{x^3}\right)x^2$$

$$= \frac{3y}{x}$$

since $c_1 = y/x^3$. Now implicit differentiation of $x^2 + 3y^2 = c_2^2$ leads to exactly the same differential equation as for $x^2 + 3y^2 = 4$; namely,

$$\frac{dy}{dx} = -\frac{x}{3y}.$$

Hence, at the point (x,y) on each curve

$$\left(\frac{dy}{dx}\right)_{\mathscr{C}_1} \cdot \left(\frac{dy}{dx}\right)_{\mathscr{C}_2} = \left(\frac{3y}{x}\right)\left(-\frac{x}{3y}\right) = -1.$$

Since the slopes of the tangent lines are negative reciprocals, the curves $\mathscr{C}_1$ and $\mathscr{C}_2$ intersect each other in an orthogonal manner.

This discussion leads to the following definition.

DEFINITION 3.1   When *all* the curves of one family of curves $G(x,y,c_1) = 0$ intersect orthogonally *all* the curves of another family $H(x,y,c_2) = 0$, then the families are said to be **orthogonal trajectories** of each other.   ☐

In other words, an orthogonal trajectory is any *one* curve which intersects every curve of another family at right angles.

EXAMPLES

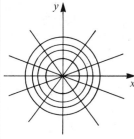

Figure 3.4

(a) The graph of $y = (-1/2)x + 1$ is an orthogonal trajectory of $y = 2x + c_1$. The families $y = (-1/2)x + c_2$ and $y = 2x + c_1$ are orthogonal trajectories.

(b) The graph of of $y = 4x^3$ is an orthogonal trajectory of $x^2 + 3y^2 = c_2^2$. The families $y = c_1 x^3$ and $x^2 + 3y^2 = c_2^2$ are orthogonal trajectories.

(c) In Figure 3.4 it is seen that the family of straight lines $y = c_1 x$ through the origin and the family $x^2 + y^2 = c_2^2$ of concentric circles with center at the origin are orthogonal trajectories.

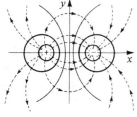

Figure 3.5

Orthogonal trajectories occur naturally in the construction of meteorological maps and in the study of electricity and magnetism. For example, in an electric field around two bodies of opposite charge the lines of force are perpendicular to the equipotential curves (that is, curves along which the potential is constant). The lines of force are indicated in Figure 3.5 by dashed lines.

The general method   To find the orthogonal trajectories of a given family of curves we first find the differential equation

$$\frac{dy}{dx} = f(x,y)$$

which describes the family. The differential equation of the second, and orthogonal, family is then

$$\frac{dy}{dx} = \frac{-1}{f(x,y)}.$$

EXAMPLE   Find the orthogonal trajectories of the family of rectangular hyperbolas

$$y = \frac{c_1}{x}.$$

Solution: The derivative of $y = c_1/x$ is

$$\frac{dy}{dx} = \frac{-c_1}{x^2}.$$

Replacing c_1 by $c_1 = xy$ yields the differential equation of the given family:

$$\frac{dy}{dx} = -\frac{y}{x}.$$

The differential equation of the orthogonal family is then

$$\frac{dy}{dx} = \frac{-1}{(-y/x)} = \frac{x}{y}.$$

We solve this last equation by separation of variables:

$$y\, dy = x\, dx$$

$$\int y\, dy = \int x\, dx$$

$$\frac{y^2}{2} = \frac{x^2}{2} + c_2'$$

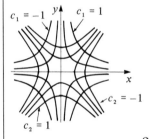

Figure 3.6

or

$$y^2 - x^2 = c_2,$$

where, for convenience we have replaced $2c_2'$ by c_2.

The graphs of the two families, for various values of c_1 and c_2, are given in Figure 3.6.

EXAMPLE   Find the orthogonal trajectories of

$$y = \frac{c_1 x}{1 + x}.$$

Solution: From the quotient rule we find

$$\frac{dy}{dx} = \frac{c_1}{(1 + x)^2}.$$

But $c_1 = y(1 + x)/x$ so that

$$\frac{dy}{dx} = \frac{y}{x(1 + x)}.$$

The differential equation of the orthogonal trajectories is then

$$\frac{dy}{dx} = -\frac{x(1 + x)}{y}.$$

Again by separating variables we have:

$$y \, dy = -x(1 + x) \, dx$$

$$\int y \, dy = -\int (x + x^2) \, dx$$

$$\frac{y^2}{2} = -\frac{x^2}{2} - \frac{x^3}{3} + c_2'$$

or $3y^2 + 3x^2 + 2x^3 = c_2$ ($6c_2'$ replaced by c_2).

EXERCISES 3.1

Answers to odd-numbered problems begin on page A-7.

In Problems 1–26 find the orthogonal trajectories of the given family of curves.

1. $y = c_1 x$

2. $3x + 4y = c_1$

3. $y = c_1 x^2$

4. $y = (x - c_1)^2$

5. $c_1 x^2 + y^2 = 1$

6. $2x^2 + y^2 = c_1^2$

7. $y = c_1 e^{-x}$

★ 8. $y = e^{c_1 x}$

9. $y^2 = c_1 x^3$

10. $y^a = c_1 x^b$, a and b constants

11. $y = \dfrac{x}{1 + c_1 x}$

12. $y = \dfrac{1 + c_1 x}{1 - c_1 x}$

13. $2x^2 + y^2 = 4c_1 x$

★14. $x^2 + y^2 = 2c_1 x$

15. $y^3 + 3x^2 y = c_1$

16. $y^2 - x^2 = c_1 x^3$

17. $y = \dfrac{c_1}{1 + x^2}$

18. $y = \dfrac{1}{c_1 + x}$

19. $4y + x^2 + 1 + c_1 e^{2y} = 0$

★20. $y = -x - 1 + c_1 e^x$

21. $y = \dfrac{1}{\ln c_1 x}$

22. $y = \ln (\tan x + c_1)$

23. $\sinh y = c_1 x$

24. $y = c_1 \sin x$

25. $x^{1/3} + y^{1/3} = c_1$

26. $x^a + y^a = c_1$, $a \neq 2$

27. Find the member of the orthogonal trajectories for $x + y = c_1 e^y$ which passes through $(0, 5)$.

★**28.** Find the member of the orthogonal trajectories for $3xy^2 = 2 + 3c_1x$ which passes through $(0, 10)$.

29. Verify that the orthogonal trajectories of the family of curves given by the parametric equations $x = c_1e^t \cos t$, $y = c_1e^t \sin t$ are

$$x = c_2e^{-t} \cos t, \qquad y = c_2e^{-t} \sin t.$$

[*Hint*: $dy/dx = (dy/dt)/(dx/dt)$.]

30. In calculus it is shown that for a graph in polar coordinates

$$\tan \psi = r\frac{d\theta}{dr}$$

where ψ is the positive counterclockwise angle between the radius vector and the tangent line. Show that two polar curves $r = f_1(\theta)$ and $r = f_2(\theta)$ are orthogonal if and only if

$$(\tan \psi_1)_{\epsilon_1}(\tan \psi_2)_{\epsilon_2} = -1.$$

In Problems 31–36 find the orthogonal trajectories of the given polar curves. Use the results of Problem 30.

EXAMPLE

Find the orthogonal trajectories of

$$r = c_1(1 - \sin \theta).$$

Solution: For the given curve we can write

$$\frac{dr}{d\theta} = -c_1 \cos \theta$$

$$= \frac{-r \cos \theta}{1 - \sin \theta}$$

so that

$$r\frac{d\theta}{dr} = -\frac{1 - \sin \theta}{\cos \theta} = \tan \psi_1.$$

Thus, by Problem 30 the differential equation of the orthogonal trajectories is

$$r\frac{d\theta}{dr} = \frac{\cos \theta}{1 - \sin \theta} = \tan \psi_2.$$

Separating variables then gives

$$\frac{dr}{r} = \frac{1 - \sin\theta}{\cos \theta} d\theta$$

$$= (\sec \theta - \tan \theta) \, d\theta$$

so that

$$\ln |r| = \ln |\sec \theta + \tan \theta| + \ln |\cos \theta| + \ln c_2$$
$$= \ln |c_2(1 + \sin \theta)|.$$

Hence

$$r = c_2(1 + \sin \theta).$$

31. $r = 2c_1 \cos \theta$

32. $r = c_1(1 + \cos \theta)$

33. $r^2 = c_1 \sin 2\theta$

★34. $r = \dfrac{c_1}{1 + \cos \theta}$

35. $r = c_1 \sec \theta$

36. $r = c_1 e^\theta$

37. A family of curves that intersects a given family of curves at a specified constant angle $\alpha \neq \pi/2$ is said to be an isogonal family. The two families are said to be **isogonal trajectories** of each other. If $dy/dx = f(x, y)$ is the differential equation of the given family, show that the differential equation of the isogonal family is

$$\frac{dy}{dx} = \frac{f(x, y) \pm \tan \alpha}{1 \mp f(x, y) \tan \alpha}.$$

In Problems 38–40 use the results of Problem 37 to find the isogonal family which intersects the one-parameter family of straight lines $y = c_1 x$ at the given angle.

38. $\alpha = 45°,$ **39.** $\alpha = 60°,$ **★40.** $\alpha = 30°$

A family of curves can be *self-orthogonal* in the sense that a member of the orthogonal trajectories is also a member of the original family. In Problems 41–43 show that the given family of curves is self-orthogonal.

41. parabolas $y^2 = 2c_1(x + c_1)$

42. cardioids $r = c_1(1 + \sin \theta)$

43. confocal conics $\dfrac{x^2}{c_1 + 1} + \dfrac{y^2}{c_1} = 1$

3.2   Applications of Linear Equations

3.2.1   Growth and Decay

The initial-value problem

$$\frac{dx}{dt} = kx$$

$$x(t_0) = x_0$$

(1)

where k is a constant, occurs in many physical theories involving either growth or decay. For example, in biology it is often observed that the rate at which certain bacteria grow is proportional to the number of bacteria present at any time. Over short intervals of time the population of small animals, such as rodents, can be predicted fairly accurately by the solution of (1). The constant k can be determined from the solution of the differential equation by using a subsequent measurement of the population at a time $t_1 > t_0$.

In physics an initial-value problem such as (1) provides a model for approximating the remaining amount of a substance which is disintegrating through radioactivity. The differential equation in (1) could also determine the temperature in a cooling body. In chemistry the amount of a substance remaining during certain reactions is also described by (1).

EXAMPLE

A culture initially has N_0 number of bacteria. At $t = 1$ hour the number of bacteria is measured to be $(3/2)N_0$. If the rate of growth is proportional to the number of bacteria present, determine the time necessary for the number of bacteria to triple.

Solution: We first solve the differential equation

$$\frac{dN}{dt} = kN \tag{2}$$

subject to $N(0) = N_0$.

After we have solved the above problem we then use the empirical condition $N(1) = (3/2)N_0$ to determine the constant of proportionality k.

Now (2) is both separable and linear. When put into the form

$$\frac{dN}{dt} - kN = 0$$

we can see by inspection that the integrating factor is e^{-kt}. Multiplying both sides of the equation by this term gives immediately

$$\frac{d}{dt}[e^{-kt}N] = 0.$$

Integrating both sides of the last equation yields

$$e^{-kt}N = c \quad \text{or} \quad N(t) = ce^{kt}.$$

At $t = 0$ it follows that $N_0 = ce^0 = c$ and so $N(t) = N_0e^{kt}$. At $t = 1$ we have

$$\frac{3}{2}N_0 = N_0e^k \quad \text{or} \quad e^k = \frac{3}{2}$$

from which we get

$$k = \ln\left(\frac{3}{2}\right)$$
$$= 0.4055.$$

Thus,

$$N(t) = N_0e^{0.4055t}.$$

To find the time at which the bacteria have tripled we solve

$$3N_0 = N_0e^{0.4055t}$$

for t.

It follows that

$$0.4055t = \ln 3$$
$$t = \frac{\ln 3}{0.4055}$$
$$\approx 2.71 \text{ hours.}$$

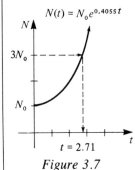

$N(t) = N_0e^{0.4055t}$

Figure 3.7

Note: We can write the function $N(t)$ obtained in the preceding example in an alternative form. From the laws of exponents

$$N(t) = N_0(e^k)^t$$

$$= N_0\left(\frac{3}{2}\right)^t$$

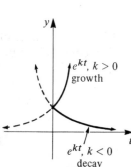

Figure 3.8

since $e^k = 3/2$. This latter solution provides a convenient method for computing $N(t)$ for small positive integral values of t; it also clearly shows the influence of the subsequent experimental observation at $t = 1$ on the solution for all time. We notice too, that the actual number of bacteria present at time $t = 0$ is quite irrelevant in finding the time required to triple the number in the culture. The necessary time to triple, say, 100, or 10,000 bacteria is still approximately 2.71 hours.

As shown in Figure 3.8, the exponential function e^{kt} increases as t increases for $k > 0$, and decreases as t increases if $k < 0$. Thus, problems describing growth, such as population, bacteria, or even capital, are characterized by a positive value of k, whereas problems involving decay, as in radioactive disintegration, will yeild a negative k value.

Half-life

In physics the half-life is a measure of the stability of a radioactive substance. The half-life is simply the time it takes for one-half of the atoms in an initial amount A_0 to disintegrate. The longer the half-life of a substance the more stable it is. For example, the half-life of highly radioactive radium is approximately 1700 years, whereas the most commonly occurring uranium isotope, U-238, has a half-life of approximately 4,500,000,000 years.

EXAMPLE

A breeder reactor converts the relatively stable uranium 238 into the isotope plutonium 239. After 15 years it is determined that 0.043% of the initial amount A_0 of the plutonium has disintegrated. Find the half-life of this isotope if the rate of disintegration is proportional to the amount remaining.

Solution: Let $A(t)$ denote the amount of plutonium remaining at any time. As in the previous example, the solution of the initial value problem

$$\frac{dA}{dt} = kA$$

$$A(0) = A_0$$

is $A(t) = A_0 e^{kt}.$

If 0.043% of the atoms of A_0 have disintegrated then 99.957% of the substance remains. To find k we must solve

$$0.99957 A_0 = A_0 e^{15k}.$$

Therefore, $e^{15k} = 0.99957$

$$15k = \ln(0.99957)$$

$$k = \frac{\ln(0.99957)}{15}$$

$$= -0.00002867.$$

Hence,
$$A(t) = A_0 e^{-0.00002867t}.$$

Now the half-life is the corresponding value of time for which $A(t) = A_0/2$. Solving for t gives,

$$\frac{A_0}{2} = A_0 e^{-0.00002867t} \qquad \text{or} \qquad \frac{1}{2} = e^{-0.00002867t},$$

$$-0.00002867t = \ln\left(\frac{1}{2}\right) = -\ln 2$$

$$t = \frac{\ln 2}{0.00002867}$$

$$\approx 24,180 \text{ years.}$$

Carbon dating

About 1950 the chemist Willard Libby devised a method of using radioactive carbon as a means of determining the approximate ages of fossils. The theory is based on the fact that the isotope carbon 14 is produced in the atmosphere by the action of cosmic radiation on nitrogen. The ratio of the amount of C-14 to ordinary carbon in the atmosphere appears to be a constant, and as a consequence the proportionate amount of the isotope present in all living organisms is the same as that in the atmosphere. When an organism dies the absorption of C-14, by either breathing or eating, ceases. Thus, by comparing the proportionate amount of C-14 present, say, in a fossil with the constant ratio found in the atmosphere it is possible to obtain a reasonable estimation of its age. The method is based upon the knowledge that the half-life of the radioactive C-14 is approximately 5600 years. For his work Libby won the Nobel Prize for chemistry in 1960.

EXAMPLE

A fossilized bone is found to contain 1/1000 the original amount of C-14. Determine the age of the fossil.

Solution: The starting point is again
$$A(t) = A_0 e^{kt}.$$

When $t = 5600$ years, $A(t) = A_0/2$ from which we can determine the value of k

$$\frac{A_0}{2} = A_0 e^{5600k}$$

$$5600k = \ln\left(\frac{1}{2}\right) = -\ln 2$$

$$k = -\frac{\ln 2}{5600}$$

$$= -0.00012378.$$

Therefore

$$A(t) = A_0 e^{-0.00012378t}.$$

When $A(t) = A_0/1000$ we have

$$\frac{A_0}{1000} = A_0 e^{-0.00012378t}$$

so that

$$-0.00012378t = \ln\left(\frac{1}{1000}\right) = -\ln 1000$$

$$t = \frac{\ln 1000}{0.00012378}$$

$$\approx 55{,}800 \text{ years.}$$

Remark: The date found in the last example is really at the border of accuracy for this method. The usual carbon 14 technique is limited to about 9 half-lives of the isotope or about 50,000 years. One reason is that the chemical analysis needed to obtain an accurate measurement of the remaining C-14 becomes somewhat formidable around the point of $A_0/1000$. Also, this analysis demands the destruction of a rather large sample of the specimen. If this measurement is accomplished indirectly, based on the actual radioactivity of the specimen, than it is very difficult to distinguish between the radiation from the fossil and the normal background radiation. But in recent developments, the use of a particle accelerator has enabled scientists to separate the C-14 from the stable C-12 directly. By computing the precise value of the ratio of C-14 to C-12 the accuracy of this method can be extended to 70,000–100,000 years. Other isotopic techniques such as using potassium 40 and argon 40 can give dates of several million years. Nonisotopic methods based on the use of amino acids are also sometimes possible.

3.2.2   Cooling, Circuits, and Chemical Mixtures

Cooling

Newton's law of cooling states that the rate at which the temperature $T(t)$ changes in a cooling body is proportional to the difference between the temperature in the body and the constant temperature T_0 of the surrounding medium. That is,

$$\frac{dT}{dt} = k(T - T_0) \tag{3}$$

where k is a constant of proportionality.

EXAMPLE   When a cake is removed from a baking oven its temperature is measured at 300°F. Three minutes later its temperature is 200°F. How long will it take to cool off to a room temperature of 70°F?

Solution: We must solve the initial-value problem

$$\frac{dT}{dt} = k(T - 70)$$

$$T(0) = 300 \tag{4}$$

and determine the value of k so that $T(3) = 200$.

Equation (3) is both linear and separable; utilizing this latter procedure yields

$$\frac{dT}{T - 70} = k\, dt$$

$$\ln |T - 70| = kt + c_1$$

$$T - 70 = c_2 e^{kt}$$

$$T = 70 + c_2 e^{kt}.$$

When $t = 0$, $T = 300$ so that $300 = 70 + c_2$ gives $c_2 = 230$ and therefore $T = 70 + 230\, e^{kt}$.

From $T(3) = 200$ we find

$$e^{3k} = \frac{13}{23}$$

or

$$k = \frac{1}{3} \ln \frac{13}{23}$$

$$= -0.19018.$$

Thus,

$$T(t) = 70 + 230 e^{-0.19018t}. \tag{5}$$

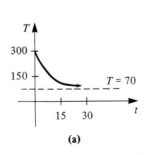

T(t)	t (minutes)
75°	20.1
74°	21.3
73°	22.8
72°	24.9
71°	28.6
70.5°	32.3

(a)   (b)

Figure 3.9

Unfortunately (5) furnishes no finite solution to $T(t) = 70$ since $\lim_{t \to \infty} T(t) = 70$. Yet intuitively we expect the cake will assume the room temperature after a reasonably long period of time. How long is long? Of course, we should not be the least bit disturbed by the fact that the model (4) does not quite live up to our physical intuition. Parts (a) and (b) of Figure 3.9 clearly show that the cake will be approximately at room temperature in about one half hour.

An *L*-*R* series circuit

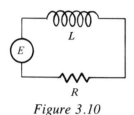

Figure 3.10

In a series circuit containing only a resistor and an inductor, Kirchoff's second law states the sum of the voltage drop across the inductor ($L(di/dt)$) and the voltage drop across the resistor (iR) is the same as the impressed voltage ($E(t)$) on the circuit. See Figure 3.10.

Thus we obtain the linear differential equation for the current $i(t)$,

$$L\frac{di}{dt} + Ri = E(t) \tag{6}$$

where L and R are constants known as the inductance and the resistance, respectively. The current $i(t)$ is sometimes called the **response** of the system.

EXAMPLE

A 12-volt battery is connected to a simple series circuit in which the inductance is 1/2 henry and the resistance is 10 ohms. Determine the current i if the initial current is zero.

Solution: We solve the initial-value problem

$$\frac{1}{2}\frac{di}{dt} + 10i = 12,$$

$$i(0) = 0.$$

First, multiply the differential equation (2) and read off the integrating factor e^{20t}.

Hence we find
$$\frac{d}{dt}[e^{20t}i] = 24e^{20t}$$

$$e^{20t}i = \frac{24}{20}e^{20t} + c$$

$$i = \frac{6}{5} + ce^{-20t}.$$

Now $i(0) = 0$ implies $0 = 6/5 + c$ or $c = -6/5$. Therefore the response is
$$i(t) = \frac{6}{5} - \frac{6}{5}e^{-20t}.$$

Transient and steady-state terms

From equation (7) of Section 2.5 we can write down a general solution of (6),

$$i(t) = \frac{e^{-(R/L)t}}{L}\int e^{(R/L)t}E(t)\,dt + ce^{-(R/L)t}. \tag{7}$$

In particular, when $E(t) = E_0$ is a constant then (7) becomes

$$i(t) = \frac{E_0}{R} + ce^{-(R/L)t}. \qquad (8)$$

Note that as $t \to \infty$, the second term in equation (8) approaches zero. Such a term is usually called a **transient term**; the remaining term(s) is called the **steady-state** part of the solution. In this case E_0/R is also called the **steady-state current**; for large time it then appears that the current in the circuit is simply governed by Ohm's law ($E = iR$).

A mixture problem

The mixing of two fluids sometimes gives rise to a linear first-order differential equation. In the next example we consider the mixture of two salt solutions of different concentrations.

EXAMPLE

Initially 50 pounds of salt is dissolved in a large tank holding 300 gallons of water. A brine solution is pumped into the tank at a rate of 3 gallons per minute, and a well-stirred solution is then pumped out at the same rate. If the concentration of the solution entering is 2 pounds per gallon, determine the amount of salt in the tank at any time. How much salt is present after 50 minutes? after a long time?

Solution: Let $A(t)$ be the amount of salt (in pounds) in the tank at any time. For problems of this sort, the net rate at which $A(t)$ changes is given by

$$\frac{dA}{dt} = \left(\begin{array}{c} \text{rate of} \\ \text{substance entering} \end{array} \right) - \left(\begin{array}{c} \text{rate of} \\ \text{substance leaving} \end{array} \right) = R_1 - R_2. \qquad (9)$$

Now the rate at which the salt enters the tank is, in pounds per minute,

$$R_1 = (3 \text{ gal/min}) \cdot (2 \text{ lb/gal}) = 6 \text{ lb/min}$$

whereas the rate at which salt is leaving is

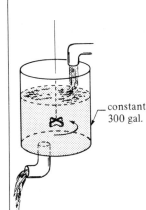
—constant
300 gal.

$$R_2 = (3 \text{ gal/min}) \cdot \left(\frac{A}{300} \text{ lb/gal} \right) = \frac{A}{100} \text{ lb/min}.$$

Thus equation (9) becomes

$$\frac{dA}{dt} = 6 - \frac{A}{100} \qquad (10)$$

which we solve subject to the initial condition $A(0) = 50$.

Figure 3.11

Since the integrating factor is $e^{t/100}$ we can write (10) as

$$\frac{d}{dt}[e^{t/100}A] = 6e^{t/100}$$

and therefore

$$e^{t/100}A = 600e^{t/100} + c$$

$$A = 600 + ce^{-t/100}. \qquad (11)$$

When $t = 0$, $A = 50$ so we find that $c = -550$. Finally, we obtain

$$A(t) = 600 - 550e^{-t/100}. \qquad (12)$$

At $t = 50$ we find $A(50) = 266.41$ lb. Also, as $t \to \infty$ it is seen from (12) and Figure 3.12 that $A \to 600$. Of course this is what we would expect; over a long period of time the number of pounds of salt in the solution must be

$$(300 \text{ gal})(2 \text{ lb/gal}) = 600 \text{ lb}.$$

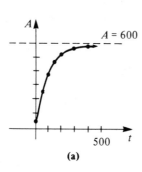

t (minutes)	A (lbs)
50	266.41
100	397.67
150	477.27
200	525.57
300	572.62
400	589.93

(a)   (b)

Figure 3.12

In the preceding example we assumed that the rate at which the solution was pumped in was the same rate as the rate at which the solution was pumped out. However, this need not be the case; the mixed brine solution could be pumped out at a rate faster or slower than the rate at which the other solution is pumped in. The resulting differential equation in this latter situation is linear with a variable coefficient.

EXAMPLE

In the preceding example, if the well-stirred solution is pumped out at a slower rate of 2 gallons per minute then the solution is *accumulating* at a rate of

$$(3 - 2) \text{ gal/min} = 1 \text{ gal/min}.$$

After t minutes there are

$$300 + t \text{ gallons}$$

of brine in the tank. The rate at which the salt is leaving is then

$$R_2 = (2 \text{ gal/min}) \cdot \left(\frac{A}{300 + t} \text{ lb/gal} \right) = \frac{2A}{300 + t} \text{ lb/min}.$$

Hence equation (9) becomes

$$\frac{dA}{dt} = 6 - \frac{2A}{300 + t} \quad \text{or} \quad \frac{dA}{dt} + \frac{2A}{300 + t} = 6.$$

Finding the integrating factor and solving the last equation gives

$$A(t) = 2(300 + t) + c(300 + t)^{-2}.$$

The initial condition $A(0) = 50$ yields $c = -4.95 \times 10^7$ and so

$$A(t) = 2(300 + t) - (4.95 \times 10^7)(300 + t)^{-2}.$$

EXERCISES 3.2   *Answers to odd-numbered problems begin on page A-8.*

[3.2.1]

1. The population of a certain community is known to increase at a rate proportional to the number of people present at any time. If the population has doubled in 5 years, how long will it take to triple? to quadruple?

★ 2. Suppose it is known that the population of the community in Problem 1 is 10,000 after 3 years. What was the initial population? What will be the population in 10 years?

3. If P_0 is the initial population of a community, show that if P is governed by (2), then

$$\left(\frac{P_1}{P_0}\right)^{t_2} = \left(\frac{P_2}{P_0}\right)^{t_1}$$

where $P_1 = P(t_1)$ and $P_2 = P(t_2)$, $t_1 < t_2$.

4. Use the results of Problem 3 to solve the following. A town has an initial population of 500 that increases by 15% in 10 years. What will be the population in 30 years?

5. In one model of the changing population $P(t)$ of a community it is assumed that

$$\frac{dP}{dt} = \frac{dB}{dt} - \frac{dD}{dt}$$

where dB/dt and dD/dt are the birth and death rates, respectively.

(a) Solve for $P(t)$ if

$$\frac{dB}{dt} = k_1 P \quad \text{and} \quad \frac{dD}{dt} = k_2 P.$$

(b) Analyze the cases $k_1 > k_2$, $k_1 = k_2$, $k_1 < k_2$.

6. Initially there were 100 milligrams of a radioactive substance present. After 6 hours the mass decreased by 3%. If the rate of decay is proportional to the amount of the substance present at any time, find the amount remaining after 24 hours.

7. Determine the half-life of the radioactive substance described in Problem 6.

★ 8. Show in general that the half-life of a radioactive substance is

$$t = \frac{(t_2 - t_1) \ln 2}{\ln A_1/A_2}$$

where $A_1 = A(t_1)$ and $A_2 = A(t_2)$, $t_1 < t_2$.

9. When a vertical beam of light passes through a transparent substance, the rate at which its intensity I decreases is proportional to $I(t)$, where t

represents the thickness of the medium in feet. In clear sea water the intensity 3 feet below the surface is 25% of the initial intensity I_0 of the incident beam. What is the intensity of the beam 15 feet below the surface?

10. When interest is compounded continuously the amount of money S increases at a rate proportional to the amount present at any time: $dS/dt = rS$ where r is the annual rate of interest (see Section 1.2).

   (a) Find the amount of money accrued at the end of 5 years when $5000 is deposited in a savings account drawing $5\frac{3}{4}\%$ annual interest compounded continuously.

   (b) In how many years will the initial sum deposited be doubled?

   (c) Use a hand calculator to compare the number obtained in part (a) with the value

$$S = 5000\left(1 + \frac{0.0575}{4}\right)^{5(4)}.$$

   This value represents the amount that would be accrued when interest is compounded quarterly.

11. In a piece of burned wood, or charcoal, it was found that 85.5% of the C-14 has decayed. What is the approximate age of the wood? (It is precisely this data that archaeologists used to date prehistoric paintings in a cave in Lascaux, France.)

[3.2.2] ★12. A thermometer is taken from an inside room to the outside where the air temperature is 5°F. After 1 minute the thermometer reads 55°F and after 5 minutes the reading is 30°F. What is the initial temperature of the room?

13. A thermometer is removed from a room where the air temperature is 70°F to the outside where the temperature is 10°F. After $\frac{1}{2}$ minute the thermometer reads 50°F. What is the reading at $t = 1$ minute? How long will it take for the thermometer to reach 15°F?

14. Formula (3) also obtains when an object absorbs heat from the surrounding medium. If a small metal bar, whose initial temperature is 20°C, is dropped into a container of boiling water, how long will it take for the bar to reach 90°C if it is known that its temperature increased 2° in 1 second? How long will it take the bar to reach 98°C?

15. A 30-volt electromotive force is applied to a series circuit in which the inductance is 0.1 henry and the resistance is 50 ohms. Find the current $i(t)$ if $i(0) = 0$. Determine the behavior of the current for large time.

16. Solve the general equation (6) under the assumption that

$$E(t) = E_0 \sin \omega t \qquad i(0) = i_0.$$

17. A tank contains 200 liters of fluid in which 30 g of salt is dissolved. Brine containing 1 g of salt per liter is then pumped into the tank at a rate of 4 liters per minute; the well-mixed solution is pumped out at the same rate. Find the number of grams of salt $A(t)$ in the tank at any time.

★**18.** Solve problem 17 under the assumption that pure water is pumped into the tank.

19. A large tank is partially filled with 100 gallons of fluid in which 10 lb of salt is dissolved. Brine containing $\frac{1}{2}$ lb of salt per gallon is pumped into the tank at a rate of 6 gallons per minute; the well-mixed solution is then pumped out at a slower rate of 4 gallons per minute. Find the number of pounds of salt in the tank after 30 minutes.

20. Beer containing 6% alcohol per gallon is pumped into a vat which initially contains 400 gallons of beer at 3% alcohol. The rate at which the beer is pumped in is 3 gallons per minute, whereas the mixed liquid is pumped out at a rate of 4 gallons per minute. Find the number of gallons of alcohol $A(t)$ in the tank at any time. What is the percentage of alcohol in the tank after 60 minutes? When is the tank empty?

21. The differential equation governing the velocity v of a falling weight w subjected to air resistance proportional to the instantaneous velocity is

$$m\frac{dv}{dt} = mg - kv$$

where k is a positive constant of proportionality. Solve the equation subject to the initial condition $v(0) = v_0$ and determine the limiting velocity of the weight. If distance s is related to velocity $ds/dt = v$, find an explicit expression for s if it is further known that $s(0) = s_0$.

22. The rate at which a drug disseminates into the bloodstream is governed by the differential equation

$$\frac{dX}{dt} = A - BX$$

where A and B are positive constants. The function $X(t)$ describes the concentration of the drug in the bloodstream at any time t. Find the limiting value of X as $t \to \infty$. At what time is the concentration one-half this limiting value? Assume that $X(0) = 0$.

3.3 Applications of Nonlinear Equations

We have seen that if a population P is described by

$$\frac{dP}{dt} = kP, \qquad k > 0 \tag{1}$$

then $P(t)$ exhibits unbounded exponential growth. In many instances this differential equation provides an unrealistic model of the growth of a population, that is, what is actually observed differs substantially from what is predicted.

Around 1840, the Belgian mathematician-biologist P. F. Verhulst was concerned with mathematical formulations for predicting the human populations of various countries. One of the equations he studied was

$$\frac{dP}{dt} = P(a - bP) \tag{2}$$

where a and b are positive constants. Equation (2) came to be known as the **logistic equation** and its solution is called the **logistic function** (the graph of which is naturally called a logistic curve).

Equation (1) does not provide a very accurate model for population growth when the population itself is very large. Overcrowded conditions with the resulting detrimental effects on the environment, such as pollution, excessive and competitive demands for food and fuel, can have an inhibitive effect on the population growth. If a, $a > 0$, is a constant average birth rate, let us assume that the average death rate is proportional to the population $P(t)$ at any time. Thus, if $\frac{1}{P}\frac{dP}{dt}$ is the rate of growth per individual in a population, then

$$\frac{1}{P}\frac{dP}{dt} = \left(\begin{array}{c}\text{average}\\\text{birth rate}\end{array}\right) - \left(\begin{array}{c}\text{average}\\\text{death rate}\end{array}\right) = a - bP \tag{3}$$

where b is a positive constant of proportionality. Cross multiplying (3) by P immediately gives (2).

As we shall now see, the solution of (2) is bounded as $t \to \infty$. If we rewrite (2) as

$$\frac{dP}{dt} = aP - bP^2$$

the term $-bP^2$, $b > 0$, can therefore be interpreted as an "inhibition" or "competition" term. Also, in most applications the positive constant a is much larger than the constant b.

Logistic curves have proved to be quite accurate in predicting the growth patterns, in a limited space, of certain types of bacteria, protozoa, water fleas (*Daphnia*), and fruit flies (*Drosophila*.). We have already seen equation (2) in the form

$$\frac{dx}{dt} = kx(n + 1 - x) \qquad k > 0.$$

This differential equation provides a reasonable model for describing the spread of an epidemic brought about by initially introducing an infected individual into a static population. The solution $x(t)$ represents the number of individuals infected with the disease at any time. (See Section 1.2) Sociologists, and even the business world, have borrowed this latter model to study the spread of information and the impact of advertising in certain centers of population.

The solution

One method for solving equation (2) is separation of variables.* By partial fractions we can write

$$\frac{dP}{P(a - bP)} = dt$$

$$\left[\frac{1/a}{P} + \frac{b/a}{a - bP}\right]dP = dt$$

$$\frac{1}{a}\frac{dP}{P} - \frac{1}{a}\frac{(-b\,dP)}{a - bP} = dt$$

$$\frac{1}{a}\ln|P| - \frac{1}{a}\ln|a - bP| = t + c$$

$$\ln\left|\frac{P}{a - bP}\right| = at + ac$$

$$\frac{P}{a - bP} = c_1 e^{at}. \tag{4}$$

It follows from the last equation that

$$P(t) = \frac{ac_1 e^{at}}{1 + bc_1 e^{at}}$$

$$= \frac{ac_1}{bc_1 + e^{-at}}. \tag{5}$$

Now if we are given the initial condition $P(0) = P_0$, $P_0 \neq a/b$,[†] equation (4) implies

$$c_1 = \frac{P_0}{a - bP_0}$$

and therefore

$$P(t) = \frac{aP_0/(a - bP_0)}{[bP_0/(a - bP_0)] + e^{-at}}$$

or

$$P(t) = \frac{aP_0}{bP_0 + (a - bP_0)e^{-at}}. \tag{6}$$

Graphs of $P(t)$

The basic shape of the graph of the logistic function $P(t)$ can be obtained without too much effort. Although the variable t usually represents time and we are seldom concerned with applications in which $t < 0$, it is nonetheless

---

*In the form

$$\frac{dP}{dt} - aP = bP^2$$

you might recognize the logistic equation as a special case of Bernoulli's equation (see Section 2.6).

[†]Notice that $P = a/b$ is a singular solution of equation (2).

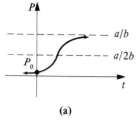

(a)

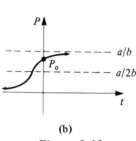

(b)

Figure 3.13

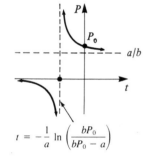

$$t = -\frac{1}{a} \ln\left(\frac{bP_0}{bP_0 - a}\right)$$

Figure 3.14

of some interest to include this interval when displaying the various graphs of P. From (6) we see for $t > 0$ that

$$P(t) \rightarrow \frac{aP_0}{bP_0} = \frac{a}{b} \qquad \text{as } t \rightarrow \infty;$$

and for $t < 0$ $\qquad P(t) \rightarrow 0 \qquad \text{as } t \rightarrow -\infty.$

Now differentiating (2) by the product rule gives

$$\frac{d^2 P}{dt^2} = P\left(-b\frac{dP}{dt}\right) + (a - bP)\frac{dP}{dt}$$

$$= \frac{dP}{dt}(a - 2bP)$$

$$= P(a - bP)(a - 2bP)$$

$$= 2b^2 P\left(P - \frac{a}{b}\right)\left(P - \frac{a}{2b}\right). \tag{7}$$

From calculus recall that the points where $d^2P/dt^2 = 0$ are possible points of inflection, but $P = 0$ and $P'' = a/b$ can obviously be ruled out. Hence, $P = a/2b$ is the only possible ordinate value at which the concavity of the graph can change. For $0 < P < a/2b$ it follows from (7) that $P'' > 0$, and $a/2b < P < a/b$ implies $P'' < 0$. Thus, reading from left to right, the graph changes from concave up to concave down at the point corresponding to $P = a/2b$. When the initial value satisfies $0 < P_0 < a/2b$ Figure 3.13(a) shows that the graph of $P(t)$ assumes the shape of an S. For $a/2b < P_0 < a/b$ the graph, as shown in Figure 3.13(b), is still S-shaped but the point of inflection occurs at a negative and, therefore, physically meaningless value of t.

If $P_0 > a/b$ equation (7) shows $P'' > 0$ for all t in the domain of $P(t)$ for which $P > 0$; when $P < 0$ equation (7) implies $P'' < 0$. However, $P = 0$ is not a point of inflection since, whenever $a - bP_0 < 0$, an inspection of (6) reveals a vertical asymptote at

$$t = -\frac{1}{a} \ln\left(\frac{bP_0}{bP_0 - a}\right).$$

The graph of $P(t)$ in this case is given in Figure 3.14.

EXAMPLE

Suppose a student carrying a flu virus returns to an isolated college campus of 1000 students. If it is assumed that the rate at which the virus spreads is proportional not only to the number x of infected students but also the number of students not infected (see Section 1.2), determine the number of infected students after 6 days if it is further observed that after 4 days $x(4) = 50$.

Solution: Assuming that no one leaves the campus throughout the duration of the disease, we must then solve the initial-value problem

$$\frac{dx}{dt} = kx(1000 - x)$$

$$x(0) = 1.$$

By making the identifications $a = 1000k$, $b = k$ we have immediately from (6) that

$$x(t) = \frac{1000k}{k + 999ke^{-1000kt}}$$

$$= \frac{1000}{1 + 999e^{-1000kt}} \qquad (8)$$

Now using the information $x(4) = 50$ we can determine k:

$$50 = \frac{1000}{1 + 999e^{-4000k}}$$

$$50 + 49{,}950e^{-4000k} = 1000$$

$$e^{-4000k} = \frac{19}{999}$$

$$k = \frac{-1}{4000} \ln \frac{19}{999}$$

$$= 0.0009906$$

Thus, (8) becomes
$$x(t) = \frac{1000}{1 + 999e^{-0.9906t}}.$$

Finally we find
$$x(6) = \frac{1000}{1 + 999e^{-5.9436}}$$

$$= 276 \text{ students.}$$

Additional calculated values of $x(t)$ are given in the table in Figure 3.15.

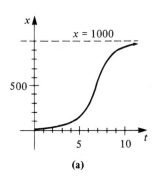

(a)

t days	x (number infected)
4	50 (observed)
5	124
6	276
7	507
8	735
9	882
10	953

(b)

Figure 3.15

Gompertz curves

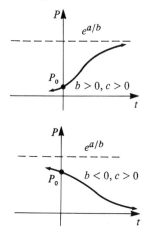

Figure 3.16

A modification of the logistic equation is

$$\frac{dP}{dt} = P(a - b \ln P) \qquad (9)$$

where a and b are constants. It is readily shown by separation of variables (see Problem 5) that a solution of (9) is

$$P(t) = e^{a/b} e^{-ce^{-bt}} \qquad (10)$$

where c is an arbitrary constant. We note when $b > 0$, $P \to e^{a/b}$ as $t \to \infty$, whereas for $b < 0$, $c > 0$, $P \to 0$ as $t \to \infty$. The graph of the function (10), called a **Gompertz curve**,* is quite similar to the graph of the logistic function. Figure 3.16 shows two possibilities for the graph of $P(t)$.

Functions such as (10) are encountered for example, in studies of the growth, or decline, of certain populations; in actuarial predictions; and in the study of growth of revenue in the sale of a commercial product.

Chemical reactions

The disintegration of a radioactive substance, governed by equation (1) of the preceding section, is said to be a **first-order reaction**. In chemistry a few reactions follow the same empirical law: If the molecules of a substance A decompose into smaller molecules it is a natural assumption to suppose that the rate at which this decomposition takes place is proportional to the amount of the first substance which has not undergone conversion. That is, if $X(t)$ is the amount of substance A remaining at any time, then

$$\frac{dX}{dt} = kX,$$

where k is negative since X is decreasing. An example of a first-order chemical reaction is the conversion of t-butyl chloride into t-butyl alcohol

$$(CH_3)_3CCl + NaOH \quad \rightarrow \quad (CH_3)_3COH + NaCl.$$

Only the concentration of the t-butyl chloride controls the rate of reaction. Now in the reaction

$$CH_3Cl + NaOH \quad \rightarrow \quad CH_3OH + NaCl$$

for every molecule of methyl chloride, one molecule of sodium hydroxide is consumed, thus forming one molecule of methyl alcohol and one molecule of sodium chloride. In this case the rate at which the reaction proceeds is proportional to the product of the remaining concentrations of CH_3Cl and of NaOH. If X deontes the amount of CH_3OH formed and α and β are the given amounts of the first two chemicals A and B, then the instantaneous amounts not converted to chemical C are α-X and β-X, respectively. Hence, the rate of formation of C is given by

---

*Named after Benjamin Gompertz (1779–1865), an English mathematician.

$$\frac{dX}{dt} = k(\alpha - X)(\beta - X) \tag{11}$$

where k is a constant of proportionality. A reaction described by equation (11) is said to be of **second-order**.

EXAMPLE

A compound C is formed when two chemicals A and B are combined. The resulting reaction between the two chemicals is such that for each gram of A, 4 grams of B are used. It is observed that 30 grams of the compound C are formed in 10 minutes. Determine the amount of C at any time if the rate of the reaction is proportional to the amounts of A and B remaining and that initially there are 50 grams of A and 32 grams of B. How much of the compound C is present at 15 minutes? Interpret the solution as $t \to \infty$.

Solution: Let $X(t)$ denote the number of grams of the compound C present at any time. Clearly $X(0) = 0$ and $X(10) = 30$.

Now for example, if there are 2 grams of compound C then we must have used, say, a grams of A and b grams of B so that

$$a + b = 2$$

$$b = 4a.$$

Thus we must use $a = 2/5 = 2(1/5)$ grams of chemical A and $b = 8/5 = 2(4/5)$ grams of B. In general, for X grams of C then we must use

$$\frac{X}{5} \text{ grams of } A \quad \text{and} \quad \frac{4}{5}X \text{ grams of } B.$$

The amounts of A and B remaining at any time are then

$$50 - \frac{X}{5} \quad \text{and} \quad 32 - \frac{4}{5}X,$$

respectively.

Now we know that the rate at which chemical C is formed satisfies

$$\frac{dX}{dt} \propto \left(50 - \frac{X}{5}\right)\left(32 - \frac{4}{5}X\right).$$

To simplify the subsequent algebra, we factor 1/5 from the first term and 4/5 from the second, and then introduce the constant of proportionality,

$$\frac{dX}{dt} = k(250 - X)(40 - X).$$

By separation of variables and partial fractions we can write

$$\frac{dX}{(250 - X)(40 - X)} = k\,dt$$

$$-\frac{1/210}{250 - X}dX + \frac{1/210}{40 - X}dX = k\,dt$$

$$\ln \left| \frac{250 - X}{40 - X} \right| = 210kt + c_1$$

$$\frac{250 - X}{40 - X} = c_2 e^{210kt}. \qquad (12)$$

When $t = 0$, $X = 0$, so it follows at this point that $c_2 = 25/4$. Using $X = 30$ at $t = 10$ we find

$$210k = \frac{1}{10} \ln \frac{88}{25}$$

$$= 0.1258.$$

Using this information we solve (12) for X

$$X(t) = 1000 \frac{1 - e^{-0.1258t}}{25 - 4e^{-0.1258t}}. \qquad (13)$$

The behavior of X as a function of time is displayed in Figure 3.17. It is clear from the accompanying table and equation (13) that $X \to 40$ as $t \to \infty$. This means there are 40 grams of compound C formed, leaving

$$50 - \tfrac{1}{5}(40) = 42 \text{ grams of chemical } A$$

and $\qquad 32 - \tfrac{4}{5}(40) = 0$ grams of chemical B.

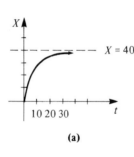

t (minutes)	X (grams)
10	30 (measured)
15	34.78
20	37.25
25	38.54
30	39.22
35	39.59

(a) (b)

Figure 3.17

Law of mass action   The preceding example can be generalized in the following manner. Suppose that a grams of substance A are combined with b grams of substance B. If there are M and N parts of A and B, respectively, in the compound formed, then

$$a - \frac{M}{M + N} X$$

represents the amount of substance A remaining at any time and

$$b - \frac{N}{M + N} X$$

is the amount of substance B remaining at any time. Thus

$$\frac{dX}{dt} \propto \left[a - \frac{M}{M + N} X \right] \left[b - \frac{N}{M + N} X \right]. \tag{14}$$

Proceeding as before, if we factor out $M/(M + N)$ from the first term and $N/(M + N)$ from the second term, the resulting differential equation is the same as (11)

$$\frac{dX}{dt} = k(\alpha - X)(\beta - X) \tag{15}$$

where $\qquad \alpha = \dfrac{a(M + N)}{M} \quad$ and $\quad \beta = \dfrac{b(M + N)}{N}.$

Chemists refer to reactions described by equation (15) as the **law of mass action**.

When $\alpha \neq \beta$, it is readily shown (see Problem 9) that a solution of (15) is

$$\frac{1}{\alpha - \beta} \ln \left| \frac{\alpha - X}{\beta - X} \right| = kt + c. \tag{16}$$

Assuming the natural initial condition $X(0) = 0$ equation (16) yields the explicit solution

$$X(t) = \frac{\alpha\beta[1 - e^{(\alpha - \beta)kt}]}{\beta - \alpha e^{(\alpha - \beta)kt}}. \tag{17}$$

Without loss of generality we assume in (17) that $\beta > \alpha$ or $\alpha - \beta < 0$. Since $X(t)$ is an increasing function we expect $k > 0$ and so it follows immediately from (17) that $X \to \alpha$ as $t \to \infty$.

EXERCISES 3.3   *Answers to odd-numbered problems begin on page A-9.*

1. The number of supermarkets $C(t)$ throughout the country that are using a computerized checkout system is described by the initial-value problem

$$\frac{dC}{dt} = C(1 - 0.0005C), \qquad t > 0$$

$$C(0) = 1.$$

How many supermarkets are using the computerized method when $t = 10$? How many companies are estimated to adopt the new procedure over a long period of time?

★ 2. The number of people $N(t)$ in a community who are exposed to a particular advertisement is governed by the logistic equation. Initially $N(0) = 500$, and it is observed that $N(1) = 1000$. If it is predicted that the limiting number of people in the community that will see the advertisement is 50,000, determine $N(t)$ at any time.

3. The population $P(t)$ at any time in a suburb of a large city is governed by the initial-value problem

$$\frac{dP}{dt} = P(10^{-1} - 10^{-7}P)$$

$$P(0) = 5000,$$

where t is measured in months. What is the limiting value of the population? At what time will the population be equal to one-half of this limiting value?

4. Find a solution of the *modified logistic equation*

$$\frac{dP}{dt} = P(a - bP)(1 - cP^{-1}), \qquad a,b,c > 0.$$

5. **(a)** Solve equation (9):

$$\frac{dP}{dt} = P(a - b \ln P).$$

   (b) Determine the value of c in equation (10) if $P(0) = P_0$.

6. Assuming $0 < P_0 < e^{a/b}$, and $a > 0$, use equation (9) to find the ordinate of the point of inflection for a Gompertz curve.

7. Two chemicals A and B are combined to form a chemical C. The rate or velocity of the reaction is proportional to the product of the instantaneous amounts of A and B not converted to chemical C. Initially there are 40 grams of A and 50 grams of B, and for each gram of B, 2 grams of A are used. It is observed that 10 grams of C are formed in 5 minutes. How much is formed in 20 minutes? What is the limiting amount of C after a long time? How much of chemicals A and B remain after a long time?

★ 8. Solve the preceding problem if there are 100 grams of chemical A present initially. At what time is chemical C half-formed?

9. Obtain a solution of the equation

$$\frac{dX}{dt} = k(\alpha - X)(\beta - X)$$

governing second-order reactions in the two cases $\alpha \neq \beta$ and $\alpha = \beta$.

10. In a third-order chemical reaction the number of grams X of a compound obtained by combining three chemicals is governed by

$$\frac{dX}{dt} = k(\alpha - X)(\beta - X)(\gamma - X).$$

Solve the equation under the assumption $\alpha \neq \beta \neq \gamma$.

EXAMPLE   A rocket is shot vertically upward from the ground with an initial velocity v_0 (see Figure 3.18). If the positive direction is taken to be upward, the student

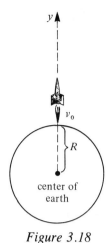

Figure 3.18

might recall from Section 1.2 (Problem 43) that in the absence of air resistance, the differential equation of motion after fuel burnout is

$$\frac{d^2y}{dt^2} = -\frac{k}{y^2}$$

where k is a positive constant. While this is not a first-order equation, we note that if we write the acceleration as

$$\frac{d^2y}{dt^2} = \frac{dv}{dt} = \frac{dv}{dy}\frac{dy}{dt} = v\frac{dv}{dy}$$

then the given equation becomes first-order in v. That is,

$$v\frac{dv}{dy} = -\frac{k}{y^2}.$$

Separating variables and integrating then gives

$$\frac{v^2}{2} = \frac{k}{y} + c.$$

Since $v = v_0$ at $y = R$ we obtain

$$\frac{v^2}{2} = \frac{k}{y} - \frac{k}{R} + \frac{v_0^2}{2}.$$

11. The reader might object that in the preceding example we really have not solved the original equation for y. Actually, the solution gives quite a bit of information. If $k = gR^2$ show that the "escape velocity" for a rocket is $v_0 = 25{,}000$ mi/hr. Use a hand calculator and the value $R = 4000$ miles.

12. In the discussion of Section 1.2 we saw that the differential equation describing the shape of a wire of constant density w hanging under its own weight is

$$\frac{d^2y}{dx^2} = \frac{w}{T_1}\sqrt{1 + \left(\frac{dy}{dx}\right)^2}$$

where T_1 is the horizontal tension in the wire at its lowest point. Using the substitution $p = dy/dx$ solve this equation subject to the initial conditions

$$y(0) = 1 \qquad \frac{dy}{dx}\bigg|_{x=0} = 0.$$

13. An equation similar to that given in the preceding problem is

$$x\frac{d^2y}{dx^2} = \frac{v_1}{v_2}\sqrt{1 + \left(\frac{dy}{dx}\right)^2}.$$

In this case the equation arises in the study of the shape of the path that a pursuer, travelling at a speed v_2, must take in order to intercept a prey travelling at speed v_1. Use the same substitution as in Problem 12 and the initial conditions

$$y(1) = 0 \qquad \left.\frac{dy}{dx}\right|_{x=1} = 0$$

to solve the equation. Consider the two cases: $v_1 = v_2$ and $v_1 \neq v_2$.

★**14.** According to **Stefan's law** of radiation, the rate of change of temperature from a body at absolute temperature T is

$$\frac{dT}{dt} = k(T^4 - T_0^4)$$

where T_0 is the absolute temperature of the surrounding medium. Find a solution of this differential equation. It can be shown that when $T - T_0$ is small compared to T_0 that this particular equation is closely approximated by Newton's law of cooling (Equation (3), Section 3.2.).

15. The height of h of water which is flowing through an orifice at the bottom of a cylindrical tank is given by

$$\frac{dh}{dt} = -\frac{A_2}{A_1}\sqrt{2gh}, \qquad g = 32\,\text{ft/sec}^2,$$

where A_1 and A_2 are the cross-sectional areas of the tank and orifice, respectively. (See Problem 34, Exercise 1.2). Solve the equation if the initial height of the water is 20 ft and $A_1 = 50\,\text{ft}^2$ and $A_2 = 1/4\,\text{ft}^2$. At what time is the tank empty?

★**16.** The nonlinear differential equation

$$\left(\frac{dr}{dt}\right)^2 = \frac{2\mu}{r} + 2h,$$

where μ and h are nonnegative constants, arises in the study of the two-body problem of celestial mechanics. Here the variable r represents the distance between the two masses. Solve the equation in the two cases $h = 0$ and $h > 0$.

17. Solve the differential equation of the **tractrix**

$$\frac{dy}{dx} = -\frac{y}{\sqrt{s^2 - y^2}}.$$

(See Problem 35, Exercise 1.2.) Assume that the initial point on the y-axis is $(0,10)$ and the length of rope is $s = 10\,\text{ft}$.

18. A body of mass m falling through a viscous medium encounters a resisting force proportional to the square of its instantaneous velocity. In this situation the differential equation for the velocity $v(t)$ at any time is

$$m\frac{dv}{dt} = mg - kv^2$$

where k is a positive constant of proportionality. Solve the equation subject to $v(0) = v_0$. What is the limiting velocity of the falling body?

19. The differential equation

$$x\left(\frac{dx}{dy}\right)^2 + 2y\frac{dx}{dy} = x,$$

where $x = x(y)$, occurs in the study of optics. The equation describes the type of plane curve that will reflect all incoming light rays to the same point. (See Problem 41, Exercise 1.2.) Show that the curve must be a parabola. [*Hint:* Use the substitution $w = x^2$ and then re-examine Section 2.6.]

★20. Solve the equation of Problem 19 with the aid of the quadratic formula.

21. The equations of Lotka and Volterra*

$$\frac{dy}{dt} = y(\alpha - \beta x)$$

$$\frac{dx}{dt} = x(-\gamma + \delta y)$$

where α, β, γ and δ are positive constants, occur in the analysis of the biological balance of two species of animals such as a predator and its prey (for example, foxes and rabbits). Here $x(t)$ and $y(t)$ denote the populations of the two species at any time. Although no explicit solutions of the system exist, solutions can be found relating the two populations at any time. Divide the first equation by the second and solve the resulting nonlinear first-order differential equation.

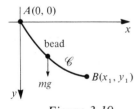

$A(0, 0)$

bead

$\mathscr{C}$

mg

$B(x_1, y_1)$

Figure 3.19

22. A classical problem in the calculus of variations is to find the shape of a curve $\mathscr{C}$ (see Figure 3.19) such that a bead, under the influence of gravity, will slide from $A(0,0)$ to $B(x_1, y_1)$ in the least time. It can be shown that the differential equation for the shape of the path is $y[1 + (y')^2] = k$ where k is a constant. First solve for dx in terms of y and dy, and then use the substitution $y = k \sin^2 \theta$ to obtain the parametric form of the solution. The curve $\mathscr{C}$ turns out to be a cycloid.

23. The initial-value problem describing the motion of a simple pendulum released from rest from an angle $\theta_0 > 0$ is

$$\frac{d^2\theta}{dt^2} + \frac{g}{l} \sin \theta = 0$$

$$\theta(0) = \theta_0 \qquad \left.\frac{d\theta}{dt}\right|_{t=0} = 0.$$

*A. J. Lotka (1880–1949) an Austrian-American biomathematician. Vito Volterra (1860–1940) an Italian mathematician.

(a) Obtain the first-order equation

$$\left(\frac{d\theta}{dt}\right)^2 = \frac{2g}{l}(\cos\theta - \cos\theta_0).$$

[*Hint*: Multiply the given equation by $2\dfrac{d\theta}{dt}$.]

(b) Use the equation in part (a) to show that the period of motion is

$$T = 2\sqrt{\frac{2l}{g}}\int_0^{\theta_0}\frac{d\theta}{\sqrt{\cos\theta - \cos\theta_0}}.$$

**CHAPTER 3
SUMMARY**

If every curve in a one-parameter family of curves $G(x,y,c_1) = 0$ is orthogonal to every curve in a second one-parameter family $H(x,y,c_2) = 0$, we say that the two families are **orthogonal trajectories.** Two curves are orthogonal if their tangent lines are perpendicular at a point of intersection. When given a family, we find its differential equation $dy/dx = f(x,y)$ by differentiating the equation $G(x,y,c_1) = 0$ and eliminating the parameter c_1. The differential equation of the second, and orthogonal family, is then $dy/dx = -1/f(x,y)$. We solve this latter equation by the methods of Chapter 2.

In the mathematical analysis of population growth, radioactive decay, or chemical mixtures, we often encounter **linear** differential equations such as

$$\frac{dx}{dt} = kx \quad \text{and} \quad \frac{dx}{dt} = a + bx$$

or **nonlinear** differential equations such as

$$\frac{dx}{dt} = x(a - bx) \quad \text{and} \quad \frac{dx}{dt} = k(\alpha - x)(\beta - x).$$

The student should be able to solve these particular equations without hesitation. It is never a good idea simply to memorize solutions of differential equations.

**CHAPTER 3
TEST**

Answers to odd-numbered problems begin on page A-10.

1. Find the orthogonal trajectories of the family of curves $y(x^3 + c_1) = 3$.

2. Find the orthogonal trajectory to the family $y = 4x + 1 + c_1e^{4x}$ passing through the point $(0,0)$.

3. Find the orthogonal trajectories of the family of parabolas opening in the y-direction with vertex at $(1,2)$.

4. If a population expands at a rate proportional to the number of people present at any time, show that the doubling time of the population is

continued

**CHAPTER 3
TEST**

then $T = (\ln 2)/k$ where k is the positive growth rate. This is known as the **Law of Malthus.** *

5. In March of 1976 the world population reached 4 billion. A popular news magazine has predicted that with an average yearly growth rate of 1.8%, the world population will be 8 billion in 45 years. How does this value compare with that predicted by the model which says that the rate of increase is proportional to the population at any time?

6. Air containing 0.06% carbon dioxide is pumped into a room whose volume is $8000 \, \text{ft}^3$. The rate at which the air is pumped in is $2000 \, \text{ft}^3/\text{min}$, and the circulated air is then pumped out at the same rate. If there is an initial concentration of 0.2% carbon dioxide, determine the subsequent amount in the room at any time. What is the concentration at 10 minutes? What is the steady-state or equilibrium concentration of carbon-dioxide?

7. The populations of two competing species of animals are described by the nonlinear system of first order differential equations

$$\frac{dx}{dt} = k_1 x (\alpha - x) \qquad \frac{dy}{dt} = k_2 xy .$$

Solve for x and y in terms of t.

8. A projectile is shot vertically into the air with an initial veolcity of $v_0 \, \text{ft/sec}$. Assuming that air resistance is proportional to the square of the instantaneous velocity, the motion is described by the pair of differential equations:

$$m \frac{dv}{dt} = -mg - kv^2, \qquad k > 0,$$

positive y-axis up, origin at ground level so that $v = v_0$ at $y = 0$,

$$m \frac{dv}{dt} = mg - kv^2, \qquad k > 0,$$

positive y-axis down, origin at the maximum height so that $v = 0$ at $y = h$. The first and second equations describe the motion of the projectile when rising and falling, respectively. Prove that the impact velocity v_i of the projectile is less than the initial velocity v_0. It can also be shown that the time t_1 needed to attain its maximum height h is less than the time t_2 that it takes to fall from this height. See Figure 3.20

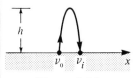

Figure 3.20

---

*Thomas R. Malthus (1766–1834) an English clergyman and economist.

NTIAL EQUATIONS SECOND EDITION A FIRST COURSE IN DIFFERENTIAL EQUATIONS
D EDITION A FIRST COURSE IN DIFFERENTIAL EQUATIONS SECOND EDITION A FIRST C
E IN DIFFERENTIAL EQUATIONS SECOND EDITION A FIRST COURSE IN DIFFERENTIAL E
IONS SECOND EDITION A FIRST COURSE IN DIFF...   ...ECOND EDITI
IRST COURSE IN DIFFERENTIAL EQUATIONS SEC...   ...SE IN DIF
NTIAL EQUATIONS SECOND EDITION A FIRST COURSE IN DIFFERENTIAL EQUATIONS S
D EDITION A FIRST COURSE IN DIFFERENTIAL EQUATIONS SECOND EDITION A FIRST C
E IN DIFFERENTIAL EQUATIONS SECOND EDITION A FIRST COURSE IN DIFFERENTIAL E
IONS SECOND EDIT...   ...FIRST COURSE IN DIFFERENTIAL EQUATIONS SECOND EDITI
IRST C...   ...ITION A FIRST COURSE IN DIFFERENTIAL EQUATIONS SECOND EDIT
NTIAL EQ...   ...ITION A FIRST COURSE IN DIFFERENTIAL EQUATIONS S
D EDIT...   ...ERENTIAL EQUATIONS SECOND EDITION A FIRST C
E IN DIFFERENTIAL EQUATIONS SECOND EDITION A FIRST COURSE IN DIFFERENTIAL E
IONS SECOND EDITION A FIRST COURSE IN DIFFERENTIAL EQUATIONS SECOND EDIT
IRST COURSE IN DIFFERENTIAL EQUATIONS SECOND EDITION A FIRST COURSE IN DIF
NTIAL EQUATIONS SECOND EDITION A FIRST COURSE IN DIFFERENTIAL EQUATIONS S
D EDITION A FIRST COURSE IN DIFFERENTIAL EQUATIONS SECOND EDITION A FIRST C

CHAPTER 4

Linear Differential Equations of Higher Order

4.1 Preliminary Theory

We begin the discussion of higher order differential equations, as we did with first-order equations, with the notion of an initial-value problem. However, we shall confine our attention to linear differential equations.

4.1.1 Initial-Value and Boundary-Value Problems

Initial-value problem

For a linear nth-order differential equation, the problem

$$Solve: \quad a_n(x)\frac{d^n y}{dx^n} + a_{n-1}(x)\frac{d^{n-1}y}{dx^{n-1}} + \cdots + a_1(x)\frac{dy}{dx} + a_0(x)y = g(x)$$

$$Subject\ to: \qquad\qquad y(x_0) = y_0$$
$$y'(x_0) = y_0'$$
$$\vdots \qquad \vdots$$
$$y^{(n-1)}(x_0) = y_0^{(n-1)}$$

(1)

where $y_0, y_0', \ldots, y_0^{(n-1)}$ are arbitrary constants, is called an **initial-value** problem. We seek a solution on some interval I containing the point $x = x_0$.

In the important case of a linear second-order equation, a solution of

$$a_2(x)\frac{d^2 y}{dx^2} + a_1(x)\frac{dy}{dx} + a_0(x)y = g(x)$$

$$y(x_0) = y_0$$

$$y'(x_0) = y_0'$$

is a function defined on I whose graph passes through (x_0, y_0) such that the slope of the curve at the point is the number y_0'.

The next theorem gives sufficient conditions for the existence of a unique solution to (1).

THEOREM 4.1 Let $a_n(x), a_{n-1}(x), \ldots, a_1(x), a_0(x)$ and $g(x)$ be continuous on an interval I and let $a_n(x) \neq 0$ for every x in this interval. If $x = x_0$ is any point in this interval, then a solution $y(x)$ of the initial-value problem (1) exists on the interval and is unique. $\square$

While we are not in a position to prove Theorem 4.1 in its full generality a demonstration of the *uniqueness* of the solution in the special case

$$a_2 y'' + a_1 y' + a_0 y = g(x)$$

$$y(0) = y_0$$

$$y'(0) = y_0',$$

where a_2, a_1, and a_0 are *positive* constants and $g(x)$ is continuous for all x, is given in the chapter appendix.

EXAMPLE

The reader should verify that $y = 3e^{2x} + e^{-2x} - 3x$ is a solution of the initial-value problem

$$y'' - 4y = 12x$$

$$y(0) = 4 \qquad y'(0) = 1.$$

Now the differential equation is linear and the constant coefficients as well as $g(x) = 12x$ are continuous on any interval containing $x = 0$. We conclude from Theorem 4.1 that the given function is the unique solution.

EXAMPLE

The initial-value problem

$$3y''' + 5y'' - y' + 7y = 0$$

$$y(1) = 0 \qquad y'(1) = 0 \qquad y''(1) = 0$$

possesses the solution $y \equiv 0$. Since the third-order equation is linear with constant coefficients, it follows that all the conditions of Theorem 4.1 are fulfilled. Hence $y \equiv 0$ is the *only* solution on any interval containing $x = 1$.

EXAMPLE

The function $y = \frac{1}{4}\sin 4x$ is a solution of the initial-value problem

$$y'' + 16y = 0$$

$$y(0) = 0 \qquad y'(0) = 1.$$

It follows from Theorem 4.1 that on any interval containing $x = 0$ the solution is unique.

The requirements in Theorem 4.1 that $a_i(x), i = 1,2,\ldots,n$ be continuous and $a_n(x) \neq 0$ for every x in I are both important. Specifically, if $a_n(x) = 0$ for some x in the interval then the solution of a linear initial-value problem may not be unique or even exist.

EXAMPLE

Verify that the function

$$y = cx^2 + x + 3$$

is a solution of the initial-value problem

$$x^2y'' - 2xy' + 2y = 6$$

$$y(0) = 3 \qquad y'(0) = 1,$$

on the interval $-\infty < x < \infty$ for any choice of the parameter c.

Solution: Since $y' = 2cx + 1$, $y'' = 2c$ it follows that

$$x^2y'' - 2xy' + 2y = x^2(2c) - 2x(2cx + 1) + 2(cx^2 + x + 3)$$

$$= 2cx^2 - 4cx^2 - 2x + 2cx^2 + 2x + 6$$

$$= 6.$$

Also, $y(0) = c(0)^2 + 0 + 3 = 3$

$$y'(0) = 2c(0) + 1 = 1.$$

Although the differential equation in the foregoing example is linear and the coefficients and $g(x) = 6$ are continuous everywhere, the obvious difficulty is that $a_2(x) = x^2$ is zero at $x = 0$.

Boundary-value problem

Another type of problem consists of solving a differential equation of order two or greater in which the dependent variable y (or its derivatives) is specified at *two different points*. A problem such as

$$\text{Solve:} \qquad a_2(x)\frac{d^2y}{dx^2} + a_1(x)\frac{dy}{dx} + a_0(x)y = g(x)$$

$$\text{Subject to:} \qquad y(a) = y_0, \qquad y(b) = y_1$$

is called a **two-point boundary-value problem** or simply a **boundary-value problem**.

EXAMPLE

For the boundary-value problem

$$x^2 y'' - 2xy' + 2y = 6$$

$$y(1) = 0 \qquad y(2) = 3$$

we seek a function defined on an interval containing $x = 1$ and $x = 2$ that satisfies the differential equation and whose graph passes through the two points $(1,0)$ and $(2,3)$.

The next examples show that even when the conditions of Theorem 4.1 are fulfilled a boundary-value problem may either have (a) several solutions, (b) a unique solution, or (c) no solution at all.

EXAMPLE

We have seen on page 6 that a two-parameter family of solutions for the differential equation $y'' + 16y = 0$ is

$$y = c_1 \cos 4x + c_2 \sin 4x.$$

Suppose we now wish to determine that solution of the equation which further satisfies the boundary conditions

$$y(0) = 0 \qquad y\left(\frac{\pi}{2}\right) = 0.$$

Observe that the first condition

$$0 = c_1 \cos 0 + c_2 \sin 0$$

implies $c_1 = 0$ so that

$$y = c_2 \sin 4x$$

But when $x = \pi/2$ we have

$$0 = c_2 \sin 2\pi$$

Since $\sin 2\pi = 0$ this latter condition is satisfied for any choice of c_2, so it follows that a solution of the problem

$$y'' + 16y = 0$$

$$y(0) = 0, \qquad y\left(\frac{\pi}{2}\right) = 0$$

is the one-parameter family

$$y = c_2 \sin 4x.$$

As Figure 4.1 on page 122 shows there are an infinite number of functions satisfying the differential equation whose graphs pass through the two points $(0,0)$ and $(\pi/2, 0)$.

Observe that had the boundary conditions been $y(0) = 0$ and $y(\pi/8) = 0$ then necessarily $c_1 = 0$ and $c_2 = 0$. Thus $y \equiv 0$ would be a solution of this new boundary-value problem. In fact, as we shall see later on in this section, it is the only solution.

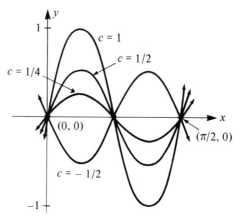

Figure 4.1

EXAMPLE

The boundary-value problem

$$y'' + 16y = 0$$

$$y(0) = 0 \qquad y\left(\frac{\pi}{2}\right) = 1$$

has no solution in the family $y = c_1 \cos 4x + c_2 \sin 4x$. As in the preceding example, the condition $y(0) = 0$ still implies that $c_1 = 0$. Thus $y = c_2 \sin 4x$ so that when $x = \pi/2$ we obtain the contradiction $1 = c_2 \sin 2\pi = c_2 \cdot 0 = 0$.

Boundary-value problems are often encountered in the applications of partial differential equations. See Chapter 10.

4.1.2   Linear Dependence and Linear Independence

The next two concepts are basic to the study of linear differential equations.

DEFINITION 4.1   A set of functions $f_1(x), f_2(x), \ldots, f_n(x)$ is said to be **linearly dependent** on an interval I if there exist constants $c_1, c_2, \ldots, c_n$, not all zero, such that

$$c_1 f_1(x) + c_2 f_2(x) + \cdots + c_n f_n(x) = 0$$

for every x in the interval. ☐

DEFINITION 4.2   A set of functions $f_1(x), f_2(x), \ldots, f_n(x)$ is said to be **linearly independent** on an interval I if it is not linearly dependent on the interval. ☐

In other words, a set of functions is linearly independent on an interval if the only constants for which

$$c_1 f_1(x) + c_2 f_2(x) + \cdots + c_n f_n(x) = 0,$$

for every x in the interval, are $c_1 = c_2 = \cdots = c_n = 0$.

It is easy to understand these definitions in the case of two functions $f_1(x)$ and $f_2(x)$. If the functions are linearly dependent on an interval then there exist constants c_1 and c_2 which are not both zero such that for every x in the interval

$$c_1 f_1(x) + c_2 f_2(x) = 0.$$

Therefore, if we assume that $c_1 \neq 0$, it follows that

$$f_1(x) = -\frac{c_2}{c_1} f_2(x).$$

That is, *if two functions are linearly dependent then one is simply a constant multiple of the other.* Conversely, if $f_1(x) = c_2 f_2(x)$, for some constant c_2, then

$$(-1) \cdot f_1(x) + c_2 f_2(x) = 0$$

for every x on some interval. Hence the functions are linearly dependent since at least one of the constants (namely, $c_1 = -1$) is not zero. We conclude that two functions are linearly independent when *neither* is a constant multiple of the other on an interval.

EXAMPLE

The functions $f_1(x) = \sin 2x$ and $f_2(x) = \sin x \cos x$ are linearly dependent on the interval $-\infty < x < \infty$ since

$$c_1 \sin 2x + c_2 \sin x \cos x = 0$$

is satisfied for every real x if we choose $c_1 = 1/2$ and $c_2 = -1$. (Recall the trigonometric identity $\sin 2x = 2 \sin x \cos x$.)

EXAMPLE

The functions $f_1(x) = x$ and $f_2(x) = |x|$ are linearly independent on the interval $-\infty < x < \infty$. Inspection of Figure 4.2 should convince the reader that neither function is a constant multiple of the other. Thus in order to have $c_1 f_1(x) + c_2 f_2(x) = 0$ for every real x we must choose $c_1 = 0$ and $c_2 = 0$.

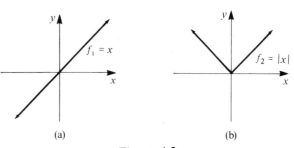

Figure 4.2

In the consideration of linear dependence or linear independence the interval on which the functions are defined is important. The functions $f_1(x) = x$ and $f_2(x) = |x|$ of the preceding example are linearly dependent on the interval $0 < x < \infty$ since

$$c_1 x + c_2 |x| = c_1 x + c_2 x = 0$$

is satisfied for any nonzero choice of c_1 and c_2 for which $c_1 = -c_2$.

EXAMPLE

The functions $f_1(x) = \sqrt{x} + 5$, $f_2(x) = \sqrt{x} + 5x$, $f_3(x) = x - 1$, $f_4(x) = x^2$ are linearly dependent on the interval $0 < x < \infty$ since

$$c_1 \cos^2 x + c_2 \sin^2 x + c_3 \sec^2 x + c_4 \tan^2 x = 0$$

when $c_1 = c_2 = 1$, $c_3 = -1$, $c_4 = 1$. We note that $\cos^2 x + \sin^2 x = 1$ and $1 + \tan^2 x = \sec^2 x$.

A set of functions $f_1(x), f_2(x), \ldots, f_n(x)$ are linearly dependent on an interval if at least one function can be expressed as a nontrivial linear combination of the remaining functions.

EXAMPLE

The functions $f_1(x) = \cos^2 x$, $f_2(x) = \sin^2 x$, $f_3(x) = \sec^2 x$, and $f_4(x) = \tan^2 x$ are linearly dependent on the interval $0 < x < \infty$ since

$$f_2(x) = 1 \cdot f_1(x) + 5 \cdot f_3(x) + 0 \cdot f_4(x)$$

for every x in the interval.

The following theorem provides a sufficient condition for the linear independence of n functions on an interval. Each function is assumed to be differentiable at least $n - 1$ times.

THEOREM 4.2   Suppose $f_1(x), f_2(x), \ldots, f_n(x)$ possess at least $n - 1$ derivatives. If

$$
\begin{vmatrix}
f_1 & f_2 & \cdots & f_n \\
f_1' & f_2' & \cdots & f_n' \\
\vdots & \vdots & \vdots & \vdots \\
f_1^{(n-1)} & f_2^{(n-1)} & \cdots & f_n^{(n-1)}
\end{vmatrix} \neq 0
$$

for at least one point in the interval I, then the functions $f_1(x)$, $f_2(x), \ldots, f_n(x)$ are linearly independent on the interval.   □

The determinant in the above theorem is denoted by

$$W(f_1(x), f_2(x), \ldots, f_n(x))$$

and is called the **Wronskian*** of the functions.

Proof: We prove Theorem 4.2 by contradiction for the case when $n = 2$. Assume that $W(f_1(x_0), f_2(x_0)) \neq 0$ for a fixed x_0 in the interval I and that $f_1(x)$ and $f_2(x)$ are linearly dependent on the interval. The fact that the functions are linearly dependent means there exist constants c_1 and c_2, not both zero, for which

$$c_1 f_1(x) + c_2 f_2(x) = 0$$

for every x in I. Differentiating this combination then gives

$$c_1 f_1'(x) + c_2 f_2'(x) = 0.$$

Thus we obtain the system of linear equations

$$c_1 f_1(x) + c_2 f_2(x) = 0.$$
$$c_1 f_1'(x) + c_2 f_2'(x) = 0. \tag{2}$$

But the linear dependence of f_1 and f_2 implies that (2) possesses a nontrivial solution for each x in the interval. Hence

$$W(f_1(x), f_2(x)) = \begin{vmatrix} f_1(x) & f_2(x) \\ f_1'(x) & f_2'(x) \end{vmatrix} = 0$$

for every x in I.† This contradicts the assumption that $W(f_1(x_0),$ $f_2(x_0)) \neq 0$. We conclude that f_1 and f_2 are linearly independent. $\square$

COROLLARY If $f_1(x), f_2(x), \ldots, f_n(x)$ possess at least $n - 1$ derivatives and are linearly dependent on I then the Wronskian of the functions is zero for every x in the interval. That is,

$$W(f_1(x), f_2(x), \ldots, f_n(x)) \equiv 0$$

on the interval. $\square$

EXAMPLE The functions $f_1(x) = \sin^2 x$ and $f_2(x) = 1 - \cos 2x$ are linearly dependent on $-\infty < x < \infty$. By the preceding corollary $W(\sin^2 x, 1 - \cos 2x) \equiv 0$ on the interval. To see this we observe

$$W(\sin^2 x, 1 - \cos 2x) = \begin{vmatrix} \sin^2 x & 1 - \cos 2x \\ 2 \sin x \cos x & 2 \sin 2x \end{vmatrix}$$

$$= 2 \sin^2 x \sin 2x - 2 \sin x \cos x + 2 \sin x \cos x \cos 2x$$

---

*Named after the Polish philosopher-mathematician Josef M. H. Wronski (1778–1853). He is remembered solely because his name is associated with the above determinant.
†Recall, a system $\qquad ax + by = 0$
$\qquad\qquad\qquad\qquad\qquad cx + dy = 0$

has nontrivial solutions for x and y if and only if the determinant of the coefficients is zero. In (2) the symbols c_1 and c_2 play the roles of x and y.

$$= \sin 2x[2\sin^2 x - 1 + \cos 2x]$$

$$= \sin 2x[2\sin^2 x - 1 + \cos^2 x - \sin^2 x]$$

$$= \sin 2x[\sin^2 x + \cos^2 x - 1]$$

$$\equiv 0.$$

EXAMPLE

For $f_1(x) = e^{m_1 x}$, $f_2(x) = e^{m_2 x}$, $m_1 \neq m_2$

$$W(e^{m_1 x}, e^{m_2 x}) = \begin{vmatrix} e^{m_1 x} & e^{m_2 x} \\ m_1 e^{m_1 x} & m_2 e^{m_2 x} \end{vmatrix}$$

$$= (m_2 - m_1)e^{(m_1 + m_2)x}$$

$$\neq 0$$

for every real value of x. Thus f_1 and f_2 are linearly independent on any interval of the x-axis.

EXAMPLE

If α and β are real numbers, $\beta \neq 0$, then $y_1 = e^{\alpha x}\cos \beta x$ and $y_2 = e^{\alpha x}\sin \beta x$ are linearly independent on any interval of the x-axis since

$$W(e^{\alpha x}\cos \beta x, e^{\alpha x}\sin \beta x) = \begin{vmatrix} e^{\alpha x}\cos \beta x & e^{\alpha x}\sin \beta x \\ -\beta e^{\alpha x}\sin \beta x + \alpha e^{\alpha x}\cos \beta x & \beta e^{\alpha x}\cos \beta x + \alpha e^{\alpha x}\sin \beta x \end{vmatrix}$$

$$= \beta e^{2\alpha x}(\cos^2 \beta x + \sin^2 \beta x)$$

$$= \beta e^{2\alpha x}$$

$$\neq 0.$$

Notice when $\alpha = 0$ we see that $\cos \beta x$ and $\sin \beta x$, $\beta \neq 0$, are also linearly independent on any interval of the x-axis.

EXAMPLE

The functions $f_1(x) = e^x$, $f_2(x) = xe^x$, and $f_3(x) = x^2 e^x$ are linearly independent on any interval of the x-axis since

$$W(e^x, xe^x, x^2 e^x) = \begin{vmatrix} e^x & xe^x & x^2 e^x \\ e^x & xe^x + e^x & x^2 e^x + 2xe^x \\ e^x & xe^x + 2e^x & x^2 e^x + 4xe^x + 2e^x \end{vmatrix} = 2e^{3x}$$

is not zero for any real value of x.

EXAMPLE

On page 123 we have seen that $f_1(x) = x$ and $f_2(x) = |x|$ are linearly independent on $-\infty < x < \infty$, however, we cannot compute the Wronskian since f_2 is not differentiable at $x = 0$.

EXAMPLE

$$W(x, xe^x) = \begin{vmatrix} x & xe^x \\ 1 & xe^x + e^x \end{vmatrix} = x^2 e^x.$$

Although $W = 0$ at $x = 0$, it is sufficient to observe that $W \neq 0$ for *at least one* other value of x on $-\infty < x < \infty$. Thus $f_1(x) = x$ and $f_2(x) = xe^x$ are linearly independent on the x-axis.

We leave it as an exercise (see Problem 30) to show that a set of functions could be linearly independent on some interval and yet have a vanishing Wronskian. In other words, $W(f_1(x), f_2(x), \ldots, f_n(x)) \equiv 0$ does not necessarily mean the functions are linearly dependent.

4.1.3 Solutions of Linear Equations

Homogeneous equations

A linear nth-order differential equation of the form

$$a_n(x)\frac{d^n y}{dx^n} + a_{n-1}(x)\frac{d^{n-1}y}{dx^{n-1}} + \cdots + a_1(x)\frac{dy}{dx} + a_0(x)y = 0 \qquad (3)$$

is said to be **homogeneous** whereas

$$a_n(x)\frac{d^n y}{dx^n} + a_{n-1}(x)\frac{d^{n-1}y}{dx^{n-1}} + \cdots + a_1(x)\frac{dy}{dx} + a_0(x)y = g(x), \qquad (4)$$

$g(x)$ not identically zero, is said to be **nonhomogeneous**.

The word "homogeneous" in this context does not refer to coefficients that are homogeneous functions (see Section 2.3).

EXAMPLES

(a) The equation
$$2y'' + 3y' - 5y = 0$$
is a homogeneous linear second-order differential equation.

(b) The equation
$$x^3 y''' - 2xy'' + 5y' + 6y = e^x$$
is a nonhomogeneous linear third-order ordinary differential equation.

We shall see in the latter part of this section, as well as in the subsequent sections of this chapter, that in order to solve a nonhomogeneous equation (4) we must first solve the associated homogeneous equation (3).

Note: To avoid needless repetition throughout the remainder of this text we shall as a matter-of-course make the following important assumptions when giving definitions and proving theorems about the linear equations (3) and (4). On some common interval I

(i) the coefficients $a_i(x)$, $i = 0, 1, \ldots , n$ are continuous;

(ii) the right-hand member $g(x)$ is continuous;

(iii) and $a_n(x) \neq 0$ for every x in the interval.

The superposition principle

The following theorem is known as the **superposition principle**.

THEOREM 4.3   Let $y_1, y_2, \ldots , y_k$ be solutions of the homogeneous linear nth-order differential equation (3) on an interval I. Then the linear combination

$$y = c_1 y_1(x) + c_2 y_2(x) + \cdots + c_k y_k(x), \tag{5}$$

where the c_i, $i = 1, 2, \ldots , k$ are arbitrary constants, is also a solution on the interval.

Proof: We prove the case when $n = k = 2$. Let $y_1(x)$ and $y_2(x)$ be solutions of

$$a_2(x)y'' + a_1(x)y' + a_0(x)y = 0.$$

If we define $y = c_1 y_1(x) + c_2 y_2(x)$ then

$$a_2(x)[c_1 y_1'' + c_2 y_2''] + a_1(x)[c_1 y_1' + c_2 y_2'] + a_0(x)[c_1 y_1 + c_2 y_2]$$

$$= c_1 \underbrace{[a_2(x)y_1'' + a_1(x)y_1' + a_0(x)y_1]}_{\text{zero}} + c_2 \underbrace{[a_2(x)y_2'' + a_1(x)y_2' + a_0(x)y_2]}_{\text{zero}}$$

$$= c_1 \cdot 0 + c_2 \cdot 0$$

$$= 0. \qquad \square$$

COROLLARIES   **(A)** A constant multiple $y = c_1 y_1(x)$ of a solution $y_1(x)$ of a homogeneous linear differential equation is also a solution.

(B) A homogeneous linear differential equation always possesses the solution $y \equiv 0$. $\qquad \square$

The superposition principle defined by (5), and its special case given in Corollary **(A)**, are properties that nonlinear differential equations, in general, do not possess (see Problems 31 and 32).

EXAMPLE

The functions $y_1 = x^2$ and $y_2 = x^2 \ln x$

both are solutions of the homogeneous third-order equation

$$x^3 y''' - 2xy' + 4y = 0$$

on the interval $0 < x < \infty$. By the superposition principle the linear combination

$$y = c_1 x^2 + c_2 x^2 \ln x$$

is also a solution of the equation on the interval.

EXAMPLE

The functions $y_1 = e^x$, $y_2 = e^{2x}$ and $y_3 = e^{3x}$ all satisfy the homogeneous equation

$$\frac{d^3 y}{dx^3} - 6\frac{d^2 y}{dx^2} + 11\frac{dy}{dx} - 6y = 0.$$

on $-\infty < x < \infty$. By Theorem 4.3 another solution is

$$y = c_1 e^x + c_2 e^{2x} + c_3 e^{3x}.$$

EXAMPLE

The function $y = x^2$ is a solution of the homogeneous linear equation

$$x^2 y'' - 3xy' + 4y = 0$$

on $0 < x < \infty$. Hence $y = cx^2$ is also a solution. For various values of c we see that $y = 3x^2$, $y = ex^2$, $y \equiv 0$, . . . are all solutions of the equation on the interval.

Linearly independent solutions

We are interested in determining when n solutions, $y_1, y_2, \ldots, y_n$, of the **homogeneous** differential equation (3) are linearly independent. Surprisingly the nonvanishing of the Wronskian of a set of n such solutions on an interval I is both necessary and sufficient for linear independence.

> **THEOREM 4.4** Let $y_1, y_2, \ldots, y_n$ be n solutions of the homogeneous linear nth-order differential equation (3) on an interval I. Then the set of solutions is linearly independent on I if and only if
>
> $$W(y_1, y_2, \ldots, y_n) \neq 0$$
>
> for every x in the interval.

Proof: We prove Theorem 4.4 for the case when $n = 2$. First, if $W(y_1, y_2) \neq 0$ for every x in I it follows immediately from Theorem 4.2 that y_1 and y_2 are linearly independent. Next, we must show that if y_1 and y_2 are linearly independent solutions of a homogeneous linear second-order differential equation then $W(y_1, y_2) \neq 0$ for every x in I. To show this, let us suppose y_1 and y_2 are linearly independent and there is some fixed x_0 in I for which $W(y_1(x_0), y_2(x_0)) = 0$. Hence there must exist c_1 and c_2, not both zero, such that

$$c_1 y_1(x_0) + c_2 y_2(x_0) = 0$$

$$c_1 y_1'(x_0) + c_2 y_2'(x_0) = 0. \tag{6}$$

If we define $\qquad y(x) = c_1 y_1(x) + c_2 y_2(x)$

then in view of (6), $y(x)$ must also satisfy

$$y(x_0) = 0$$

$$y'(x_0) = 0. \tag{7}$$

But the identically zero function satisfies both the differential equation and the initial conditions (7), and thus by Theorem 4.1 it is the only solution. In other words $y \equiv 0$ or

$$c_1 y_1(x) + c_2 y_2(x) = 0$$

for every x in I. This contradicts the assumption that y_1 and y_2 are linearly independent on the interval. $\qquad\square$

From the foregoing discussion we conclude that when $y_1, y_2, \ldots, y_n$ are n solutions of (3) on an interval I, either the Wronskian is identically zero or is never zero on the interval.

DEFINITION 4.3   Any set $y_1, y_2, \ldots, y_n$ of n linearly independent solutions of the homogeneous linear nth-order differential equation (3) on an interval I is said to be a **fundamental set of solutions** on the interval. $\qquad\square$

THEOREM 4.5   Let $y_1, y_2, \ldots, y_n$ be a fundamental set of solutions of the homogeneous linear nth-order differential equation (3) on an interval I. Then for any solution $Y(x)$ of (3) on I, constants $C_1, C_2, \ldots, C_n$ can be found so that

$$Y = C_1 y_1(x) + C_2 y_2(x) + \cdots + C_n y_n(x).$$

Proof: We prove the case when $n = 2$. Let Y be a solution, and let y_1 and y_2 be linearly independent solutions of

$$a_2(x)y'' + a_1(x)y' + a_0(x)y = 0$$

on an interval I. Suppose $x = t$ is a point in this interval for which $W(y_1(t), y_2(t)) \neq 0$. Suppose also that the values of $Y(t)$ and $Y'(t)$ are given by

$$Y(t) = k_1$$

$$Y'(t) = k_2.$$

If we now examine the system of equations

$$C_1 y_1(t) + C_2 y_2(t) = k_1$$

$$C_1 y_1'(t) + C_2 y_2'(t) = k_2,$$

it follows that we can determine C_1 and C_2 uniquely provided the determinant of the coefficients satisfies

$$\begin{vmatrix} y_1(t) & y_2(t) \\ y_1'(t) & y_2'(t) \end{vmatrix} \neq 0.$$

But this latter determinant is simply the Wronskian evaluated at $x = t$, and by assumption $W \neq 0$. If we now define the function

$$G(x) = C_1 y_1(x) + C_2 y_2(x)$$

we then observe:

(i) $G(x)$ satisfies the differential equation since it is the superposition of two known solutions y_1 and y_2,

(ii) $G(x)$ satisfies the initial conditions

$$G(t) = C_1 y_1(t) + C_2 y_2(t) = k_1$$

$$G'(t) = C_1 y_1'(t) + C_2 y_2'(t) = k_2,$$

(iii) $Y(x)$ satisfies the *same* linear equation and the *same* initial conditions.

Since the solution of this linear initial-value problem is unique (Theorem 4.1) we have

$$Y(x) \equiv G(x)$$

$$Y(x) = C_1 y_1(x) + C_2 y_2(x). \qquad \square$$

The basic question of whether a fundamental set of solutions exists for a linear equation is answered in the next theorem.

THEOREM 4.6 There exists a fundamental set of solutions for the homogeneous linear nth-order differential equation (3) on an interval I. $\qquad \square$

The proof of this result follows from Theorem 4.1. The justification of Theorem 4.6 in the special case of second-order equations is left as an exercise (see Problem 46).

Since we have shown that any solution of (3) is obtained from a linear combination of functions in a fundamental set of solutions we are able to make the following definition.

DEFINITION 4.4 Let $y_1, y_1, \ldots, y_n$ be a fundamental set of solutions of the homogeneous linear nth-order differential equation (3) on an interval I. The **general solution** of the equation on the interval is defined to be

$$y = c_1 y_1(x) + c_2 y_2(x) + \cdots + c_n y_n(x),$$

where the c_i, $i = 1, 2, \ldots, n$ are arbitrary constants. $\qquad \square$

Recall, the general solution as defined in Section 1.1 is also called the **complete solution** of the differential equation.

EXAMPLE

The second-order equation $y'' - 9y = 0$ possesses two solutions

$$y_1 = e^{3x} \quad \text{and} \quad y_2 = e^{-3x}.$$

Since

$$W(e^{3x}, e^{-3x}) = \begin{vmatrix} e^{3x} & e^{-3x} \\ 3e^{3x} & -3e^{-3x} \end{vmatrix}$$

$$= -6$$

$$\neq 0$$

for every value of x, y_1 and y_2 form a fundamental set of solutions on $-\infty < x < \infty$. The general solution of the differential equation on the interval is

$$y = c_1 e^{3x} + c_2 e^{-3x}.$$

EXAMPLE

The reader should verify that the function $y = 4 \sinh 3x - 5e^{-3x}$ also satisfies the differential equation in the preceding example. By choosing $c_1 = 2$, $c_2 = -7$ in the general solution $y = c_1 e^{3x} + c_2 e^{-3x}$ we obtain

$$y = 2e^{3x} - 7e^{-3x}$$

$$= 2e^{3x} - 2e^{-3x} - 5e^{-3x}$$

$$= 4\left(\frac{e^{3x} - e^{-3x}}{2}\right) - 5e^{-3x}$$

$$= 4 \sinh 3x - 5e^{-3x}.$$

EXAMPLE

The functions $y_1 = e^x$, $y_2 = e^{2x}$, and $y_3 = e^{3x}$ satisfy the third-order equation

$$\frac{d^3y}{dx^3} - 6\frac{d^2y}{dx^2} + 11\frac{dy}{dx} - 6y = 0.$$

Since

$$W(e^x, e^{2x}, e^{3x}) = \begin{vmatrix} e^x & e^{2x} & e^{3x} \\ e^x & 2e^{2x} & 3e^{3x} \\ e^x & 4e^{2x} & 9e^{3x} \end{vmatrix}$$

$$= e^x(18e^{5x} - 12e^{5x}) - e^{2x}(9e^{4x} - 3e^{4x}) + e^{3x}(4e^{3x} - 2e^{3x})$$

$$= 2e^{6x}$$

$$\neq 0$$

for every real value of x, y_1, y_2 and y_3 form a fundamental set of solutions on $-\infty < x < \infty$. We conclude

$$y = c_1 e^x + c_2 e^{2x} + c_3 e^{3x}$$

is the general solution of the differential equation on the interval.

EXAMPLE

We have seen that $W(\cos \beta x, \sin \beta x) = \beta$. Since $y_1 = \cos \beta x$ and $y_2 = \sin \beta x$, $\beta \neq 0$, are two linearly independent solutions of the second-order differential equation

$$y'' + \beta^2 y = 0$$

its general solution on $-\infty < x < \infty$ must be

$$y = c_1 \cos \beta x + c_2 \sin \beta x.$$

Nonhomogeneous equations

We now turn our attention to defining the general solution of a **nonhomogeneous** linear equation. Any function y_p, free of arbitrary parameters, which satisfies (4) is said to be a **particular solution** of the equation (sometimes also called a **particular integral**).

EXAMPLES

(a) A particular solution of

$$y'' + 9y = 27$$

is $y_p = 3$ since $y_p'' = 0$, and $0 + 9y_p = 9(3) = 27$.

(b) $y_p = x^3 - x$ is a particular solution of

$$x^2 y'' + 2xy' - 8y = 4x^3 + 6x$$

since $y_p' = 3x^2 - 1$, $y_p'' = 6x$, and

$$x^2 y_p'' + 2x y_p' - 8y_p = x^2(6x) + 2x(3x^2 - 1) - 8(x^3 - x)$$

$$= 4x^3 + 6x.$$

THEOREM 4.7 Let y_1, y_2, . . . , y_k be solutions of the homogeneous linear nth-order differential equation (3) on an interval I and let y_p be any solution of the nonhomogeneous equation (4) on the same interval. Then

$$y = c_1 y_1(x) + c_2 y_2(x) + \cdots + c_k y_k(x) + y_p(x)$$

is also a solution of the nonhomogeneous equation on the interval for any constants c_1, c_2, . . . , c_k. ☐

We can now prove the following analogue of Theorem 4.5 for nonhomogeneous differential equations.

THEOREM 4.8   Let y_p be a given solution of the nonhomogeneous linear nth-order differential equation (4) on an interval I and let $y_1, y_2, \ldots, y_n$ be a fundamental set of solutions of the associated homogeneous equation (3) on the interval. Then for any solution $Y(x)$ of (4) on I, constants $C_1, C_2, \ldots, C_n$ can be found so that

$$Y = C_1 y_1(x) + C_2 y_2(x) + \cdots + C_n y_n(x) + y_p(x)$$

Proof: We prove the case when $n = 2$. Suppose Y and y_p are both solutions of

$$a_2(x)y'' + a_1(x)y' + a_0(x)y = g(x).$$

If we define a function u by

$$u(x) = Y(x) - y_p(x)$$

then $a_2(x)u'' + a_1(x)u' + a_0(x)u$

$$= a_2(x)[Y'' - y_p''] + a_1(x)[Y' - y_p'] + a_0(x)[Y - y_p]$$

$$= a_2(x)Y'' + a_1(x)Y' + a_0(x)Y - [a_2(x)y_p'' + a_1(x)y_p' + a_0(x)y_p]$$

$$= g(x) - g(x)$$

$$= 0.$$

Therefore, in view of Definition 4.4 and Theorem 4.5 we can write

$$u(x) = C_1 y_1(x) + C_2 y_2(x) + \cdots + C_n y_n(x)$$

$$Y(x) - y_p(x) = C_1 y_1(x) + C_2 y_2(x) + \cdots + C_n y_n(x)$$

or $$Y(x) = C_1 y_1(x) + C_2 y_2(x) + \cdots + C_n y_n(x) + y_p(x). \quad \square$$

Thus we come to the last definition of this section.

DEFINITION 4.5   Let y_p be a given solution of the nonhomogeneous linear nth-order differential equation (4) on an interval I and let

$$y_c = c_1 y_1(x) + c_2 y_2(x) + \cdots + c_n y_n(x)$$

denote the general solution of the associated homogeneous equation (3) on the interval. The **general solution** of the nonhomogeneous equation on the interval is defined to be

$$y = c_1 y_1(x) + c_2 y_2(x) + \cdots + c_n y_n(x) + y_p(x)$$

$$= y_c(x) + y_p(x). \quad \square$$

Complementary function

In Definition 4.5 the linear combination

$$y_c(x) = c_1 y_1(x) + c_2 y_2(x) + \cdots + c_n y_n(x),$$

which is the general solution of (3), is called the **complementary function** for equation (4). In other words, the general solution of a nonhomogeneous linear differential equation is

$$y = \text{complementary function} + \text{any particular solution}.$$

EXAMPLE

The function
$$y_p = -\frac{11}{12} - \frac{1}{2}x$$

can be shown to be a particular solution of the nonhomogeneous equation

$$\frac{d^3y}{dx^3} - 6\frac{d^2y}{dx^2} + 11\frac{dy}{dx} - 6y = 3x. \tag{8}$$

In order to write down the general solution of (8) we must also be able to solve the associated homogeneous equation

$$\frac{d^3y}{dx^3} - 6\frac{d^2y}{dx^2} + 11\frac{dy}{dx} - 6y = 0.$$

But on page 133 we saw that the general solution of this latter equation on the interval $-\infty < x < \infty$ was

$$y_c = c_1e^x + c_2e^{2x} + c_3e^{3x}.$$

Hence the general solution of (8) on the interval is

$$y = y_c + y_p$$

$$= c_1e^x + c_2e^{2x} + c_3e^{3x} - \frac{11}{12} - \frac{1}{2}x.$$

Before we actually start solving homogeneous and nonhomogeneous linear differential equations we need the one additional bit of theory presented in the next section.

EXERCISES 4.1

Answers to odd-numbered problems begin on page A-10.

[4.1.1]

1. Given that
$$y = c_1e^x + c_2e^{-x}$$
is a two-parameter family of solutions of
$$y'' - y = 0.$$
on the interval $-\infty < x < \infty$. Find a member of the family satisfying the initial conditions

$$y(0) = 0 \qquad y'(0) = 1.$$

2. Find a solution of the differential equation in Problem 1 satisfying the boundary conditions

$$y(0) = 0 \qquad y(1) = 1.$$

3. Given that
$$y = c_1e^{4x} + c_2e^{-x}$$
is a two-parameter family of solutions of
$$y'' - 3y' - 4y = 0.$$
on the interval $-\infty < x < \infty$. Find a member of the family satisfying the initial conditions

$$y(0) = 1 \qquad y'(0) = 2.$$

★ **4.** Given that
$$y = c_1 + c_2 \cos x + c_3 \sin x$$
is a three-parameter family of solutions of
$$y''' + y' = 0.$$
on the interval $-\infty < x < \infty$. Find a member of the family satisfying the initial conditions
$$y(\pi) = 0 \qquad y'(\pi) = 2 \qquad y''(\pi) = -1.$$

5. Given that
$$y = c_1 x + c_2 x \ln x$$
is two-parameter family of solutions of
$$x^2 y'' - xy' + y = 0$$
on the interval $\infty < x < \infty$. Find a member of the family satisfying the initial conditions
$$y(1) = 3 \qquad y'(1) = -1.$$

6. Given that
$$y = c_1 + c_2 x^2$$
is a two-parameter family of solutions of
$$xy'' - y' = 0.$$
on the interval $-\infty < x < \infty$. Show that constants c_1 and c_2 cannot be found so that a member of the family satisfies the initial conditions
$$y(0) = 0 \qquad y'(0) = 1.$$
Explain why this does not violate Theorem 4.1.

7. Find two members of the family of solutions of $xy'' - y' = 0$ given in Problem 6 satisfying the initial conditions
$$y(0) = 0 \qquad y'(0) = 0.$$

★ **8.** Find a member of the family of solutions of $xy'' - y' = 0$ given in Problem 6 satisfying the conditions
$$y(0) = 1 \qquad y'(1) = 6.$$
Does Theorem 4.1 guarantee that this solution is unique?

9. Given that
$$y = c_1 e^x \cos x + c_2 e^x \sin x$$
is a two-parameter family of solutions of
$$y'' - 2y' + 2y = 0.$$
on the interval $-\infty < x < \infty$. Determine whether a member of the family can be found that satisfies the conditions
(a) $y(0) = 1, \quad y'(0) = 0$
(b) $y(0) = 1, \quad y(\pi) = -1$
(c) $y(0) = 1, \quad y(\pi/2) = 1$
(d) $y(0) = 0, \quad y(\pi) = 0.$

10. Given that $y = c_1 x^2 + c_2 x^4 + 3$

is a two-parameter family of solutions of

$$x^2y'' - 5xy' + 8y = 24$$

on the interval $-\infty < x < \infty$. Determine whether a member of the family can be found that satisfies the conditions

(a) $y(-1) = 0, \quad y(1) = 4$ (c) $y(0) = 3, \quad y(1) = 0$
(b) $y(0) = 1, \quad y(1) = 2$ (d) $y(1) = 3, \quad y(2) = 15$.

In Problems 11 and 12 find an interval around $x = 0$ for which the given initial-value problem has a unique solution.

11. $(x - 2)y'' + 3y = x, \ y(0) = 0, \ y'(0) = 1$

12. $y'' + (\tan x)y = e^x, \ y(0) = 1, \ y'(0) = 0$

13. Given that $y = c_1 \cos \lambda x + c_2 \sin \lambda x$ is a family of solutions of the differential equation $y'' + \lambda^2 y = 0$. Determine the values of the parameter λ for which the boundary-value problem

$$y'' + \lambda^2 y = 0, \qquad y(0) = 0, \qquad y(\pi) = 0$$

has nonzero solutions.

★**14.** Determine the values of the parameter λ for which the boundary-value problem

$$y'' + \lambda^2 y = 0, \qquad y(0) = 0, \qquad y(5) = 0$$

has nonzero solutions. (See Problem 13.)

[4.1.2]   In Problems 15–22 determine whether the given functions are linearly independent or dependent on $-\infty < x < \infty$.

15. $f_1(x) = x, \quad f_2(x) = x^2, \quad f_3(x) = 4x - 3x^2$

★**16.** $f_1(x) = 0, \quad f_2(x) = x, \quad f_3(x) = e^x$

17. $f_1(x) = 5, \quad f_2(x) = \cos^2 x, \quad f_3(x) = \sin^2 x$

18. $f_1(x) = \cos 2x, \quad f_2(x) = \sin 2x, \quad f_3(x) = \sin^2 x$

19. $f_1(x) = x, \quad f_2(x) = x - 1, \quad f_3(x) = x + 3$

20. $f_1(x) = 2 + x, \quad f_2(x) = 2 + |x|$

21. $f_1(x) = 1 + x, \quad f_2(x) = x, \quad f_3(x) = x^2$

22. $f_1(x) = e^x, \quad f_2(x) = e^{-x}, \quad f_3(x) = \sinh x$

In Problems 23–28 show by computing the Wronskian that the given functions are linearly independent on the indicated interval.

23. $x^{1/2}, x^2; \quad 0 < x < \infty$ **24.** $1 + x, x^3; \quad -\infty < x < \infty$

25. $\sin x, \csc x; \quad 0 < x < \pi$ **26.** $\tan x, \cot x; \quad 0 < x < \pi/2$

27. $e^x, e^{-x}, e^{4x}; \quad -\infty < x < \infty$ **28.** $x, x \ln x, x^2 \ln x; \quad 0 < x < \infty$

29. Observe that for the functions $f_1(x) = 2$ and $f_2(x) = e^x$

$$1 \cdot f_1(0) - 2 \cdot f_2(0) = 0.$$

Does this imply that f_1 and f_2 are linearly dependent on any interval containing $x = 0$?

★30. (a) Show graphically that

$$f_1(x) = x^2 \quad \text{and} \quad f_2(x) = x|x|$$

are linearly independent on $-\infty < x < \infty$.

(b) Show that

$$W(f_1(x), f_2(x)) \equiv 0,$$

that is, $W = 0$ for every real value of x.

[4.1.3]   **31. (a)** Verify that $y = 1/x$ is a solution of the nonlinear differential equation $y'' = 2y^3$ on the interval $0 < x < \infty$.

(b) Show that the constant multiple $y = c/x$ is not a solution of the equation when $c \neq 0, \pm 1$.

★32. (a) Verify that $y_1 = 1$ and $y_2 = \ln x$ are solutions of the nonlinear differential equation $y'' + (y')^2 = 0$ on the interval $0 < x < \infty$.

(b) Is $y_1 + y_2$ a solution of the equation? Is $c_1 y_1 + c_2 y_2$, c_1 and c_2 arbitrary, a solution of the equation?

In Problems 33–40 verify that the given functions form a fundamental set of solutions of the differential equation on the indicated interval. Form the general solution.

33. $y'' - y' - 12y = 0;$ $e^{-3x}, e^{4x}, -\infty < x < \infty$

34. $y'' - 4y = 0;$ $\cosh 2x, \sinh 2x, -\infty < x < \infty$

35. $y'' - 2y' + 5y = 0;$ $e^x \cos 2x, e^x \sin 2x, -\infty < x < \infty$

36. $4y'' - 4y' + y = 0;$ $e^{x/2}, xe^{x/2}, -\infty < x < \infty$

37. $x^2 y'' - 6xy' + 12y = 0;$ $x^3, x^4, 0 < x < \infty$

★38. $x^2 y'' + xy' + y = 0;$ $\cos(\ln x), \sin(\ln x), 0 < x < \infty$

39. $x^3 y''' + 6x^2 y'' + 4xy' - 4y = 0;$ $x, x^{-2}, x^{-2} \ln x, 0 < x < \infty$

40. $y^{(4)} + y'' = 0;$ $1, x, \cos x, \sin x, -\infty < x < \infty$

In Problems 41–44 verify that the given two-parameter family of functions is the general solution of the nonhomogeneous differential equation on the indicated interval.

41. $y'' - 7y' + 10y = 24e^x;$
$y = c_1 e^{2x} + c_2 e^{5x} + 6e^x, -\infty < x < \infty$

42. $y'' + y = \sec x;$
$y = c_1 \cos x + c_2 \sin x + x \sin x + (\cos x) \ln(\cos x), -\pi/2 < x < \pi/2$

43. $y'' - 4y' + 4y = 2e^{2x} + 4x - 12$;
$y = c_1 e^{2x} + c_2 x e^{2x} + x^2 e^{2x} + x - 2, \quad -\infty < x < \infty$

44. $2x^2 y'' + 5xy' + y = x^2 - x$;

$y = c_1 x^{-1/2} + c_2 x^{-1} + \dfrac{1}{15} x^2 - \dfrac{1}{6} x, \quad 0 < x < \infty$

45. (a) Verify that $\qquad y_1 = x^3 \qquad$ and $\qquad y_2 = |x|^3$

are linearly independent solutions of the differential equation $x^2 y'' - 4xy' + 6y = 0$ on $-\infty < x < \infty$.

(b) Show that $W(y_1, y_2) \equiv 0$.

(c) Does the result of part (b) violate Theorem 4.4?

(d) Verify that $\qquad Y_1 = x^3 \qquad$ and $\qquad Y_2 = x^2$

are also linearly independent solutions of the differential equation on $-\infty < x < \infty$.

(e) Find a solution of the equation satisfying $y(0) = 0$, $y'(0) = 0$.

(f) By the superposition principle both linear combinations

$$y = c_1 y_1 + c_2 y_2 \qquad \text{and} \qquad y = c_1 Y_1 + c_2 Y_2,$$

are solutions of the differential equation. Is one, both, or neither the general solution of the differential equation on $-\infty < x < \infty$?

46. Consider the second-order differential equation

$$a_2(x)y'' + a_1(x)y' + a_0(x)y = 0 \tag{9}$$

where $a_2(x)$, $a_1(x)$, and $a_0(x)$ are continuous on an interval I and $a_2(x) \neq 0$ for every x in the interval. From Theorem 4.1 there exists only one solution y_1 of the equation satisfying $y(x_0) = 1$ and $y'(x_0) = 0$, where x_0 is a point in I. Similarly there exists a unique solution y_2 of the equation satisfying $y(x_0) = 0$ and $y'(x_0) = 1$. Show that y_1 and y_2 form a fundamental set of solutions of the differential equation on the interval I.

Miscellaneous problems

47. Let y_1 and y_2 be two solutions of (9).

(a) If $W(y_1, y_2)$ is the Wronskian of y_1 and y_2, show that

$$a_2(x) \dfrac{dW}{dx} + a_1(x)W = 0.$$

(b) Derive **Abel's formula***

$$W = ce^{-\int [a_1(x)/a_2(x)]\, dx}$$

where c is a constant.

(c) Using an alternative form of Abel's formula

$$W = ce^{-\int_{x_0}^{x} [a_1(t)/a_2(t)]\, dt},$$

---

*Niels Henrik Abel (1802–1829) a brilliant Norwegian mathematician whose tragic death at age 26 was an inestimable loss for mathematics.

for x_0 in I show that

$$W(y_1, y_2) = W(x_0)e^{-\int_{x_0}^{x}[a_1(t)/a_2(t)]\, dt}.$$

(d) Show that if $W(x_0) = 0$, then $W = 0$ for every x in I whereas if $W(x_0) \neq 0$, then $W \neq 0$ for every x in the interval.

In Problems 48 and 49 use the results of Problem 47.

★**48.** If y_1 and y_2 are two solutions of

$$(1 - x^2)y'' - 2xy' + n(n + 1)y = 0$$

on $-1 < x < 1$, show that

$$W(y_1, y_2) = \frac{c}{1 - x^2}$$

where c is a constant.

49. In Chapter 6 we shall see that the solutions y_1 and y_2 of $xy'' + y' + xy = 0$, $0 < x < \infty$, are infinite series. Suppose we consider initial conditions

$$y_1(x_0) = k_1 \qquad y_1'(x_0) = k_2$$

and

$$y_2(x_0) = k_3 \qquad y_2'(x_0) = k_4$$

for $x_0 > 0$. Show that

$$W(y_1, y_2) = \frac{(k_1 k_4 - k_2 k_3)x_0}{x}.$$

50. Suppose f is a differentiable function on an interval I and $f(x) \neq 0$ for every x in the interval. Prove that $f(x)$ and $xf(x)$ are linearly independent on I.

4.2   Constructing a Second Solution from a Known Solution

Reduction of order   It is one of the more interesting, as well as important, facts of life in the study of linear *second-order* differential equations that we can construct a second solution from a *known* solution. Suppose $y_1(x)$ is a nonzero solution of the equation*

$$a_2(x)y'' + a_1(x)y' + a_0(x)y = 0. \tag{1}$$

The process we shall use to find a second solution $y_2(x)$ consists of **reducing the order** of equation (1) to a first-order equation. For example, it is easily verified that $y_1 = e^x$ satisfies the differential equation $y'' - y = 0$. If we try to determine a solution of the form $y = u(x)e^x$ then

---

*We naturally assume, as we did in the preceding section, that the coefficients in (1) are continuous and $a_2(x) \neq 0$ for every x in some interval I.

$$y' = ue^x + e^x u'$$

$$y'' = ue^x + 2e^x u' + e^x u''$$

and so
$$y'' - y = e^x(u'' + 2u') = 0.$$

Since $e^x \neq 0$ this last equation requires that $u'' + 2u' = 0$.

If we let $w = u'$, then the latter equation is recognized as a linear first-order equation in $w:w' + 2w = 0$. Using the integrating factor e^{2x} we can write

$$\frac{d}{dx}[e^{2x}w] = 0$$

and thus
$$w = c_1 e^{-2x}$$

$$\frac{du}{dx} = c_1 e^{-2x}$$

$$u = -\frac{c_1}{2}e^{-2x} + c_2.$$

Hence we obtain
$$y = u(x)e^x$$

$$= -\frac{c_1}{2}e^{-x} + c_2 e^x.$$

By picking $c_2 = 0$ and $c_1 = -2$, we obtain the second solution $y_2 = e^{-x}$. Since $W(e^x, e^{-x}) \neq 0$ for every x the solutions are linearly independent on $-\infty < x < \infty$ and thus the expression for y is actually the general solution of the given equation.

EXAMPLE

Given that $y_1 = x^3$ is a solution of $x^2 y'' - 6y = 0$. Use reduction of order to find a second solution on the interval $0 < x < \infty$.

Solution: Define
$$y_2 = u(x)x^3$$

so that
$$y_2' = 3x^2 u + x^3 u'$$

$$y_2'' = x^3 u'' + 6x^2 u' + 6xu$$

$$x^2 y_2'' - 6y_2 = x^2(x^3 u'' + 6x^2 u' + 6xu) - 6ux^3$$

$$= x^5 u'' + 6x^4 u'$$

$$= 0$$

provided $u(x)$ is a solution of

$$x^5 u'' + 6x^4 u' = 0 \quad \text{or} \quad u'' + \frac{6}{x}u' = 0.$$

If $w = u'$, we obtain the linear first-order equation

$$w' + \frac{6}{x}w = 0$$

which possesses the integrating factor $e^{6\int dx/x} = e^{6\ln x} = x^6$. It follows that

$$\frac{d}{dx}[x^6 w] = 0$$

$$x^6 w = c$$

$$w = u' = \frac{c_1}{x^6}$$

$$u = -\frac{c_1}{5x^5} + c_2.$$

Hence

$$y = u(x)x^3$$

$$y = -\frac{c_1}{5x^2} + c_2 x^3.$$

Choosing $c_2 = 0$ and $c_1 = -5$ yields the second solution $y_2 = 1/x^2$.

The general case

Suppose we divide by $a_2(x)$ in order to put equation (1) in the form

$$y'' + P(x)y' + Q(x)y = 0. \tag{2}$$

where $P(x)$ and $Q(x)$ are continuous on some interval I. Let us suppose further that $y_1(x)$ is a known solution of (2) on I and that $y_1(x) \neq 0$ for every x in the interval. If we define $y = u(x)y_1(x)$, it follows that

$$y' = uy_1' + y_1 u'$$

$$y'' = uy_1'' + 2y_1' u' + y_1 u''$$

$$y'' + Py' + Qy = u\underbrace{[y_1'' + Py_1' + Qy_1]}_{\text{zero}} + y_1 u'' + (2y_1' + Py_1)u' = 0.$$

This implies we must have

$$y_1 u'' + (2y_1' + Py_1)u' = 0$$

or

$$y_1 w' + (2y_1' + Py_1)w = 0 \tag{3}$$

where we have let $w = u'$. Observe that equation (3) is both linear and separable. Applying the latter technique we obtain

$$\frac{dw}{w} + 2\frac{y_1'}{y_1}dx + P\,dx = 0$$

$$\ln|w| + 2\ln|y_1| = -\int P\,dx + c$$

$$\ln|wy_1^2| = -\int P\,dx + c$$

$$wy_1^2 = c_1 e^{-\int P\,dx}$$

$$w = c_1 \frac{e^{-\int P\, dx}}{y_1^2}$$

$$\frac{du}{dx} = c_1 \frac{e^{-\int P\, dx}}{y_1^2}.$$

Integrating again gives

$$u = c_1 \int \frac{e^{-\int P\, dx}}{y_1^2}\, dx + c_2$$

and therefore

$$y = u(x) y_1(x)$$

$$= c_1 y_1(x) \int \frac{e^{-\int P(x)dx}}{y_1^2(x)}\, dx + c_2 y_1(x).$$

By choosing $c_2 = 0$ and $c_1 = 1$, we find that a second solution of equation (2) is

$$y_2 = y_1(x) \int \frac{e^{-\int P(x)dx}}{y_1^2(x)}\, dx . \tag{4}$$

It makes a good review exercise in differentiation to start with formula (4) and actually verify that equation (2) is satisfied.

Now $y_1(x)$ and $y_2(x)$ are linearly independent since

$$W(y_1(x),\, y_2(x)) = \begin{vmatrix} y_1 & y_1 \int \dfrac{e^{-\int P\, dx}}{y_1^2}\, dx \\[4mm] y_1' & \dfrac{e^{-\int P\, dx}}{y_1} + y_1' \int \dfrac{e^{-\int P\, dx}}{y_1^2}\, dx \end{vmatrix} = e^{-\int P\, dx}$$

is not zero on any interval on which $y_1(x)$ is not zero.*

EXAMPLE

The function $y_1 = x^2$ is a solution of $x^2 y'' - 3xy' + 4y = 0$. Find the general solution on the interval $0 < x < \infty$.

Solution: Since the equation has the alternative form

$$y'' - \frac{3}{x} y' + \frac{4}{x^2} y = 0$$

we find from (4)

$$y_2 = x^2 \int \frac{e^{3\int dx/x}}{x^4}\, dx \qquad [e^{3\int dx/x} = e^{\ln x^3} = x^3]$$

$$= x^2 \int \frac{dx}{x}$$

$$= x^2 \ln x.$$

---

*Alternatively, if $y_2 = u(x) y_1$ then $W(y_1, y_2) = u'(y_1)^2 \neq 0$, since $y_1 \neq 0$ for every x on some interval. If $u' = 0$, then $u = $ constant.

The general solution on $0 < x < \infty$ is given by

$$y = c_1 y_1 + c_2 y_2$$

$$= c_1 x^2 + c_2 x^2 \ln x.$$

EXAMPLE

It can be verified that
$$y_1 = \frac{\sin x}{\sqrt{x}}$$
is a solution of $x^2 y'' + xy' + (x^2 - \tfrac{1}{4})y = 0$ on $0 < x < \pi$. Find a second solution.

Solution: First put the equation into the form

$$y'' + \frac{1}{x} y' + \left(1 - \frac{1}{4x^2}\right) y = 0.$$

Then by formula (4) we have

$$y_2 = \frac{\sin x}{\sqrt{x}} \int \frac{e^{-\int dx/x}}{\left(\dfrac{\sin x}{\sqrt{x}}\right)^2} \, dx$$

$$= \frac{\sin x}{\sqrt{x}} \int \csc^2 x \, dx \qquad [e^{-\int dx/x} = e^{\ln x^{-1}} = x^{-1}]$$

$$= \frac{\sin x}{\sqrt{x}} (-\cot x)$$

$$= -\frac{\cos x}{\sqrt{x}}.$$

Since the differential equation is homogeneous, we can disregard the negative sign and take the second solution to be $y_2 = \cos x / \sqrt{x}$.

Observe that $y_1(x)$ and $y_2(x)$ of the previous example are linearly independent solutions of the given differential equation on the larger interval $0 < x < \infty$.

EXERCISES 4.2   *Answers to odd-numbered problems begin on page A-12.*

In Problems 1–30 find a second solution of each differential equation. Assume an appropriate interval of validity.

1. $y'' + 4y' = 0; \quad y_1 = 1$ **2.** $y'' - y' = 0; \quad y_1 = 1$

3. $y'' - 4y' + 4y = 0; \quad y_1 = e^{2x}$

4. $y'' + 2y' + y = 0; \quad y_1 = xe^{-x}$

5. $y'' + 16y = 0;$ $y_1 = \cos 4x$ **6.** $y'' + 9y = 0;$ $y_1 = \sin 3x$

7. $y'' - y = 0;$ $y_1 = \cosh x$ **8.** $y'' - 25y = 0;$ $y_1 = e^{5x}$

9. $9y'' - 12y' + 4y = 0;$ $y_1 = e^{2x/3}$

10. $6y'' + y' - y = 0;$ $y_1 = e^{x/3}$

11. $x^2y'' - 7xy' + 16y = 0;$ $y_1 = x^4$

12. $x^2y'' + 2xy' - 6y = 0;$ $y_1 = x^2$

13. $xy'' + y' = 0;$ $y_1 = \ln x$

★**14.** $4x^2y'' + y = 0;$ $y_1 = x^{1/2}\ln x$

15. $(1 - 2x - x^2)y'' + 2(1 + x)y' - 2y = 0;$ $y_1 = x + 1$

16. $(1 - x^2)y'' - 2xy' = 0;$ $y_1 = 1$

17. $x^2y'' - xy' + 2y = 0;$ $y_1 = x \sin (\ln x)$

18. $x^2y'' - 3xy' + 5y = 0;$ $y_1 = x^2 \cos (\ln x)$

19. $(1 + 2x)y'' + 4xy' - 4y = 0;$ $y_1 = e^{-2x}$

★**20.** $(1 + x)y'' + xy' - y = 0;$ $y_1 = x$

21. $x^2y'' - xy' + y = 0;$ $y_1 = x$ **22.** $x^2y'' - 20y = 0;$ $y_1 = x^{-4}$

23. $x^2y'' - 5xy' + 9y = 0;$ $y_1 = x^3 \ln x$

★**24.** $x^2y'' + xy' + y = 0;$ $y_1 = \cos (\ln x)$

25. $x^2y'' - 4xy' + 6y = 0;$ $y_1 = x^2 + x^3$

26. $x^2y'' - 7xy' - 20y = 0;$ $y_1 = x^{10}$

27. $(3x + 1)y'' - (9x + 6)y' + 9y = 0;$ $y_1 = e^{3x}$

28. $xy'' - (x + 1)y' + y = 0;$ $y_1 = e^x$

29. $y'' - 3(\tan x)y' = 0;$ $y_1 = 1$ **30.** $xy'' - (2 + x)y' = 0;$ $y_1 = 1$

In Problems 31–34 use the method of reduction of order to find a solution of the given nonhomogeneous equation. The indicated function $y_1(x)$ is a solution of the associated homogeneous equation. Determine a second solution of the homogeneous equation and a particular solution of the nonhomogeneous equation.

31. $y'' - 4y = 2;$ $y_1 = e^{-2x}$ **32.** $y'' + y' = 1;$ $y_1 = 1$

33. $y'' - 3y' + 2y = 5e^{3x};$ $y_1 = e^x$

★**34.** $y'' - 4y' + 3y = x;$ $y_1 = e^x$

Miscellaneous problem

35. Verify by direct substitution that formula (4) satisfies equation (2).

4.3   Homogeneous Linear Equations with Constant Coefficients

We have seen that the linear first-order equation $dy/dx + ay = 0$, where a is a constant, has the exponential solution $y = c_1 e^{-ax}$ on $-\infty < x < \infty$. Therefore it is natural to seek to determine whether exponential solutions exist on $-\infty < x < \infty$ for higher order equations such as

$$a_n y^{(n)} + a_{n-1} y^{(n-1)} + \cdots + a_2 y'' + a_1 y' + a_0 y = 0, \qquad (1)$$

where the a_i, $i = 0, 1, \ldots, n$ are constants. The surprising fact is that *all* solutions of (1) are exponential functions or constructed out of exponential functions. We shall begin by considering the special case of the second-order equation

$$ay'' + by' + cy = 0. \qquad (2)$$

The auxiliary equation

If we try a solution of the form $y = e^{mx}$, then $y' = me^{mx}$ and $y'' = m^2 e^{mx}$ so that equation (2) becomes

$$am^2 e^{mx} + bm e^{mx} + ce^{mx} = 0 \qquad \text{or} \qquad e^{mx}[am^2 + bm + c] = 0.$$

Because e^{mx} is never zero for real values of x, it is apparent that the only way that this exponential function can satisfy the differential equation is to choose m so that it is a root of the quadratic equation

$$am^2 + bm + c = 0. \qquad (3)$$

This latter equation is called the **auxiliary equation** or **characteristic equation** of the differential equation (2). We shall consider three cases, namely, the solutions of the auxiliary equation corresponding to real distinct roots, real but equal roots, and lastly, complex conjugate roots.

CASE I   Under the assumption that the auxiliary equation (3) has two unequal real roots m_1 and m_2 we find two solutions

$$y_1 = e^{m_1 x} \qquad \text{and} \qquad y_2 = e^{m_2 x}.$$

We have already seen that these functions are linearly independent on $-\infty < x < \infty$ (see page 126) and hence form a fundamental set. It follows that the general solution of (2) on this interval is

$$y = c_1 e^{m_1 x} + c_2 e^{m_2 x}. \qquad \square (4)$$

CASE II   When $m_1 = m_2$ we necessarily obtain only one exponential solution $y_1 = e^{m_1 x}$. However, it follows immediately from the discussion of Section 4.2 that a second solution is

$$y_2 = e^{m_1 x} \int \frac{e^{-(b/a)x}}{e^{2m_1 x}} \, dx \qquad (5)$$

But from the quadratic formula we have

$$m = \frac{-b \pm \sqrt{b^2 - 4ac}}{2a} = -\frac{b}{2a}$$

since the only way to have $m_1 = m_2$ is to have $b^2 - 4ac = 0$. In view of the fact that $2m_1 = -b/a$, (5) becomes

$$y_2 = e^{m_1 x} \int \frac{e^{2m_1 x}}{e^{2m_1 x}} \, dx = e^{m_1 x} \int dx$$

$$= xe^{m_1 x}.$$

The general solution of (2) is then

$$y = c_1 e^{m_1 x} + c_2 x e^{m_1 x}. \qquad \square \ (6)$$

CASE III If m_1 and m_2 are complex then we can write

$$m_1 = \alpha + i\beta \qquad \text{and} \qquad m_2 = \alpha - i\beta$$

where α and β are real and $i^2 = -1$. Formally there is no difference between this case and Case I, and hence

$$y = C_1 e^{(\alpha + i\beta)x} + C_2 e^{(\alpha - i\beta)x}. \qquad (7)$$

However, in practice we would prefer to work with real functions instead of complex exponentials. Now we can write (7) in a more practical form by using Euler's formula*

$$e^{i\theta} = \cos\theta + i\sin\theta,$$

where θ is any real number. From this result we can write

$$e^{i\beta x} = \cos\beta x + i\sin\beta x \quad \text{and} \quad e^{-i\beta x} = \cos\beta x - i\sin\beta x,$$

where we have used $\cos(-\beta x) = \cos\beta x$ and $\sin(-\beta x) = -\sin\beta x$. Thus (7) becomes

$$y = e^{\alpha x}[C_1 e^{i\beta x} + C_2 e^{-i\beta x}]$$

$$= e^{\alpha x}[C_1\{\cos\beta x + i\sin\beta x\} + C_2\{\cos\beta x - i\sin\beta x\}]$$

$$= e^{\alpha x}[(C_1 + C_2)\cos\beta x + (C_1 i - C_2 i)\sin\beta x].$$

Since $e^{\alpha x}\cos\beta x$ and $e^{\alpha x}\sin\beta x$ themselves form a fundamental set of solutions of the given differential equation on $-\infty < x < \infty$ we can simply relabel $C_1 + C_2$ as c_1 and $C_1 i - C_2 i$ as c_2 and use the superposition principle to write the general solution

---

*In case you have never seen this result before, it is worth going through the formal proof at least once. Write down the MacLaurin series for e^x and then substitute $x = i\theta$. Separate the real and imaginary parts and look closely at the resulting two series.

$$y = c_1 e^{\alpha x} \cos \beta x + c_2 e^{\alpha x} \sin \beta x$$
$$= e^{\alpha x}[c_1 \cos \beta x + c_2 \sin \beta x].$$
$\qquad\qquad\qquad\qquad\qquad\qquad\qquad\qquad\qquad\qquad$ $\square$ (8)

EXAMPLES

Solve the following differential equations.

(a) $2y'' - 5y' - 3y = 0$,
(b) $y'' - 10y' + 25y = 0$,
(c) $y'' + y' + y = 0$.

Solutions: (a) $2m^2 - 5m - 3 = 0$
$$(2m + 1)(m - 3) = 0$$

$$m_1 = -\frac{1}{2}, \qquad m_2 = 3$$

$$y = c_1 e^{-x/2} + c_2 e^{3x}.$$

(b) $m^2 - 10m + 25 = 0$
$$(m - 5)^2 = 0$$

$$m_1 = m_2 = 5$$

$$y = c_1 e^{5x} + c_2 x e^{5x}.$$

(c) $m^2 + m + 1 = 0$

$$m = \frac{-1 \pm \sqrt{-3}}{2}$$

$$m_1 = -\frac{1}{2} + \frac{\sqrt{3}}{2} i, \qquad m_2 = -\frac{1}{2} - \frac{\sqrt{3}}{2} i$$

$$y = e^{-x/2}\left[c_1 \cos \frac{\sqrt{3}}{2} x + c_2 \sin \frac{\sqrt{3}}{2} x\right].$$

EXAMPLE

The two equations $\qquad\qquad y'' + k^2 y = 0$ $\qquad\qquad\qquad$ (9)

and $\qquad\qquad\qquad\qquad y'' - k^2 y = 0$ $\qquad\qquad\qquad$ (10)

are frequently encountered in the study of applied mathematics. For the former differential equation the auxiliary equation $m^2 + k^2 = 0$ has the roots $m_1 = ki$ and $m_2 = -ki$. It follows from (8) that the general solution of (9) is

$$y = c_1 \cos kx + c_2 \sin kx.$$ $\qquad\qquad\qquad$ (11)

The latter differential equation has the auxiliary equation $m^2 - k^2 = 0$ with real roots $m_1 = k$ and $m_2 = -k$, so that its general solution is

$$y = c_1 e^{kx} + c_2 e^{-kx}$$ $\qquad\qquad\qquad$ (12)

Notice that if we choose $c_1 = c_2 = 1/2$ in (12), then,

$$y = \frac{e^{kx} + e^{-kx}}{2}$$

$$= \cosh kx$$

is also a solution of (10). Furthermore, when $c_1 = 1/2$, $c_2 = -1/2$ then (12) becomes

$$y = \frac{e^{kx} - e^{-kx}}{2}$$

$$= \sinh kx.$$

Since $\cosh kx$ and $\sinh kx$ are linearly independent on any interval of the x-axis they form a fundamental set. Thus an alternative form for the general solution of (10) is

$$y = c_1 \cosh kx + c_2 \sinh kx. \tag{13}$$

Higher order equations

In general, to solve an nth-order differential equation

$$a_n y^{(n)} + a_{n-1} y^{(n-1)} + \cdots + a_2 y'' + a_1 y' + a_0 y = 0, \tag{14}$$

where the a_i, $= 0, 1, \ldots, n$ are real constants, we must solve an nth-degree polynominal equation

$$a_n m^n + a_{n-1} m^{n-1} + \cdots + a_2 m^2 + a_1 m + a_0 = 0. \tag{15}$$

If all the roots of (15) are real and distinct, then the general solution of (14) is

$$y = c_1 e^{m_1 x} + c_2 e^{m_2 x} + \cdots + c_n e^{m_n x}. \tag{16}$$

It is somewhat harder to summarize the analogues of Case II and Case III because the roots of an auxiliary equation of degree greater than two can occur in many combinations. For example, a fifth degree equation could have five distinct real roots, or three distinct real and two complex roots, or one real and four complex roots, or five real but equal roots, or five real roots but two of them equal, and so on. When m_1 is a root of multiplicity k of an nth-degree auxiliary equation (that is, k roots are equal to m_1) then it can be shown that the linearly independent solutions are

$$e^{m_1 x}, xe^{m_1 x}, x^2 e^{m_1 x}, \ldots, x^{k-1} e^{m_1 x},$$

and the general solution must contain the linear combination

$$c_1 e^{m_1 x} + c_2 x e^{m_1 x} + c_3 x^2 e^{m_1 x} + \cdots + c_k x^{k-1} e^{m_1 x}.$$

Lastly, it should be remembered that when the coefficients are real, complex roots of an auxiliary equation will always appear in pairs. Thus for example, a cubic polynomial equation can have at most two complex roots.

EXAMPLE   Solve
$$y''' + 3y'' - 4y = 0.$$

Solution: It should be apparent from inspection of
$$m^3 + 3m^2 - 4 = 0$$
that one root is $m_1 = 1$. Now if we divide* $m^3 + 3m^2 - 4$ by $m - 1$ we find
$$m^3 + 3m^2 - 4 = (m - 1)(m^2 + 4m + 4)$$
$$= (m - 1)(m + 2)^2,$$
and so the other roots are $m_2 = m_3 = -2$. Thus the general solution is
$$y = c_1 e^x + c_2 e^{-2x} + c_3 x e^{-2x}.$$

Of course, the most difficult aspect of solving constant-coefficient equations is finding the roots of auxiliary equations of degree greater than two. As illustrated in the preceding example, one way to solve an equation is to guess[†] a root m_1 and then divide by $m - m_1$ to obtain the factorization $(m - m_1)Q(m)$. We then try to find the roots of $Q(m)$.

EXAMPLE   Solve
$$3y''' - 19y'' + 36y' - 10y = 0.$$

Solution: It is easily verified that $m_1 = 1/3$ is one root of
$$3m^3 - 19m^2 + 36m - 10 = 0.$$
By division we find that
$$3m^3 - 19m^2 + 36m - 10 = \left(m - \frac{1}{3}\right)(3m^2 - 18m + 30)$$
$$= (3m - 1)(m^2 - 6m + 10).$$
From the quadratic formula we find $m_2 = 3 + i$ and $m_3 = 3 - i$. The general solution is
$$y = c_1 e^{x/3} + e^{3x}[c_2 \cos x + c_3 \sin x].$$

---

*Synthetic division provides a quick way of testing for roots. In the above case

1	3	0	− 4	$\lfloor +1$
	1	4	4	
1	4	4	$\lfloor 0 = R$	

If $R \neq 0$ the number tested is not a root. For example, dividing $m + 1$ is equivalent to

1	3	0	− 4	$\lfloor - 1$
	− 1	− 2	2	
1	2	− 2	$\lfloor - 2 = R.$	

Therefore $m = -1$ is not a root and so $m + 1$ is not a factor.

†If the equation $a_n m^n + \cdots + a_1 m + a_0 = 0$ has a *rational* real root $m_1 = p/q$, where p and q are integers, then q must be a factor of a_n and p must be a factor of a_0. We can test all the possible ratios by synthetic division.

EXAMPLE   Solve
$$\frac{d^4y}{dx^4} + 2\frac{d^2y}{dx^2} + y = 0.$$

Solution: The auxiliary equation
$$m^4 + 2m^2 + 1 = 0$$

is equivalent to $(m^2 + 1)^2 = 0.$

Therefore, the roots are $m_1 = m_3 = i$ and $m_2 = m_4 = -i$. From Case II the formal solution would be
$$y = C_1 e^{ix} + C_2 e^{-ix} + C_3 x e^{ix} + C_4 x e^{-ix}.$$

By Euler's formula the grouping $C_1 e^{ix} + C_2 e^{-ix}$ can be rewritten as $c_1 \cos x + c_2 \sin x$ after   a   relabeling   of   constants.   Similarly, $x(C_3 e^{ix} + C_4 e^{-ix})$ can be expressed as $x(c_3 \cos x + c_4 \sin x)$. Hence the general solution is
$$y = c_1 \cos x + c_2 \sin x + c_3 x \cos x + c_4 x \sin x.$$

When $m_1 = \alpha + i\beta$ is a complex root of multiplicity k of an auxiliary equation with real coefficients, its complex conjugate $m_2 = \alpha - i\beta$ is also a root of multiplicity k. The general solution of the corresponding differential equation must then contain a linear combination of the linearly independent solutions
$$e^{\alpha x}\cos \beta x, \ xe^{\alpha x}\cos \beta x, \ x^2 e^{\alpha x}\cos \beta x, \ \ldots, \ x^{k-1}e^{\alpha x}\cos \beta x,$$

$$e^{\alpha x}\sin \beta x, \ xe^{\alpha x}\sin \beta x, \ x^2 e^{\alpha x}\sin \beta x, \ \ldots, \ x^{k-1}e^{\alpha x}\sin \beta x.$$

In the last example we identify $k = 2$, $\alpha = 0$, and $\beta = 1$.

EXAMPLE   Solve $y'' - 4y' + 13y = 0$

subject to $y(0) = -1 \qquad y'(0) = 2.$

Solution: The roots of the auxiliary equation
$$m^2 - 4m + 13 = 0$$

are $m_1 = 2 + 3i$ and $m_2 = 2 - 3i$

so that $y = e^{2x}[c_1 \cos 3x + c_2 \sin 3x]$

The condition $y(0) = -1$ implies
$$-1 = e^0[c_1 \cos 0 + c_2 \sin 0]$$
$$= c_1$$

from which we can write
$$y = e^{2x}[-\cos 3x + c_2 \sin 3x].$$

Differentiating this latter expression and using the second condition gives
$$y' = e^{2x}[3 \sin 3x + 3c_2 \cos 3x] + 2e^{2x}[-\cos 3x + c_2 \sin 3x]$$
$$2 = 3c_2 - 2$$

so that $c_2 = 4/3$. Hence

$$y = e^{2x}\left[-\cos 3x + \frac{4}{3}\sin 3x \right].$$

EXERCISES 4.3   *Answers to odd-numbered problems begin on page A-12.*

In Problems 1-36 find the general solution of the given differential equation.

1. $3y'' - y' = 0$ **2.** $2y'' + 5y' = 0$

3. $y'' - 16y = 0$ **4.** $y'' - 8y = 0$

5. $y'' + 9y = 0$ **6.** $4y'' + y = 0$

7. $y'' - 3y' + 2y = 0$ **8.** $y'' - y' - 6y = 0$

9. $\dfrac{d^2 y}{dx^2} + 8\dfrac{dy}{dx} + 16y = 0$ **10.** $\dfrac{d^2 y}{dx^2} - 10\dfrac{dy}{dx} + 25y = 0$

11. $y'' + 3y' - 5y = 0$ **12.** $y'' + 4y' - y = 0$

13. $12y'' - 5y' - 2y = 0$ **14.** $8y'' + 2y' - y = 0$

15. $y'' - 4y' + 5y = 0$ **★16.** $2y'' - 3y' + 4y = 0$

17. $3y'' + 2y' + y = 0$ **18.** $2y'' + 2y' + y = 0$

19. $y''' - 4y'' - 5y' = 0$ **20.** $4y''' + 4y'' + y' = 0$

21. $y''' - y = 0$ **22.** $y''' + 5y'' = 0$

23. $y''' - 5y'' + 3y' + 9y = 0$ **24.** $y''' + 3y'' - 4y' - 12y = 0$

25. $y''' + y'' - 2y = 0$ **★26.** $y''' - y'' - 4y = 0$

27. $y''' + 3y'' + 3y' + y = 0$ **28.** $y''' - 6y'' + 12y' - 8y = 0$

29. $\dfrac{d^4 y}{dx^4} + \dfrac{d^3 y}{dx^3} + \dfrac{d^2 y}{dx^2} = 0$ **30.** $\dfrac{d^4 y}{dx^4} - 2\dfrac{d^2 y}{dx^2} + y = 0$

31. $16\dfrac{d^4 y}{dx^4} + 24\dfrac{d^2 y}{dx^2} + 9y = 0$ **★32.** $\dfrac{d^4 y}{dx^4} - 7\dfrac{d^2 y}{dx^2} - 18y = 0$

33. $\dfrac{d^5 y}{dx^5} - 16\dfrac{dy}{dx} = 0$ **34.** $\dfrac{d^5 y}{dx^5} - 2\dfrac{d^4 y}{dx^4} + 17\dfrac{d^3 y}{dx^3} = 0$

35. $\dfrac{d^5 y}{dx^5} + 5\dfrac{d^4 y}{dx^4} - 2\dfrac{d^3 y}{dx^3} - 10\dfrac{d^2 y}{dx^2} + \dfrac{dy}{dx} + 5y = 0$

36. $2\dfrac{d^5 y}{dx^5} - 7\dfrac{d^4 y}{dx^4} + 12\dfrac{d^3 y}{dx^3} + 8\dfrac{d^2 y}{dx^2} = 0$

37. The roots of the auxiliary equation are
$$m_1 = 4 \qquad m_2 = m_3 = -5.$$
What is the corresponding differential equation?

★38. The roots of the auxiliary equation are

$$m_1 = -\frac{1}{2}, \qquad m_2 = 3 + i, \qquad m_3 = 3 - i.$$

What is the corresponding differential equation?

In Problems 39–54 solve the given differential equation subject to the indicated initial conditions.

39. $y'' + 16y = 0, \quad y(0) = 2, \quad y'(0) = -2$

40. $y'' - y = 0, \quad y(0) = y'(0) = 1$

41. $y'' + 6y' + 5y = 0, \quad y(0) = 0, \quad y'(0) = 3$

42. $y'' - 8y' + 17y = 0, \quad y(0) = 4, \quad y'(0) = -1$

43. $2y'' - 2y' + y = 0, \quad y(0) = -1, \quad y'(0) = 0$

44. $y'' - 2y' + y = 0, \quad y(0) = 5, \quad y'(0) = 10$

45. $y'' + y' + 2y = 0, \quad y(0) = y'(0) = 0$

★46. $4y'' - 4y' - 3y = 0, \quad y(0) = 1, \quad y'(0) = 5$

47. $y'' - 3y' + 2y = 0, \quad y(1) = 0, \quad y'(1) = 1$

48. $y'' + y = 0, \quad y(\pi/3) = 0, \quad y'(\pi/3) = 2$

49. $y''' + 12y'' + 36y' = 0, \quad y(0) = 0, \quad y'(0) = 1, \quad y''(0) = -7$

50. $y''' + 2y'' - 5y' - 6y = 0, \quad y(0) = y'(0) = 0, \quad y''(0) = 1$

51. $y''' - 8y = 0, \quad y(0) = 0, \quad y'(0) = -1, \quad y''(0) = 0$

52. $\dfrac{d^4y}{dx^4} = 0, \quad y(0) = 2, \quad y'(0) = 3, \quad y''(0) = 4, \quad y'''(0) = 5$

53. $\dfrac{d^4y}{dx^4} - 3\dfrac{d^3y}{dx^3} + 3\dfrac{d^2y}{dx^2} - \dfrac{dy}{dx} = 0, \quad y(0) = y'(0) = 0,$
$y''(0) = y'''(0) = 1$

54. $\dfrac{d^4y}{dx^4} - y = 0, \quad y(0) = y'(0) = y''(0) = 0, \quad y'''(0) = 1$

In Problems 55 and 56 solve the given differential equation subject to the indicated boundary-conditions.

55. $y'' - 10y' + 25y = 0, \quad y(0) = 1, \quad y(1) = 0$

56. $y'' + 4y = 0, \quad y(0) = 0, \quad y(\pi) = 0$

57. Use the fact that

$$i = \left(\frac{\sqrt{2}}{2} + \frac{\sqrt{2}}{2}i\right)^2 \qquad \text{and} \qquad -i = \left(\frac{\sqrt{2}}{2} - \frac{\sqrt{2}}{2}i\right)^2$$

to solve the differential equation

$$\frac{d^4y}{dx^4} + y = 0.$$

[*Hint:* Write the auxiliary equation $m^4 + 1 = 0$ as $(m^2 + 1)^2 - 2m^2 = 0$. See what happens when you factor.]

4.4 Undetermined Coefficients

4.4.1 Differential Operators

In calculus the symbol D^n is often used to denote the nth-order derivative of a function:

$$D^n y = \frac{d^n y}{dx^n}.$$

Hence a linear differential equation with constant coefficients

$$a_n y^{(n)} + a_{n-1} y^{(n-1)} + \cdots + a_2 y'' + a_1 y' + a_0 y = g(x)$$

can be written as

$$a_n D^n y + a_{n-1} D^{n-1} y + \cdots + a_2 D^2 y + a_1 D y + a_0 y = g(x)$$

or $\quad (a_n D^n + a_{n-1} D^{n-1} + \cdots + a_2 D^2 + a_1 D + a_0)y = g(x).$

The expression

$$a_n D^n + a_{n-1} D^{n-1} + \cdots + a_2 D^2 + a_1 D + a_0 \qquad (1)$$

is called an **linear nth-order differential operator**. Since (1) is a polynomial in the symbol D, it is often abbreviated by $P(D)$.

It can be shown that when the a_i, $i = 0, 1, \ldots, n$ are constants

(a) $P(D)$ can possibly be factored into differential operators of lower order.* This is accomplished by treating $P(D)$ as an ordinary polynomial.

(b) The factors of $P(D)$ commute.

EXAMPLES

(a) The operators $D^2 + D$ and $D^2 - 1$ can be factored as
$$D(D + 1) \quad \text{and} \quad (D + 1)(D - 1),$$
respectively.

(b) The operator $D^2 + 1$ does not factor using real numbers.

(c) The operator $D^2 + 5D + 6$ can be written as $(D + 2)(D + 3)$ or $(D + 3)(D + 2)$. The following example shows that these factors commute.

---

*If one is willing to use complex numbers, then a differential operator with constant coefficients can *always* be factored. We are primarily concerned with writing differential equations in operator form with real coefficients.

EXAMPLE   If $y = f(x)$ possesses a second derivative then
$$(D^2 + 5D + 6)y = (D + 2)(D + 3)y$$
$$= (D + 3)(D + 2)y.$$

To show this let $w = (D + 3)y = y' + 3y$

$$(D + 2)w = Dw + 2w$$

$$= \frac{d}{dx}[y' + 3y] + 2[y' + 3y]$$

$$= y'' + 3y' + 2y' + 6y$$

$$= y'' + 5y' + 6y.$$

Similarly, if we let $w = (D + 2)y = y' + 2y$ then

$$(D + 3)w = Dw + 3w$$

$$= \frac{d}{dx}[y' + 2y] + 3[y' + 2y]$$

$$= y'' + 2y' + 3y' + 6y$$

$$= y'' + 5y' + 6y.$$

EXAMPLE   The operator $D^2 + 4D + 4$ can be written as $(D + 2)(D + 2)$ or $(D + 2)^2$.

Annihilator operator   Let $y = f(x)$ be a function possessing at least n derivatives.
If

$$(a_n D^n + a_{n-1}D^{n-1} + \cdots + a_1 D + a_0)f(x) = 0$$

then the differential operator $a_n D^n + a_{n-1}D^{n-1} + \cdots + a_1 D + a_0$ is said to **annihilate** f. For example, if $f(x) = k$, a constant, then $Dk = 0$. Also, $D^2 x = 0$, $D^3 x^2 = 0$, and so on.

> The differential operator D^n annihilates
> each of the functions   (2)
> $$1, x, x^2, \ldots, x^{n-1}.$$

As an immediate consequence of (2), and the fact that differentiation can be done termwise, a polynomial

$$c_0 + c_1 x + \cdots + c_{n-1}x^{n-1}$$

can be annihilated by finding an operator that annihilates the highest power of x.

EXAMPLE

Find a differential operator that annihilates
$$1 - 5x^2 + 8x^3.$$

Solution: From (2) we know that $D^4 x^3 = 0$ and so it follows that
$$D^4(1 - 5x^2 + 8x^3) = 0.$$

The differential operator $(D - \alpha)^n$ annihilates each of the functions   (3)
$$e^{\alpha x}, xe^{\alpha x}, x^2 e^{\alpha x}, \ldots, x^{n-1} e^{\alpha x}.$$

To see this, note that the auxiliary equation of the homogeneous equation $(D - \alpha)^n y = 0$ is $(m - \alpha)^n = 0$. Since α is a root of multiplicity n the general solution is
$$y = c_1 e^{\alpha x} + c_2 x e^{\alpha x} + \cdots + c_n x^{n-1} e^{\alpha x}. \qquad (4)$$

EXAMPLES

Find an annihilator operator for (a) e^{5x}, (b) $4e^{2x} - 6xe^{2x}$.

Solution: **(a)**   From (3) with $\alpha = 5$ and $n = 1$ we see
$$(D - 5)e^{5x} = 0.$$

(b)   From (3) and (4) with $\alpha = 2$ and $n = 2$,
$$(D - 2)^2(4e^{2x} - 6xe^{2x}) = 0.$$

EXAMPLE

Find a differential operator that annihilates
$$e^{-3x} + xe^x.$$

Solution: From (3),
$$(D + 3)e^{-3x} = 0 \qquad \text{and} \qquad (D - 1)^2 xe^x = 0.$$

The product of the two operators $(D + 3)(D - 1)^2$ will annihilate the given linear combination. Since this may not be obvious a verification is in order:
$$(D + 3)(D - 1)^2(e^{-3x} + xe^x) = (D + 3)[(D - 1)^2 e^{-3x} + (D - 1)^2 xe^x]$$
$$= (D + 3)[16e^{-3x} + 0]$$
$$= 16(D + 3)e^{-3x}$$
$$= 0.$$

When α and β are real numbers, the quadratic formula reveals that $[m^2 - 2\alpha m + (\alpha^2 + \beta^2)]^n = 0$ has complex roots $\alpha + i\beta$, $\alpha - i\beta$, both of multiplicity n. From the discussion of Section 4.3 (page 151) we have the next result.

> The differential operator $[D^2 - 2\alpha D + (\alpha^2 + \beta^2)]^n$ annihilates each of the functions
>
> $$e^{\alpha x}\cos \beta x, \ xe^{\alpha x}\cos \beta x, \ x^2 e^{\alpha x}\cos \beta x, \ \ldots, \ x^{n-1}e^{\alpha x}\cos \beta x,$$
>
> $$e^{\alpha x}\sin \beta x, \ xe^{\alpha x}\sin \beta x, \ x^2 e^{\alpha x}\sin \beta x, \ \ldots, \ x^{n-1}e^{\alpha x}\sin \beta x.$$
>
> (5)

EXAMPLE

From (5), with $\alpha = -1$, $\beta = 2$, and $n = 1$ we see that

$$(D^2 + 2D + 5)e^{-x}\cos 2x = 0 \quad \text{and} \quad (D^2 + 2D + 5)e^{-x}\sin 2x = 0.$$

EXAMPLE

From (5), with $\alpha = 0$, $\beta = 1$, and $n = 2$ it is seen that the differential operator $(D^2 + 1)^2$ or $D^4 + 2D^2 + 1$ will annihilate $\cos x$, $x \cos x$, $\sin x$, and $x \sin x$. Moreover, $(D^2 + 1)^2$ will annihilate any linear combination of these functions.

When $\alpha = 0$ and $n = 1$ a special case of (5) is

$$(D^2 + \beta^2)\begin{cases} \cos \beta x \\ \sin \beta x \end{cases} = 0. \tag{6}$$

EXAMPLE

Find a differential operator that annihilates

$$7 - x + 6 \sin 3x.$$

Solution: From (2) and (6) we have respectively,

$$D^2(7 - x) = 0 \quad \text{and} \quad (D^2 + 9)\sin 3x = 0.$$

Hence the operator $D^2(D^2 + 9)$ will annihilate the given linear combination.

The concept of an annihilator operator will be of immediate importance in the discussion that follows.

4.4.2 Solving a Nonhomogeneous Linear Equation

To obtain the general solution of a nonhomogeneous differential equation with constant coefficients we must do two things: find the complementary function y_c, and then find *any* particular solution y_p of the nonhomogeneous equation. Recall from the discussion of Section 4.1 that a particular solution is any function, free of arbitrary constants, which satisfies the equation identically. The general solution of a nonhomogeneous equation is the sum of y_c and y_p.

Method of undetermined coefficients

If $P(D)$ denotes the differential operator (1), then a nonhomogeneous linear differential equation with constant coefficients can be written simply as

$$P(D)y = g(x). \tag{7}$$

When $g(x)$ consists of either

    (a)   a constant k,

    (b)   a polynomial in x,

    (c)   an exponential function $e^{\alpha x}$,

    (d) $\sin \beta x$, $\cos \beta x$

or finite sums and products of these functions it is always possible to find another differential operator $P_1(D)$ that annihilates $g(x)$. Applying $P_1(D)$ to (7) yields

$$P_1(D)P(D)y = P_1(D)g(x) = 0.$$

By solving the homogeneous equation $P_1(D)P(D)y = 0$ it is possible to discover the *form* of a particular solution y_p for the nonhomogeneous equation (7).

    The next several examples illustrate the so-called **method of undetermined coefficients** for finding y_p. The general solution of each equation is defined on the interval $-\infty < x < \infty$.

EXAMPLE

Solve

$$\frac{d^2y}{dx^2} + 3\frac{dy}{dx} + 2y = 4x^2. \tag{8}$$

Solution: In view of (2), (8) can be rendered homogeneous by taking three derivatives of each side of the equation. In other words,

$$D^3(D^2 + 3D + 2)y = 4D^3x^2 = 0 \tag{9}$$

since $D^3x^2 = 0$. The auxiliary equation of (9) is

$$m^3(m^2 + 3m + 2) \qquad \text{or} \qquad m^3(m + 1)(m + 2) = 0,$$

and so its general solution must be

$$y = \underbrace{c_1e^{-x} + c_2e^{-2x}}_{y_c} + \underbrace{c_3 + c_4x + c_5x^2}_{y_p}. \tag{10}$$

Formally we can then argue that every solution of equation (8) should also be a solution of equation (9). Now $y_c = c_1e^{-x} + c_2e^{-2x}$ is recognized as the complementary function of (8) and therefore y_p must have the basic structure indicated in (10):

$$y_p = A + Bx + Cx^2 \tag{11}$$

where we have replaced c_3, c_4, and c_5 by A, B, and C, respectively. For (11) to be a particular solution of (8) it is necessary to find *specific* coefficients A, B, and C. Differentiating (11) and substituting (8) gives

$$y_p'' + 3y_p' + 2y_p = 2A + 3B + 2C + (2B + 6C)x + 2Cx^2$$
$$= 4x^2.$$

By equating coefficients in the last identity we obtain the system of equations

$$2A + 3B + 2C = 0$$
$$2B + 6C = 0 \qquad (12)$$
$$2C = 4$$

Solving (12) gives $A = 7$, $B = -6$, and $C = 2$. Thus, $y_p = 7 - 6x + 2x^2$ and the general solution of (8) is $y = y_c + y_p$, or

$$y = c_1 e^{-x} + c_2 e^{-2x} + 7 - 6x + 2x^2.$$

EXAMPLE   Solve $\qquad\qquad y'' - 3y' = 8e^{3x} + 4 \sin x.$ (13)

Solution: Since $(D - 3)e^{3x} = 0$ and $(D^2 + 1)\sin x = 0$ we apply the operator $(D - 3)(D^2 + 1)$ to both sides of (13):

$$(D - 3)(D^2 + 1)(D^2 - 3D)y = 0. \qquad (14)$$

The auxiliary equation of (14) is

$$(m - 3)(m^2 + 1)(m^2 - 3m) = 0 \qquad \text{or} \qquad m(m - 3)^2(m^2 + 1) = 0.$$

Thus $\qquad y = \underbrace{c_1 + c_2 e^{3x}}_{y_c} + \underbrace{c_3 x e^{3x} + c_4 \cos x + c_5 \sin x}_{y_p}.$ (15)

As shown in (15) the complementary function of (13) is $y_c = c_1 + c_2 e^{3x}$ and so we conclude that a particular solution is given by

$$y_p = Axe^{3x} + B \cos x + C \sin x.$$

Substituting y_p in (13) and simplifying yields

$$y_p'' - 3y_p' = 3Ae^{3x} + (-B - 3C)\cos x + (3B - C)\sin x$$
$$= 8e^{3x} + 4 \sin x.$$

Equating coefficients gives

$$3A = 8$$
$$-B - 3C = 0$$
$$3B - C = 4.$$

We find $A = 8/3$, $B = 6/5$, $C = -2/5$ and

$$y_p = \frac{8}{3} x e^{3x} + \frac{6}{5} \cos x - \frac{2}{5} \sin x.$$

The general solution of (13) is then

$$y = c_1 + c_2 e^{3x} + \frac{8}{3} x e^{3x} + \frac{6}{5} \cos x - \frac{2}{5} \sin x.$$

EXAMPLE

Solve $$y'' + 8y = 5x + 2e^{-x}. \tag{16}$$

***Solution*:** From (2) and (3) we know $D^2x = 0$ and $(D + 1)e^{-x} = 0$, respectively. Hence we apply $D^2(D + 1)$ to (16):
$$D^2(D + 1)(D^2 + 8)y = 0.$$

It is seen that
$$y = \underbrace{c_1 \cos 2\sqrt{2}x + c_2 \sin 2\sqrt{2}x}_{y_c} + \underbrace{c_3 + c_4x + c_5e^{-x}}_{y_p}.$$

and so $$y_p = A + Bx + Ce^{-x}.$$

Subsitituting y_p into (16) yields
$$y_p'' + 8y_p = 8A + 8Bx + 9Ce^{-x}$$
$$= 5x + 2e^{-x}.$$

This implies $A = 0$, $B = 5/8$, and $C = 2/9$ so that the general solution of (16) is

$$y = c_1 \cos 2\sqrt{2}x + c_2 \sin 2\sqrt{2}x + \frac{5}{8}x + \frac{2}{9}e^{-x}.$$

EXAMPLE

Solve $$y'' + y = x \cos x - \cos x. \tag{17}$$

***Solution*:** It was seen in an earlier example that $x \cos x$ and $\cos x$ are annihilated by the operator $(D^2 + 1)^2$. Thus
$$(D^2 + 1)^2(D^2 + 1)y = 0 \quad \text{or} \quad (D^2 + 1)^3y = 0.$$

Since i and $-i$ are both complex roots of multiplicity 3 of the auxiliary equation of the last differential equation we conclude
$$y = \underbrace{c_1 \cos x + c_2 \sin x}_{y_c} + \underbrace{c_3x \cos x + c_4x \sin x + c_5x^2 \cos x + c_6x^2 \sin x}_{y_p}.$$

Substitute
$$y_p = Ax \cos x + Bx \sin x + Cx^2\cos x + Ex^2\sin x$$

into (17) and simplify:
$$y_p'' + y_p = 4Ex \cos x - 4Cx \sin x + (2B + 2C)\cos x + (-2A + 2E)\sin x$$
$$= x \cos x - \cos x.$$

Equating coefficients gives the equations
$$4E = 1$$
$$-4C = 0$$
$$2B + 2C = -1$$
$$-2A + 2E = 0.$$

from which we find $E = 1/4$, $C = 0$, $B = -1/2$, and $A = 1/4$. Hence the general solution of (17) is

$$y = c_1 \cos x + c_2 \sin x + \frac{1}{4}x \cos x - \frac{1}{2}x \sin x + \frac{1}{4}x^2\sin x.$$

EXAMPLE

Determine the form of a particular solution for
$$y'' - 2y' + y = 10e^{-2x}\cos x. \tag{18}$$

Solution: From (5) with $\alpha = -2$, $\beta = 1$, and $n = 1$ we know that
$$(D^2 + 4D + 5)e^{-2x}\cos x = 0.$$

Applying the operator $D^2 + 4D + 5$ to (18) gives
$$(D^2 + 4D + 5)(D^2 - 2D + 1)y = 0. \tag{19}$$

Since the roots of the auxiliary equation of (19) are $-2 - i, -2 + i, 1, 1$,
$$y = \underbrace{c_1e^x + c_2xe^x}_{y_c} + \underbrace{c_3e^{-2x}\cos x + c_4e^{-2x}\sin x}_{y_p}.$$

Hence a particular solution of (18) can be found having the form
$$y_p = Ae^{-2x}\cos x + Be^{-2x}\sin x.$$

EXAMPLE

Determine the form of a particular solution for
$$y''' - 4y'' + 4y' = 5x^2 - 6x + 4x^2e^{2x} + 3e^{5x}. \tag{20}$$

Solution: Observe that
$$D^3(5x^2 - 6x) = 0, \quad (D - 2)^3x^2e^{2x} = 0, \quad \text{and} \quad (D - 5)e^{5x} = 0.$$

Therefore $D^3(D - 2)^3(D - 5)$ applied to (20) gives
$$D^3(D - 2)^3(D - 5)(D^3 - 4D^2 + 4D)y = 0$$

or $$D^4(D - 2)^5(D - 5)y = 0.$$

Hence
$$y = c_1 + c_2x + c_3x^2 + c_4x^3 + c_5e^{2x} + c_6xe^{2x} + c_7x^2e^{2x} + c_8x^3e^{2x}$$
$$+ c_9x^4e^{2x} + c_{10}e^{5x}.$$

Since the linear combination $c_1 + c_5e^{2x} + c_6xe^{2x}$ can be taken as the complementary function of (20) we see that a particular solution of the equation has the form
$$y_p = Ax + Bx^2 + Cx^3 + Ex^2e^{2x} + Fx^3e^{2x} + Gx^4e^{2x} + He^{5x}.$$

EXERCISES 4.4   *Answers to odd-numbered problems begin on page A-12.*

[4.4.1]   In Problems 1–10, if possible, factor the given differential operator.

1. $4D^2 - 9$ 2. $D^2 - 5$

3. $D^2 - 4D - 12$ 4. $2D^2 - 3D - 2$

5. $D^3 + 10D^2 + 25D$ 6. $D^3 + 4D$

7. $D^3 + 2D^2 - 13D + 10$ 8. $D^3 + 4D^2 + 3D$

9. $D^4 + 8D$ ★10. $D^4 - 8D^2 + 16$

In Problems 11–20 find a differential operator that annihilates the given function.

11. $1 + 6x - 2x^2$ 12. $x^3(1 - 5x)$

13. $1 + 7e^{2x}$ 14. $x + 3xe^{6x}$

15. $13x + 9x^2 - \sin 4x$ 16. $8x - \sin x + 10 \cos 5x$

17. $e^{-x} + 2xe^x - x^2e^x$ 18. $(2 - e^x)^2$

19. $3 + e^x\cos 2x$ 20. $e^{-x}\sin x - e^{2x}\cos x$

[4.4.2]   In Problems 21–52 solve the given differential equation by undetermined coefficients.

21. $y'' - 9y = 54$ ★22. $2y'' - 7y' + 5y = -29$

23. $y'' + y' = 3$ 24. $y''' + 2y'' + y' = 10$

25. $y'' + 4y' + 4y = 2x + 6$ 26. $y'' + 3y' = 4x - 5$

27. $y''' + y'' = 8x^2$ 28. $y'' - 2y' + y = x^3 + 4x$

29. $y'' - y' - 12y = e^{4x}$ 30. $y'' + 2y' + 2y = 5e^{6x}$

31. $y'' - 2y' - 3y = 4e^x - 9$ 32. $y'' + 6y' + 8y = 3e^{-2x} + 2x$

33. $y'' + 25y = 6 \sin x$

★34. $y'' + 4y = 4 \cos x + 3 \sin x - 8$

35. $y'' + 6y' + 9y = -xe^{4x}$ 36. $y'' + 3y' - 10y = x(e^x + 1)$

37. $y'' - y = x^2e^x + 5$ 38. $y'' + 2y' + y = x^2e^{-x}$

39. $y'' - 2y' + 5y = e^x\sin x$

40. $y'' + y' + \dfrac{1}{4}y = e^x(\sin 3x - \cos 3x)$

41. $y^{(4)} - 2y''' + y'' = e^x + 1$ 42. $y^{(4)} - 4y'' = 5x^2 - e^{2x}$

43. $16y^{(4)} - y = e^{x/2}$

★44. $y^{(4)} - 5y'' + 4y = 2 \cosh x - 6$

45. $y''' - 3y'' + 3y' - y = e^x - x + 16$

46. $y''' - y'' + y' - y = xe^x - e^{-x} + 7$

47. $y'' + 25y = 20 \sin 5x$ **48.** $y'' + y = 4 \cos x - \sin x$

49. $y'' + y' + y = x \sin x$ **★50.** $y'' + 4y = \cos^2 x$

51. $y'' - y' = e^x(1 - e^{-x})^2$

52. $2y''' - 3y'' - 3y' + 2y = (e^x + e^{-x})^2$

In Problems 53–60 solve the given differential equation subject to the indicated initial conditions.

53. $y'' - 64y = 16;$ $y(0) = 1, y'(0) = 0$

54. $y'' + y' = x;$ $y(0) = 1, y'(0) = 0$

55. $y'' - 5y' = x - 2;$ $y(0) = 0, y'(0) = 2$

56. $y'' + 5y' - 6y = 10e^{2x};$ $y(0) = 1, y'(0) = 1$

57. $y'' + y = 8 \cos 2x - 4 \sin x;$ $y(\pi/2) = -1, y'(\pi/2) = 0$

58. $y''' - 2y'' + y' = xe^x + 5;$ $y(0) = 2, y'(0) = 2, y''(0) = -1$

59. $y'' - 4y' + 8y = x^3;$ $y(0) = 2, y'(0) = 4$

★60. $y^{(4)} - y''' = x + e^x;$ $y(0) = 0, y'(0) = 0, y''(0) = 0, y'''(0) = 0$

In Problems 61 and 62 determine the form of a particular solution for the given differential equation.

61. $y'' - y = e^x(2 + 3x \cos 2x)$ **62.** $y'' + y' = 9 - e^{-x} + x^2 \sin x$

Miscellaneous problems

63. Determine whether the operator $(xD - 1)(D + 4)$ is the same as the operator $(D + 4)(xD - 1)$.

64. Prove that the differential equation

$$a_n y^{(n)} + a_{n-1} y^{(n-1)} + \cdots + a_1 y' + a_0 y = k,$$

k a constant, $a_0 \neq 0$, has the particular solution $y_p = k/a_0$.

4.5   Variation of Parameters

The linear first-order equation revisited

In Chapter 2 we have seen that the general solution of the linear equation

$$\frac{dy}{dx} + P(x)y = f(x), \tag{1}$$

where $P(x)$ and $f(x)$ are continuous on an interval I, is

$$y = e^{-\int P(x)\,dx} \int e^{\int P(x)\,dx} f(x)\,dx + c_1 e^{-\int P(x)\,dx} \tag{2}$$

Now it is easily verified that the solution (2) has the form

$$y = y_c + y_p$$

where

$$y_c = c_1 e^{-\int P(x)\, dx}$$

is a solution of

$$\frac{dy}{dx} + P(x)y = 0 \tag{3}$$

and

$$y_p = e^{-\int P(x)\, dx} \int e^{\int P(x)\, dx} f(x)\, dx$$

is a particular solution of equation (1). As a means of *motivating* an additional method for solving nonhomogeneous second-order linear equations we shall rederive equation (2) by a method known as **variation of parameters**. The basic procedure is similar to that employed in Section 4.1. For second-order equations we shall see that variation of parameters has the distinct advantage that it will always yield a particular solution y_p, provided the related homogeneous equation can be solved.

Suppose y_1 is a known solution of equation (3), that is,

$$\frac{dy_1}{dx} + P(x)y_1 = 0.$$

We have already proved in Section 2.4 that $y_1 = e^{-\int P(x)\, dx}$ is a solution, and since the differential equation is linear its general solution is

$$y = c_1 y_1(x). \tag{4}$$

Variation of parameters consists of finding a function u_1 such that $y = u_1(x)y_1(x)$ is a particular solution of (1). In other words, we replace the parameter c_1 in (4) by a variable u_1.

Substituting $y = u_1 y_1$ into equation (1) gives

$$\frac{d}{dx}[u_1 y_1] + P(x)u_1 y_1 = f(x)$$

$$u_1 \frac{dy_1}{dx} + y_1 \frac{du_1}{dx} + P(x)u_1 y_1 = f(x)$$

$$u_1 \underbrace{\left[\frac{dy_1}{dx} + P(x)y_1\right]}_{\text{zero}} + y_1 \frac{du_1}{dx} = f(x)$$

so that

$$y_1 \frac{du_1}{dx} = f(x).$$

By separating variables we find

$$du_1 = \frac{f(x)}{y_1(x)}\, dx \qquad \text{and} \qquad u_1 = \int \frac{f(x)}{y_1(x)}\, dx + c_1$$

from which it follows that

$$y = u_1 y_1 = y_1 \int \frac{f(x)}{y_1(x)}\, dx + c_1 y_1. \tag{5}$$

From the definition of y_1 we see that (5) is identical to (2).

Second-order equations

Throughout this discussion we shall put the linear second-order equation

$$a_2(x)y'' + a_1(x)y' + a_0(x)y = g(x)$$

into the form $\qquad y'' + P(x)y' + Q(x)y = f(x) \qquad\qquad$ (6)

by dividing through by $a_2(x)$. Here we shall assume that $P(x), Q(x)$, and $f(x)$ are continuous on some interval I. Equation (6) is the analogue of (1). As we know, when $P(x)$ and $Q(x)$ are constants there is absolutely no difficulty in writing down y_c. Although at this time we shall confine our attention to second-order equations with constant coefficients, it should be noted that the method of variation of parameters can be generalized to higher order equations and is also applicable to differential equations with variable coefficients.

Suppose y_1 and y_2 form a fundamental set of solutions on I of the associated homogeneous form of (6). That is,

$$y_1'' + P(x)y_1' + Q(x)y_1 = 0 \qquad \text{and} \qquad y_2'' + P(x)y_2' + Q(x)y_2 = 0.$$

Now we ask: Can two functions u_1 and u_2 be found so that

$$y_p = u_1(x)y_1(x) + u_2(x)y_2(x)$$

is a particular solution of (1)? Notice our assumption for y_p is the same as $y_c = c_1 y_1 + c_2 y_2$, but we have replaced c_1 and c_2 by the "variable parameters" u_1 and u_2. Using the product rule to differentiate y_p gives

$$y_p' = u_1 y_1' + y_1 u_1' + u_2 y_2' + y_2 u_2'. \qquad\qquad (7)$$

If we make the further demand that u_1 and u_2 be functions for which

$$y_1 u_1' + y_2 u_2' = 0 \qquad\qquad (8)$$

then (7) becomes

$$y_p' = u_1 y_1' + u_2 y_2'.$$

Continuing, we find

$$y_p'' = u_1 y_1'' + y_1' u_1' + u_2 y_2'' + y_2' u_2'$$

and hence

$$y_p'' + Py_p' + Qy_p = u_1 y_1'' + y_1' u_1' + u_2 y_2'' + y_2' u_2' + Pu_1 y_1' + Pu_2 y_2' + Qu_1 y_1 + Qu_2 y_2$$

$$= u_1 \underbrace{[y_1'' + Py_1' + Qy_1]}_{\text{zero}} + u_2 \underbrace{[y_2'' + Py_2' + Qy_2]}_{\text{zero}} + y_1' u_1' + y_2' u_2'.$$

$$= f(x).$$

In other words, u_1 and u_2 must be functions which also satisfy the condition

$$y_1' u_1' + y_2' u_2' = f(x). \qquad\qquad (9)$$

Equations (8) and (9) constitute a linear system of equations for determining the derivatives u_1' and u_2'. That is, we must solve

$$y_1 u_1' + y_2 u_2' = 0$$

$$y_1' u_1' + y_2' u_2' = f(x).$$

By Cramer's rule we obtain

$$u_1' = \frac{\begin{vmatrix} 0 & y_2 \\ f(x) & y_2' \end{vmatrix}}{\begin{vmatrix} y_1 & y_2 \\ y_1' & y_2' \end{vmatrix}} = -\frac{y_2 f(x)}{W}$$

(10)

and

$$u_2' = \frac{\begin{vmatrix} y_1 & 0 \\ y_1' & f(x) \end{vmatrix}}{\begin{vmatrix} y_1 & y_2 \\ y_1' & y_2' \end{vmatrix}} = \frac{y_1 f(x)}{W}$$

where we recognize the determinant in the denominator as the Wronskian W of y_1 and y_2. By linear independence of y_1 and y_2 on I we know that $W(y_1(x), y_2(x)) \neq 0$ for every x in the interval.

A summary of the method

Usually it is not a good idea to memorize formulas in lieu of understanding a procedure. However, the foregoing procedure is much too long and complicated to use each time we wish to solve a differential equation. In this case it is more efficient to simply use the formulas in (10). Thus to solve $ay'' + by' + cy = g(x)$ where a, b and c are constants, first find the complementary function $y_c = c_1 y_1 + c_2 y_2$, and then compute the Wronskian

$$W = \begin{vmatrix} y_1 & y_2 \\ y_1' & y_2' \end{vmatrix}.$$

By dividing by a we put the equation into the form $y'' + Py' + Qy = f(x)$ to determine $f(x)$. We find u_1 and u_2 by integrating

$$u_1' = -\frac{y_2 f(x)}{W} \quad \text{and} \quad u_2' = \frac{y_1 f(x)}{W}.$$

Finally, form the particular solution

$$y_p = u_1 y_1 + u_2 y_2.$$

EXAMPLE   Solve

$$y'' - 4y' + 4y = (x + 1)e^{2x}.$$

Solution: Since the auxiliary equation is $m^2 - 4m + 4 = (m - 2)^2 = 0$ we have

$$y_c = c_1 e^{2x} + c_2 x e^{2x}.$$

Identifying $y_1 = e^{2x}$ and $y_2 = xe^{2x}$ we next compute the Wronskian:

$$W(e^{2x}, xe^{2x}) = \begin{vmatrix} e^{2x} & xe^{2x} \\ 2e^{2x} & 2xe^{2x} + e^{2x} \end{vmatrix} = e^{4x}.$$

Because the coefficient of y'' is already unity we can then write

$$u_1' = -\frac{xe^{2x}(x + 1)e^{2x}}{e^{4x}}$$

$$= -x^2 - x$$

and

$$u_2' = \frac{e^{2x}(x + 1)e^{2x}}{e^{4x}}$$

$$= x + 1.$$

It follows that

$$u_1 = -\frac{x^3}{3} - \frac{x^2}{2}$$

$$u_2 = \frac{x^2}{2} + x$$

and therefore

$$y_p = \left(-\frac{x^3}{3} - \frac{x^2}{2}\right)e^{2x} + \left(\frac{x^2}{2} + x\right)xe^{2x}$$

$$= \left(\frac{x^3}{6} + \frac{x^2}{2}\right)e^{2x}.$$

Thus we have

$$y = y_c + y_p$$

$$= c_1 e^{2x} + c_2 xe^{2x} + \left(\frac{x^3}{6} + \frac{x^2}{2}\right)e^{2x}.$$

EXAMPLE   Solve $\qquad\qquad 4y'' + 36y = \csc 3x.$

Solution: We first write the equation in the form

$$y'' + 9y = \frac{1}{4}\csc 3x.$$

Since the roots of the auxiliary equation $m^2 + 9 = 0$ are $m_1 = 3i$ and $m_2 = -3i$ we have

$$y_c = c_1 \cos 3x + c_2 \sin 3x$$

and $\qquad W(\cos 3x, \sin 3x) = \begin{vmatrix} \cos 3x & \sin 3x \\ -3\sin 3x & 3\cos 3x \end{vmatrix} = 3.$

Therefore $\qquad u_1' = -\dfrac{(\sin 3x)(\frac{1}{4}\csc 3x)}{3}$

$$= -\frac{1}{12} \qquad \left[\csc 3x = \frac{1}{\sin 3x}\right]$$

$$u_2' = \frac{(\cos 3x)(\frac{1}{4}\csc 3x)}{3}$$

$$= \frac{1}{12}\frac{\cos 3x}{\sin 3x}$$

from which we obtain

$$u_1 = -\frac{1}{12}x \qquad u_2 = \frac{1}{36}\ln|\sin 3x|.$$

Hence

$$y_p = -\frac{1}{12}x\cos 3x + \frac{1}{36}(\sin 3x)\ln|\sin 3x|$$

and

$$y = y_c + y_p$$

$$= c_1\cos 3x + c_2\sin 3x - \frac{1}{12}x\cos 3x + \frac{1}{36}(\sin 3x)\ln|\sin 3x|. \qquad (11)$$

Equation (11) represents the general solution of the differential equation on, say, $0 < x < \pi/6$.

Constants of integration

When computing the indefinite integrals of u_1' and u_2' we need not introduce any constants. This is because

$$y = y_c + y_p$$

$$= c_1y_1 + c_2y_2 + (u_1 + a_1)y_1 + (u_2 + b_1)y_2$$

$$= (c_1 + a_1)y_1 + (c_2 + b_1)y_2 + u_1y_1 + u_2y_2$$

$$= C_1y_1 + C_2y_2 + u_1y_1 + u_2y_2.$$

EXAMPLE

Solve

$$y'' - y = \frac{1}{x}.$$

Solution:

$$m^2 - 1 = 0$$

$$m_1 = 1, \qquad m_2 = -1$$

$$y_c = c_1e^x + c_2e^{-x}$$

$$W(e^x, e^{-x}) = \begin{vmatrix} e^x & e^{-x} \\ e^x & -e^{-x} \end{vmatrix} = -2$$

$$u_1' = -\frac{e^{-x}(1/x)}{-2}, \qquad u_1 = \frac{1}{2}\int_{x_0}^x \frac{e^{-t}}{t}\,dt$$

$$u_2' = \frac{e^x(1/x)}{-2}, \qquad u_2 = -\frac{1}{2}\int_{x_0}^x \frac{e^t}{t}\,dt.$$

It is well known that the integrals defining u_1 and u_2 cannot be expressed in terms of elementary functions. Hence we write

$$y_p = \frac{1}{2} e^x \int_{x_0}^x \frac{e^{-t}}{t}\, dt - \frac{1}{2} e^{-x} \int_{x_0}^x \frac{e^t}{t}\, dt$$

and therefore

$$y = y_c + y_p$$

$$= c_1 e^x + c_2 e^{-x} + \frac{1}{2} e^x \int_{x_0}^x \frac{e^{-t}}{t}\, dt - \frac{1}{2} e^{-x} \int_{x_0}^x \frac{e^t}{t}\, dt .$$

In the above example, we can integrate on any interval $x_0 \le t \le x$ not containing the origin.

EXERCISES 4.5   *Answers to odd-numbered problems begin on page A-13.*

In Problems 1–20 solve each differential equation by variation of parameters. State an interval on which the general solution is defined.

1. $y'' + y = \sec x$ **2.** $y'' + y = \tan x$

3. $y'' + y = \sin x$ ★ **4.** $y'' + y = \sec x \tan x$

5. $y'' + y = \cos^2 x$ **6.** $y'' + y = \sec^2 x$

7. $y'' - y = \cosh x$ **8.** $y'' - y = \sinh 2x$

9. $y'' - 4y = e^{2x}/x$ **10.** $y'' - 9y = 9x/e^{3x}$

11. $y'' + 3y' + 2y = 1/(1 + e^x)$ **12.** $y'' - 3y' + 2y = e^{3x}/(1 + e^x)$

13. $y'' + 3y' + 2y = \sin e^x$ ★ **14.** $y'' - 2y' + y = e^x \arctan x$

15. $y'' - 2y' + y = e^x/(1 + x^2)$ **16.** $y'' - 2y' + 2y = e^x \sec x$

17. $y'' + 2y' + y = e^{-x}\ln x$ **18.** $y'' + 10y' + 25y = e^{-10x}/x^2$

19. $4y'' - 4y' + y = 8e^{-x} + x$

20. $4y'' - 4y' + y = e^{x/2}\sqrt{1 - x^2}$

In Problems 21–24 solve each differential equation by variations of parameters subject to the initial conditions $y(0) = 1$, $y'(0) = 0$.

21. $y'' - y = xe^x$ **22.** $2y'' + y' - y = x + 1$

23. $y'' + 2y' - 8y = 2e^{-2x} - e^{-x}$

24. $y'' - 4y' + 4y = (12x^2 - 6x)e^{2x}$

25. Given that $y_1 = x$ and $y_2 = x \ln x$ form a fundamental set of solutions of

$$x^2 y'' - xy' + y = 0$$

on $0 < x < \infty$. Find the general solution of
$$x^2y'' - xy' + y = 4x \ln x.$$

★26. Given that $y_1 = x^2$ and $y_2 = x^3$ form a fundamental set of solutions of
$$x^2y'' - 4xy' + 6y = 0$$
on $0 < x < \infty$. Find the general solution of
$$x^2y'' - 4xy' + 6y = 1/x.$$

27. Given that $y_1 = x^{-1/2}\cos x$ and $y_2 = x^{-1/2}\sin x$ form a fundamental set of solutions of
$$x^2y'' + xy' + \left(x^2 - \frac{1}{4}\right)y = 0$$
on $0 < x < \infty$. Find the general solution of
$$x^2y'' + xy' + \left(x^2 - \frac{1}{4}\right)y = x^{3/2}.$$

28. Given that $y_1 = \cos (\ln x)$ and $y_2 = \sin (\ln x)$ are known linearly independent solutions of
$$x^2y'' + xy' + y = 0$$
on $0 < x < \infty$. **(a)** Find a particular solution of
$$x^2y'' + xy' + y = \sec (\ln x).$$

(b) Give the general solution of the equation and state an interval of validity. [*Hint:* It is *not* $0 < x < \infty$. Why?]

Miscellaneous problems

29. Solve the third-order equation
$$y''' - y' = xe^x$$
[*Hint*: $(d/dx)(y'' - y) = y''' - y'$, integrate and then use variation of parameters.]

30. Solve $y''' + 4y' = \sin x \cos x$ by the method outlined in Problem 29.

31. The complementary function for
$$y''' - 2y'' - y' + 2y = e^{3x}$$
is
$$y_c = c_1e^{-x} + c_2e^x + c_3e^{2x}.$$
Find variable parameters u_1, u_2, and u_3 such that
$$y_p = u_1(x)e^{-x} + u_2(x)e^x + u_3(x)e^{2x}$$
is a particular solution of the differential equation.

CHAPTER 4 SUMMARY

We summarize the salient information for **linear** second-order differential equations.

The equation
$$a_2(x)y'' + a_1(x)y' + a_0(x)y = 0 \qquad (1)$$
is said to be **homogeneous**, whereas

**CHAPTER 4
SUMMARY**

$$a_2(x)y'' + a_1(x)y' + a_0(x)y = g(x), \qquad (2)$$

$g(x)$ not identically zero, is **nonhomogeneous**. In the consideration of the linear equations (1) and (2) we assume that $a_2(x)$, $a_1(x)$, $a_0(x)$ and $g(x)$ are continuous on an interval I and that $a_2(x) \neq 0$ for every x in the interval. Under these assumptions there exists a unique solution of (2) satisfying the **initial conditions** $y(x_0) = y_0$, $y'(x_0) = y_0'$, where x_0 is a point in I.

The **Wronskian** of two differentiable functions $f_1(x)$ and $f_2(x)$ is the determinant

$$W(f_1(x), f_2(x)) = \begin{vmatrix} f_1(x) & f_2(x) \\ f_1'(x) & f_2'(x) \end{vmatrix}.$$

When $W \neq 0$ for at least one point in an interval the functions are **linearly independent** on the interval. If the functions are **linearly dependent** on the interval then $W \equiv 0$.

In solving the homogeneous equation (1) we want linearly independent solutions. A necessary and sufficient condition that two solutions y_1 and y_2 are linearly independent on I is that $W(y_1, y_2) \neq 0$ for every x in I. We say y_1 and y_2 form a **fundamental set** on I when they are linearly independent solutions of (1) on the interval. For *any* two solutions y_1 and y_2 the **superposition principle** states that the linear combination $c_1y_1 + c_2y_2$ is also a solution of (1). When y_1 and y_2 form a fundamental set, the function $y = c_1y_1 + c_2y_2$ is called the **general solution** of (1). The **general solution** of (2) is $y = y_c + y_p$ where y_c is the **complementary function**, or general solution of (1), and y_p is any **particular solution** of (2).

To solve (1) in the case $ay'' + by' + cy = 0$, a, b, and c constants, we first solve the **auxiliary equation** $am^2 + bm + c = 0$. There are three forms of the general solution depending on the three possible ways in which the roots of the auxiliary equation can occur.

Roots	General Solution
1. m_1 and m_2: real and distinct	$y = c_1e^{m_1x} + c_2e^{m_2x}$
2. m_1 and m_2: real but $m_1 = m_2$	$y = c_1e^{m_1x} + c_2xe^{m_1x}$
3. m_1 and m_2: complex $m_1 = \alpha + i\beta$ $m_2 = \alpha - i\beta$	$y = e^{\alpha x}[c_1 \cos \beta x + c_2 \sin \beta x]$

To solve the nonhomogeneous equation $ay'' + by' + cy = g(x)$ we use either the **method of undetermined coefficients** or **variation of parameters** to find a particular solution y_p. The former procedure is limited to the cases where $g(x)$ is a constant, a polynomial, an exponential function $e^{\alpha x}$, the trigonometric functions $\cos \beta x$ or $\sin \beta x$, or finite sums and products of these functions.

**CHAPTER 4
TEST**

Answers to odd-numbered problems begin on page A-13.

Answer Problems 1–10 without referring back to the text. Fill in the blank or answer true/false. In some cases there can be more than one correct answer.

1. The only solution to $y'' + x^2 y = 0$, $y(0) = 0$, $y'(0) = 0$ is _____.

2. If two differentiable functions $f_1(x)$ and $f_2(x)$ are linearly independent on an interval then $W(f_1(x), f_2(x)) \neq 0$ for at least one point in the interval. _____

3. Two functions $f_1(x)$ and $f_2(x)$ are linearly independent on an interval if one is not a constant multiple of the other. _____

4. The functions $f_1(x) = x^2$, $f_2(x) = 1 - x^2$, and $f_3(x) = 5$ are linearly _____ on $-\infty < x < \infty$.

5. The functions $f_1(x) = x^2$ and $f_2(x) = x\,|x|$ are linearly independent on the interval _____ whereas they are linearly dependent on the interval _____.

6. Two solutions y_1 and y_2 of $y'' + y' + y = 0$ are linearly dependent if $W(y_1, y_2) = 0$ for every real value of x. _____

7. A constant multiple of a solution of a differential equation is also a solution. _____

8. A fundamental set of two solutions of $(x - 2)y'' + y = 0$ exists on any interval not containing the point _____.

9. Using the method of undetermined coefficients the assumed form of the particular solution y_p for $y'' - y = 1 + e^x$ is _____.

10. A differential operator that annihilates $e^{2x}(x + \sin x)$ is _____.

In Problems 11 and 12 find a second solution for the differential equation given that $y_1(x)$ is a known solution.

11. $y'' + 4y = 0$; $y_1 = \cos 2x$

12. $xy'' - 2(x + 1)y' + (x + 2)y = 0$; $y_1 = e^x$

In Problems 13–18 find the general solution of each differential equation.

13. $y'' - 2y' - 2y = 0$

14. $2y'' + 2y' + 3y = 0$

15. $y''' + 10y'' + 25y' = 0$

16. $2\dfrac{d^4 y}{dx^4} + 3\dfrac{d^3 y}{dx^3} + 2\dfrac{d^2 y}{dx^2} + 6\dfrac{dy}{dx} - 4y = 0$

CHAPTER 4
TEST

17. $3y''' + 10y'' + 15y' + 4y = 0$

18. $2y''' + 9y'' + 12y' + 5y = 0$

In Problems 19–21 solve each differential equation by the method of undetermined coefficients.

19. $y'' - 3y' + 5y = 4x^3 - 2x$

20. $y'' - 2y' + y = x^2 e^x$

21. $y''' - 5y'' + 6y' = 2 \sin x + 8$

22. Solve $y''' - y'' = 6$ subject to $y(0) = 0$, $y'(0) = 1$, $y''(0) = 4$.

In Problems 23 and 24 solve each differential equation by the method of variation of parameters.

23. $y'' - 2y' + 2y = e^x \tan x$

24. $y'' - y = 2e^x/(e^x + e^{-x})$

25. Solve $y'' + y = \sec^3 x$ subject to $y(0) = 1$, $y'(0) = 1/2$.

Chapter 4 Appendix   A Uniqueness Proof

Consider the initial-value problem

$$a_2 y'' + a_1 y' + a_0 y = g(x) \tag{1}$$

$$y(0) = y_0$$
$$y'(0) = y_0' \tag{2}$$

where a_2, a_1, and a_0 are positive constants and $g(x)$ is continuous for all x. To show that this problem has a unique solution, we first prove that $y \equiv 0$ is the only solution of the problem

$$a_2 y'' + a_1 y' + a_0 y = 0 \tag{3}$$

$$y(0) = 0$$
$$y'(0) = 0. \tag{4}$$

Multiplying (3) by y' and integrating over the interval $0 \le x \le t$ gives

$$a_2 \int_0^t y'' y' \, dx + a_1 \int_0^t (y')^2 \, dx + a_0 \int_0^t y y' \, dx = \int_0^t 0 \, dx$$

$$\frac{a_2}{2} [y'(t)^2 - y'(0)^2] + a_1 \int_0^t (y')^2 \, dx + \frac{a_0}{2} [y(t)^2 - y(0)^2] = 0.$$

In view of the imposed initial conditions (4) we find

$$\frac{a_2}{2} (y')^2 + a_1 \int_0^t (y')^2 \, dx + \frac{a_0}{2} y^2 = 0$$

for every t. It follows that the sum of three nonnegative quantities can be zero only when $y \equiv 0$.

To return to the original problem, let us now suppose that y_1 and y_2 are two *different* solutions of equation (1) and that both solutions satisfy the initial conditions (2). If we define the function

$$u = y_1 - y_2$$

then
$$u(0) = y_1(0) - y_2(0) = y_0 - y_0 = 0$$

$$u'(0) = y_1'(0) - y_2'(0) = y_0' - y_0' = 0$$

and

$$a_2 u'' + a_1 u' + a_0 u = a_2[y_1'' - y_2''] + a_1[y_1' - y_2'] + a_0[y_1 - y_2]$$

$$= (a_2 y_1'' + a_1 y_1' + a_0 y_1) - (a_2 y_2'' + a_1 y_2' + a_0 y_2)$$

$$= g(x) - g(x)$$

$$= 0.$$

Hence u satisfies the initial-value problem

$$a_2 u'' + a_1 u' + a_0 u = 0$$

$$u(0) = 0$$

$$u'(0) = 0.$$

Since the only solution of this last problem is $u \equiv 0$, we have $y_1 - y_2 \equiv 0$ or $y_1 \equiv y_2$.

EDITION A FIRST COURSE IN DIFFERENTIAL EQUATIONS SECOND EDITION A FIRST CC
IN DIFFERENTIAL EQUATIONS SECOND EDITION A FIRST COURSE IN DIFFERENTIAL EC
ONS SECOND EDITION A FIRST COURSE IN DIFFERE **CHAPTER 5** ONS EDITIO
RST COURSE IN DIFFERENTIAL EQUATIONS SECON... ...RSE IN DIF
ITIAL EQUATIONS SECOND EDITION A FIRST COURSE IN DIFFERENTIAL EQUATIONS SE
D EDITION A FIRST COURSE IN DIFFERENTIAL EQUATIONS SECOND EDITION A FIRST CC
IN DIFFERENTIAL EQUATIONS SECOND EDITION A FIRST COURSE IN DIFFERENTIAL EC
ONS SEC
RST COU
ITIAL EQ
D EDITION
IN DIFF
ONS SECOND
RST COURSE IN DIFFERENTIAL EQUATIONS SECOND EDITION A FIRST COURSE IN DIF
ITIAL EQUATIONS SECOND EDITION A FIRST COURSE IN DIFFERENTIAL EQUATIONS SE

Applications of Second-Order Differential Equations: Vibrational Models

5.1 Simple Harmonic Motion

Hooke's law

Suppose, as in Figure 5.1(b), a mass m_1 is attached to a flexible spring suspended from a rigid support (such as a ceiling). When m_1 is replaced with a different mass m_2, the amount of stretch, or elongation of the spring, will of course be different.

By Hooke's law, the spring itself exerts a restoring force F opposite to the direction of elongation and proportional to the amount of elongation s. Simply stated, $F = ks$ where k is a constant of proportionality. Although masses with different weights stretch a spring by different amounts, the spring is essentially characterized by the number k. For example, if a mass weighing 10 lb stretches a spring by $1/2$ ft then

$$10 = k(\tfrac{1}{2})$$

$$k = 20 \text{ lb/ft.} \tag{1}$$

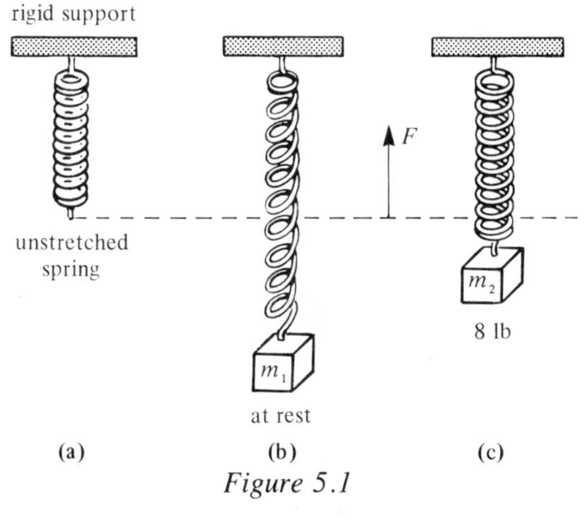

rigid support

unstretched spring

F

m_2

8 lb

m_1

at rest

(a)   (b)   (c)

Figure 5.1

Necessarily then a mass weighing 8 lb stretches the same spring 2/5 ft.

Newton's second law   After a mass m is attached to a spring it will stretch the spring by an amount s and attain a position of equilibrium at which its weight W is balanced by the restoring force ks. Recall from Section 1.3 that weight is defined by

$$W = mg \tag{2}$$

where the mass is measured in slugs, kilograms, or grams and $g = 32$ ft/sec^2, 9.8 m/sec^2, or 980 cm/sec^2, respectively. As indicated in Figure 5.2(b), the

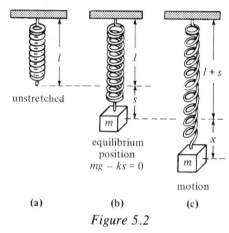

unstretched

equilibrium
position
$mg - ks = 0$

motion

(a)   (b)   (c)

Figure 5.2

condition of equilibrium is $mg = ks$ or $mg - ks = 0$. If the mass is now displaced by an amount x from its equilibrium position and released the net force F in this dynamic case is given by **Newton's second law of motion** $F = ma$ where a is the acceleration d^2x/dt^2. Assuming that there are no retarding forces acting on the system and assuming that the mass vibrates free of other external influencing forces (**free motion**), we can then equate F to the resultant force of the weight and the restoring force:

$$m\frac{d^2x}{dt^2} = -k(s + x) + mg$$

$$= -kx + \underbrace{mg - ks}_{\text{zero}}$$

$$= -kx. \tag{3}$$

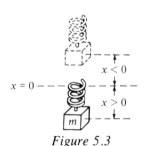

Figure 5.3

The negative sign in (3) indicates that the restoring force of the spring acts opposite to the direction of motion. Furthermore, we shall adopt the convention that displacements measured below the equilibrium position are positive. See Figure 5.3.

Differential equation of free undamped motion   By dividing (3) by the mass m we obtain the second-order differential equation

$$\frac{d^2x}{dt^2} + \frac{k}{m}x = 0 \tag{4}$$

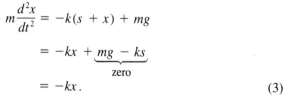

or
$$\frac{d^2x}{dt^2} + \omega^2 x = 0 \tag{5}$$

where $\omega^2 = k/m$. Equation (5) is said to describe **simple harmonic motion**, or **free undamped motion**. Associated with equation (5) there are two obvious initial conditions:

$$x(0) = \alpha \qquad \frac{dx}{dt}\bigg|_{t=0} = \beta \tag{6}$$

representing the amount of initial displacement and the initial velocity, respectively. For example, if $\alpha > 0$, $\beta < 0$, the mass would start from a point *below* the equilibrium position with an imparted *upward* velocity. If $\alpha < 0$, $\beta = 0$, the mass would be released from *rest* from a point $|\alpha|$ units *above* the equilibrium position, and so on.

The solution and the equation of motion

To solve equation (5) we note that the solutions of the auxiliary equation $m^2 + \omega^2 = 0$ are the complex numbers

$$m_1 = \omega i \qquad m_2 = -\omega i .$$

Thus, from equation (8) of Section 4.3 we find the general solution of (5) to be

$$x(t) = c_1 \cos \omega t + c_2 \sin \omega t . \tag{7}$$

The **period** of free vibrations described by (7) is $T = 2\pi/\omega$ and the **frequency** is $f = 1/T = \omega/2\pi$.* For example, for $x(t) = 2 \cos 3t - 4 \sin 3t$, the period is $2\pi/3$ and the frequency is $3/2\pi$. The former number means that the graph of $x(t)$ repeats every $2\pi/3$ units; the latter number means that there are 3 cycles of the graph every 2π units, or equivalently, the mass undergoes $3/2\pi$ complete vibrations per unit time. In addition, it can be shown that the period $2\pi/\omega$ is the time interval between two successive maxima of $x(t)$.[†] Finally, when the initial conditions (6) are used to determine the constants c_1 and c_2 in (7) we say that the resulting particular solution is the **equation of motion**.

EXAMPLE

Solve and interpret the initial-value problem

$$\frac{d^2x}{dt^2} + 16x = 0$$

$$x(0) = 10 \qquad \frac{dx}{dt}\bigg|_{t=o} = 0.$$

---

*Sometimes the number ω is called the *circular frequency* of vibrations. For free undamped motion the numbers $2\pi/\omega$ and $\omega/2\pi$ are also referred to as the *natural period* and *natural frequency*, respectively.

[†]Keep in mind that a maximum of $x(t)$ is a positive displacement corresponding to the mass attaining a maximum distance *below* the equilibrium position whereas a minimum of $x(t)$ is a negative displacement corresponding to the mass attaining a maximum height *above* the equilibrium position. We shall refer to either case as an *extreme displacement* of the mass.

Solution: The problem is equivalent to pulling a mass on a spring down 10 units below the equilibrium position, holding it until $t = 0$, and then releasing it from rest. Applying the initial conditions to the solution

$$x(t) = c_1 \cos 4t + c_2 \sin 4t$$

gives

$$x(0) = 10 = c_1 \cdot 1 + c_2 \cdot 0$$

so that $c_1 = 10$, and hence

$$x(t) = 10 \cos 4t + c_2 \sin 4t$$

$$\frac{dx}{dt} = -40 \sin 4t + 4c_2 \cos 4t$$

$$\left.\frac{dx}{dt}\right|_{t=0} = 0 = 4c_2 \cdot 1.$$

The latter equation implies $c_2 = 0$ so therefore the equation of motion is $x(t) = 10 \cos 4t$.

The solution clearly shows that once the system is set in motion, it stays in motion with the mass bouncing back and forth 10 units on either side of the equilibrium position $x = 0$. As shown in Figure 5.4(b), the period of oscillation is $2\pi/4 = \pi/2$ sec.

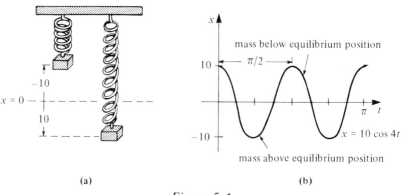

(a)

(b)

Figure 5.4

EXAMPLE

A mass weighing 2 lb stretches a spring 6 inches. At $t = 0$ the mass is released from a point 8 inches below the equilibrium position with an upward velocity of 4/3 ft/sec. Determine the function $x(t)$ which describes the subsequent free motion.

Solution: Since we are using the engineering system of units, the measurements given in terms of inches must be converted to feet: 6 inches = 6/12 = 1/2 foot; 8 inches = 8/12 = 2/3 foot. In addition we must convert the units of weight given in pounds to units of mass. From (2), $m = W/g$ and so

$$m = \frac{2}{32} = \frac{1}{16} \text{ slug.}$$

Also from Hooke's law

$$2 = k \cdot \frac{1}{2}$$

we find $k = 4$ lb/ft.

Hence the analogue of equation (3) is

$$\frac{1}{16} \frac{d^2x}{dt^2} = -4x$$

or $$\frac{d^2x}{dt^2} + 64x = 0. \qquad (8)$$

The initial displacement and initial velocity are given by

$$x(0) = \frac{2}{3} \qquad \frac{dx}{dt}\bigg|_{t=0} = -\frac{4}{3}$$

where the negative sign in the last condition is a consequence of the fact that the mass is given an initial velocity in the negative or upward direction.

Now $\omega^2 = 64$ or $\omega = 8$ so that the general solution of (8) is

$$x(t) = c_1 \cos 8t + c_2 \sin 8t. \qquad (9)$$

Applying the initial conditions to (9) we obtain

$$x(0) = \frac{2}{3} = c_1 \cdot 1 + c_2 \cdot 0,$$

$c_1 = 2/3$,

$$x(t) = \frac{2}{3} \cos 8t + c_2 \sin 8t$$

$$x'(t) = -\frac{16}{3} \sin 8t + 8c_2 \cos 8t$$

$$x'(0) = -\frac{4}{3} = -\frac{16}{3} \cdot 0 + 8c_2 \cdot 1,$$

$c_2 = -1/6$. Thus the equation of motion is

$$x(t) = \frac{2}{3} \cos 8t - \frac{1}{6} \sin 8t. \qquad (10)$$

Note: It is common, though regrettable, practice to blur the distinction between weight and mass. Thus one often speaks of both the motion of a mass on a spring and the motion of a weight on a spring.

Alternative form of $x(t)$

When $c_1 \neq 0$ and $c_2 \neq 0$, the actual amplitude A of free vibrations is not obvious from inspection of equation (7). For example, although the mass in the preceding example is initially displaced 2/3 ft beyond the equilibrium position, the amplitude of vibrations is a number larger than 2/3. Hence it is often convenient to convert a solution of form (7) to the simpler form

$$x(t) = A \, \sin (\omega t + \phi). \tag{11}$$

Here

$$A = \sqrt{c_1^2 + c_2^2}$$

and ϕ is a **phase angle** defined by

$$\left. \begin{array}{l} \sin \phi = \dfrac{c_1}{A} \\[2mm] \cos \phi = \dfrac{c_2}{A} \end{array} \right\} \quad \tan \phi = \dfrac{c_1}{c_2}. \tag{12}$$

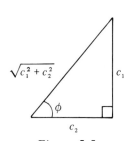

Figure 5.5

To verify this we expand (11) by the addition formula for the sine function:

$$A \sin \omega t \cos \phi + A \cos \omega t \sin \phi = (A \sin \phi) \cos \omega t + (A \cos \phi) \sin \omega t. \tag{13}$$

If follows from Figure 5.5 that if ϕ is defined by

$$\sin \phi = \frac{c_1}{\sqrt{c_1^2 + c_2^2}} = \frac{c_1}{A}, \qquad \cos \phi = \frac{c_2}{\sqrt{c_1^2 + c_2^2}} = \frac{c_2}{A}$$

then (13) becomes

$$A \frac{c_1}{A} \, \cos \omega t + A \frac{c_2}{A} \, \sin \omega t = c_1 \cos \omega t + c_2 \sin \omega t = x(t).$$

EXAMPLE

In view of the foregoing discussion, we can write the solution to the preceding example

$$x(t) = \frac{2}{3} \, \cos 8t - \frac{1}{6} \sin 8t$$

alternatively as $x(t) = A \, \sin (8t + \phi).$

The amplitude is given by

$$A = \sqrt{\left(\frac{2}{3}\right)^2 + \left(-\frac{1}{6}\right)^2}$$

$$= \sqrt{\frac{4}{9} + \frac{1}{36}}$$

$$= \frac{\sqrt{17}}{6} \approx 0.69 \text{ ft.}$$

One should exercise some care when finding the phase angle ϕ defined by (12). In this case

$$\tan \phi = \frac{2/3}{-1/6} = -4$$

and a scientific hand calculator would give

$$\tan^{-1}(-4) = -1.326 \text{ radians.}*$$

Unfortunately, this angle is located in the fourth quadrant and therefore contradicts the fact than $\sin \phi > 0$ and $\cos \phi < 0$ (recall, $c_1 > 0$ and $c_2 < 0$). Hence we must take ϕ to be the second quadrant angle

$$\phi = \pi + (-1.326)$$

$$= 1.816 \text{ radians.}$$

Thus we have

$$x(t) = \frac{\sqrt{17}}{6} \sin(8t + 1.816). \tag{14}$$

Form (11) is very useful since it is easy to find the values of time for which the graph of $x(t)$ crosses the positive t-axis (the line $x = 0$). We observe that $\sin(\omega t + \phi) = 0$ when

$$\omega t + \phi = n\pi$$

where n is a nonnegative integer.

EXAMPLE

For motion described by $x(t) = (\sqrt{17}/6) \sin(8t + 1.816)$ find the first value of time for which the mass passes through the equilibrium position heading downward.

Solution: The values $t_1, t_2, t_3, \ldots,$ for which $\sin(8t + 1.816) = 0$ are determined from

$$8t_1 + 1.816 = \pi, \qquad 8t_2 + 1.816 = 2\pi, \qquad 8t_3 + 1.816 = 3\pi, \ldots.$$

We find $t_1 = 0.166$, $t_2 = 0.558$, $t_3 = 0.951$, . . . , respectively.

Inspection of Figure 5.6 shows that the mass passes through $x = 0$ heading downward (namely, toward $x > 0$) the first time at $t_2 = 0.558$ sec.

---

*The range of the inverse tangent is $-\pi/2 < \tan^{-1} x < \pi/2$.

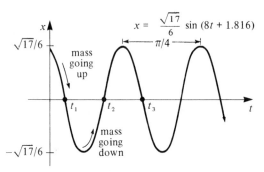

$$x = \frac{\sqrt{17}}{6} \sin (8t + 1.816)$$

Figure 5.6

EXERCISES 5.1   *Answers to odd-numbered problems begin on page A-14.*

In Problems 1 and 2 state in words a possible physical interpretation of the given problem.

1. $\dfrac{4}{32}x'' + 3x = 0$

$x(0) = -3, \quad x'(0) = -2$

2. $\dfrac{1}{16}x'' + 4x = 0$

$x(0) = 0.7, \quad x'(0) = 0$

In Problems 3–8 write the solution of the given initial-value problem in form (11).

3. $x'' + 25x = 0$

$x(0) = -2, \quad x'(0) = 10$

★ 4. $\dfrac{1}{2}x'' + 8x = 0$

$x(0) = 1, \quad x'(0) = -2$

5. $x'' + 2x = 0$

$x(0) = -1, \quad x'(0) = -2\sqrt{2}.$

6. $\dfrac{1}{4}x'' + 16x = 0$

$x(0) = 4, \quad x'(0) = 16$

7. $0.1x'' + 10x = 0$

$x(0) = 1, \quad x'(0) = 1$

8. $x'' + x = 0$

$x(0) = -4, \quad x'(0) = 3$

9. The period of free undamped oscillations of a mass on a spring is $\pi/4$ sec. If the spring constant is 16 lb/ft, what is the numerical value of the weight?

10. A spring is suspended from a ceiling. When a mass weighing 60 lb is attached, the spring is stretched 1/2 ft. The mass is removed and a person grabs the end of the spring and procedes to bounce up and down with a period of 1 sec. How much does the person weigh?

11. A 4-lb weight is attached to a spring whose spring constant is 16 lb/ft. What is the period of simple harmonic motion?

★12. A 20-kg mass is attached to a spring. If the frequency of simple harmonic motion is $2/\pi$ vibrations/sec, what is the spring constant k? What is the frequency of simple harmonic motion if the original mass is replaced with an 80-kg mass?

13. A 24-lb weight, attached to the end of a spring, stretches it 4 in. Find the equation of motion if the weight is released from rest from a point 3 in. above the equilibrium position.

14. Determine the equation of motion if the weight in Problem 13 is released from the equilibrium position with an initial downward velocity of 2 ft/sec.

15. A 20-lb weight stretches a spring 6 in. If the weight is released from rest 6 in. below the equilibrium position,

(a) find the position of the spring at $t = \pi/12, \pi/8, \pi/6, \pi/4, 9\pi/32$ sec.
(b) What is the velocity of the weight when $t = 3\pi/16$ sec? In which direction is the weight heading at this instant?
(c) At what times does the weight pass through the equilibrium position?

16. A force of 400 newtons stretches a spring 2m. A mass of 50 kg is attached to the end of the spring and released from the equilibrium position with an upward velocity of 10 m/sec. Find the equation of motion.

17. Another spring whose constant is 20 nt/m is suspended from the same rigid support but parallel to the spring-mass system in Problem 16. A mass of 20 kg is attached to the second spring and both masses are released from the equilibrium position with an upward velocity of 10 m/sec.

(a) Which mass exhibits the greater amplitude of motion?
(b) Which mass is moving faster at $t = \pi/4$ sec? at $\pi/2$ sec?
(c) At what times are the two masses in the same position? Where are the masses at these times? In which directions are they moving?

★18. A 32-lb weight stretches a spring 2 ft. Determine the amplitude and period of motion if the weight is released 1 ft above the equilibrium position with an initial upward velocity of 2 ft/sec. How many complete vibrations will the weight have completed at the end of 4π sec?

19. An 8-lb weight attached to a spring exhibits simple harmonic motion. Determine the equation of motion if the spring constant is 1 lb/ft and if the weight is released 6 in. below the equilibrium position with a downward velocity of 3/2 ft/sec. Express the solution in form (11).

20. A mass weighing 10 lb stretches a spring 1/4 ft. This mass is removed and replaced with a mass of 1.6 slugs which is released 1/3 ft above the equilibrium position with a downward velocity of 5/4 ft/sec. Express the solution in form (11). At what times does the mass attain a displacement below the equilibrium position numerically equal to one-half the amplitude?

21. A 64-lb weight attached to the end of a spring stretches it 0.32 ft. From a position 8 in. above the equilibrium position the weight is given a downward velocity of 5 ft/sec.

(a) Find the equation of motion.

(b) What is the amplitude and period of motion?

(c) How many complete vibrations will the weight have completed at the end of 3π sec?

(d) At what time does the weight pass through the equilibrium position heading downward for the second time?

(e) At what times does the weight attain its extreme displacement on either side of the equilibrium position?

(f) What is the position of the weight at $t = 3$ sec?

(g) What is the instantaneous velocity at $t = 3$ sec?

(h) What is the acceleration at $t = 3$ sec?

(i) What is the instantaneous velocity at the times when the weight passes through the equilibrium position?

(j) At what times is the weight 5 in. below the equilibrium position?

(k) At what times is the weight 5 in. below the equilibrium position heading in the upward direction?

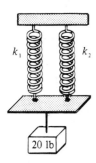

k_1 k_2

20 lb

Figure 5.7

★22. A mass of 1 slug is suspended from a spring whose characteristic spring constant is 9 lb/ft. Initially the mass starts from a point 1 ft above the equilibrium position with an upward velocity of $\sqrt{3}$ ft/sec. Find the times for which the mass is heading downward at a velocity of 3 ft/sec.

23. When two parallel springs, with constants k_1 and k_2, support a single weight W, the effective spring constant is given by $k = k_1 + k_2$. A 20-lb weight stretches one spring 6 in. and another spring 2 in. The springs are attached to a common rigid support, and the 20-lb weight is then attached to the double spring arrangement as shown in Figure 5.7. Determine the equation of motion if the weight is released from the equilibrium position with a downward velocity of 2 ft/sec.

★24. A certain weight stretches one spring $1/3$ ft and another spring $1/2$ ft. The two springs are attached to a common rigid support in a manner indicated in Problem 23 and Figure 5.7. The first weight is set aside, and an 8-lb weight is attached to the double spring arrangement and the system is set in motion. If the period of motion is $\pi/15$ sec, determine the numerical value of the first weight.

Miscellaneous problems

25. If x_0 and v_0 are the initial position and velocity, respectively, of a weight exhibiting simple harmonic motion, show that the amplitude of vibrations is

$$A = \sqrt{x_0^2 + \left(\frac{v_0}{\omega}\right)^2}.$$

26. Show that any linear combination $x(t) = c_1 \cos \omega t + c_2 \sin \omega t$ can also be written in the form

$$x(t) = A \; \cos \; (\omega t + \phi)$$

where
$$A = \sqrt{c_1^2 + c_2^2}$$

and
$$\sin \phi = -\frac{c_2}{A} \qquad \cos \phi = \frac{c_1}{A}.$$

27. Express the solution of Problem 3 in the form of the cosine function given in Problem 26.

28. Prove that when a weight attached to a spring exhibits simple harmonic motion the maximum value of the speed (that is, $|v(t)|$) occurs when the weight is passing through the equilibrium position.

29. Use (11) to prove that the time interval between two successive maxima of $x(t)$ is $2\pi/\omega$.

5.2   Damped Motion

The discussion of free harmonic motion is somewhat unrealistic since the motion described by equation (5) of Section 5.1 assumes that no retarding forces are acting on the moving mass. Unless the mass is suspended in a perfect vacuum, there will be at least a resisting force due to the surrounding medium. For example, as Figure 5.8 shows, the mass m could be suspended in a viscous medium or connected to a dashpot damping device.

Differential equation of motion with damping
In the study of mechanics, damping forces acting on a body are considered to be proportional to a power of the instantaneous velocity; in particular we shall assume throughout the subsequent discussion that this force is given by a constant multiple of dx/dt.* When no other external forces are impressed on the system it follows from Newton's second law that

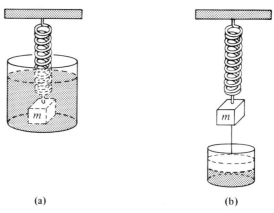

(a)   (b)

Figure 5.8

---
*In many instances, such as problems in hydrodynamics, the damping force is proportional to $(dx/dt)^2$.

$$m\frac{d^2x}{dt^2} = -kx - \beta\frac{dx}{dt} \qquad (1)$$

where β is a positive *damping constant* and the negative sign is a consequence of the fact that the damping force acts in a direction opposite to the motion.

Dividing (1) by the mass m, the differential equation of **free, damped motion** then is

$$\frac{d^2x}{dt^2} + \frac{\beta}{m}\frac{dx}{dt} + \frac{k}{m}x = 0 \qquad (2)$$

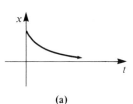

(a)

or

$$\frac{d^2x}{dt^2} + 2\lambda\frac{dx}{dt} + \omega^2 x = 0. \qquad (3)$$

In equation (3) we make the identifications

$$2\lambda = \frac{\beta}{m} \qquad \omega^2 = \frac{k}{m}. \qquad (4)$$

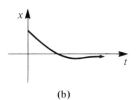

(b)

Figure 5.9

The symbol 2λ is used only for algebraic convenience since the auxiliary equation is $m^2 + 2\lambda m + \omega^2 = 0$ and the corresponding roots are then

$$m_1 = -\lambda + \sqrt{\lambda^2 - \omega^2} \qquad m_2 = -\lambda - \sqrt{\lambda^2 - \omega^2}.$$

We can now distinguish three possible cases depending on the algebraic sign of $\lambda^2 - \omega^2$. Since each solution will contain the *damping factor* $e^{-\lambda t}$, $\lambda > 0$, the displacements of the mass will become negligible for large time.

CASE I $\lambda^2 - \omega^2 > 0$.   In this situation the system is said to be **overdamped** since the damping coefficient β is large when compared to the spring constant k. The corresponding solution of (3) is

$$x(t) = c_1 e^{m_1 t} + c_2 e^{m_2 t}$$

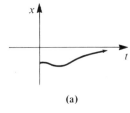

(a)

or

$$x(t) = e^{-\lambda t}[c_1 e^{\sqrt{\lambda^2 - \omega^2}\,t} + c_2 e^{-\sqrt{\lambda^2 - \omega^2}\,t}]. \qquad (5)$$

This equation represents a smooth and nonoscillatory motion. Figure 5.9 shows two possible graphs of $x(t)$.   □

CASE II $\lambda^2 - \omega^2 = 0$.   The system is said to be **critically damped** since any slight decrease in the damping force would result in oscillatory motion. The general solution of (3) is

$$x(t) = c_1 e^{m_1 t} + c_2 t e^{m_1 t}$$

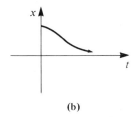

(b)

Figure 5.10

or

$$x(t) = e^{-\lambda t}[c_1 + c_2 t]. \qquad (6)$$

Some graphs of typical motion are given in Figure 5.10. Notice that the motion is quite similar to that of an overdamped system. It is also apparent from (6) that the mass can pass through the equilibrium position at most one time.* □

CASE III $\lambda^2 - \omega^2 < 0$. In this case the system is said to be **underdamped** since the damping coefficient is small compared to the spring constant. The roots m_1 and m_2 are now complex,

$$m_1 = -\lambda + \sqrt{\omega^2 - \lambda^2}\, i \qquad m_2 = -\lambda - \sqrt{\omega^2 - \lambda^2}\, i,$$

and so the general solution of equation (3) is

$$x(t) = e^{-\lambda t}[c_1 \cos \sqrt{\omega^2 - \lambda^2}\, t + c_2 \sin \sqrt{\omega^2 - \lambda^2}\, t]. \qquad (7)$$

Figure 5.11

As indicated in Figure 5.11 the motion described by (7) is oscillatory but because of the coefficient $e^{-\lambda t}$ the amplitudes of vibration $\to 0$ as $t \to \infty$. □

EXAMPLE

It is readily verified that the solution of the initial-value problem

$$\frac{d^2x}{dt^2} + 5\frac{dx}{dt} + 4x = 0$$

$$x(0) = 1 \qquad \frac{dx}{dt}\bigg|_{t=0} = 1$$

is

$$x(t) = \frac{5}{3}e^{-t} - \frac{2}{3}e^{-4t}. \qquad (8)$$

The problem can be interpreted as representing the overdamped motion of a mass on a spring. The mass starts from a position 1 unit *below* the equilibrium position with a *downward* velocity of 1 ft/sec.

To graph $x(t)$ we find the value of t for which the function has an extremum, that is, the value of time for which the first derivative (velocity) is zero. Differentiating (8) gives

$$x'(t) = -\frac{5}{3}e^{-t} + \frac{8}{3}e^{-4t}$$

so that $x'(t) = 0$ implies

$$\frac{5}{3}e^{-t} = \frac{8}{3}e^{-4t}$$

$$e^{3t} = \frac{8}{5}$$

$$t = \frac{1}{3}\ln\frac{8}{5} = 0.157.$$

---

*An examination of the derivatives of (5) and (6) would show that these functions can have at most one relative maximum or one relative minimum for $t > 0$.

It follows from the first derivative test, as well as our physical intuition, that $x(0.157) = 1.069$ ft is actually a maximum. In other words, the mass attains an extreme displacement of 1.069 ft below the equilibrium position.

We should also check to see whether the graph crosses the t-axis, that is, whether the mass passes through the equilibrium position. This cannot happen in this instance since the equation $x(t) = 0$, or

$$\frac{5}{3}e^{-t} = \frac{2}{3}e^{-4t}$$

$$e^{3t} = \frac{2}{5}$$

has the physically irrelevant solution

$$t = \frac{1}{3}\ln\frac{2}{5} = -0.305.$$

The graph of $x(t)$, along with some other pertinent data, is given in Figure 5.12.

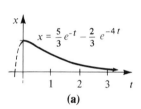

t	$x(t)$
1	0.601
1.5	0.370
2	0.225
2.5	0.137
3	0.083

(a)

(b)

Figure 5.12

EXAMPLE

An 8-lb weight stretches a spring 2 ft. Assuming a damping force numerically equal to two times the instantaneous velocity acts on the system, determine the equation of motion if the weight is released from the equilibrium position with an upward velocity of 3 ft/sec.

Solution: From Hooke's law we have

$$8 = k(2)$$

$$k = 4 \text{ lb/ft}$$

and from $m = W/g$

$$m = \frac{8}{32} = \frac{1}{4}\text{slug}.$$

Thus, the differential equation of motion is

$$\frac{1}{4}\frac{d^2x}{dt^2} = -4x - 2\frac{dx}{dt} \quad \text{or} \quad \frac{d^2x}{dt^2} + 8\frac{dx}{dt} + 16x = 0. \tag{9}$$

The initial conditions are

$$x(0) = 0 \qquad \frac{dx}{dt}\bigg|_{t=0} = -3.$$

Now the auxiliary equation for (9) is

$$m^2 + 8m + 16 = (m + 4)^2 = 0$$

so that $m_1 = m_2 = -4$. Hence the system is critically damped and

$$x(t) = c_1 e^{-4t} + c_2 t e^{-4t}. \tag{10}$$

The initial condition $x(0) = 0$ immediately demands that $c_1 = 0$, whereas using $x'(0) = -3$ gives $c_2 = -3$. Thus, the equation of motion is

$$x(t) = -3t e^{-4t}. \tag{11}$$

To graph $x(t)$ we proceed as in the preceding example:

$$x'(t) = -3(-4t e^{-4t} + e^{-4t})$$

$$= -3e^{-4t}(1 - 4t).$$

Clearly $x'(t) = 0$ when $t = 1/4$. The corresponding extreme displacement is

$$x\left(\frac{1}{4}\right) = -3\left(\frac{1}{4}\right)e^{-1} = -0.276 \text{ ft.}$$

As shown in Figure 5.13, we interpret this value to mean that the weight reaches a maximum height of 0.276 ft above the equilibrium position.

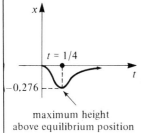

$t = 1/4$

-0.276

maximum height
above equilibrium position

Figure 5.13

EXAMPLE

A 16-lb weight is attached to a 5-ft-long spring. At equilibrium the spring measures 8.2 ft. If the weight is pushed up and released from rest at a point 2 ft above the equilibrium position, find the displacements $x(t)$ if it is further known that the surrounding medium offers a resistance numerically equal to the instantaneous velocity.

Solution: The elongation of the spring after the weight is attached is $8.2 - 5 = 3.2$ ft so it follows from Hooke's law that

$$16 = k(3.2)$$

$$k = 5 \text{ lb/ft.}$$

In addition we have

$$m = \frac{16}{32} = \frac{1}{2} \text{slug,}$$

so that the differential equation is given by

$$\frac{1}{2}\frac{dx^2}{dt^2} = -5x - \frac{dx}{dt}$$

or

$$\frac{d^2x}{dt^2} + 2\frac{dx}{dt} + 10x = 0. \tag{12}$$

This latter equation is solved subject to the conditions

$$x(0) = -2 \qquad \left. \frac{dx}{dt} \right|_{t=0} = 0.$$

Proceeding, we find that the roots of $m^2 + 2m + 10 = 0$ are $m_1 = -1 + 3i$ and $m_2 = -1 - 3i$ which then implies the system is underdamped and

$$x(t) = e^{-t}[c_1 \cos 3t + c_2 \sin 3t]. \qquad (13)$$

Now

$$x(0) = -2 = c_1$$

$$x(t) = e^{-t}[-2 \cos 3t + c_2 \sin 3t]$$

$$x'(t) = e^{-t}[6 \sin 3t + 3c_2 \cos 3t] - e^{-t}[-2 \cos 3t + c_2 \sin 3t]$$

$$x'(0) = 0 = 3c_2 + 2$$

which gives $c_2 = -2/3$. Thus we finally obtain

$$x(t) = e^{-t}\left[-2 \cos 3t - \frac{2}{3} \sin 3t \right]. \qquad (14)$$

Alternative form of the solution

In a manner identical to the procedure used in Section 5.1 we can write any solution of form (7)

$$x(t) = e^{-\lambda t}[c_1 \cos \sqrt{\omega^2 - \lambda^2}\, t + c_2 \sin \sqrt{\omega^2 - \lambda^2}\, t]$$

in the alternative form

$$x(t) = Ae^{-\lambda t} \sin \left[\sqrt{\omega^2 - \lambda^2}\, t + \phi \right] \qquad (15)$$

Where $A = \sqrt{c_1^2 + c_2^2}$ and the phase angle ϕ is determined from the equations

$$\sin \phi = \frac{c_1}{A} \qquad \cos \phi = \frac{c_2}{A} \qquad \tan \phi = \frac{c_1}{c_2}.$$

The coefficient $Ae^{-\lambda t}$ is sometimes called the *damped amplitude* of vibrations. Because (15) is not a periodic function the number $2\pi/\sqrt{\omega^2 - \lambda^2}$ is called the *quasi period* and $\sqrt{\omega^2 - \lambda^2}/2\pi$ is the *quasi frequency*. The quasi period is the time interval between two successive maxima of $x(t)$.

To graph an equation such as (15), we first find the intercepts t_1, $t_2, \ldots, t_k, \ldots$. That is, for some integer n we must solve

$$\sqrt{\omega^2 - \lambda^2}\, t + \phi = n\pi$$

for t. It follows

$$t = \frac{n\pi - \phi}{\sqrt{\omega^2 - \lambda^2}}. \qquad (16)$$

In addition, we note that $|x(t)| \leq Ae^{-\lambda t}$ since

$$\left|\sin\left[\sqrt{\omega^2 - \lambda^2}\, t + \phi\right]\right| \leq 1.$$

Indeed, the graph of (15) touches the graphs of $\pm Ae^{-\lambda t}$ at the values $t_1^*, t_2^*, \ldots, t_k^*, \ldots$ for which

$$\sin\left[\sqrt{\omega^2 - \lambda^2}\, t + \phi\right] = \pm 1.$$

This means $\sqrt{\omega^2 - \lambda^2}\, t + \phi$ must be an odd multiple of $\pi/2$,

$$\sqrt{\omega^2 - \lambda^2}\, t + \phi = (2n + 1)\frac{\pi}{2}$$

$$t = \frac{(2n + 1)\pi/2 - \phi}{\sqrt{\omega^2 - \lambda^2}}. \tag{17}$$

For example, were we asked to graph $x(t) = e^{-0.5t} \sin(2t - \pi/3)$, we find the intercepts on the *positive* t-axis by solving

$$2t_1 - \frac{\pi}{3} = 0, \qquad 2t_2 - \frac{\pi}{3} = \pi, \qquad 2t_3 - \frac{\pi}{3} = 2\pi, \ldots$$

which gives respectively

$$t_1 = \frac{\pi}{6}, \qquad t_2 = \frac{4\pi}{6}, \qquad t_3 = \frac{7\pi}{6}, \ldots.$$

Notice that even though $x(t)$ is not periodic, the difference between the successive roots is $t_k - t_{k-1} = \pi/2$ units or one-half the quasi period of $2\pi/2 = \pi$ sec. Also, $\sin(2t - \pi/3) = \pm 1$ at the solutions of

$$2t_1^* - \frac{\pi}{3} = \frac{\pi}{2}, \qquad 2t_2^* - \frac{\pi}{3} = \frac{3\pi}{2}, \qquad 2t_3^* - \frac{\pi}{3} = \frac{5\pi}{2}, \ldots$$

or

$$t_1^* = \frac{5\pi}{12}, \qquad t_2^* = \frac{11\pi}{12}, \qquad t_3^* = \frac{17\pi}{12}, \ldots.$$

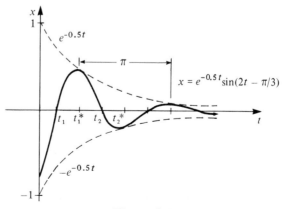

Figure 5.14

It is readily shown that the difference between the successive t_k^* is also $\pi/2$.[†]
The graph of $x(t)$ is given in Figure 5.14.

EXAMPLE

Using (15) we can write the solution of the preceding example

$$\frac{d^2x}{dt^2} + 2\frac{dx}{dt} + 10x = 0$$

$$x(0) = -2 \qquad \frac{dx}{dt}\bigg|_{t=0} = 0$$

in the form $x(t) = Ae^{-t}\sin(3t + \phi)$.

Now from (14) we have $c_1 = -2$, $c_2 = -2/3$ so that

$$A = \sqrt{4 + \frac{4}{9}} = \frac{2}{3}\sqrt{10}$$

and $\qquad\qquad \tan\phi = \dfrac{-2}{-2/3} = 3$

$$\tan^{-1}(3) = 1.249 \text{ radians.}$$

But, since $\sin\phi < 0$ and $\cos\phi < 0$ we take ϕ to be the third quadrant angle
$\phi = \pi + 1.249 = 4.391$ radians. Hence

$$x(t) = \frac{2}{3}\sqrt{10}e^{-t}\sin(3t + 4.391).$$

The graph of this function is given in Figure 5.15. The values of t_k and
t_k^* given in the accompanying table are the intercepts and the points where the
graph of $x(t)$ touches the graphs of $\pm (2/3)\sqrt{10}e^{-t}$, respectively. In this
example the quasi period is $2\pi/3$ sec and so the difference between the
successive t_k (and the successive t_k^*) is $\pi/3$ units.

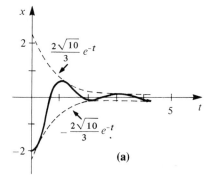

(a)

k	t_k	t_k^*	$x(t_k^*)$
1	0.631	1.154	0.665
2	1.678	2.202	-0.233
3	2.725	3.249	0.082
4	3.772	4.296	-0.029

(b)

Figure 5.15

[†]We note that the values of t for which the graph of $x(t)$ touches the exponential graphs are
not the values for which the function attains its relative extrema.

EXERCISES 5.2   *Answers to odd-numbered problems begin on page A-14.*

In Problems 1 and 2 give a possible physical interpretation of the given initial-value problem.

1. $\dfrac{1}{16}x'' + 2x' + x = 0$

$x(0) = 0$ $\quad \dfrac{dx}{dt}\bigg|_{t=0} = -1.5$

2. $\dfrac{16}{32}x'' + x' + 2x = 0$

$x(0) = -2$ $\quad x'(0) = 1$

3. A 4-lb weight is attached to a spring whose constant is 2 lb/ft. The medium offers a resistance to the motion of the weight numerically equal to the instantaneous velocity. If the weight is released from a point 1 ft above the equilibrium position with a downward velocity of 8 ft/sec, determine the time that the weight passes through the equilibrium position. Find the time for which the weight attains its extreme displacement from the equilibrium position. What is the position of the weight at this instant?

★ **4.** A 4-ft spring measures 8 ft long after an 8-lb weight is attached to it. The medium through which the weight moves offers a resistance numerically equal to $\sqrt{2}$ times the instantaneous velocity. Find the equation of motion if the weight is released from the equilibrium position with a downward velocity of 5 ft/sec. Find the time for which the weight attains its extreme displacement from the equilibrium position. What is the position of the weight at this instant?

5. A 1-kg mass is attached to a spring whose constant is 16 nt/m and the entire system is then submerged in a liquid which imparts a damping force numerically equal to 10 times the instantaneous velocity. Determine the equations of motion if

(a) the weight is released from rest 1 m below the equilibrium position;
(b) the weight is released 1 m below the equilibrium position with an upward velocity of 12 m/sec.

6. In parts (a) and (b) of Problem 5, determine whether the weight passes through the equilibrium position. In each case, find the time at which the weight attains its extreme displacement from the equilibrium position. What is the position of the weight at this instant?

7. A force of 2 lb stretches a spring 1 ft. A 3.2-lb weight is attached to the spring and the system is then immersed in a medium which imparts a damping force numerically equal to 0.4 times the instantaneous velocity.

(a) Find the equation of motion if the weight is released from rest 1 ft above the equilibrium position.
(b) Express the equation of motion in the form given in (15).
(c) Find the first time for which the weight passes through the equilibrium position heading upward.

8. After a 10-lb weight is attached to a 5 ft spring the spring measures 7 ft long. The 10-lb weight is removed and replaced with an 8-lb weight and the entire system is placed in a medium offering a resistance numerically equal to the instantaneous velocity.

   (a) Find the equation of motion if the weight is released 1/2 ft below the equilibrium position with a downward velocity of 1 ft/sec.
   (b) Express the equation of motion in the form given in (15).
   (c) Find the times for which the weight passes through the equilibrium position heading downward.
   (d) Graph the equation of motion.

9. A 10-lb weight attached to a spring stretches it 2 ft. The weight is attached to a dashpot damping device which offers a resistance numerically equal to $\beta(\beta > 0)$ times the instantaneous velocity. Determine the values of the damping constant β so that the subsequent motion is **(a)** overdamped, **(b)** critically damped, and **(c)** underdamped.

★**10.** A 24-lb weight stretches a spring 4 ft. The subsequent motion takes place in a medium offering a resistance numerically equal to $\beta(\beta > 0)$ times the instantaneous velocity. If the weight starts from the equilibrium position with an upward velocity of 2 ft/sec, show that if $\beta > 3\sqrt{2}$, the equation of motion is

$$x(t) = \frac{-3}{\sqrt{\beta^2 - 18}} e^{-2\beta t/3} \sinh \frac{2}{3}\sqrt{\beta^2 - 18}\, t.$$

11. A mass of 40 g stretches a spring 10 cm. A damping device imparts a resistance to motion numerically equal to 560 times the instantaneous velocity. Find the equation of motion if the mass is released from the equilibrium position with a downward velocity of 2 cm/sec.

12. Find the equation of motion for the mass in Problem 11 if the damping constant is doubled.

13. A mass of 1 slug is attached to a spring whose constant is 9 lb/ft. The medium offers a resistance to the motion numerically equal to 6 times the instantaneous velocity. The mass is released from a point 8 in. above the equilibrium position with a downward velocity of v_0 ft/sec. Determine the values of v_0 such that the mass will subsequently pass through the equilibrium position.

★**14.** The quasi period of an underdamped, vibrating 1-slug mass on a spring is $\pi/2$ sec. If the spring constant is 25 lb/ft, find the damping constant β.

Miscellaneous problems

15. In the case of underdamped motion show that the difference in times between two successive positive maxima of the equation of motion is $2\pi/\sqrt{\omega^2 - \lambda^2}$.

16. Use (16) to show that the time interval between successive intercepts of (15) is one-half the quasi period.

17. Use (17) to show that the time interval between successive values of t for which the graph of (15) touches the graphs of $\pm Ae^{-\lambda t}$ is one-half the quasi period.

18. Use equation (17) to show that the intercepts of the graph of $x(t) = Ae^{-\lambda t} \sin(\sqrt{\omega^2 - \lambda^2}\,t + \phi)$ are halfway between the values of t for which the graph of $x(t)$ touches the graphs of $\pm Ae^{-\lambda t}$. The values of t for which $x(t)$ is a maximum or minimum are not located halfway between the intercepts of the graph of $x(t)$. Verify this last statement by considering the function $x(t) = e^{-t} \sin(t + \pi/4)$.

19. In the case of underdamped motion show that the ratio between two consecutive maximum (or minimum) displacements x_n and x_{n+2} is the constant

$$x_n/x_{n+2} = e^{2\pi\lambda/\sqrt{\omega^2 - \lambda^2}}$$

The number $\delta = \ln(x_n/x_{n+2}) = 2\pi\lambda/\sqrt{\omega^2 - \lambda^2}$ is called the **logarithmic decrement.**

20. The logarithmic decrement defined in Problem 19 is an indicator of the rate at which the motion is damped out.

    (a) Describe the motion of an underdamped system if δ is very small positive number.
    (b) Compute the logarithmic decrement for the motion described in Problem 8.

5.3 Forced Motion

With damping

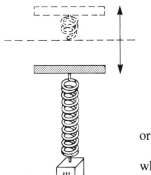

Figure 5.16

Suppose we now take into consideration an external force $f(t)$ acting on a vibrating mass on a spring. For example, $f(t)$ could represent a driving force causing an oscillatory vertical motion of the support of the spring (Figure 5.16). The inclusion of $f(t)$ in the formulation of Newton's second law gives

$$m\frac{d^2x}{dt^2} = -kx - \beta\frac{dx}{dt} + f(t), \tag{1}$$

$$\frac{d^2x}{dt^2} + \frac{\beta}{m}\frac{dx}{dt} + \frac{k}{m}x = \frac{f(t)}{m}, \tag{2}$$

or

$$\frac{d^2x}{dt^2} + 2\lambda\frac{dx}{dt} + \omega^2x = F(t), \tag{3}$$

where $F(t) = f(t)/m$ and, as in the preceding section, $2\lambda = \beta/m$, $\omega^2 = k/m$. To solve the latter nonhomogeneous equation, we can employ either the method of undetermined coefficients or variation of parameters.

EXAMPLE   Interpret and solve the initial-value problem

$$\frac{1}{5}\frac{d^2x}{dt^2} + 1.2\frac{dx}{dt} + 2x = 5\cos 4t \tag{4}$$

$$x(0) = \frac{1}{2} \qquad \frac{dx}{dt}\bigg|_{t=0} = 0.$$

Solution: We can interpret the problem to represent a vibrational system consisting of a mass ($m = 1/5$ slug or kilogram) attached to a spring ($k = 2$ lb/ft or nt/m). The mass is released from rest 1/2 unit (foot or meter) below the equilibrium position. Although the motion is damped ($\beta = 1.2$), the system is also being driven by an external periodic ($T = \pi/2$ sec) force beginning at $t = 0$. Intuitively we would expect that even with damping the system will remain in motion until such time as the forcing function is "turned off," in which case the amplitudes would gradually diminish. However, as the problem is given, $f(t) = 5\cos 4t$ will remain "on" forever.

We first multiply (4) by 5 and solve the homogeneous equation

$$\frac{dx^2}{dt^2} + 6\frac{dx}{dt} + 10x = 0$$

by the usual methods. Since $m_1 = -3 + i$, $m_2 = -3 - i$ it follows that

$$x_c(t) = e^{-3t}(c_1 \cos t + c_2 \sin t).$$

Using the method of undetermined coefficients, we assume a particular solution of the form $x_p(t) = A\cos 4t + B\sin 4t$. Now

$$x_p' = -4A\sin 4t + 4B\cos 4t$$

$$x_p'' = -16A\cos 4t - 16B\sin 4t$$

so that

$$x_p'' + 6x_p' + 10x_p = -16A\cos 4t - 16B\sin 4t - 24A\sin 4t$$

$$+ 24B\cos 4t + 10A\cos 4t + 10B\sin 4t$$

$$= (-6A + 24B)\cos 4t + (-24A - 6B)\sin 4t$$

$$= 25\cos 4t.$$

The resulting system of equations

$$-6A + 24B = 25$$

$$-24A - 6B = 0$$

yields $A = -25/102$ and $B = 50/51$. It follows that

$$x(t) = e^{-3t}(c_1 \cos t + c_2 \sin t) - \frac{25}{102}\cos 4t + \frac{50}{51}\sin 4t. \tag{5}$$

When we set $t = 0$ in the above equation we immediately obtain $c_1 = 38/51$. By differentiating the expression and then setting $t = 0$ we also find that $c_2 = -86/51$. Therefore, the equation of motion is

$$x(t) = e^{-3t}\left(\frac{38}{51}\cos t - \frac{86}{51}\sin t\right) - \frac{25}{102}\cos 4t + \frac{50}{51}\sin 4t. \quad (6)$$

Transient and steady-state terms

Notice that the complementary function

$$x_c(t) = e^{-3t}\left(\frac{38}{51}\cos t - \frac{86}{51}\sin t\right)$$

in the preceding example possesses the property that

$$\lim_{t\to\infty} x_c(t) = 0.$$

Since $x_c(t)$ becomes negligible (namely, $\to 0$) as $t \to \infty$ it is said to be a **transient term** or **transient solution**. Thus, for large time, the displacements of the weight in the preceding problem are closely approximated by the particular solution $x_p(t)$. This latter function is also called the **steady-state solution**. When F is a periodic function, such as $F(t) = F_0 \sin \gamma t$ or $F(t) = F_0 \cos \gamma t$, the general solution of (3) consists of

$$x(t) = \text{transient} + \text{steady-state}.$$

EXAMPLE

The solution to the initial-value problem

$$\frac{d^2x}{dt^2} + 2\frac{dx}{dt} + 2x = 4\cos t + 2\sin t$$

$$x(0) = 0 \qquad \frac{dx}{dt}\bigg|_{t=0} = 3$$

is readily shown to be

$$x = x_c + x_p$$

$$= \underbrace{e^{-t}\sin t}_{\text{transient}} + \underbrace{2\sin t}_{\text{steady-state}}$$

Inspection of Figure 5.17 shows that the effect of the transient term on the solution is, in this case, neglibible for about $t > 2\pi$.

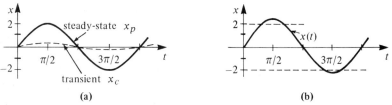

Figure 5.17

Without damping

In the absence of a damping force there will be no transient term in the solution of a problem. Also, we shall see that a periodic impressed force with a frequency near, or the same as, the frequency of free, undamped vibrations can cause a severe problem in any oscillatory mechanical system.

EXAMPLE   Solve the initial-value problem

$$\frac{d^2x}{dt^2} + \omega^2 x = F_0 \sin \gamma t, \qquad F_0 = \text{constant}, \tag{7}$$

$$x(0) = 0 \qquad \frac{dx}{dt}\bigg|_{t=0} = 0.$$

Solution: The complementary function is $x_c(t) = c_1 \cos \omega t + c_2 \sin \omega t$. To obtain a particular solution we assume $x_p(t) = A \cos \gamma t + B \sin \gamma t$, so that

$$x_p' = -A\gamma \sin \gamma t + B\gamma \cos \gamma t$$

$$x_p'' = -A\gamma^2 \cos \gamma t - B\gamma^2 \sin \gamma t$$

$$x_p'' + \omega^2 x_p = A(\omega^2 - \gamma^2)\cos \gamma t + B(\omega^2 - \gamma^2)\sin \gamma t$$

$$= F_0 \sin \gamma t.$$

It follows $A = 0$

$$B = \frac{F_0}{\omega^2 - \gamma^2} \qquad (\gamma \neq \omega)$$

and therefore $x_p(t) = \dfrac{F_0}{\omega^2 - \gamma^2} \sin \gamma t.$

Applying the given initial conditions to the general solution

$$x(t) = c_1 \cos \omega t + c_2 \sin \omega t + \frac{F_0}{\omega^2 - \gamma^2} \sin \gamma t$$

yields $c_1 = 0$ and $c_2 = -\gamma F_0 / \omega(\omega^2 - \gamma^2)$. Thus the solution is

$$x(t) = \frac{-\gamma F_0}{\omega(\omega^2 - \gamma^2)} \sin \omega t + \frac{F_0}{\omega^2 - \gamma^2} \sin \gamma t$$

$$= \frac{F_0}{\omega(\omega^2 - \gamma^2)}[-\gamma \sin \omega t + \omega \sin \gamma t] \qquad \gamma \neq \omega. \tag{8}$$

Pure resonance   Although equation (8) is not defined for $\gamma = \omega$, it is interesting to observe that its limiting value as $\gamma \rightarrow \omega$ can be obtained by applying L'Hopital's rule. This limiting process is analogous to "tuning in" the frequency of the driving force ($\gamma/2\pi$) to the frequency of free vibrations ($\omega/2\pi$). Intuitively we expect that over a length of time we should be able to substantially increase the amplitudes of vibration.* For $\gamma = \omega$ we define the solution to be

---

*Forgetting about damping effects of shock absorbers, the situation is roughly equivalent to a number of passengers jumping up and down in the back of a bus in time with the natural vertical motion caused by equally spaced faults (such as cracks) in the road. Theoretically these passengers could upset the bus—assuming they are not kicked off first.

$$x(t) = \lim_{\gamma \to \omega} F_0 \frac{-\gamma \sin \omega t + \omega \sin \gamma t}{\omega(\omega^2 - \gamma^2)}$$

$$= F_0 \lim_{\gamma \to \omega} \frac{\dfrac{d}{d\gamma}[-\gamma \sin \omega t + \omega \sin \gamma t]}{\dfrac{d}{d\gamma}[\omega^3 - \omega\gamma^2]}$$

$$= F_0 \lim_{\gamma \to \omega} \frac{-\sin \omega t + \omega t \cos \gamma t}{-2\omega\gamma}$$

$$= F_0 \frac{-\sin \omega t + \omega t \cos \omega t}{-2\omega^2}$$

$$= \frac{F_0}{2\omega^2}[\sin \omega t - \omega t \cos \omega t]$$

$$= \frac{F_0}{2\omega^2} \sin \omega t - \frac{F_0}{2\omega} t \cos \omega t. \tag{9}$$

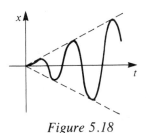

Figure 5.18

As suspected, when $t \to \infty$ the displacements become large, in fact, $|x(t)| \to \infty$. The phenomenon we have just described is known as **pure resonance**. The graph in Figure 5.18 displays typical motion in this case.

In conclusion, it should be noted that there is no actual need to use a limiting process on (8) to obtain the solution for $\gamma = \omega$. Alternatively, equation (9) follows by solving the initial-value problem

$$\frac{d^2x}{dt^2} + \omega^2 x = F_0 \sin \omega t$$

$$x(0) = 0 \qquad \frac{dx}{dt}\bigg|_{t=0} = 0$$

directly by conventional methods.

Remark: If a mechanical system were actually described by a function such as (9) of this section, it would necessarily fail; large oscillations of a weight on a spring would eventually force the spring beyond its elastic limit. One might argue too that the resonating model presented (Figure 5.18) is completely unrealistic since it ignores the retarding effects of ever-present damping forces. While it is true that pure resonance cannot occur when the smallest amount of damping is taken into consideration, nevertheless, large and equally destructive amplitudes of vibration (though bounded as $t \to \infty$) could take place (see Problem 11).

If you have ever looked out a window while in flight you have probably observed that wings on an airplane are not perfectly rigid. A reasonable amount of flutter is not only tolerated but necessary to prevent the wing from snapping like a piece of peppermint stick candy. In late 1959 and early 1960 two commercial plane crashes occurred, with a then relatively new model of prop-jet, which illustrate the destructive effects of large mechanical oscillations.

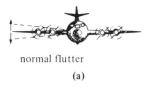

normal flutter

(a)

wing snaps

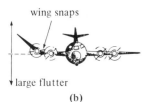

large flutter

(b)

Figure 5.19

The unusual aspect of these crashes was that they both happened while the planes were in mid-flight. Barring mid-air collisions, the safest period during any flight is when the plane has attained its cruising altitude. It is well known that a plane is most vulnerable to an accident when it is least manuverable, namely, either during take-off or landing. So having two planes simply fall out of the sky was at least an embarassment to the aircraft industry as well as a thoroughly puzzling problem to aerodynamic engineers. In crashes of this sort, a structural failure of some kind is immediately suspected. After a subsequent and massive technical investigation the problem was eventually traced in each case to an outboard engine and engine housing. Roughly, it was determined that when each plane surpassed a critical speed of approximately 400 mph, a propeller and engine began to wobble, causing a gyroscopic force which would not be quelled or damped by the engine housing. This external vibrational force was then transferred to the already oscillating wing. This in itself need not have been destructively dangerous since the aircraft wings are designed to withstand the stress of unusual and excessive forces. (In fact, the particular wing in question was so incredibly strong that test engineers and pilots, who were deliberately trying, failed to snap a wing under every conceivable flight condition.) Unfortunately, after a short period of time during which the engine wobbled rapidly, the frequency of the impressed force actually slowed to a point where it approached and finally coincided with the maximum frequency of wing flutter (around 3 cycles per second). The resulting resonance situation finally accomplished what the test engineers could not do, namely, the amplitudes of wing flutter became large enough to snap the wing. The problem was solved in two steps. All models of this particular plane were required to fly at speeds substantially below 400 mph until each plane could be modified by considerably strengthening (or stiffening) the engine housings. A strengthened engine housing was shown to be able to impart a damping effect capable of preventing the critical resonance phenomenon even in the unlikely event of a subsequent engine wobble.*

You may be aware that soldiers usually do not march in step across bridges. The reason for breaking stride is simply to avoid any possibility of resonance occuring between the natural vibrations inherent in the bridges structure and the frequency of the external force of a multitude of feet stomping in unison on the bridge.

Bridges are good examples of vibrating mechanical systems which are constantly being subjected to external forces, either from people walking, cars and trucks driving on them, water pushing against their foundations, or wind blowing against their superstructures. On November 7, 1940, the Tacoma Narrows Bridge at Puget Sound in Washington collapsed. However, the crash came as no surprise since this particular bridge was famous in the local community for a vertical undulating motion of its roadway which gave many motorists a very exciting crossing. On November 7, only four months after

*For a fascinating nontechnical account of the investigation see Robert J. Serling, *Loud and Clear*, New York: Dell, 1970), Chapter 5.

its grand opening the amplitudes of these undulations became so large that the bridge failed and a substantial portion was sent splashing into the water below. In the investigation that followed, it was found that a poorly designed superstructure caused the wind blowing across it to vortex in a periodic manner. When the frequency of this periodic force approached the natural frequency of the bridge, large upheavals of the road resulted. In a word, the bridge was another victim of the destructive effect of mechanical resonance. Since this disaster developed over a matter of months, there was sufficient opportunity to record on film the strange and frightening phenomenon of a bucking and heaving bridge and its ultimate collapse.*

Acoustic vibrations can be as equally destructive as large mechanical vibrations. Operatic and jazz singers sometimes take pride in their ability to inflict destruction on the lowly water glass. The sounds from organs and piccolos have been known to crack windows.

> "As the horns blew, the people began to shout. When they heard the signal horn, they raised a tremendous shout. The wall collapsed. . . ."   *Joshua* 6:20

Did the power of acoustic resonance cause the walls of Jericho to tumble down? This is the conjecture of some contemporary scholars.

EXERCISES 5.3   *Answers to odd-numbered problems begin on page A-15.*

1. A 16-lb weight stretches a spring 8/3 ft. Initially the weight starts from rest 2 ft below the equilibrium position and the subsequent motion takes place in a medium that offers a damping force numerically equal to 1/2 the instantaneous velocity. Find the equation of motion if the weight is driven by an external force equal to $f(t) = 10 \cos 3t$.

2. A mass of 1 slug is attached to a spring whose constant is 5 lb/ft. Initially the mass is released 1 ft below the equilibrium position with an downward velocity of 5 ft/sec and the subsequent motion takes place in a medium that offers a damping force numerically equal to 2 times the instantaneous velocity.

   (a) Find the equation of motion if the mass is driven by an external force equal to $f(t) = 12 \cos 2t + 3 \sin 2t$.
   (b) Graph the transient and steady-state solutions on the same coordinate axes.
   (c) Graph the equation of motion.

3. A mass of 1 slug when attached to a spring stretches it 2 ft and then comes to rest in the equilibrium position. Starting at $t = 0$ an external force

---

*National Committee for Fluid Mechanics Films, Educational Services, Inc., Watertown, Mass. See also, *American Society of Civil Engineers: Proceedings*, "Failure of The Tacoma Narrows Bridge", Vol. 69, pp. 1555–86, Dec. 1943.

equal to $f(t) = 8 \sin 4t$ is applied to the system. Find the equation of motion if the surrounding medium offers a damping force numerically equal to 8 times the instantaneous velocity.

★ **4.** In Problem 3 determine the equation of motion if the external force is $f(t) = e^{-t} \sin 4t$. Analyze the displacements for $t \to \infty$.

5. When a mass of 2 kilograms is attached to a spring whose constant is 32 nt/m it comes to rest in the equilibrium position. Starting at $t = 0$ a force equal to $f(t) = 68e^{-2t} \cos 4t$ is applied to the system. Find the equation of motion in the absence of damping.

6. In Problem 5, write the equation of motion in the form $x(t) = A \sin(\omega t + \phi) + Be^{-2t} \sin(4t + \theta)$. What is the amplitude of vibrations after a very long time?

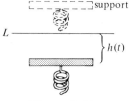

Figure 5.20

7. A mass m is attached to the end of a spring whose constant is k. After the mass reaches equilibrium, its support begins to oscillate vertically about a horizontal line L according to a formula $h(t)$. The value of h represents the distance in feet measured from L (see Figure 5.20). Determine the differential equation of motion if the entire system moves through a medium offering a damping force numerically equal to $\beta(dx/dt)$.

★ **8.** Solve the differential equation of the preceding problem if the spring is stretched 4 ft by a weight of 16 lb, and $\beta = 2$, $h(t) = 5 \cos t$, $x(0) = x'(0) = 0$.

9. A mass of 100 g is attached to a spring whose constant is 1600 dynes/cm. After the mass reaches equilibrium, its support oscillates according to the formula $h(t) = \sin 8t$, where h represents displacement from its original position (See Problem 7 and Figure 5.20).

  (a) In the absence of damping, determine the equation of motion if the mass starts from rest from the equilibrium position.
  (b) At what times does the mass pass through the equilibrium position?
  (c) At what times does the mass attain its extreme displacements?
  (d) What are the maximum and minimum displacements?
  (e) Graph the equation of motion.

10. In the case of underdamped vibrations show that the general solution of the differential equation

$$\frac{d^2x}{dt^2} + 2\lambda \frac{dx}{dt} + \omega^2 x = F_0 \sin \gamma t$$

is

$$x(t) = Ae^{-\lambda t} \sin[\sqrt{\omega^2 - \lambda^2}\,t + \phi] + \frac{F_0}{\sqrt{(\omega^2 - \gamma^2)^2 + 4\lambda^2\gamma^2}} \sin(\gamma t + \theta)$$

where $A = \sqrt{c_1^2 + c_2^2}$ and the phase angles ϕ and θ are respectively defined by

$$\sin \phi = \frac{c_1}{A}$$

$$\cos \phi = \frac{c_2}{A}$$

$$\sin \theta = \frac{-2\lambda\gamma}{\sqrt{(\omega^2 - \gamma^2)^2 + 4\lambda^2\gamma^2}}$$

$$\cos \theta = \frac{\omega^2 - \gamma^2}{\sqrt{(\omega^2 - \gamma^2)^2 + 4\lambda^2\gamma^2}}.$$

EXAMPLE

Inspection of Problem 10 shows that $x_c(t)$ is transient when damping is present and hence for large values of time the solution is closely approximated by the steady-state

$$x_p(t) = g(\gamma) \sin(\gamma t + \theta)$$

where we define

$$g(\gamma) = \frac{F_0}{\sqrt{(\omega^2 - \gamma^2)^2 + 4\lambda^2\gamma^2}} \tag{10}$$

Although the amplitude of x_p is bounded as $t \to \infty$ it is easily shown that the maximum oscillations will occur at the value $\gamma_1 = \sqrt{\omega^2 - 2\lambda^2}$. See Problem 11. Thus when the frequency of the external force is $\sqrt{\omega^2 - 2\lambda^2}/2\pi$ the system is said to be in **resonance**.

In the specific case $k = 4$, $m = 1$, $F_0 = 2$, $g(\gamma)$ becomes

$$g(\gamma) = \frac{2}{\sqrt{(4 - \gamma^2)^2 + \beta^2\gamma^2}}. \tag{11}$$

Figure 5.21(a) shows the graph of (11) for various values of the damping coefficient β. This family of graphs is called the **resonance curve** of the system. Observe the behavior of the amplitudes $g(\gamma)$ as $\beta \to 0$, that is, as the system approaches pure resonance.

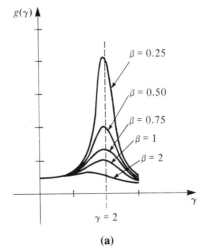

β	γ_1	$g(\gamma_1)$
2	1.41	0.58
1	1.87	1.03
0.75	1.93	1.36
0.50	1.97	2.02
0.25	1.99	4.01

(a)   *Figure 5.21*   (b)

11.(a) Prove that $g(\gamma)$ given in (10) of the preceding example has a maximum value at $\gamma_1 = \sqrt{\omega^2 - 2\lambda^2}$. [*Hint:* Differentiate with respect to γ.]

(b) What is the maximum value of $g(\gamma)$ at resonance?

★**12.(a)** If $k = 3$ lb/ft and $m = 1$ slug, use the information in the above example to show that the system is underdamped when the damping coefficient β satisfies $0 < \beta < 2\sqrt{3}$ but that resonance can occur only if $0 < \beta < \sqrt{6}$.

(b) Construct the resonance curve of the system when $F_0 = 3$.

13. A mass of $1/2$ slug is suspended on a spring whose constant is 6 lb/ft. The system is set in motion in a medium offering a damping force numerically equal to twice the instantaneous velocity. Find the steady-state solution if an external force $f(t) = 40 \sin 2t$ is applied to the system starting at $t = 0$. Write this solution in the form of a constant multiple of $\sin(2t + \theta)$.

14. Verify that the mechanical system described in Problem 13 is in resonance. Show that the amplitude of the steady-state solution is the maximum value of $g(\gamma)$ described in Problem 11.

15.(a) Show that the solution of the initial-value problem

$$\frac{d^2x}{dt^2} + \omega^2 x = F_0 \cos \gamma t$$

$$x(0) = 0 \qquad \frac{dx}{dt}\bigg|_{t=o} = 0$$

is

$$x(t) = \frac{F_0}{\omega^2 - \gamma^2}[\cos \gamma t - \cos \omega t].$$

(b) Evaluate

$$\lim_{\gamma \to \omega} \frac{F_0}{\omega^2 - \gamma^2}[\cos \gamma t - \cos \omega t].$$

16. Compare the result obtained in part (b) of the preceding problem with the solution obtained using variation of parameters when the external force is $F_0 \cos \omega t$.

In Problems 17 and 18 solve the given initial-value problem.

17. $\dfrac{d^2x}{dt^2} + 4x = -5 \sin 2t + 3 \cos 2t$

$x(0) = -1 \qquad \dfrac{dx}{dt}\bigg|_{t=0} = 1.$

18. $\dfrac{d^2x}{dt^2} + 9x = 5 \sin 3t$

$x(0) = 2 \qquad \dfrac{dx}{dt}\bigg|_{t=0} = 0$

19.(a) Show that $x(t)$ given in part (a) of Problem 15 can be written in the form

$$x(t) = \frac{-2F_0}{\omega^2 - \gamma^2} \sin \frac{1}{2}(\gamma - \omega)t \, \sin\frac{1}{2}(\gamma + \omega)t.$$

(b) If we define $\varepsilon = \frac{1}{2}(\gamma - \omega)$, show that when ε is small that an *approximate* solution is

$$x(t) = \frac{F_0}{2\varepsilon\gamma} \sin \varepsilon t \sin \gamma t.$$

(c) Evaluate
$$\lim_{\varepsilon \to 0} \frac{F_0}{2\varepsilon\gamma} \sin \varepsilon t \sin \gamma t.$$

EXAMPLE   In part (b) of the preceding problem, when ε is small, the frequency $\gamma/2\pi$ of the impressed force is close to the frequency $\omega/2\pi$ of free vibrations. When this occurs, the motion is as indicated in Figure 5.22. Oscillations of this kind are called *beats* and are due to the fact that the frequency of $\sin \varepsilon t$ is quite small in comparison to the frequency of $\sin \gamma t$. The dotted curves, or *envelope* of the graph of $x(t)$, are obtained from the graphs of $\pm(F_0/2\varepsilon\gamma) \sin \varepsilon t$.

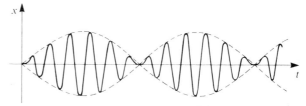

Figure 5.22

---

20. Show that the solution of

$$\frac{d^2x}{dt^2} + 25x = 10 \cos 7t$$

$$x(0) = 0 \qquad \frac{dx}{dt}\bigg|_{t=0} = 0$$

is $x(t) = \dfrac{5}{6} \sin t \sin 6t$.

[O] 5.4 Analogous Systems

One differential equation can serve as a mathematical model for many different physical phenomena.

In Section 1.3 it was seen that the angular displacements θ of a simple pendulum are described by the nonlinear second-order equation

$$\frac{d^2\theta}{dt^2} + \frac{g}{l}\sin\theta = 0 \tag{1}$$

where l is the length of the pendulum rod. For small displacements $\sin\theta$ is replaced by θ and the resulting differential equation,

$$\frac{d^2\theta}{dt^2} + \frac{g}{l}\theta = 0, \tag{2}$$

indicates that the pendulum exhibits simple harmonic motion. Inspection of the solution (2) reveals that the period of small oscillations is given by the familiar formula $T = 2\pi\sqrt{l/g}$.

The series circuit analogue

You may also recall from Section 1.3 that when the various voltage drops in an L–R–C series electrical circuit (Figure 5.23) are added (Kirchoff's second law) we obtain the following differential equation

$$L\frac{d^2q}{dt^2} + R\frac{dq}{dt} + \frac{1}{C}q = E(t) \tag{3}$$

where $q(t)$ is the instantaneous charge on the capacitor and $E(t)$ is the impressed voltage or electromotive force (emf), on the circuit.

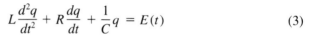

Figure 5.23

EXAMPLE

A series circuit contains only a capacitor and inductor. If the capacitor has an initial charge q_0, determine the subsequent charge $q(t)$.

Solution: From equation (3) we can write

$$L\frac{d^2q}{dt^2} + \frac{1}{C}q = 0$$

subject to $q(0) = q_0$. Assuming that no current flows initially (for example, a switch could be open) then $q'(0) = 0$ since $q'(t) = i(t)$. The general solution of the equation is

$$q(t) = c_1 \cos\frac{1}{\sqrt{LC}}t + c_2 \sin\frac{1}{\sqrt{LC}}t.$$

Now the initial conditions imply $c_1 = q_0$ and $c_2 = 0$ so that

$$q(t) = q_0 \cos\frac{1}{\sqrt{LC}}t.$$

The twisted shaft

Similarly, it can be shown that the differential equation governing the torsional motion of a weight suspended from the end of an elastic shaft is

$$I\frac{d^2\theta}{dt^2} + c\frac{d\theta}{dt} + k\theta = T(t). \tag{4}$$

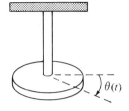

Figure 5.24

As shown in Figure 5.24 the function $\theta(t)$ represents the amount of twist of the weight at any time.

By comparing equations (3) and (4) with the differential equation of forced motion with damping

$$m\frac{d^2x}{dt^2} + \beta\frac{dx}{dt} + kx = f(t) \tag{5}$$

we see, that with the exception of terminology, there is absolutely no difference between the mathematics of vibrating springs, torsional vibrations, and simple series circuits. The following table gives a comparison of the analogous parts of these three kinds of systems.

Mechanical	Series electrical	Torsional
m (mass)	L (inductance)	I (moment of inertia)
β (damping)	R (resistance)	c (damping)
k (spring constant)	$\dfrac{1}{C}$ (reciprocal of capacitance— called elastance)	k (elastic shaft constant)
$f(t)$ (applied force)	$E(t)$ (impressed voltage)	$T(t)$ (applied torque)

A second-order linear differential equation such as (5) appears in the mathematical analysis of many other problems in physics, engineering, and even chemistry. Although a substantial number of these probems deal directly with classical vibrational phenomena such as oscillations of tuning forks or oscillations of atomic particles, many new and interesting applications are being found in the field of mathematical biology.* Our goal, of course, is not to study all possible applications, but to acquaint the student with the mathematical procedures which are common to these problems.

EXERCISES 5.4   *Answers to odd-numbered problems begin on page A-16.*

1. Find the equation of motion describing the small displacements $\theta(t)$ of a simple pendulum of length 2 ft released at $t = 0$ with a displacement of 1/2 radians to the right of the vertical and angular velocity of $2\sqrt{3}$ ft/sec to the right. What are the amplitude, period, and frequency of motion?

*See M. Braun, *Differential Equations and Their Applications*, (N.Y.: Springer-Verlag, 1975), pp. 257–268, for an application of $x'' + 2\lambda x' + \omega^2 x = 0$ to "A model for the detection of diabetes."

2. In Problem 1, at what times does the pendulum pass through its equilibrium position? At what times does the pendulum attain its extreme angular displacements on either side of its equilibrium position?

3. Show that the current $i(t)$ in a series circuit such as illustrated in Figure 5.23 satisfies the differential equation

$$L\frac{d^2i}{dt^2} + R\frac{di}{dt} + \frac{1}{C}i = E'(t).$$

4. Use undetermined coefficients to find a particular solution of the differential equation in Problem 3 when $L = 1$ farad, $R = 2$ ohms, $C = 0.25$ farad, $E(t) = 50 \cos t$ volts. The particular solution i_p in this case is called the *steady-state current*.

5. Use the differential equation in Problem 3 to find the steady-state current in an L—C circuit when $L = 0.1$ henry, $C = 0.1$ farad, and $E(t) = 120 \sin kt$ volts, $k \neq 10$.

6. Solve equation (3) of Section 5.4 for $q(t)$ if $L = 5/3$ henry, $R = 10$ ohms, $C = 1/30$ farad, $E = 300$ volts, and $q(0) = 0$, $i(0) = 0$. Determine the current $i(t)$. What is the maximum charge on the capacitor?

7. A series L—R—C circuit contains $L = 1/2$ henry, $R = 10$ ohms, $C = 1/100$ farad, and $E(t) = 150$ volts. Determine the instantaneous charge $q(t)$ on the capacitor for $t > 0$ if $q(0) = 1$ and $i(0) = 0$. What is the charge on the capacitor after a long time?

**CHAPTER 5
SUMMARY**

When a mass is attached to a spring it stretches to a position where the restoring force ks of the spring is balanced by the weight mg. Any subsequent motion is then measured x units (feet in the engineering system) above or below this **equilibrium position**. When the mass is above the equilibrium position, we adopt the convention that $x < 0$, whereas when the mass is below the equilibrium position we take $x > 0$.

The differential equation of motion is obtained by equating Newton's second law $F = ma = m(d^2x/dt^2)$ with the net force acting on the mass at any time. We distinguish three cases.

CASE I   The equation

$$m\frac{d^2x}{dt^2} = -kx$$

or

$$\frac{d^2x}{dt^2} + \omega^2x = 0, \qquad \omega^2 = \frac{k}{m} \tag{1}$$

describes the motion under the assumptions that no damping force and no external impressed forces are acting on the system. The solution of (1) is $x(t) = c_1 \cos \omega t + c_2 \sin \omega t$ and the mass is said to exhibit **simple harmonic motion**. The constants c_1 and c_2 are determined by the initial position $x(0)$ and the initial velocity $x'(0)$ of the mass.   □

**CHAPTER 5
SUMMARY**

CASE II   When a damping force is present the differential equation becomes

$$m\frac{d^2x}{dt^2} = -kx - \beta\frac{dx}{dt}, \qquad \beta > 0$$

or $$\frac{d^2x}{dt^2} + 2\lambda\frac{dx}{dt} + \omega^2x = 0, \qquad 2\lambda = \frac{\beta}{m}, \ \omega^2 = \frac{k}{m}. \tag{2}$$

The resulting motion is said to be **overdamped, critically damped,** or **underdamped** accordingly as $\lambda^2 - \omega^2 > 0$, $\lambda^2 - \omega^2 = 0$, $\lambda^2 - \omega^2 < 0$.

The respective solutions of equation (2) are then

$$x(t) = c_1e^{m_1t} + c_2e^{m_2t},$$

where $m_1 = -\lambda + \sqrt{\lambda^2 - \omega^2}$, $m_2 = -\lambda - \sqrt{\lambda^2 - \omega^2}$;

$$x(t) = c_1e^{m_1t} + c_2te^{m_1t},$$

where $m_1 = -\lambda$; and

$$x(t) = e^{-\lambda t}[c_1 \cos \sqrt{\omega^2 - \lambda^2}\, t + c_2 \sin \sqrt{\omega^2 - \lambda^2}\, t].$$

In each case, the damping force is responsible for the displacements becoming negligible for large time, that is, $x \to 0$ as $t \to \infty$.   □

The motion described in Cases I and II is said to be **free motion**.

CASE III   When an external force is impressed on the system for $t > 0$, the differential equation becomes

$$m\frac{d^2x}{dt^2} = -kx - \beta\frac{dx}{dt} + f(t)$$

or $$\frac{d^2x}{dt^2} + 2\lambda\frac{dx}{dt} + \omega^2x = F(t), \tag{3}$$

where λ and ω^2 are defined in (2). The solution of the nonhomogeneous equation (3) is $x(t) = x_c + x_p$.

Since the complementary function x_c always contains the factor $e^{-\lambda t}$, it will be **transient**, that is, $x_c \to 0$ as $t \to \infty$. If $F(t)$ is periodic then x_p will be a **steady-state solution**.   □

In the absence of a damping force, an impressed periodic force can cause the amplitudes of vibration to become very large. If the frequency of the external force is the same as the frequency $\omega/2\pi$ of free vibrations, we say that the system is in a state of **pure resonance**. In this case the amplitudes of vibrations become unbounded as $t \to \infty$. In the presence of a damping force, amplitudes of oscillatory motion are always bounded. However, large and potentially destructive amplitudes can occur.

When a series circuit containing an inductor, resistor, and capacitor is driven by an electromotive force $E(t)$, the resulting differential equations for the charge $q(t)$ or the current $i(t)$ are quite similar in structure to equation (3). Hence the analysis of such circuits is the same as outlined above.

CHAPTER 5
TEST

Answers to odd-numbered problems begin on page A-16.

Answer Problems 1–9 without referring back to the text. Fill in the blank or answer true/false.

1. If a 10-lb weight stretches a spring 2.5 ft, a 32-lb weight will stretch it _____ ft.

2. The period of simple harmonic motion of an 8-lb weight attached to a spring whose constant is 6.25 lb/ft is _____ sec.

3. The differential equation of a weight on a spring is $x'' + 16x = 0$. If the weight is released at $t = 0$ from 1 m above the equilibrium position with a downward velocity of 3 m/sec, the amplitude of vibrations is _____ m.

4. Pure resonance cannot take place in the presence of a damping force. _____

5. In the presence of damping the displacements of a weight on a spring will always approach zero as $t \to \infty$. _____

6. A weight on a spring whose motion is critically damped can possibly pass through the equilibrium position twice. _____

7. At critical damping any increase in damping will result in an _____ system.

8. In describing simple harmonic motion by $x = \sqrt{2}/2 \sin(2t + \phi)$, the phase angle ϕ is _____ when $x(0) = -1/2$ and $x'(0) = 1$.

9. A 16-lb weight attached to a spring exhibits simple harmonic motion. If the frequency of oscillations is $3/2\pi$ vibrations/sec, the spring constant is _____ .

10. A 12-lb weight stretches a spring 2 ft. The weight is released from a point 1 ft below the equilibrium position with an upward velocity of 4 ft/sec.

   (a) Find the equation describing the resulting simple harmonic motion.

   (b) What is the amplitude, period, and frequency of motion?

   (c) At what times does the weight return to the point 1 ft below the equilibrium position?

**CHAPTER 5
TEST**

(d) At what times does the weight pass through the equilibrium position moving upward? Moving downward?

(e) What is the velocity of weight at $t = 3\pi/16$?

(f) At what times is the velocity zero?

11. A force of 2 lb stretches a spring 1 ft. With one end held fixed, an 8-lb weight is attached to the other end and the system lies on a table which imparts a frictional force numerically equal to 3/2 times the instantaneous velocity. Initially the weight is displaced 4 in. above the equilibrium position and released from rest. Find the equation of motion if the motion takes place along a horizontal straight line which is taken as the x-axis.

12. A 32-lb weight stretches a spring 6 in. The weight moves through a medium offering a damping force numerically equal to β times the instantaneous velocity. Determine the values of β for which the system will exhibit oscillatory motion.

13. A spring with constant $k = 2$ is suspended in a liquid that offers a damping force numerically equal to 4 times the instantaneous velocity. If a mass m is suspended from the spring, determine the values of m for which the subsequent free motion is nonoscillatory.

14. The vertical motion of a weight attached to a spring is described by the initial-value problem

$$\frac{1}{4}\frac{d^2x}{dt^2} + \frac{dx}{dt} + x = 0$$

$$x(0) = 4 \qquad x'(0) = 2.$$

Determine the maximum vertical displacement.

15. A 4-lb weight stretches a spring 18 in. A periodic force equal to $f(t) = \cos \gamma t + \sin \gamma t$ is impressed on the system starting at $t = 0$. In the absence of a damping force, for what value of γ will the system be in a state of pure resonance?

16. Find a particular solution for

$$\frac{d^2x}{dt^2} + 2\lambda\frac{dx}{dt} + \omega^2 x = A,$$

where A is a constant force.

17. A 4-lb weight is suspended from a spring whose constant is 3 lb/ft. The entire system is immeresed in a fluid offering a damping force numerically equal to the instantaneous velocity. Beginning at $t = 0$,
continued

an external force equal to $f(t) = e^{-t}$ is impressed on the system. Determine the equation of motion if the weight is released from rest at a point 2 ft below the equilibrium position.

18. A series circuit contains an inductance of $L = 1$ henry, a capacitance of $C = 10^{-4}$ farad, and an electromotive force of $E(t) = 100 \sin 50t$ volts. Initially the charge q and current i are zero.

    (a) Find the equation for the charge at any time.

    (b) Find the equation for the current at any time.

    (c) Find the times for which the charge on the capacitor is zero.

Differential Equations With Variable Coefficients

Introduction

The same ease with which we solved differential equations with constant coefficients does not usually carry over to equations with variable coefficients. In fact, we cannot expect to be able to express the solutions of even a simple linear equation such as $y'' - xy = 0$ in terms of the usual sines, cosines, logarithms, exponentials, and other elementary functions. Although it is easily verified that

$$(1 - x^2)y'' - 2xy' + 2y = 0 \qquad \text{and} \qquad x^2y'' + xy' + (x^2 - \tfrac{1}{4})y = 0$$

have elementary solutions

$$y = x \qquad \text{and} \qquad y = \frac{\sin x}{\sqrt{x}}$$

respectively, the best that we can *usually* expect from equations of this sort is an infinite series solution. On the other hand, we shall now consider one important type of equation with variable coefficients whose general solution can always be written in terms of elementary functions.

6.1 The Cauchy-Euler Equation

Any differential equation of the form

$$a_n x^n \frac{d^n y}{dx^n} + a_{n-1} x^{n-1} \frac{d^{n-1}y}{dx^{n-1}} + \cdots + a_1 x \frac{dy}{dx} + a_0 y = g(x)$$

where $a_n, a_{n-1}, \ldots, a_0$ are constants is said to be a **Cauchy-Euler equation.**[*] The obvious characteristic of this type of equation is that the

---

[*]Named after the famous mathematicians Leonhard Euler (Swiss, 1707–1783) and Augustin Louis Cauchy (French, 1789–1857). The Cauchy-Euler equation is sometimes referred to as the **equidimensional equation**.

the polynomial coefficients x^k matches the order of differentiation in the terms $d^k y / dx^k$ for $k = 1, 2, \ldots, n$.

For the sake of discussion we shall confine our attention to solving the homogeneous second-order equation

$$ax^2 \frac{d^2 y}{dx^2} + bx \frac{dy}{dx} + cy = 0.$$

The solution of higher order equations follows analogously. Also, we can solve the nonhomogeneous equation

$$ax^2 \frac{d^2 y}{dx^2} + bx \frac{dy}{dx} + cy = g(x)$$

by variation of parameters once we have determined the complementary function $y_c(x)$.

Note: The coefficient of $d^2 y/dx^2$ is zero at $x = 0$. Hence, in order to guarantee that the fundamental results of Theorem 4.1 are applicable to the Cauchy-Euler equation, we shall confine our attention to finding the general solution on the interval $0 < x < \infty$. Solutions on the interval $-\infty < x < 0$ can be obtained by substituting $t = -x$ in the differential equation.

The method of solution

We try a solution of the form $y = x^m$ where m is to be determined. The first and second derivatives are

$$\frac{dy}{dx} = mx^{m-1}$$

$$\frac{d^2 y}{dx^2} = m(m-1)x^{m-2},$$

so the differential equation becomes

$$ax^2 \frac{d^2 y}{dx^2} + bx \frac{dy}{dx} + cy = ax^2 \cdot m(m-1)x^{m-2} + bx \cdot mx^{m-1} + cx^m$$

$$= am(m-1)x^m + bmx^m + cx^m$$

$$= x^m[am(m-1) + bm + c].$$

Thus $y = x^m$ will be a solution of the differential equation whenever m is a solution of the **auxiliary equation**

$$am(m-1) + bm + c = 0 \tag{1}$$

or $\qquad\qquad\quad am^2 + (b - a)m + c = 0.$

There are three different cases to be considered depending on whether the roots of this quadratic equation are real and distinct, real and equal, or complex conjugates.

CASE I   Let m_1 and m_2 be the real roots of (1) such that $m_1 \neq m_2$.

Then

$$y_1 = x^{m_1} \quad \text{and} \quad y_2 = x^{m_2}$$

form a fundamental set of solutions. Hence the general solution is

$$y = c_1 x^{m_1} + c_2 x^{m_2}. \tag{2}$$

□

EXAMPLE   Solve

$$x^2 \frac{d^2 y}{dx^2} - 2x \frac{dy}{dx} - 4y = 0.$$

Solution: Rather than just memorizing equation (1) it is preferable to assume $y = x^m$ as the solution a few times in order to understand the origin and the difference between this new form of the auxiliary equation and that obtained in Chapter 4. Differentiate twice and substitute back into the equation:

$$\frac{dy}{dx} = mx^{m-1}$$

$$\frac{d^2 y}{dx^2} = m(m-1)x^{m-2}$$

$$x^2 \frac{d^2 y}{dx^2} - 2x \frac{dy}{dx} - 4y = x^2 \cdot m(m-1)x^{m-2} - 2x \cdot mx^{m-1} - 4x^m$$

$$= x^m[m(m-1) - 2m - 4]$$

$$= x^m[m^2 - 3m - 4]$$

$$= 0$$

if $m^2 - 3m - 4 = 0$. Now $(m+1)(m-4) = 0$ implies $m_1 = -1, m_2 = 4$ so that

$$y = c_1 x^{-1} + c_2 x^4.$$

CASE II If $m_1 = m_2$ then we obtain only one solution, namely, $y = x^{m_1}$. When the roots of the quadratic equation $am^2 + (b - a)m + c = 0$ are equal, the discriminant of the coefficients is necessarily zero. It follows from the quadratic formula that the root must be $m_1 = -(b - a)/2a$.

Now we can construct a second solution y_2 using formula (4) of Section 4.2. We first write the Cauchy-Euler equation in the form

$$\frac{d^2 y}{dx^2} + \frac{b}{ax} \frac{dy}{dx} + \frac{c}{ax^2} y = 0$$

and make the identification $P(x) = b/ax$. Thus

$$y_2 = x^{m_1} \int \frac{e^{-\int (b/ax)\, dx}}{(x^{m_1})^2}\, dx$$

$$= x^{m_1} \int \frac{e^{-(b/a)\ln x}}{x^{2m_1}}\, dx$$

$$= x^{m_1} \int x^{-b/a} \cdot x^{-2m_1}\, dx \qquad [e^{-(b/a)\ln x} = e^{\ln x^{-b/a}} = x^{-b/a}]$$

$$= x^{m_1} \int x^{-b/a} \cdot x^{(b-a)/a}\, dx \qquad \left[2m_1 = -\frac{b-a}{a} \right]$$

$$= x^{m_1} \int \frac{dx}{x}$$

$$= x^{m_1} \ln x.$$

The general solution is then

$$y = c_1 x^{m_1} + c_2 x^{m_1} \ln x. \qquad \qquad \Box(3)$$

EXAMPLE Solve $$4x^2 \frac{d^2y}{dx^2} + 8x \frac{dy}{dx} + y = 0.$$

Solution: The substitution $y = x^m$ yields

$$4x^2 \frac{d^2y}{dx^2} + 8x \frac{dy}{dx} + y = x^m[4m(m-1) + 8m + 1]$$

$$= x^m[4m^2 + 4m + 1]$$

$$= 0$$

when $$4m^2 + 4m + 1 = 0 \qquad \text{or} \qquad (2m+1)^2 = 0.$$

Since $m_1 = -1/2$ the general solution is

$$y = c_1 x^{-1/2} + c_2 x^{-1/2} \ln x.$$

CASE III If m_1 and m_2 are complex conjugates, say,

$$m_1 = \alpha + i\beta \qquad m_2 = \alpha - i\beta$$

where α and β are real, then a formal solution is

$$y = C_1 x^{\alpha+i\beta} + C_2 x^{\alpha-i\beta}$$

But as in the case of equations with constant coefficients, when the roots of the auxiliary equation are complex we wish to write the solution in terms of real functions only. We note the identity

$$x^{i\beta} = (e^{\ln x})^{i\beta} = e^{i\beta \ln x}$$

which, by Euler's formula, is the same as

$$x^{i\beta} = \cos(\beta \ln x) + i \sin(\beta \ln x).$$

Therefore

$$y = C_1 x^{\alpha + i\beta} + C_2 x^{\alpha - i\beta}$$

$$= x^{\alpha}[C_1 x^{i\beta} + C_2 x^{-i\beta}]$$

$$= x^{\alpha}[C_1\{\cos(\beta \ln x) + i \sin(\beta \ln x)\} + C_2\{\cos(\beta \ln x) - i \sin(\beta \ln x)\}]$$

$$= x^{\alpha}[(C_1 + C_2)\cos(\beta \ln x) + C_1 i - C_2 i)\sin(\beta \ln x)]$$

$$= x^{\alpha}[(C_1 + C_2)\cos(\beta \ln x) + (C_1 i - C_2 i)\sin(\beta \ln x)]$$

On the interval $0 < x < \infty$ it can be verified that

$$y_1 = x^{\alpha} \cos(\beta \ln x) \quad \text{and} \quad y_2 = x^{\alpha} \sin(\beta \ln x)$$

constitute a fundamental set of solutions of the differential equation. It follows that the general solution is

$$y = x^{\alpha}[c_1 \cos(\beta \ln x) + c_2 \sin(\beta \ln x)] \qquad \Box(4)$$

EXAMPLE

Solve
$$x^2 \frac{d^2 y}{dx^2} + 3x \frac{dy}{dx} + 3y = 0.$$

Solution: We have

$$x^2 \frac{d^2 y}{dx^2} + 3x \frac{dy}{dx} + 3y = x^m[m(m-1) + 3m + 3]$$

$$= x^m[m^2 + 2m + 3]$$

$$= 0$$

when $m^2 + 2m + 3 = 0$.

From the quadratic formula we find $m_1 = -1 + \sqrt{2}i$ and $m_2 = -1 - \sqrt{2}i$. If we make the identifications $\alpha = -1$ and $\beta = \sqrt{2}$, we see from (4) that the general solution is

$$y = x^{-1}[c_1 \cos(\sqrt{2} \ln x) + c_2 \sin(\sqrt{2} \ln x)].$$

EXAMPLE

Solve the third-order Cauchy–Euler equation

$$x^3 \frac{d^3 y}{dx^3} + 5x^2 \frac{d^2 y}{dx^2} + 7x \frac{dy}{dx} + 8y = 0.$$

Solution: The first three derivatives of $y = x^m$ are

$$\frac{dy}{dx} = mx^{m-1}$$

$$\frac{d^2y}{dx^2} = m(m-1)x^{m-2}$$

$$\frac{d^3y}{dx^3} = m(m-1)(m-2)x^{m-3}$$

so that the given differential equation becomes

$$x^3\frac{d^3y}{dx^3} + 5x^2\frac{d^2y}{dx^2} + 7x\frac{dy}{dx} + 8y$$

$$= x^3m(m-1)(m-2)x^{m-3} + 5x^2m(m-1)x^{m-2} + 7xmx^{m-1} + 8x^m$$

$$= x^m[m(m-1)(m-2) + 5m(m-1) + 7m + 8]$$

$$= x^m[m^3 + 2m^2 + 4m + 8].$$

In this case we see that $y = x^m$ will be a solution of the differential equation provided m is a root of the cubic equation

$$m^3 + 2m^2 + 4m + 8 = 0$$

or

$$m^2(m+2) + 4(m+2) = 0$$

$$(m+2)(m^2+4) = 0.$$

The roots are: $m_1 = -2$, $m_2 = 2i$, $m_3 = -2i$. Hence the general solution is

$$y = c_1x^{-2} + c_2\cos(2\ln x) + c_3\sin(2\ln x).$$

EXAMPLE

Solve the nonhomogeneous equation

$$x^2y'' - 3xy' + 3y = 2x^4e^x.$$

Solution: The substitution $y = x^m$ leads to the auxiliary equation

$$m(m-1) - 3m + 3 = 0 \quad \text{or} \quad (m-1)(m-3) = 0.$$

Thus

$$y_c = c_1x + c_2x^3.$$

Before using variation of parameters, recall that the formulas $u_1' = -y_2f(x)/W$ and $u_2' = y_1f(x)/W$ were derived under the assumption that the differential equation has been put into the form $y'' + P(x)y' + Q(x)y = f(x)$. Therefore we divide the given equation by x^2 and then make the identification that $f(x) = 2x^2e^x$.

Now

$$W = \begin{vmatrix} x & x^3 \\ 1 & 3x^2 \end{vmatrix} = 3x^3 - x^3 = 2x^3$$

so that

$$u_1' = -\frac{x^3(2x^2e^x)}{2x^3} = -x^2e^x \quad \text{and} \quad u_2' = \frac{x(2x^2e^x)}{2x^3} = e^x.$$

The integral of the latter function is immediate, but in the case of u_1' we integrate by parts twice. The results are

$$u_1 = -x^2e^x + 2xe^x - 2e^x \quad \text{and} \quad u_2 = e^x.$$

Hence $\qquad y_p = u_1 y_1 + u_2 y_2$

$$= (-x^2 e^x + 2xe^x - 2e^x)x + e^x x^3$$

$$= 2x^2 e^x - 2xe^x.$$

Finally we have $\qquad y = y_c + y_p$

$$= c_1 x + c_2 x^3 + 2x^2 e^x - 2xe^x.$$

EXERCISES 6.1 *Answers to odd-numbered problems begin on page A-16.*

In Problems 1–20 solve the given differential equation.

1. $x^2 y'' - 2y = 0$ $\qquad\qquad$ **2.** $4x^2 y'' + y = 0$

3. $xy'' + y' = 0$ $\qquad\qquad$ **4.** $xy'' - y' = 0$

5. $x^2 y'' + xy' + 4y = 0$ $\qquad\qquad$ **6.** $x^2 y'' + 5xy' + 3y = 0$

7. $x^2 y'' - 3xy' - 2y = 0$ $\qquad\qquad$ ★ **8.** $x^2 y'' + 3xy' - 4y = 0$

8. $25x^2 y'' + 25xy' + y = 0$ $\qquad\qquad$ **10.** $4x^2 y'' + 4xy' - y = 0$

11. $x^2 y'' + 5xy' + 4y = 0$ $\qquad\qquad$ **12.** $x^2 y'' + 8xy' + 6y = 0$

13. $x^2 y'' - xy' + 2y = 0$ $\qquad\qquad$ **14.** $x^2 y'' - 7xy' + 41y = 0$

15. $3x^2 y'' + 6xy' + y = 0$ $\qquad\qquad$ **16.** $2x^2 y'' + xy' + y = 0$

17. $x^3 y''' - 6y = 0$ $\qquad\qquad$ ★**18.** $x^3 y''' + xy' - y = 0$

19. $x^3 \dfrac{d^3 y}{dx^3} - 2x^2 \dfrac{d^2 y}{dx^2} - 2x \dfrac{dy}{dx} + 8y = 0$

20. $x^3 \dfrac{d^3 y}{dx^3} - 2x^2 \dfrac{d^2 y}{dx^2} + 4x \dfrac{dy}{dx} - 4y = 0$

In Problems 21–24 solve the given differential equation subject to the indicated initial conditions.

21. $x^2 y'' + 3xy' = 0$, $\quad y(1) = 0$, $\quad y'(1) = 4$

★**22.** $x^2 y'' - 5xy' + 8y = 0$, $\quad y(2) = 32$, $\quad y'(2) = 0$

23. $x^2 y'' + xy' + y = 0$, $\quad y(1) = 1$, $\quad y'(1) = 2$

24. $x^2 y'' - 3xy' + 4y = 0$, $\quad y(1) = 5$, $\quad y'(1) = 3$

Solve Problems 25–30 by variation of parameters.

25. $xy'' + y' = x$ $\qquad\qquad$ **26.** $xy'' - 4y' = x^4$

27. $2x^2 y'' + 5xy' + y = x^2 - x$ $\qquad$ ★**28.** $x^2 y'' - 2xy' + 2y = x^4 e^x$

29. $x^2 y'' - xy' + y = 2x$ $\qquad\qquad$ **30.** $x^2 y'' - 2xy' + 2y = x^3 \ln x$

The Cauchy-Euler equation can be reduced to an equation with constant coefficients by means of the substitution $x = e^t$. Solve Problems 31–36 using this change of variables.

EXAMPLE   Solve
$$x^2\frac{d^2y}{dx^2} - x\frac{dy}{dx} + y = \ln x.$$

Solution: With the substitution $x = e^t$ or $t = \ln x$ it follows from the chain rule that

$$\frac{dy}{dx} = \frac{dy}{dt}\frac{dt}{dx} = \frac{1}{x}\frac{dy}{dt}$$

$$\frac{d^2y}{dx^2} = \frac{1}{x}\frac{d}{dx}\left(\frac{dy}{dt}\right) + \frac{dy}{dt}\left(-\frac{1}{x^2}\right)$$

$$= \frac{1}{x}\left(\frac{d^2y}{dt^2}\frac{1}{x}\right) + \frac{dy}{dt}\left(-\frac{1}{x^2}\right)$$

$$= \frac{1}{x^2}\left[\frac{d^2y}{dt^2} - \frac{dy}{dt}\right].$$

Substituting in the given differential equation and simplifying yields

$$\frac{d^2y}{dt^2} - 2\frac{dy}{dt} + y = t.$$

Since this last equation has constant coefficients, its auxiliary equation is $m^2 - 2m + 1 = 0$ or $(m - 1)^2 = 0$. Thus we obtain

$$y_c = c_1e^t + c_2te^t.$$

By undetermined coefficients we try a particular solution of the form $y_p = A + Bt$. This assumption leads to $-2B + A + Bt = t$ so that $A = 2$ and $B = 1$. Hence

$$y = y_c + y_p$$
$$= c_1e^t + c_2te^t + 2 + t$$

and so the general solution of the original differential equation on the interval $0 < x < \infty$ is

$$y = c_1x + c_2x \ln x + 2 + \ln x.$$

31. $x^2\frac{d^2y}{dx^2} + 10x\frac{dy}{dx} + 8y = x^2$ **32.** $x^2y'' - 4xy' + 6y = \ln x^2$

33. $x^2y'' - 3xy' - 13y = 4 + 3x$

★**34.** $2x^2y'' - 3xy' - 3y = 1 + 2x + x^2$ **35.** $x^2y'' + 9xy' - 20y = 5/x^3$

36. $x^3\frac{d^3y}{dx^3} - 3x^2\frac{d^2y}{dx^2} + 6x\frac{dy}{dx} - 6y = 3 + \ln x^3$

Miscellaneous problems

In Problems 37–39 solve the given differential equation.

37. $(x - 1)^2\frac{d^2y}{dx^2} - 2(x - 1)\frac{dy}{dx} - 4y = 0$ [*Hint:* Let $t = x - 1$.]

★**38.** $(3x + 4)^2 y'' + 10(3x + 4)y' + 9y = 0$

39. $(x + 2)^2 y'' + (x + 2)y' + y = 0$

40. Show that the auxiliary equation for the fourth-order Cauchy-Euler equation

$$x^4 \frac{d^4 y}{dx^4} + 6x^3 \frac{d^3 y}{dx^3} + 9x^2 \frac{d^2 y}{dx^2} + 3x \frac{dy}{dx} + y = 0$$

has repeated complex roots. Find the general solution of the differential equation.

6.2 Power Series Solutions

6.2.1 The Procedure

A review

In the discussion that follows we shall be concerned with the use of *power series*. The reader should recall the following facts from calculus.

(a) A power series in $(x - a)$ is an infinite series of the form $\sum_{n=0}^{\infty} c_n (x - a)^n$.

(b) Every power series has an *interval of convergence*. The interval of convergence is the set of all numbers for which the series converges.

(c) A power series *converges absolutely* at a number x_1 if $\sum_{n=0}^{\infty} |c_n||x_1 - a|^n$ converges.

(d) Every interval of convergence has a *radius of convergence R*.

(e) A power series converges absolutely for $|x - a| < R$ and diverges for $|x - a| > R$. When $R = 0$ the interval of convergence consists of the single number a; when $R = \infty$ the power series converges for all numbers x.

(f) The radius of convergence is usually obtained from the *ratio test:* *

$$\lim_{n \to \infty} \left| \frac{c_{n+1}}{c_n} \right| |x - a|$$

The series will converge absolutely for those values of x for which the limit is strictly less than one.

(g) If R is not 0 or ∞ then the interval of convergence may include the endpoints $a - R$ and $a + R$.

(h) A power series represents a continuous function within its interval of convergence.

---

*The *root test* is sometimes applicable.

(i) A power series can be differentiated termwise within its interval of convergence.

(j) A power series can be integrated termwise within its interval of convergence.

(k) Two power series with a common interval of convergence can be added term by term.

Solution of a differential equation

We have already seen in Section 1.1 that the function $y = e^{x^2}$ is an explicit solution of the linear first-order differential equation

$$\frac{dy}{dx} - 2xy = 0. \tag{1}$$

But e^x has the well known power series representation

$$e^x = \sum_{n=0}^{\infty} \frac{x^n}{n!} \tag{2}$$

so that the solution can then be written as

$$y = e^{x^2} = \sum_{n=0}^{\infty} \frac{x^{2n}}{n!} \tag{3}$$

Both series (2) and (3) converge for all real values of x.

In other words, knowing the solution in advance, we were able to find an infinite series solution of the differential equation. We now propose to obtain the power series solution of (1) directly; the method of attack is similar to the technique of undetermined coefficients.

If we assume that a solution exists in the form of power series in x,

$$y = \sum_{n=0}^{\infty} c_n x^n \tag{4}$$

we pose the question: Can we determine coefficients c_n for which (4) converges to a function satisfying (1)? Formal* term-by-term differentiation of (4) gives

$$\frac{dy}{dx} = \sum_{n=0}^{\infty} n c_n x^{n-1}$$

$$= \sum_{n=1}^{\infty} n c_n x^{n-1}. \qquad \text{[the first term corresponding} \atop \text{to } n = 0 \text{ is zero]}$$

Using this last result and the assumption (4) we find

$$\frac{dy}{dx} - 2xy = \sum_{n=1}^{\infty} n c_n x^{n-1} - \sum_{n=0}^{\infty} 2 c_n x^{n+1}. \tag{5}$$

We would like to add the two series in (5), but in order to do this we must have both summation indices start at the same value. In addition, it is desirable that the numerical values of the powers of x be "in phase" in each

---

*At this point we do not know the interval of convergence.

summation. That is, if one series starts with a multiple of, say, x to the first power, then we want the other series to start with this same power. To this end we write (5) as

$$\frac{dy}{dx} - 2xy = 1 \cdot c_1 x^0 + \sum_{n=2}^{\infty} nc_n x^{n-1} - \sum_{n=0}^{\infty} 2c_n x^{n+1} \tag{6}$$

and let $k = n - 1$ in the first series and $k = n + 1$ in the second.* The right side of equation (6) then becomes

$$c_1 + \sum_{k=1}^{\infty} (k + 1)c_{k+1} x^k - \sum_{k=1}^{\infty} 2c_{k-1} x^k.$$

By adding the series termwise it follows that

$$\frac{dy}{dx} - 2xy = c_1 + \sum_{k=1}^{\infty} [(k + 1)c_{k+1} - 2c_{k-1}]x^k = 0. \tag{7}$$

Hence, in order to have (7) identically zero it is necessary that the coeffients satisfy $\qquad\qquad\qquad c_1 = 0$

and $\qquad\qquad (k + 1)c_{k+1} - 2c_{k-1} = 0, \ k = 1,2,3 \dots. \tag{8}$

Equation (8) provides a *recurrence relation* that determines the c_k. Since $k + 1 \neq 0$ for all of the indicated values of k we can write (8) as

$$c_{k+1} = \frac{2c_{k-1}}{k + 1}. \tag{9}$$

Iteration of this last formula then gives

$$k = 1, \quad c_2 = \frac{2}{2}c_0 = c_0$$

$$k = 2, \quad c_3 = \frac{2}{3}c_1 = 0$$

$$k = 3, \quad c_4 = \frac{2}{4}c_2 = \frac{1}{2}c_0 = \frac{1}{2!}c_0$$

$$k = 4, \quad c_5 = \frac{2}{5}c_3 = 0$$

$$k = 5, \quad c_6 = \frac{2}{6}c_4 = \frac{1}{3 \cdot 2!}c_0 = \frac{1}{3!}c_0$$

$$k = 6, \quad c_7 = \frac{2}{7}c_5 = 0$$

$$k = 7, \quad c_8 = \frac{2}{8}c_6 = \frac{1}{4 \cdot 3!}c_0 = \frac{1}{4!}c_0$$

---

*Recall the summation index is a "dummy" variable. The fact that $k = n - 1$ in one case and $k = n + 1$ in the other should cause no confusion if you keep in mind that it is the *value* of the summation index which is important. In both cases k takes on the successive values 1, 2, 3, . . . for $n = 2, 3, 4, \dots$ (for $k = n - 1$) and $n = 0, 1, 2, \dots$ (for $k = n + 1$) respectively.

and so on. Thus from the original assumption (4) we find

$$y = \sum_{n=0}^{\infty} c_n x^n$$

$$= c_0 + c_1 x + c_2 x^2 + c_3 x^3 + c_4 x^4 + c_5 x^5 + \cdots$$

$$= c_0 + 0 + c_0 x^2 + 0 + \frac{1}{2!} c_0 x^4 + 0 + \frac{1}{3!} c_0 x^6 + 0 + \cdots$$

$$= c_0 \left[1 + x^2 + \frac{1}{2!} x^4 + \frac{1}{3!} x^6 + \cdots \right]$$

$$= c_0 \sum_{n=0}^{\infty} \frac{x^{2n}}{n!}. \tag{10}$$

Since the iteration of (9) leaves c_0 completely undetermined, we have in fact found the general solution of (1).

6.2.2   Solutions Around Ordinary Points

Ordinary and singular points

Suppose the linear second-order differential equation

$$a_2(x)y'' + a_1(x)y' + a_0(x)y = 0 \tag{11}$$

is put into the form

$$y'' + P(x)y' + Q(x)y = 0 \tag{12}$$

where $P(x) = a_1(x)/a_2(x)$ and $Q(x) = a_0(x)/a_2(x)$. We make the following definition.

> **DEFINITION 6.1**   A point $x = x_0$ is said to be an **ordinary point** of equation (11) if both $P(x)$ and $Q(x)$ are **analytic** at x_0, that is, both $P(x)$ and $Q(x)$ have a power series in $(x - x_0)$ with a positive radius of convergence. A point that is not an ordinary point is said to be a **singular point** of the equation.   □

EXAMPLE

Every finite value of x is an ordinary point of

$$y'' + (e^x)y' + (\sin x)y = 0.$$

In particular we see that $x = 0$ is an ordinary point since

$$e^x = 1 + \frac{x}{1!} + \frac{x^2}{2!} + \cdots\cdot$$

and

$$\sin x = x - \frac{x^3}{3!} + \frac{x^5}{5!} + \cdots$$

converge for all finite values of x.

EXAMPLE

The differential equation $xy'' + (\sin x)y = 0$ has an ordinary point at $x = 0$ since it can be shown that $Q(x) = (\sin x)/x$ possesses the power series expansion

$$Q(x) = 1 - \frac{x^2}{3!} + \frac{x^4}{5!} - \frac{x^6}{7!} + \cdots$$

that converges for all finite values of x.

EXAMPLE

The differential equation

$$y'' + (\ln x)y = 0$$

has a singular point at $x = 0$ because $Q(x) = \ln x$ possesses no power series in x.

Polynomial coefficients

We shall be primarily concerned with the case when (11) has *polynomial* coefficients. As a consequence of Definition 6.1 we note that when $a_2(x)$, $a_1(x)$ and $a_0(x)$ are polynomials with *no common factors*, a point $x = x_0$ is

 (i) an ordinary point if $a_2(x_0) \neq 0$, whereas it is

 (ii) a singular point if $a_2(x_0) = 0$.

EXAMPLES

(a)   The singular points of the equation $(x^2 - 1)y'' + 2xy' + 6y = 0$ are the solutions of $x^2 - 1 = 0$ or $x = \pm 1$. All other finite values of x are ordinary points.

(b)   Singular points need not be real numbers. The equation $(x^2 + 1)y'' + xy' - y = 0$ has singular points at the solutions of $x^2 + 1 = 0$, namely, $x = \pm i$. All other finite values of x, real or complex, are ordinary points.

EXAMPLE

The Cauchy-Euler equation $ax^2y'' + bxy' + cy = 0$, where a, b, and c are constants has a singular point at $x = 0$. All other finite values of x, real or complex, are ordinary points.

In the remaining discussion of this section our goal is to find power series solutions about ordinary points for differential equations of type (11) in which the coefficients are polynomials.

Note: For our purposes ordinary points and singular points will always be finite. It is possible for a differential equation to have, say, a singular point at infinity (see Exercises 6.3).

We state the following theorem without proof.

THEOREM 6.1 If $x = x_0$ is an ordinary point of equation (11), we can always find two distinct power series solutions of the form

$$y = \sum_{n=0}^{\infty} c_n(x - x_0)^n.$$

A series solution will converge at least for $|x - x_0| < R_1$, where R_1 is the distance to the closest singular point.* $\qquad\square$

To solve an equation such as (11) we find two different sets of coefficients c_n so that we have two linearly independent power series $y_1(x)$ and $y_2(x)$, both expanded about the same ordinary point $x = x_0$. On a common interval of convergence not containing the origin the general solution of the equation is then $y = C_1 y_1(x) + C_2 y_2(x)$. The procedure used to solve a second-order equation is the same as illustrated in solving $y' - 2xy = 0$, namely, we assume a solution $y = \sum_{n=0}^{\infty} c_n(x - x_0)^n$ and then determine the c_n. In fact, we shall find that $C_1 = c_0$ and $C_2 = c_1$ where c_0 and c_1 are arbitrary.

Note: For the sake of simplicity we shall assume an ordinary point is always located at $x = 0$, since, if not, the substitution $t = x - x_0$ translates the value $x = x_0$ to $t = 0$.

EXAMPLE

Solve the second-order equation

$$y'' - 2xy = 0.$$

Solution: We see that $x = 0$ is an ordinary point of the equation. Since there are no finite singular points, Theorem 6.1 guarantees two solutions of the form $y = \sum_{n=0}^{\infty} c_n x^n$ convergent for $|x| < \infty$. Proceeding, we write

$$y' = \sum_{n=0}^{\infty} nc_n x^{n-1} = \sum_{n=1}^{\infty} nc_n x^{n-1}$$

$$y'' = \sum_{n=1}^{\infty} n(n - 1)c_n x^{n-2} = \sum_{n=2}^{\infty} n(n - 1)c_n x^{n-2}$$

where we have used the fact that the first term in each series, corresponding to $n = 0$ and $n = 1$, respectively, is zero. Therefore

---

*For example, there could exist polynomial solutions in which case the solution is valid for all finite values of x.

$$y'' - 2xy = \sum_{n=2}^{\infty} n(n-1)c_n x^{n-2} - \sum_{n=0}^{\infty} 2c_n x^{n+1} = 2 \cdot 1c_2 x^0 + \underbrace{\sum_{n=3}^{\infty} n(n-1)c_n x^{n-2} - \sum_{n=0}^{\infty} 2c_n x^{n+1}}_{\text{both series start with } x^1.}$$

Letting $k = n - 2$ in the first series and $k = n + 1$ in the second gives

$$y'' - 2xy = 2c_2 + \sum_{k=1}^{\infty} (k+2)(k+1)c_{k+2} x^k - \sum_{k=1}^{\infty} 2c_{k-1} x^k$$

$$= 2c_2 + \sum_{k=1}^{\infty} [(k+2)(k+1)c_{k+2} - 2c_{k-1}] x^k = 0.$$

We must then have

$$2c_2 = 0 \qquad \text{and} \qquad (k+2)(k+1)c_{k+2} - 2c_{k-1} = 0.$$

The last expression is the same as

$$c_{k+2} = \frac{2c_{k-1}}{(k+2)(k+1)}, \qquad k = 1, 2, 3 \ldots .$$

Iterating

$$c_3 = \frac{2c_0}{3 \cdot 2}$$

$$c_4 = \frac{2c_1}{4 \cdot 3}$$

$$c_5 = \frac{2c_2}{5 \cdot 4} = 0$$

$$c_6 = \frac{2c_3}{6 \cdot 5} = \frac{2^2}{6 \cdot 5 \cdot 3 \cdot 2} c_0$$

$$c_7 = \frac{2c_4}{7 \cdot 6} = \frac{2^2}{7 \cdot 6 \cdot 4 \cdot 3} c_1$$

$$c_8 = \frac{2c_5}{8 \cdot 7} = 0$$

$$c_9 = \frac{2c_6}{9 \cdot 8} = \frac{2^3}{9 \cdot 8 \cdot 6 \cdot 5 \cdot 3 \cdot 2} c_0$$

$$c_{10} = \frac{2c_7}{10 \cdot 9} = \frac{2^3}{10 \cdot 9 \cdot 7 \cdot 6 \cdot 4 \cdot 3} c_1$$

$$c_{11} = \frac{2c_8}{11 \cdot 10} = 0$$

and so on. It should be apparent that both c_0 and c_1 are arbitrary. Now

$$y = c_0 + c_1 x + c_2 x^2 + c_3 x^3 + c_4 x^4 + c_5 x^5 + c_6 x^6 + c_7 x^7 + c_8 x^8$$
$$+ c_9 x^9 + c_{10} x^{10} + c_{11} x^{11} + \cdots$$

$$= c_0 + c_1 x + 0 + \frac{2}{3 \cdot 2} c_0 x^3 + \frac{2}{4 \cdot 3} c_1 x^4 + 0 + \frac{2^2}{6 \cdot 5 \cdot 3 \cdot 2} c_0 x^6$$

$$+ \frac{2^2}{7 \cdot 6 \cdot 4 \cdot 3} c_1 x^7 + 0 + \frac{2^3}{9 \cdot 8 \cdot 6 \cdot 5 \cdot 3 \cdot 2} c_0 x^9$$

$$+ \frac{2^3}{10 \cdot 9 \cdot 7 \cdot 6 \cdot 4 \cdot 3} c_1 x^{10} + 0 + \cdots$$

$$= c_0 \left[1 + \frac{2}{3 \cdot 2} x^3 + \frac{2^2}{6 \cdot 5 \cdot 3 \cdot 2} x^6 + \frac{2^3}{9 \cdot 8 \cdot 6 \cdot 5 \cdot 3 \cdot 2} x^9 + \cdots \right]$$

$$+ c_1 \left[x + \frac{2}{4 \cdot 3} x^4 + \frac{2^2}{7 \cdot 6 \cdot 4 \cdot 3} x^7 + \frac{2^3}{10 \cdot 9 \cdot 7 \cdot 6 \cdot 4 \cdot 3} x^{10} + \cdots \right].$$

Although the pattern of the coefficients in the preceding example should be clear it is sometimes useful to write the solutions in terms of summation notation. By using the properties of the factorial we can write

$$y_1(x) = c_0 \left[1 + \sum_{k=1}^{\infty} \frac{2^k [1 \cdot 4 \cdot 7 \cdots (3k - 2)]}{(3k)!} x^{3k} \right]$$

and

$$y_2(x) = c_1 \left[x + \sum_{k=1}^{\infty} \frac{2^k [2 \cdot 5 \cdot 8 \cdots (3k - 1)]}{(3k + 1)!} x^{3k+1} \right].$$

In this form the ratio test can be used to show that each series converges for $|x| < \infty$.

EXAMPLE   Solve $(x^2 + 1) y'' + xy' - y = 0$

Solution: Since the singular points are $x = \pm i$ a power series solution will converge at least for $|x| < 1.$* The assumption $y = \sum_{n=0}^{\infty} c_n x^n$ leads to

$$(x^2 + 1) \sum_{n=2}^{\infty} n(n - 1) c_n x^{n-2} + x \sum_{n=1}^{\infty} n c_n x^{n-1} - \sum_{n=0}^{\infty} c_n x^n$$

$$= \sum_{n=2}^{\infty} n(n - 1) c_n x^n + \sum_{n=2}^{\infty} n(n - 1) c_n x^{n-2} + \sum_{n=1}^{\infty} n c_n x^n - \sum_{n=0}^{\infty} c_n x^n$$

$$= 2c_2 x^0 - c_0 x^0 + 6c_3 x + c_1 x - c_1 x + \underbrace{\sum_{n=2}^{\infty} n(n - 1) c_n x^n}_{k \,=\, n}$$

---

*The modulus or magnitude of the complex number $x = i$ is $|x| = 1$. If $x = a + bi$ is a singular point then $|x| = \sqrt{a^2 + b^2}$.

$$+ \sum_{\underbrace{n=4}_{k \,=\, n \,-\, 2}}^{\infty} n(n-1)c_n x^{n-2} + \sum_{\underbrace{n=2}_{k \,=\, n}}^{\infty} nc_n x^n - \sum_{\underbrace{n=2}_{k \,=\, n}}^{\infty} c_n x^n$$

$$= 2c_2 - c_0 + 6c_3 x$$

$$+ \sum_{k=2}^{\infty} [k(k-1)c_k + (k+2)(k+1)c_{k+2} + kc_k - c_k]x^k$$

$$= 2c_2 - c_0 + 6c_3 x$$

$$+ \sum_{k=2}^{\infty} [(k+1)(k-1)c_k + (k+2)(k+1)c_{k+2}]x^k = 0.$$

Thus we must have

$$2c_2 - c_0 = 0$$

$$c_3 = 0$$

$$(k+1)(k-1)c_k + (k+2)(k+1)c_{k+2} = 0$$

or, after dividing by $k + 1$,

$$c_3 = 0$$

$$c_2 = \frac{1}{2}c_0$$

$$c_{k+2} = \frac{1-k}{k+2}c_k, \qquad k = 2, 3, 4 \ldots$$

Iteration of the last formula gives

$$c_4 = -\frac{1}{4}c_2 = -\frac{1}{2 \cdot 4}c_0 = -\frac{1}{2^2 2!}c_0$$

$$c_5 = -\frac{2}{5}c_3 = 0$$

$$c_6 = -\frac{3}{6}c_4 = \frac{3}{2 \cdot 4 \cdot 6}c_0 = \frac{1 \cdot 3}{2^3 3!}c_0$$

$$c_7 = -\frac{4}{7}c_5 = 0$$

$$c_8 = -\frac{5}{8}c_6 = -\frac{3 \cdot 5}{2 \cdot 4 \cdot 6 \cdot 8}c_0 = -\frac{1 \cdot 3 \cdot 5}{2^4 4!}c_0$$

$$c_9 = -\frac{6}{9}c_7 = 0$$

$$c_{10} = -\frac{7}{10}c_8 = \frac{3 \cdot 5 \cdot 7}{2 \cdot 4 \cdot 6 \cdot 8 \cdot 10}c_0 = \frac{1 \cdot 3 \cdot 5 \cdot 7}{2^5 5!}c_0$$

and so on. Therefore

$$y = c_0 + c_1 x + c_2 x^2 + c_3 x^3 + c_4 x^4 + c_5 x^5 + c_6 x^6 + c_7 x^7 + c_8 x^8 + \cdots$$

$$= c_1 x + c_0 \left[1 + \frac{1}{2} x^2 - \frac{1}{2^2 2!} x^4 \right.$$

$$\left. + \frac{1 \cdot 3}{2^3 3!} x^6 - \frac{1 \cdot 3 \cdot 5}{2^4 4!} x^8 + \frac{1 \cdot 3 \cdot 5 \cdot 7}{2^5 5!} x^{10} + \cdots \right].$$

The solutions are

$$y_1(x) = c_0 \left[1 + \frac{1}{2} x^2 + \sum_{n=2}^{\infty} (-1)^{n-1} \frac{1 \cdot 3 \cdot 5 \cdots (2n-3)}{2^n n!} x^{2n} \right], \quad |x| < 1,$$

$$y_2(x) = c_1 x.$$

EXAMPLE

If we seek a solution $y = \sum_{n=0}^{\infty} c_n x^n$ for the equation

$$y'' - (1 + x)y = 0$$

we obtain $c_2 = c_0/2$ and the three-term recurrence relation

$$c_{k+2} = \frac{c_k + c_{k-1}}{(k+1)(k+2)}, \qquad k = 1, 2, 3 \ldots.$$

To simplify the iteration we can first choose $c_0 \neq 0$, $c_1 = 0$; this will yield one solution. The other solution follows from next choosing $c_0 = 0$, $c_1 \neq 0$. With the first assumption we find

$$c_2 = \frac{1}{2} c_0$$

$$c_3 = \frac{c_1 + c_0}{2 \cdot 3} = \frac{c_0}{2 \cdot 3} = \frac{1}{6} c_0$$

$$c_4 = \frac{c_2 + c_1}{3 \cdot 4} = \frac{c_0}{2 \cdot 3 \cdot 4} = \frac{1}{24} c_0$$

$$c_5 = \frac{c_3 + c_2}{4 \cdot 5} = \frac{c_0}{4 \cdot 5} \left[\frac{1}{2 \cdot 3} + \frac{1}{2} \right] = \frac{1}{30} c_0$$

and so on. Thus one solution is

$$y_1(x) = c_0 \left[1 + \frac{1}{2} x^2 + \frac{1}{6} x^3 + \frac{1}{24} x^4 + \frac{1}{30} x^5 + \cdots \right].$$

Similarly if we choose $c_0 = 0$ then

$$c_2 = 0$$

$$c_3 = \frac{c_1 + c_0}{2 \cdot 3} = \frac{c_1}{2 \cdot 3} = \frac{1}{6} c_1$$

$$c_4 = \frac{c_2 + c_1}{3 \cdot 4} = \frac{c_1}{3 \cdot 4} = \frac{1}{12} c_1$$

$$c_5 = \frac{c_3 + c_2}{4 \cdot 5} = \frac{c_1}{2 \cdot 3 \cdot 4 \cdot 5} = \frac{1}{120} c_1$$

and so on. Hence another solution is

$$y_2(x) = c_1 \left[x + \frac{1}{6} x^3 + \frac{1}{12} x^4 + \frac{1}{120} x^5 + \cdots \right].$$

Each series converges for all finite values of x.

EXERCISES 6.2   *Answers to odd-numbered problems begin on page A-17.*

[6.2.1]   In Problems 1–10 solve each differential equation in the manner of the previous chapters and then compare the results with the solutions obtained by assuming a power series solution $y = \sum_{n=0}^{\infty} c_n x^n$.

1. $y' + y = 0$

2. $y' = 2y$

3. $y' - x^2 y = 0$

4. $y' + x^3 y = 0$

5. $(1 - x)y' - y = 0$

6. $(1 + x)y' - 2y = 0$

7. $y'' + y = 0$

★ 8. $y'' - y = 0$

9. $y'' = y'$

10. $2y'' + y' = 0$

[6.2.2]   In Problems 11–24 for each differential equation find two linearly independent power series solutions about the ordinary point $x = 0$.

11. $y'' = xy$

12. $y'' + x^2 y = 0$

13. $y'' - 2xy' + y = 0$

14. $y'' - xy' + 2y = 0$

15. $y'' + x^2 y' + xy = 0$

16. $y'' + 2xy' + 2y = 0$

17. $(x - 1)y'' + y' = 0$

★18. $(x + 2)y'' + xy' - y = 0$

19. $(x^2 - 1)y'' + 4xy' + 2y = 0$

20. $(x^2 + 1)y'' - 6y = 0$

21. $(x^2 + 2)y'' + 3xy' - y = 0$

22. $(x^2 - 1)y'' + xy' - y = 0$

23. $y'' - (x + 1)y' - y = 0$

★24. $y'' - xy' - (x + 2)y = 0$

In Problems 25–28 use the power series method to solve the given differential equation subject to the indicated initial conditions.

25. $(x - 1)y'' - xy' + y = 0$, $y(0) = -2$, $y'(0) = 6$

26. $(x + 1)y'' - (2 - x)y' + y = 0$, $y(0) = 2$, $y'(0) = -1$

27. $y'' - 2xy' + 8y = 0$, $y(0) = 3$, $y'(0) = 0$

★28. $(x^2 + 1)y'' + 2xy' = 0$, $y(0) = 0$, $y'(0) = 1$

The series method of this section can be used when the coefficients are not polynomials. In Problems 29–32 find two power series solutions about the ordinary point $x = 0$.

EXAMPLE
$$y'' + (\cos x)y = 0.$$

Solution: Since $\cos x = 1 - \dfrac{x^2}{2!} + \dfrac{x^4}{4!} - \dfrac{x^2}{6!} + \cdots$, it is seen that $x = 0$ is an ordinary point. Thus the assumption $y = \sum_{n=0}^{\infty} c_n x^n$ leads to

$$y'' + (\cos x)y = \sum_{n=2}^{\infty} n(n-1)c_n x^{n-2} + \left(1 - \frac{x^2}{2!} + \frac{x^4}{4!} + \cdots\right)\sum_{n=0}^{\infty} c_n x^n$$

$$= (2c_2 + 6c_3 x + 12c_4 x^2 + 20c_5 x^3 + \cdots)$$

$$+ \left(1 - \frac{x^2}{2} + \frac{x^4}{24} + \cdots\right)(c_0 + c_1 x + c_2 x^2 + c_3 x^3 + \cdots)$$

$$= 2c_2 + c_0 + (6c_3 + c_1)x + \left(12c_4 + c_2 - \frac{1}{2}c_0\right)x^2$$

$$+ \left(20c_5 + c_3 - \frac{1}{2}c_1\right)x^3 + \cdots.$$

Since the last line is to be identically zero we must have

$$2c_2 + c_0 = 0$$

$$6c_3 + c_1 = 0$$

$$12c_4 + c_2 - \frac{1}{2}c_0 = 0$$

$$20c_5 + c_3 - \frac{1}{2}c_1 = 0$$

and so on. Since c_0 and c_1 are arbitrary we find

$$y_1(x) = c_0\left[1 - \frac{1}{2}x^2 + \frac{1}{12}x^4 + \cdots\right]$$

and
$$y_2(x) = c_1\left[x - \frac{1}{6}x^3 + \frac{1}{30}x^5 + \cdots\right].$$

Both series converge for all finite values of x.

---

29. $y'' + (\sin x)y = 0$

30. $xy'' + (\sin x)y = 0$
[*Hint:* See page 225.]

31. $y'' + e^{-x}y = 0$

★32. $y'' + e^x y' - y = 0$

In Problems 33 and 34 use the power series method to solve the non-homogeneous equations.

33. $y'' - xy = 1$

★34. $y'' - 4xy' - 4y = e^x$

6.3   Solutions around Singular Points

6.3.1   Regular Singular Points; The Method of Frobenius—Case I

We saw in the preceding section that there is no basic problem in finding a power series solution of

$$a_2(x)y'' + a_1(x)y' + a_0(x)y = 0 \qquad (1)$$

around an ordinary point $x = x_0$. However, when $x = x_0$ is a singular point it is not always possible to find a solution of the form $y = \sum_{n=0}^{\infty} c_n (x - x_0)^n$; it turns out that we may be able to find a solution of the form $y = \sum_{n=0}^{\infty} c_n (x - x_0)^{n+r}$ where r is a constant that must be determined.

Regular and irregular singular points

Singular points are further classified as either **regular** or **irregular.** To define these concepts we again put (1) into the form

$$y'' + P(x)y' + Q(x)y = 0. \qquad (2)$$

> **DEFINITION 6.2**   A singular point $x = x_0$ of equation (1) is said to be a **regular singular point** if both $(x - x_0)P(x)$ and $(x - x_0)^2 Q(x)$ are analytic at x_0, that is, both $(x - x_0)P(x)$ and $(x - x_0)^2 Q(x)$ have a power series in $(x - x_0)$ with a positive radius of convergence. A singular point that is not regular is said to be an **irregular singular point** of the equation.   □

Polynomial coefficients

In the case when coefficients in (1) are polynomials with no common factors, Definition 6.2 is equivalent to the following.

> Let $a_2(x_0) = 0$. Form $P(x)$ and $Q(x)$ by reducing $a_1(x)/a_2(x)$ and $a_0(x)/a_2(x)$ to lowest terms, respectively. If the factor $(x - x_0)$ appears *at most* to the first power in the denominator of $P(x)$ and *at most* to the second power in the denominator of $Q(x)$ then $x = x_0$ is a regular singular point.

EXAMPLE

It should be clear that $x = -2$ and $x = 2$ are singular points of the equation

$$(x^2 - 4)^2 y'' + (x - 2)y' + y = 0.$$

Dividing the equation by $(x^2 - 4)^2 = (x - 2)^2(x + 2)^2$ we find that

$$P(x) = \frac{1}{(x - 2)(x + 2)^2} \qquad \text{and} \qquad Q(x) = \frac{1}{(x - 2)^2(x + 2)^2}.$$

We now test $P(x)$ and $Q(x)$ at each singular point.

In order that $x = -2$ be a regular singular point, the factor $x + 2$ can appear at most to the first power in the denominator of $P(x)$, and can appear at most to the second power in the denominator of $Q(x)$. Inspection of $P(x)$

and $Q(x)$ shows that the first condition does not obtain, and so we conclude that $x = -2$ is an irregular singular point.

In order that $x = 2$ be a regular singular point, the factor $x - 2$ can appear at most to the first power in the denominator of $P(x)$, and can appear at most to the second power in the denominator of $Q(x)$. Further inspection of $P(x)$ and $Q(x)$ shows that both these conditions are satisfied, so $x = 2$ is a regular singular point.

EXAMPLE

Both $x = 0$ and $x = -1$ are singular points of the differential equation
$$x^2(x + 1)^2 y'' + (x^2 - 1)y' + 2y = 0.$$

Inspection of

$$P(x) = \frac{x - 1}{x^2(x + 1)} \quad \text{and} \quad Q(x) = \frac{2}{x^2(x + 1)^2}$$

shows that $x = 0$ is an irregular singular point since $(x - 0)$ appears to the second power in the denominator of $P(x)$. Note however that $x = -1$ is a regular singular point.

EXAMPLE

(a) $x = 1$ and $x = -1$ are regular singular points of
$$(1 - x^2)y'' - 2xy' + 30y = 0.$$

(b) $x = 0$ is an irregular singular point of
$$x^3 y'' - 2xy' + 5y = 0$$
since
$$Q(x) = 5/x^3$$

(c) $x = 0$ is a regular singular point of
$$xy'' - 2xy' + 5y = 0$$
since
$$P(x) = -2 \quad \text{and} \quad Q(x) = 5/x.$$

In part **(c)** of the preceding example notice that $(x - 0)$ and $(x - 0)^2$ do not even appear in the denominators of $P(x)$ and $Q(x)$. Remember these factors can appear at most in this fashion. For a singular point $x = x_0$, any nonnegative power of $(x - x_0)$ less than one (namely, zero) and nonnegative power less than two (namely, zero and one) in the denominators of $P(x)$ and $Q(x)$, respectively, implies x_0 is a regular singular point.

Also, recall that singular points could be complex numbers. It should be apparent that both $x = 3i$ and $x = -3i$ are regular singular points of the equation $(x^2 + 9)y'' - 3xy' + (1 - x)y = 0$ since

$$P(x) = \frac{-3x}{(x - 3i)(x + 3i)} \quad \text{and} \quad Q(x) = \frac{1 - x}{(x - 3i)(x + 3i)}.$$

EXAMPLE

From our discussion of the Cauchy–Euler equation in Section 6.1, we can show that $y_1 = x^2$ and $y_2 = x^2 \ln x$ are solutions of the equation $x^2 y'' - 3xy + 4y = 0$ on the interval $0 < x < \infty$. If the procedure of Theorem 6.1 is formally attempted at the regular singular point $x = 0$ (that is, an assumed solution of the form $y = \sum_{n=0}^{\infty} c_n x^n$) we would succeed in obtaining only the solution $y_1 = x^2$. The fact that we would not obtain the second solution is not really surprising since $y_2 = x^2 \ln x$ does not possess a Taylor series expansion about $x = 0$.

EXAMPLE

The differential equation

$$6x^2 y'' + 5xy' + (x^2 - 1)y = 0$$

has a regular singular point at $x = 0$, but does not possess *any* solution of the form $y = \sum_{n=0}^{\infty} c_n x^n$. By the procedure that we shall now consider it can be shown, however, that there exist two series solutions of the form

$$y = \sum_{n=0}^{\infty} c_n x^{n+1/2} \quad \text{and} \quad y = \sum_{n=0}^{\infty} c_n x^{n-1/3}.$$

The method of Frobenius

To solve a differential equation such as (1) about a regular singular point we employ the following theorem due to Georg Frobenius.*

THEOREM 6.2   If $x = x_0$ is a regular singular point of equation (1) then there exists at least one series solution of the form

$$y = (x - x_0)^r \sum_{n=0}^{\infty} c_n (x - x_0)^n = \sum_{n=0}^{\infty} c_n (x - x_0)^{n+r} \qquad (3)$$

where the number r is a constant which must be determined. The series will converge at least on some interval $0 < x - x_0 < R$.[†]   ☐

As an example, we note that $x = 0$ is a regular singular point of the equation

$$3xy'' + y' - y = 0 \qquad (4)$$

so we try a solution of the form

$$y = \sum_{n=0}^{\infty} c_n x^{n+r}.$$

Now

$$y' = \sum_{n=0}^{\infty} (n + r)c_n x^{n+r-1}$$

$$y'' = \sum_{n=0}^{\infty} (n + r)(n + r - 1)c_n x^{n+r-2}$$

---

*Frobenius, a German mathematician (1884–1917), published this method in 1873.
[†]As noted in the foregoing section, for the sake of simplicity we shall always assume $x_0 = 0$.

so that

$$3xy'' + y' - y$$

$$= 3 \sum_{n=0}^{\infty} (n + r)(n + r - 1)c_n x^{n+r-1} + \sum_{n=0}^{\infty} (n + r)c_n x^{n+r-1} - \sum_{n=0}^{\infty} c_n x^{n+r}$$

$$= \sum_{n=0}^{\infty} (n + r)(3n + 3r - 2)c_n x^{n+r-1} - \sum_{n=0}^{\infty} c_n x^{n+r}$$

$$= x^r \left[r(3r - 2)c_0 x^{-1} + \underbrace{\sum_{n=1}^{\infty} (n + r)(3n + r - 2)c_n x^{n-1}}_{k = n - 1} - \underbrace{\sum_{n=0}^{\infty} c_n x^n}_{k = n} \right]$$

$$= x^r \left[r(3r - 2)c_0 x^{-1} + \sum_{k=0}^{\infty} [(k + r + 1)(3k + 3r + 1)c_{k+1} - c_k] x^k \right]$$

$$= 0$$

which implies

$$r(3r - 2)c_0 = 0$$

$$(k + r + 1)(3k + 3r + 1)c_{k+1} - c_k = 0, \qquad k = 0, 1, 2 \ldots . \quad (5)$$

Since nothing is gained by taking $c_0 = 0$ we must then have

$$r(3r - 2) = 0$$

and

$$c_{k+1} = \frac{c_k}{(k + r + 1)(3k + 3r + 1)}, \qquad k = 0, 1, 2 \ldots . \quad (7)$$

The two values of r that satisfy (6), $r_1 = 2/3$ and $r_2 = 0$, when substituted in (7) gives two different recurrence relations

$$r_1 = \frac{2}{3}, \qquad c_{k+1} = \frac{c_k}{(3k + 5)(k + 1)}, \qquad (8)$$

and

$$r_2 = 0, \qquad c_{k+1} = \frac{c_k}{(k + 1)(3k + 1)}. \qquad (9)$$

Iteration of (8) gives

$$c_1 = \frac{c_0}{5 \cdot 1}$$

$$c_2 = \frac{c_1}{8 \cdot 2} = \frac{c_0}{2!5 \cdot 8}$$

$$c_3 = \frac{c_2}{11 \cdot 3} = \frac{c_0}{3!5 \cdot 8 \cdot 11}$$

$$c_4 = \frac{c_3}{14 \cdot 4} = \frac{c_0}{4!5 \cdot 8 \cdot 11 \cdot 14}$$

$$\vdots$$

$$c_n = \frac{c_0}{n!5 \cdot 8 \cdot 11 \cdots (3n + 2)}, \qquad n = 1, 2, 3 \ldots,$$

whereas iteration of (9) yields

$$c_1 = \frac{c_0}{1 \cdot 1}$$

$$c_2 = \frac{c_1}{2 \cdot 4} = \frac{c_0}{2!1 \cdot 4}$$

$$c_3 = \frac{c_2}{3 \cdot 7} = \frac{c_0}{3!1 \cdot 4 \cdot 7}$$

$$c_4 = \frac{c_3}{4 \cdot 10} = \frac{c_0}{4!1 \cdot 4 \cdot 7 \cdot 10}$$

$$\vdots$$

$$c_n = \frac{c_0}{n!1 \cdot 4 \cdot 7 \cdots (3n - 2)}, \qquad n = 1, 2, 3 \ldots.$$

Thus we obtain two series solutions

$$y_1 = c_0 x^{2/3} \left[1 + \sum_{n=1}^{\infty} \frac{1}{n!5 \cdot 8 \cdot 11 \cdots (3n + 2)} x^n \right] \qquad (10)$$

and

$$y_2 = c_0 x^0 \left[1 + \sum_{n=1}^{\infty} \frac{1}{n!1 \cdot 4 \cdot 7 \cdots (3n - 2)} x^n \right]. \qquad (11)$$

By the ratio test it can be demonstrated that both (10) and (11) converge for all finite values of x. Also, it should be clear from the form of (10) and (11) neither series is a constant multiple of the other and therefore $y_1(x)$ and $y_2(x)$ are linearly independent solutions on the x-axis. Hence, by the superposition principle

$$y = C_1 y_1(x) + C_2 y_2(x)$$

$$= C_1 \left[x^{2/3} + \sum_{n=1}^{\infty} \frac{1}{n!5 \cdot 8 \cdot 11 \cdots (3n + 2)} x^{n+2/3} \right]$$

$$+ C_2 \left[1 + \sum_{n=1}^{\infty} \frac{1}{n!1 \cdot 4 \cdot 7 \cdots (3n - 2)} x^n \right], \qquad |x| < \infty,$$

is another solution of (4). On any interval not containing the origin (such as, $0 < x < \infty$) this combination represents the general solution of the differential equation.

Although the foregoing example illustrates the general procedure for using the method of Frobenius, we hasten to point out that we may not always be able to find two solutions so readily, or for that matter, that we can find two solutions which are infinite series consisting entirely of powers of x.

Indicial equation

Equation (6) is called the **indicial equation** of the problem and the values $r_1 = 2/3$ and $r_2 = 0$ are called the **indicial roots** or **exponents** of the singularity ($x = 0$). In general, when using the assumption $y = \sum_{n=0}^{\infty} c_n x^{n+r}$ in an attempt to solve a linear second-order differential equation, the indicial equation is a quadratic equation in r resulting from equating to zero the *total coefficient of the lowest power of x*. We then solve for the two values of the exponents and substitute these values into a corresponding recurrence relation such as (7). Theorem 6.2 guarantees that we can always find at least one solution of the assumed series form.

EXAMPLE

The method of Frobenius applied to

$$xy'' + 3y' - y = 0 \tag{12}$$

yields

$$xy'' + 3y' - y$$

$$= x^r \left[r(r+2)c_0 x^{-1} + \sum_{k=0}^{\infty} [(k+r+1)(k+r+3)c_{k+1} - c_k]x^k \right] = 0$$

so that the indicial equation and exponents are $r(r+2) = 0$ and $r_1 = 0$, $r_2 = -2$ respectively.

Since

$$(k+r+1)(k+r+3)c_{k+1} - c_k = 0, \qquad k = 0, 1, 2 \ldots \tag{13}$$

it follows that when $r_1 = 0$

$$c_{k+1} = \frac{c_k}{(k+1)(k+3)}$$

$$c_1 = \frac{c_0}{1 \cdot 3}$$

$$c_2 = \frac{c_1}{2 \cdot 4} = \frac{2c_0}{2!4!}$$

$$c_3 = \frac{c_2}{3 \cdot 5} = \frac{2c_0}{3!5!}$$

$$c_4 = \frac{c_3}{4 \cdot 6} = \frac{2c_0}{4!6!}$$

$$\vdots$$

$$c_n = \frac{2c_0}{n!(n+2)!}, \qquad n = 1, 2, 3 \ldots.$$

Thus one series solution is

$$y_1 = c_0 x^0 \left[1 + \sum_{n=1}^{\infty} \frac{2}{n!(n+2)!} x^n \right].$$

$$= c_0 \sum_{n=0}^{\infty} \frac{2}{n!(n+2)!} x^n, \qquad |x| < \infty. \tag{14}$$

Now when $r_2 = -2$, (13) becomes

$$(k - 1)(k + 1)c_{k+1} - c_k = 0, \tag{15}$$

but note here that we *do not divide* by $(k - 1)(k + 1)$ immediately since this term is zero for $k = 1$. However, we use the recurrence relation (15) for the cases $k = 0$ and $k = 1$:

$$-1 \cdot 1c_1 - c_0 = 0 \quad \text{and} \quad 0 \cdot 2c_2 - c_1 = 0.$$

The latter equation implies that $c_1 = 0$ and so the former equation implies $c_0 = 0$. Continuing we find

$$c_{k+1} = \frac{c_k}{(k - 1)(k + 1)}, \qquad k = 2, 3, \ldots$$

and so

$$c_3 = \frac{c_2}{1 \cdot 3}$$

$$c_4 = \frac{c_3}{2 \cdot 4} = \frac{2c_2}{2!4!}$$

$$c_5 = \frac{c_4}{3 \cdot 5} = \frac{2c_2}{3!5!}$$

$$\vdots$$

$$c_n = \frac{2c_2}{(n - 2)!n!}, \qquad n = 2, 3, 4, \ldots .$$

Thus we can formally write

$$y_2 = c_2 x^{-2} \sum_{n=2}^{\infty} \frac{2}{(n - 2)!n!} x^n. \tag{16}$$

However, close inspection of (16) reveals that y_2 is simply a constant multiple of (14). To see this, let $k = n - 2$ in (16). We conclude that the method of Frobenius gives only one series solution of equation (12).

Cases of indicial roots

When using the method of Frobenius we usually distinguish three possible cases corresponding to the nature of the indicial roots. For the sake of discussion let us suppose that r_1 and r_2 are the real solutions of the indicial equation and that, when appropriate, r_1 *denotes the largest root*.

CASE I If r_1 and r_2 are distinct and *do not* differ by an integer, then there exist two linearly independent solutions of equation (1) of the form

$$y_1 = \sum_{n=0}^{\infty} c_n x^{n+r_1}, \qquad c_0 \neq 0 \tag{17a}$$

$$y_2 = \sum_{n=0}^{\infty} b_n x^{n+r_2}, \qquad b_0 \neq 0. \tag{17b}$$

CASE I   Roots Not Differing by an Integer

EXAMPLE   Solve $2xy'' + (1 + x)y' + y = 0.$ (18)

Solution: If $y = \displaystyle\sum_{n=0}^{\infty} c_n x^{n+r}$ then

$2xy'' + (1 + y)y' + y$

$$= 2 \sum_{n=0}^{\infty} (n + r)(n + r - 1)c_n x^{n+r-1} + \sum_{n=0}^{\infty} (n + r)c_n x^{n+r-1}$$

$$+ \sum_{n=0}^{\infty} (n + r)c_n x^{n+r} + \sum_{n=0}^{\infty} c_n x^{n+r}$$

$$= \sum_{n=0}^{\infty} (n + r)(2n + 2r - 1)c_n x^{n+r-1} + \sum_{n=0}^{\infty} (n + r + 1)c_n x^{n+r}$$

$$= x^r \Bigg[r(2r - 1)c_0 x^{-1} + \underbrace{\sum_{n=1}^{\infty} (n + r)(2n + 2r - 1)c_n x^{n-1}}_{k = n - 1}$$

$$+ \underbrace{\sum_{n=0}^{\infty} (n + r + 1)c_n x^{n}}_{k = n} \Bigg]$$

$$= x^r \Bigg[r(2r - 1)c_0 x^{-1} + \sum_{k=0}^{\infty} [(k + r + 1)(2k + 2r + 1)c_{k+1}$$

$$+ (k + r + 1)c_k]x^k \Bigg] = 0$$

which implies $r(2r - 1) = 0$ (19)

$$(k + r + 1)(2k + 2r + 1)c_{k+1} + (k + r + 1)c_k = 0, \ 1, \ 2. \ldots \quad (20)$$

For $r_1 = 1/2$ we can divide by $k + 3/2$ in (20) to obtain

$$c_{k+1} = \frac{-c_k}{2(k + 1)}$$

$$c_1 = \frac{-c_0}{2 \cdot 1}$$

$$c_2 = \frac{-c_1}{2 \cdot 2} = \frac{c_0}{2^2 \cdot 2!}$$

$$c_3 = \frac{-c_2}{2 \cdot 3} = \frac{-c_0}{2^3 \cdot 3!}$$

$$\vdots$$

$$c_n = \frac{(-1)^n c_0}{2^n n!}, \qquad n = 1, 2, 3 \ldots .$$

Thus we have

$$y_1 = c_0^{1/2} \left[1 + \sum_{n=0}^{\infty} \frac{(-1)^n}{2^n n!} x^n \right]$$

$$= c_0 \sum_{n=0}^{\infty} \frac{(-1)^n}{2^n 2!} x^{n+1/2} \qquad (21)$$

which converges for $x \geq 0$. As given, the series is not meaningful for $x < 0$ because of the presence of $x^{1/2}$.

Now for $r_2 = 0$, (20) becomes

$$c_{k+1} = \frac{-c_k}{2k + 1}$$

$$c_1 = \frac{-c_0}{1}$$

$$c_2 = \frac{-c_1}{3} = \frac{c_0}{1 \cdot 3}$$

$$c_3 = \frac{-c_2}{5} = \frac{-c_0}{1 \cdot 3 \cdot 5}$$

$$c_4 = \frac{-c_3}{7} = \frac{c_0}{1 \cdot 3 \cdot 5 \cdot 7}$$

$$\vdots$$

$$c_n = \frac{(-1)^n c_0}{1 \cdot 3 \cdot 5 \cdot 7 \cdots (2n - 1)}, \qquad n = 1, 2, 3. \ldots .$$

We conclude that a second solution to (18) is

$$y_2 = c_0 \left[1 + \sum_{n=1}^{\infty} \frac{(-1)^n}{1 \cdot 3 \cdot 5 \cdot 7 \cdots (2n - 1)} x^n \right], \qquad |x| < \infty. \quad (22)$$

On the interval $0 < x < \infty$ the general solution is $y = C_1 y_1(x) + C_2 y_2(x)$.

6.3.2 The Method of Frobenius—Cases II and III

When the roots of the indicial equation differ by a positive integer we may or may not be able to find two solutions of (1) having form (3). If not, then one solution, corresponding to the smaller root, contains a logarithmic term. When the exponents are equal a second solution will *always* contain a logarithm. This latter situation is analogous to the solutions of the Cauchy–Euler differential equation when the roots of the auxiliary equation are equal. We have the next two cases.

CASE II   If $r_1 - r_2 = N$, where N is a positive integer, then there exist two linearly independent solutions of equation (1) of the form

$$y_1 = \sum_{n=0}^{\infty} c_n x^{n+r_1} \qquad c_0 \neq 0 \tag{23a}$$

$$y_2 = Cy_1(x) \ln x + \sum_{n=0}^{\infty} b_n x^{n+r_2} \qquad b_0 \neq 0, \tag{23b}$$

where C is a constant that could be zero.   □

CASE III   If $r_1 = r_2$ there always exist two linearly independent solutions of equation (1) of the form

$$y_1 = \sum_{n=0}^{\infty} c_n x^{n+r_1} \qquad c_0 \neq 0, \tag{24a}$$

$$y_2 = y_1(x) \ln x + \sum_{n=1}^{\infty} b_n x^{n+r_1}. \qquad □ \tag{24b}$$

CASE II   Roots Differing by a Positive Integer

EXAMPLE   Solve $xy'' + (x - 6)y' - 3y = 0.$ (25)

Solution: The assumption $y = \sum_{n=0}^{\infty} c_n x^{n+r}$ leads to

$xy'' + (x - 6)y' - 3y$

$= \sum_{n=0}^{\infty} (n + r)(n + r - 1)c_n x^{n+r-1} - 6 \sum_{n=0}^{\infty} (n + r)c_n x^{n+r-1} + \sum_{n=0}^{\infty} (n + r)c_n x^{n+r} - 3 \sum_{n=0}^{\infty} c_n x^{n+r}$

$= x^r \left[r(r - 7)c_0 x^{-1} + \underbrace{\sum_{n=1}^{\infty} (n + r)(n + r - 7)c_n x^{n-1}}_{k = n - 1} + \underbrace{\sum_{n=0}^{\infty} (n + r - 3)c_n x^n}_{k = n} \right]$

$= x^r \left[r(r - 7)c_0 x^{-1} + \sum_{k=0}^{\infty} [(k + r + 1)(k + r - 6)c_{k+1} + (k + r - 3)c_k]x^k \right] = 0.$

Thus $r(r - 7) = 0$ so that $r_1 = 7$, $r_2 = 0$, $r_1 - r_2 = 7$, and

$$(k + r + 1)(k + r - 6)c_{k+1} + (k + r - 3)c_k = 0, \qquad k = 0, 1, 2. \ldots \tag{26}$$

For the smaller root $r_2 = 0$, (26) becomes

$$(k + 1)(k - 6)c_{k+1} + (k - 3)c_k = 0. \tag{27}$$

Since $k - 6 = 0$ when $k = 6$ we do not divide by this term until $k > 6$. We find

$$1 \cdot (-6)c_1 + (-3)c_0 = 0$$

$$2 \cdot (-5)c_2 + (-2)c_1 = 0$$

$$3 \cdot (-4)c_3 + (-1)c_2 = 0$$

$$4 \cdot (-3)c_4 + 0 \cdot c_3 = 0$$

$$5 \cdot (-2)c_5 + 1 \cdot c_4 = 0$$

$$6 \cdot (-1)c_6 + 2 \cdot c_5 = 0$$

$$7 \cdot 0c_7 + 3 \cdot c_6 = 0$$

implies $c_4 = c_5 = c_6 = 0$ but c_0 and c_7 can be chosen arbitrarily

Thus

$$c_1 = -\frac{1}{2}c_0$$

$$c_2 = -\frac{1}{5}c_1 = \frac{1}{10}c_0 \tag{28}$$

$$c_3 = -\frac{1}{12}c_2 = -\frac{1}{120}c_0$$

and for $k \geq 7$

$$c_{k+1} = \frac{-(k-3)c_k}{(k+1)(k-6)}$$

$$c_8 = \frac{-4}{8 \cdot 1}c_7$$

$$c_9 = \frac{-5}{9 \cdot 2}c_8 = \frac{4 \cdot 5}{2!8 \cdot 9}c_7$$

$$c_{10} = \frac{-6}{10 \cdot 3}c_9 = \frac{-4 \cdot 5 \cdot 6}{3!8 \cdot 9 \cdot 10}c_7$$

$$\vdots$$

$$c_n = \frac{(-1)^{n+1}4 \cdot 5 \cdot 6 \cdots (n-4)}{(n-7)!8 \cdot 9 \cdot 10 \cdots n}c_7, \qquad n = 8, 9, 10. \ldots \tag{29}$$

If we choose $c_7 = 0$ and $c_0 \neq 0$, we obtain the polynomial solution

$$y_1 = c_0 \left[1 - \frac{1}{2}x + \frac{1}{10}x^2 - \frac{1}{120}x^3 \right], \tag{30}$$

but when $c_7 \neq 0$ and $c_0 = 0$, it follows that a second, though infinite series, solution is

$$y_2 = c_7 \left[x^7 + \sum_{n=8}^{\infty} \frac{(-1)^{n+1}4 \cdot 5 \cdot 6 \cdots (n-4)}{(n-7)!8 \cdot 9 \cdot 10 \cdots n} x^n \right]$$

$$= c_7 \left[x^7 + \sum_{k=1}^{\infty} \frac{(-1)^k 4 \cdot 5 \cdot 6 \cdots (k+3)}{k!8 \cdot 9 \cdot 10 \cdots (k+7)} x^{k+7} \right], \qquad |x| < \infty \tag{31}$$

Finally, for $x > 0$ a general solution of equation (25) is

$$y = C_1 y_1(x) + C_2 y_2(x)$$

$$= C_1\left[1 - \frac{1}{2}x + \frac{1}{10}x^2 - \frac{1}{120}x^3\right] + C_2\left[x^7 + \sum_{k=1}^{\infty} \frac{(-1)^k 4 \cdot 5 \cdot 6 \cdots (k+3)}{k! 8 \cdot 9 \cdot 10 \cdots (k+7)} x^{k+7}\right].$$

It is interesting to observe that in the preceding example the larger root $r_1 = 7$ was not used. Had we done so, we would have obtained a series solution of the form*

$$\sum_{n=0}^{\infty} c_n x^{n+7} \tag{32}$$

where the c_n are defined by the formula (equation (26) with $r_1 = 7$)

$$c_{k+1} = \frac{-(k+4)}{(k+8)(k+1)} c_k, \qquad k = 0, 1, 2. \ldots$$

Iteration of this latter recurrence relation then would yield only *one* solution, namely, the solution given by (31).

When the roots of the indicial equation differ by a positive integer the second solution *may* contain a logarithm. In practice this is something we do not know in advance, but is determined after we have found the indicial roots and have carefully examined the recurrence relation which defines the coefficients c_n. As the foregoing example shows, we just may be lucky enough to find two solutions which involve only powers of x. On the other hand, if we fail to find a second series-type solution, we can always formally use the fact that

$$y_2 = y_1(x) \int \frac{e^{-\int P(x)\,dx}}{y_1^2(x)}\,dx \tag{33}$$

is also a solution of the equation $y'' + P(x)y' + Q(x)y = 0$ whenever y_1 is a known solution (see Section 4.2).

EXAMPLE

Find the general solution of

$$xy'' + 3y' - y = 0.$$

Solution: Recall from page 238 that the method of Frobenius provides only one solution to this equation, namely,

$$y_1 = \sum_{n=0}^{\infty} \frac{2}{n!(n+2)!} x^n$$

$$= 1 + \frac{1}{3}x + \frac{1}{24}x^2 + \frac{1}{360}x^3 + \cdots \tag{34}$$

From (33) we obtain a second solution

---

*Observe that both (31) and (32) start with the power x^7. In Case II it is always a good idea to work with the smaller root first.

$$y_2 = y_1(x) \int \frac{e^{-\int 3/x\, dx}}{y_1^2(x)} dx$$

$$= y_1(x) \int \frac{dx}{x^3 \left[1 + \frac{1}{3}x + \frac{1}{24}x^2 + \frac{1}{360}x^3 + \cdots \right]^2}$$

$$= y_1(x) \int \frac{dx}{x^3 \left[1 + \frac{2}{3}x + \frac{7}{36}x^2 + \frac{1}{30}x^3 + \cdots \right]} \qquad \text{(squaring)}$$

$$= y_1(x) \int \frac{1}{x^3}\left[1 - \frac{2}{3}x + \frac{1}{4}x^2 - \frac{19}{270}x^3 + \cdots \right] dx \qquad \text{(long division)}$$

$$= y_1(x) \int \left[\frac{1}{x^3} - \frac{2}{3x^2} + \frac{1}{4x} - \frac{19}{270} + \cdots \right] dx$$

$$= y_1(x)\left[-\frac{1}{2x^2} + \frac{2}{3x} + \frac{1}{4}\ln x - \frac{19}{270}x + \cdots \right]$$

or $\qquad y_2 = \frac{1}{4}y_1(x)\ln x + y_1(x)\left[-\frac{1}{2x^2} + \frac{2}{3x} - \frac{19}{270}x \cdots \right].$ (35)

Hence, on the interval $0 < x < \infty$ the general solution is

$$y = C_1 y_1(x) + C_2\left[\frac{1}{4} y_1(x)\ln x + y_1(x)\left(-\frac{1}{2x^2} + \frac{2}{3x} - \frac{19}{270}x + \cdots \right) \right]$$ (36)

where $y_1(x)$ is defined by (34).

**An alternative
procedure**

There are several alternative procedures to formula (33) when the method of Frobenius fails to provide a second series solution. Although the next method is somewhat tedious, it is nonetheless straightforward. The basic idea is to assume a solution either of the form (23b) or (24b) and determine coefficients b_n in terms of the coefficients c_n which define the known solution $y_1(x)$.

EXAMPLE

The smaller of the two indicial roots for the equation $xy'' + 3y' - y = 0$ is $r_2 = -2$. From (23b) we now assume a second solution

$$y_2 = y_1 \ln x + \sum_{n=0}^{\infty} b_n x^{n-2}$$ (37)

where

$$y_1 = \sum_{n=0}^{\infty} \frac{2}{n!(n+2)!} x^n$$ (38)

Differentiation of (37) gives

$$y_2' = \frac{y_1}{x} + y_1' \ln x + \sum_{n=0}^{\infty} (n-2)b_n x^{n-3}$$

$$y_2'' = -\frac{y_1}{x^2} + \frac{2y_1'}{x} + y_1'' \ln x + \sum_{n=0}^{\infty} (n-2)(n-3)b_n x^{n-4}$$

so that

$$xy_2'' + 3y_2' - y_2 = \ln x \underbrace{[xy_1'' + 3y_1' - y_1]}_{\text{zero}} + 2y_1' + \frac{2y_1}{x}$$

$$+ \sum_{n=0}^{\infty} (n-2)(n-3)b_n x^{n-3}$$

$$+ 3 \sum_{n=0}^{\infty} (n-2)b_n x^{n-3} - \sum_{n=0}^{\infty} b_n x^{n-2}$$

$$= 2y_1' + \frac{2y_1}{x} + \sum_{n=0}^{\infty} (n-2)nb_n x^{n-3} - \sum_{n=0}^{\infty} b_n x^{n-2} \qquad (39)$$

where we have combined the first two summations and have used the fact that $xy_1'' + 3y_1' - y_1 = 0$.

By differentiating (38) we can write (39) as

$$\sum_{n=0}^{\infty} \frac{4n}{n!(n+2)!} x^{n-1} + \sum_{n=0}^{\infty} \frac{4}{n!(n+2)!} x^{n-1} + \sum_{n=0}^{\infty} (n-2)nb_n x^{n-3} - \sum_{n=0}^{\infty} b_n x^{n-2}$$

$$= 0(-2)b_0 x^{-3} + (-b_0 - b_1)x^{-2} + \underbrace{\sum_{n=0}^{\infty} \frac{4(n+1)}{n!(n+2)!} x^{n-1}}_{k=n} + \underbrace{\sum_{n=2}^{\infty} (n-2)nb_n x^{n-3}}_{k=n-2} - \underbrace{\sum_{n=0}^{\infty} b_n x^{n-2}}_{k=n-1}$$

$$= -(b_0 + b_1)x^{-2} + \sum_{k=0}^{\infty} \left[\frac{4(k+1)}{k!(k+2)!} + k(k+2)b_{k+2} - b_{k+1} \right]x^{k-1} \qquad (40)$$

Setting (40) equal to zero then gives $b_1 = -b_0$ and

$$\frac{4(k+1)}{k!(k+2)!} + k(k+2)b_{k+2} - b_{k+1} = 0, \qquad \text{for } k = 0, 1, 2, \ldots \qquad (41)$$

When $k = 0$ in (41) we have $2 + 0 \cdot 2b_2 - b_1 = 0$ so that $b_1 = 2, b_0 = -2$, but b_2 is arbitrary.

Rewriting (41) as

$$b_{k+2} = \frac{b_{k+1}}{k(k+2)} - \frac{4(k+1)}{k!(k+2)!k(k+2)} \qquad (42)$$

and evaluating for $k = 1, 2, \ldots$, gives

$$b_3 = \frac{b_2}{3} - \frac{4}{9}$$

$$b_4 = \frac{1}{8}b_3 - \frac{1}{32} = \frac{1}{24}b_2 - \frac{25}{288}$$

and so on. Thus we can finally write

$$y_2 = y_1 \ln x + b_0 x^{-2} + b_1 x^{-1} + b_2 + b_3 x + \cdots$$

$$= y_1 \ln x - 2x^{-2} + 2x^{-1} + b_2 + \left(\frac{b_2}{3} - \frac{4}{9}\right)x + \cdots \quad (43)$$

where b_2 is arbitrary.

Equivalent solutions   At this point you may be wondering whether (35) and (43) are really equivalent. If we choose $C_2 = 4$ in (36) then

$$y_2 = y_1 \ln x + y_1\left(-\frac{2}{x^2} + \frac{8}{3x} - \frac{38}{135}x + \cdots\right)$$

$$= y_1 \ln x + \left(1 + \frac{1}{3}x + \frac{1}{24}x^2 + \frac{1}{360}x^3 + \cdots\right)$$

$$\times \left(-\frac{2}{x^2} + \frac{8}{3x} - \frac{38}{135}x + \cdots\right)$$

$$= y_1 \ln x - 2x^{-2} + 2x^{-1} + \frac{29}{36} - \frac{19}{108}x + \cdots \quad (44)$$

which is precisely what we obtain from (43) if b_2 is chosen as 29/36.

CASE III   Equal Indicial Roots

EXAMPLE   Find the general solution of

$$xy'' + y' - 4y = 0 \quad (45)$$

Solution: The assumption $y = \displaystyle\sum_{n=0}^{\infty} c_n x^{n+r}$ leads to

$$xy'' + y' - 4y = \sum_{n=0}^{\infty} (n + r)(n + r - 1)c_n x^{n+r-1}$$

$$+ \sum_{n=0}^{\infty} (n + r)c_n x^{n+r-1} - 4\sum_{n=0}^{\infty} c_n x^{n+r}$$

$$= \sum_{n=0}^{\infty} (n + r)^2 c_n x^{n+r-1} - 4\sum_{n=0}^{\infty} c_n x^{n+r}$$

$$= x^r\left[r^2 c_0 x^{-1} + \underbrace{\sum_{n=1}^{\infty} (n + r)^2 c_n x^{n-1}}_{k = n - 1} - 4\underbrace{\sum_{n=0}^{\infty} c_n x^n}_{k = n}\right]$$

$$= x^r\left[r^2 c_0 x^{-1} + \sum_{k=0}^{\infty} [(k + r + 1)^2 c_{k+1} - 4c_k]x^k\right] = 0.$$

Therefore, $r^2 = 0$, $r_1 = r_2 = 0$, and

$$(k + r + 1)^2 c_{k+1} + 1 - 4c_k = 0, \quad k = 0, 1, 2 \dots. \quad (46)$$

Clearly, the root $r_1 = 0$ will only yield one solution corresponding to the coefficients defined by the iteration of

$$c_{k+1} = \frac{4c_k}{(k + 1)^2}, \qquad k = 0, 1, 2 \ldots$$

The result is
$$y_1 = c_0 \sum_{n=0}^{\infty} \frac{4^n}{(n!)^2} x^n, \qquad |x| < \infty. \tag{47}$$

To obtain the second linearly independent solution we set $c_0 = 1$ in (47) and then use formula (33)

$$y_2 = y_1(x) \int \frac{e^{-\int (1/x)\,dx}}{y_1^2(x)}\,dx$$

$$= y_1(x) \int \frac{dx}{x\left[1 + 4x + 4x^2 + \dfrac{16}{9}x^3 + \cdots\right]^2}$$

$$= y_1(x) \int \frac{dx}{x\left[1 + 8x + 24x^2 + \dfrac{16}{9}x^3 + \cdots\right]}$$

$$= y_1(x) \int \frac{1}{x}\left[1 - 8x + 40x^2 - \frac{1472}{9}x^3 + \cdots\right] dx$$

$$= y_1(x) \int \left[\frac{1}{x} - 8 + 40x - \frac{1472}{9}x^2 + \cdots\right] dx$$

$$= y_1(x)\left[\ln x - 8x + 20x^2 - \frac{1472}{27}x^3 + \cdots\right]. \tag{48}$$

Thus, on the interval $0 < x < \infty$ the general solution of (45) is

$$y = C_1 y_1(x) + C_2\left[y_1(x) \ln x + y_1(x)\left(-8x + 20x^2 - \frac{1472}{27}x^3 + \cdots\right)\right] \tag{49}$$

where $y_1(x)$ is defined by (47).

Alternative procedure

As in Case II, we can determine $y_2(x)$ of the preceding example directly from the assumption (24b).

EXAMPLE

For equation (45) we know $r_1 = r_2 = 0$ so that (24b) becomes

$$y_2 = y_1 \ln x + \sum_{n=1}^{\infty} b_n x^n \tag{50}$$

and therefore

$$y_2' = \frac{y_1}{x} + y_1' \ln x + \sum_{n=1}^{\infty} b_n n x^{n-1}$$

$$y_2'' = -\frac{y_1}{x^2} + \frac{2y_1'}{x} + y_1'' \ln x + \sum_{n=1}^{\infty} b_n n(n - 1)x^{n-2}$$

$$xy_2'' + y_2' - 4y_2 = \ln x \underbrace{[xy_1'' + y_1' - 4y_1]}_{\text{zero}} + 2y_1' + \sum_{n=1}^{\infty} n(n-1)b_n x^{n-1}$$

$$+ \sum_{n=1}^{\infty} nb_n x^{n-1} - 4 \sum_{n=1}^{\infty} b_n x^n$$

$$= 2y_1' + \sum_{n=1}^{\infty} n^2 b_n x^{n-1} - 4 \sum_{n=1}^{\infty} b_n x^n. \tag{51}$$

Using the fact that
$$y_1 = \sum_{n=0}^{\infty} \frac{4^n}{(n!)^2} x^n$$

(51) then becomes

$$2 \sum_{n=0}^{\infty} \frac{4^n n}{(n!)^2} x^{n-1} + \sum_{n=1}^{\infty} n^2 b_n x^{n-1} - 4 \sum_{n=1}^{\infty} b_n x^n$$

$$= 8 + b_1 + 2 \underbrace{\sum_{n=2}^{\infty} \frac{4^n n}{(n!)^2} x^{n-1}}_{k=n-1} + \underbrace{\sum_{n=2}^{\infty} n^2 b_n x^{n-1}}_{k=n-1} - 4 \underbrace{\sum_{n=1}^{\infty} b_n x^n}_{k=n}$$

$$= 8 + b_1 + \sum_{k=1}^{\infty} \left[\frac{2 \cdot 4^{k+1}(k+1)}{[(k+1)!]^2} + (k+1)^2 b_{k+1} - 4b_k \right] x^k \tag{52}$$

Equating (52) to zero gives $b_1 = -8$ and

$$b_{k+1} = \frac{4}{(k+1)^2} b_k - \frac{2 \cdot 4^{k+1}}{(k+1)[(k+1)!]^2}, \qquad k = 1, 2, 3, \ldots, \tag{53}$$

which implies
$$b_2 = b_1 - 4 = -12$$

$$b_3 = \frac{4}{9} b_2 - \frac{32}{27} = -\frac{176}{27}$$

and so on. Hence, it follows that

$$y_2 = y_1 \ln x - 8x - 12x^2 - \frac{176}{27} x^3 - \cdots. \tag{54}$$

Equivalent solutions   If we examine (48) in a little more detail, we find that

$$y_2 = y_1 \ln x + y_1 \left(-8x + 20x^2 - \frac{1472}{27} x^3 + \cdots \right)$$

$$= y_1 \ln x + \left(1 + 4x + 4x^2 + \frac{16}{9} x^3 + \cdots \right)$$

$$\times \left(-8x + 20x^2 - \frac{1472}{27} x^3 + \cdots \right)$$

$$= y_1 \ln x - 8x - 12x^2 - \frac{176}{27}x^3 - \cdots$$

which, of course, is the same as (54).

Remark: We purposely have not considered two further complications when solving a differential equation such as (1) about a point x_0 for which $a_2(x_0) = 0$. When using (3), it is quite possible that the roots of the indicial equation could turn out to be complex numbers. When the exponents r_1 and r_2 are complex, the statement $r_1 > r_2$ is meaningless and must be replaced with $Re(r_1) > Re(r_2)$ (for example, if $r = \alpha + i\beta$, then $Re(r) = \alpha$). In particular, when the indicial equation has real coefficients, the complex roots will be a conjugate pair

$$r_1 = \alpha + i\beta \qquad r_2 = \alpha - i\beta$$

and $r_1 - r_2 = i\beta \neq$ integer. Thus, for $x_0 = 0$, there will always exist two solutions

$$y_1 = \sum_{n=0}^{\infty} c_n x^{n+r_1} \qquad \text{and} \qquad y_2 = \sum_{n=0}^{\infty} b_n x^{n+r_2}.$$

Unfortunately, both solutions will give complex values of y for each real choice of x. This latter difficulty can be surmounted by the superposition principle. Since a combination of solutions is also a solution to the differential equation, we could form appropriate combinations of $y_1(x)$ and $y_2(x)$ to yield real solutions. (See Case III of the solution of the Cauchy–Euler Equation.)

Lastly, if $x = 0$ is an irregular singular point, it should be noted that we may not be able to find *any* solution of the form $y = \sum_{n=0}^{\infty} c_n x^{n+r}$.

EXERCISES 6.3

Answers to odd-numbered problems begin on page A-17.

[6.3.1]

In Problems 1–10 determine the singular points of each differential equation. Classify each singular point as regular or irregular.

1. $x^3 y'' + 4x^2 y' + 3y = 0$ ★**2.** $xy'' - (x + 3)^{-2}y = 0$

3. $(x^2 - 9)^2 y'' + (x + 3)y' + 2y = 0$

4. $y'' - \dfrac{1}{x}y' + \dfrac{1}{(x - 1)^3}y = 0$

5. $(x^3 + 4x)y'' - 2xy' + 6y = 0$

6. $x^2(x - 5)^2 y'' + 4xy' + (x^2 - 25)y = 0$

7. $(x^2 + x - 6)y'' + (x + 3)y' + (x - 2)y = 0$

8. $x(x^2 + 1)^2 y'' + y = 0$

9. $x^3(x^2 - 25)(x - 2)^2 y'' + 3x(x - 2)y' + 7(x + 5)y = 0$

10. $(x^3 - 2x^2 - 3x)^2 y'' + x(x - 3)^2 y' - (x + 1)y = 0$

In the Problems 11–22 show that the indicial roots do not differ by an integer. Use the method of Frobenius to obtain two linearly independent series solutions about the regular singular point $x_0 = 0$. Form the general solution on $0 < x < \infty$.

11. $2xy'' - y' + 2y = 0$ **12.** $2xy'' + 5y' + xy = 0$

13. $4xy'' + \dfrac{1}{2}y' + y = 0$ **14.** $2x^2y'' - xy' + (x^2 + 1)y = 0$

15. $3xy'' + (2 - x)y' - y = 0$ **★16.** $x^2y'' - \left(x - \dfrac{2}{9}\right)y = 0$

17. $2xy'' - (3 + 2x)y' + y = 0$ **18.** $x^2y'' + xy' + \left(x^2 - \dfrac{4}{9}\right)y = 0$

19. $9x^2y'' + 9x^2y' + 2y = 0$

20. $2x^2y'' + 3xy' + (2x - 1)y = 0$

21. $2x^2y'' - x(x - 1)y' - y = 0$ **★22.** $x(x - 2)y'' + y' - 2y = 0$

[6.3.2] In Problems 23–34 show that the indicial roots differ by an integer. Use the method of Frobenius to obtain two linearly independent series solutions about the regular singular point $x_0 = 0$. Form the general solution on $0 < x < \infty$.

23. $xy'' + 2y' - xy = 0$ **24.** $x^2y'' + xy' + \left(x^2 - \dfrac{1}{4}\right)y = 0$

25. $x(x - 1)y'' + 3y' - 2y = 0$ **26.** $y'' + \dfrac{3}{x}y' - 2y = 0$

27. $xy'' + (1 - x)y' - y = 0$ **★28.** $xy'' + y = 0$

29. $xy'' + y' + y = 0$ **30.** $xy'' - xy' + y = 0$

31. $x^2y'' + x(x - 1)y' + y = 0$ **32.** $xy'' + y' - 4xy = 0$

33. $xy'' + (x - 1)y' - 2y = 0$ **34.** $xy'' - y' + x^3y = 0$

In Problems 35 and 36 note that $x_0 = 0$ is an irregular singular point of each equation. In each case determine whether the method of Frobenius will yield a solution.

35. $x^3y'' + y = 0$ **★36.** $x^2y'' - y' + y = 0$

Miscellaneous problems

A differential equation is said to have a singular point at ∞ if, after the substitution $w = 1/x$, the resulting equation has a singular point at $w = 0$. In Problems 37–40 determine whether the given equation has a singular point at ∞, if so, state whether it is regular or irregular.

EXAMPLE

The equation $y'' + xy = 0$ has no singular points in the finite plane. Using the substitution $w = 1/x$, it follows from the chain rule that

$$\frac{dy}{dx} = \frac{dy}{dw}\frac{dw}{dx}$$

$$= -\frac{1}{x^2}\frac{dy}{dw}$$

$$= -w^2\frac{dy}{dw},$$

$$\frac{d^2y}{dx^2} = -w^2\frac{d}{dx}\left[\frac{dy}{dw}\right] - \frac{dy}{dw}\frac{d}{dx}[w^2]$$

$$= -w^2\frac{d}{dw}\left(\frac{dy}{dw}\right)\frac{dw}{dx} - \frac{dy}{dw}\left(2w\frac{dw}{dx}\right)$$

$$= w^4\frac{d^2y}{dw^2} + 2w^3\frac{dy}{dw}$$

so that the original equation transforms into

$$w^4\frac{d^2y}{dw^2} + 2w^3\frac{dy}{dw} + \frac{1}{w}y = 0.$$

Inspection of $P(w) = \dfrac{2}{w}$ and $Q(w) = \dfrac{1}{w^5}$

indicates that $w = 0$ is an irregular singular point. Hence ∞ is an irregular singular point.

---

37. $x^2y'' - 4y = 0$

38. $(1 - x)y'' + xy' - y = 0$

39. $x^3y'' + 2x^2y' + 3y = 0$

40. $x^2y'' + (2x + 1)y' + 5y = 0$

41. Solve the Cauchy-Euler equation
$$x^2y'' + 3xy' - 8y = 0.$$
by the method of Frobenius.

6.4 Two Special Equations

The two equations

$$x^2y'' + xy' + (x^2 - \nu^2)y = 0 \tag{1}$$

$$(1 - x^2)y'' - 2xy' + n(n + 1)y = 0 \tag{2}$$

occur frequently in advanced studies in applied mathematics, physics, and engineering. They are called **Bessel's equation** and **Legendre's equation,** respectively.* In solving (1) we shall assume $\nu \geq 0$, whereas in (2) we shall

---

*Named after Friedrich Wilhelm Bessel (1784–1846) and Adrien Marie Legendre (1752–1833). Bessel was a German astronomer who was the first to measure the distance to a star. Legendre, a French mathematician, is best remembered for spending almost 40 years of his

consider only the case when n is a nonnegative integer. Since we seek series solutions of each equation about $x = 0$, we observe that the origin is a regular singular point of Bessel's equation, but it is an ordinary point of Legendre's equation.

6.4.1 Solution of Bessel's Equation

If we assume $y = \sum_{n=0}^{\infty} c_n x^{n+r}$ then

$$x^2 y'' + xy' + (x^2 - v^2) = \sum_{n=0}^{\infty} c_n (n + r)(n + r - 1)x^{n+r} + \sum_{n=0}^{\infty} c_n (n + r)x^{n+r} + \sum_{n=0}^{\infty} c_n x^{n+r+2}$$

$$- v^2 \sum_{n=0}^{\infty} c_n x^{n+r}$$

$$= c_0(r^2 - r + r - v^2)x^r + x^r \sum_{n=1}^{\infty} c_n [(n + r)(n + r - 1)$$

$$+ (n + r) - v^2]x^n + x^r \sum_{n=0}^{\infty} c_n x^{n+2}$$

$$= c_0(r^2 - v^2)x^r + x^r \sum_{n=1}^{\infty} c_n \times [(n + r)^2 - v^2]x^n$$

$$+ x^r \sum_{n=0}^{\infty} c_n x^{n+2} \tag{3}$$

From (3) we see that the indicial equation is $r^2 - v^2 = 0$ so that the indicial roots are $r_1 = v$ and $r_2 = -v$. When $r_1 = v$, (3) becomes

$$x^v \sum_{n=1}^{\infty} c_n n(n + 2v)x^n + x^v \sum_{n=0}^{\infty} c_n x^{n+2}$$

$$= x^v \left[(1 + 2v)c_1 + \underbrace{\sum_{n=2}^{\infty} c_n n(n + 2v)x^n}_{k = n - 2} + \underbrace{\sum_{n=0}^{\infty} c_n x^{n+2}}_{k = n} \right]$$

$$= x^v \left[(1 + 2v)c_1 + \sum_{k=0}^{\infty} [(k + 2)(k + 2 + 2v)c_{k+2} + c_k]x^{k+2} \right] = 0.$$

Therefore by the usual argument we can write

$$(1 + 2v)c_1 = 0$$

$$(k + 2)(k + 2 + 2v)c_{k + 2} + c_k = 0$$

life studying and calculating elliptic integrals. However, the particular polynomial solutions of the equation which bears his name were encountered in his studies of gravitation.

or
$$c_{k+2} = \frac{-c_k}{(k+2)(k+2+2\nu)}, \qquad k = 0, 1, 2 \ldots . \qquad (4)$$

The choice $c_1 = 0$ in (4) implies $c_3 = c_5 = c_7 = \cdots = 0$ so that for $k = 0, 2, 4, \ldots$, we find, after letting $k + 2 = 2n$, $n = 1, 2, 3, \ldots$, that

$$c_{2n} = -\frac{c_{2n-2}}{2^2 n(n+\nu)}. \qquad (5)$$

Thus, $c_2 = -\dfrac{c_0}{2^2 \cdot 1 \cdot (1+\nu)}$

$$c_4 = -\frac{c_2}{2^2 \cdot 2(2+\nu)} = \frac{c_0}{2^4 \cdot 1 \cdot 2(1+\nu)(2+\nu)}$$

$$c_6 = -\frac{c_4}{2^2 \cdot 3(3+\nu)} = \frac{c_0}{2^6 \cdot 1 \cdot 2 \cdot 3(1+\nu)(2+\nu)(3+\nu)}$$

$$\vdots$$

$$c_{2n} = \frac{(-1)^n c_0}{2^{2n} n!(1+\nu)(2+\nu)\cdots(n+\nu)}, \qquad n = 1, 2, 3 \ldots . \qquad (6)$$

It is standard practice to choose c_0 to be a specific value, namely,

$$c_0 = \frac{1}{2^\nu \Gamma(1+\nu)}$$

where $\Gamma(1+\nu)$ is the Gamma function (see the Chapter Appendix). Since this latter function possesses the convenient property $\Gamma(1+\alpha) = \alpha\Gamma(\alpha)$, we can collapse the indicated product in the denominator of (6) into one term. For example,

$$\Gamma(1+\nu+1) = (1+\nu)\Gamma(1+\nu)$$

$$\Gamma(1+\nu+2) = (2+\nu)\Gamma(2+\nu)$$

$$= (2+\nu)(1+\nu)\Gamma(1+\nu).$$

Hence we can write (6) as

$$c_{2n} = \frac{(-1)^n}{2^{2n+\nu} n!(1+\nu)(2+\nu)\cdots(n+\nu)\Gamma(1+\nu)}$$

$$= \frac{(-1)^n}{2^{2n+\nu} n!\Gamma(1+\nu+n)}, \qquad n = 0, 1, 2 \ldots .$$

It follows that one solution

$$y = \sum_{n=0}^{\infty} c_{2n} x^{2n+\nu}$$

$$= \sum_{n=0}^{\infty} \frac{(-1)^n}{n!\Gamma(1+\nu+n)} \left(\frac{x}{2}\right)^{2n+\nu} \qquad (7)$$

If $\nu \geq 0$, the series will converge at least on the interval $0 \leq x < \infty$.

Bessel functions

The series given in (7) is usually denoted by $J_\nu(x)$:

$$J_\nu(x) = \sum_{n=0}^{\infty} \frac{(-1)^n}{n!\,\Gamma(1+\nu+n)} \left(\frac{x}{2}\right)^{2n+\nu}$$

Also, for the second exponent $r_2 = -\nu$ we obtain, in exactly the same manner,

$$J_{-\nu}(x) = \sum_{n=0}^{\infty} \frac{(-1)^n}{n!\,\Gamma(1-\nu+n)} \left(\frac{x}{2}\right)^{2n-\nu} \tag{8}$$

The functions $J_\nu(x)$ and $J_{-\nu}(x)$ are called **Bessel functions of the first kind** of order ν and $-\nu$, respectively. Depending on the value of ν, (8) may contain negative powers of x and hence converges on $0 < x < \infty$*.

Now some care must be taken in writing the general solution of (1). When $\nu = 0$ it is apparent that (7) and (8) are the same. If $\nu > 0$ and $r_1 - r_2 = \nu - (-\nu) = 2\nu$ is not a positive integer it follows from Case I of Section 6.3 that $J_\nu(x)$ and $J_{-\nu}(x)$ are linearly independent solutions of (1) on $0 < x < \infty$ and so the general solution of the interval would be $y = c_1 J_\nu(x) + c_2 J_{-\nu}(x)$. But we also know from Case II of Section 6.3 that when $r_1 - r_2 = 2\nu$ is a positive integer, a second series solution of (1) *may* exist. In this second case, we distinguish two possibilities. When $\nu = m = $ positive integer, $J_{-m}(x)$ defined by (8) and $J_m(x)$ are not linearly independent solutions. It can be shown J_{-m} is a constant multiple of J_m, specifically, $J_{-m}(x) = (-1)^m J_m(x)$ (see Problem 37). In addition, $r_1 - r_2 = 2\nu$ can be a positive integer when ν is half an odd positive integer. It can be shown in this latter event that $J_\nu(x)$ and $J_{-\nu}(x)$ are linearly independent. In other words, the general solution of (1) on $0 < x < \infty$ is

$$y = c_1 J_\nu(x) + c_2 J_{-\nu}(x) \tag{9}$$

provided $\nu \neq$ integer.

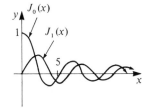

Figure 6.1

The graphs of $y = J_0(x)$ and $y = J_1(x)$ are given in Figure 6.1. Observe that the graphs of J_0 and J_1 resemble damped cosine and sine graphs, respectively.[†]

EXAMPLE

Find the general solution of the equation

$$x^2 y'' + xy' + (x^2 - \tfrac{1}{4})y = 0$$

on $0 < x < \infty$.

Solution: We identify $\nu^2 = 1/4$ and so $\nu = 1/2$. From (9) we see that the general solution of the differential equation on $0 < x < \infty$ is

$$y = c_1 J_{1/2}(x) + c_2 J_{-1/2}(x).$$

---

*By replacing x by $|x|$, the series given in (7) and (8) converge for $0 < |x| < \infty$.
[†]Bessel functions belong to a class of functions that are called "almost-periodic."

EXAMPLE

Find the general solution of the equation

$$x^2y'' + xy' + (x^2 - 9)y = 0.$$

on $0 < x < \infty$.

Solution: Since $\nu^2 = 9$, we identify $\nu = 3$. Thus one solution is $y_1 = J_3(x)$. Dividing the equation by x^2 yields $P(x) = 1/x$, and hence a second solution can be constructed by the method of Section 4.2:

$$y_2 = J_3(x) \int \frac{e^{-\int dx/x}}{J_3^2(x)} \, dx$$

$$= J_3(x) \int \frac{dx}{x J_3^2(x)}.$$

The general solution on $0 < x < \infty$ is then

$$y = c_1 J_3(x) + c_2 J_3(x) \int \frac{dx}{x J_3^2(x)}.$$

When $\nu = 0$, $r_1 = r_2$ and so a second solution of Bessel's equation on $0 < x < \infty$ can be found in the form

$$y_2 = J_0(x) \ln x + \sum_{n=1}^{\infty} b_n x^n.$$

Similarly, when $\nu = m$ is a positive integer, a second solution is given by

$$y_2 = C J_m(x) \ln x + \sum_{n=0}^{\infty} b_n x^{n-m}, \qquad b_0 \neq 0, \qquad C \neq 0, \qquad x > 0.$$

With some struggle it can be verified that the second solution $y_2 = J_3(x) \int dx/x J_3^2(x)$ of the preceding example contains a logarithm.

6.4.2   Solution of Legendre's Equation

Since $x = 0$ is an ordinary point of equation (2) we assume a solution of the form $y = \sum_{k=0}^{\infty} c_k x^k$. Therefore

$$(1 - x^2)y'' - 2xy' + n(n + 1)y$$

$$= (1 - x^2) \sum_{k=0}^{\infty} c_k k(k - 1)x^{k-2} - 2 \sum_{k=0}^{\infty} c_k k x^k + n(n + 1) \sum_{k=0}^{\infty} c_k x^k$$

$$= \sum_{k=2}^{\infty} c_k k(k-1) x^{k-2} - \sum_{k=2}^{\infty} c_k k(k-1) x^k - 2 \sum_{k=1}^{\infty} c_k k x^k + n(n+1) \sum_{k=0}^{\infty} c_k x^k$$

$$= [n(n+1)c_0 + 2c_2] x^0 + [n(n+1)c_1 - 2c_1 + 6c_3] x$$

$$+ \underbrace{\sum_{k=4}^{\infty} c_k k(k-1) x^{k-2}}_{j = k-2} - \underbrace{\sum_{k=2}^{\infty} c_k k(k-1) x^k}_{j = k}$$

$$- 2 \underbrace{\sum_{k=2}^{\infty} c_k k x^k}_{j = k} + n(n+1) \underbrace{\sum_{k=2}^{\infty} c_k x^k}_{j = k}$$

$$= [n(n+1)c_0 + 2c_2] + [(n-1)(n+2)c_1 + 6c_3]x$$

$$+ \sum_{j=2}^{\infty} [(j+2)(j+1)c_{j+2} + (n-j)(n+j+1)c_j] x^j = 0$$

implies that

$$n(n+1)c_0 + 2c_2 = 0$$

$$(n-1)(n+2)c_1 + 6c_3 = 0$$

$$(j+2)(j+1)c_{j+2} + (n-j)(n+j+1)c_j = 0$$

or

$$c_2 = -\frac{n(n+1)}{2!} c_o$$

$$c_3 = -\frac{(n-1)(n+2)}{3!} c_1$$

$$c_{j+2} = -\frac{(n-j)(n+j+1)}{(j+2)(j+1)} c_j, \qquad j = 2, 3, 4 \dots . \qquad (10)$$

Iterating (10) gives

$$c_4 = -\frac{(n-2)(n+3)}{4 \cdot 3} c_2 = \frac{(n-2)n(n+1)(n+3)}{4!} c_0$$

$$c_5 = -\frac{(n-3)(n+4)}{5 \cdot 4} c_3 = \frac{(n-3)(n-1)(n+2)(n+4)}{5!} c_1$$

$$c_6 = -\frac{(n-4)(n+5)}{6 \cdot 5} c_4 = -\frac{(n-4)(n-2)n(n+1)(n+3)(n+5)}{6!} c_0$$

$$c_7 = -\frac{(n-5)(n+6)}{7 \cdot 6} c_5$$

$$= -\frac{(n-5)(n-3)(n-1)(n+2)(n+4)(n+6)}{7!} c_1$$

and so on. Thus for at least $|x| < 1$ we obtain two formal linearly independent power series solutions

$$y_1(x) = c_o \left[1 - \frac{n(n + 1)}{2!}x^2 + \frac{(n - 2)n(n + 1)(n + 3)}{4!}x^4 \right.$$

$$\left. - \frac{(n - 4)(n - 2)n(n + 1)(n + 3)(n + 5)}{6!}x^6 + \cdots \right]$$

$$y_2(x) = c_1 \left[x - \frac{(n - 1)(n + 2)}{3!}x^3 + \frac{(n - 3)(n - 1)(n + 2)(n + 4)}{5!}x^5 \right.$$

$$\left. - \frac{(n - 5)(n - 3)(n - 1)(n + 2)(n + 4)(n + 6)}{7!}x^7 + \cdots \right]$$

(11)

Notice that if n is an even integer the first series terminates, whereas $y_2(x)$ is an infinite series. For example, if $n = 4$ then

$$y_1(x) = c_0 \left[1 - \frac{4 \cdot 5}{2!}x_2 + \frac{2 \cdot 4 \cdot 5 \cdot 7}{4!}x^4 \right]$$

$$= c_0 \left[1 - 10x^2 + \frac{35}{3}x^4 \right].$$

Similarly, when n is an odd integer the series for $y_2(x)$ terminates with x^n. That is, *when n is a nonnegative integer we obtain an nth degree polynomial solution* of Legendre's equation.

Since we know that a constant multiple of a solution of Legendre's equation is also a solution, it is traditional to choose specific values for c_0 and c_1 depending on whether n is an even or odd positive integer, respectively. For $n = 0$ we choose

$$c_0 = 1,$$

and for $n = 2, 4, 6, \ldots$,

$$c_0 = (-1)^{n/2}\frac{1 \cdot 3 \cdots (n - 1)}{2 \cdot 4 \cdots n},$$

whereas for $n = 1$ we choose

$$c_1 = 1,$$

and for $n = 3, 5, 7, \ldots$,

$$c_1 = (-1)^{(n-1)/2}\frac{1 \cdot 3 \cdots n}{2 \cdot 4 \cdots (n - 1)}.$$

For example, when $n = 4$ we have

$$y_1(x) = (-1)^{4/2}\frac{1 \cdot 3}{2 \cdot 4}\left[1 - 10x^2 + \frac{35}{3}x^4 \right]$$

$$= \frac{3}{8} - \frac{30}{8}x^2 + \frac{35}{8}x^4$$

$$= \frac{1}{8}(35x^4 - 30x^2 + 3).$$

Legendre polynomials

These specific nth degree polynomial solutions are called **Legendre polynomials** and are denoted by $P_n(x)$. From the series for $y_1(x)$ and $y_2(x)$ and from the above choices of c_0 and c_1, it is easily verified that the first several Legendre polynomials are

$$P_0(x) = 1$$

$$P_1(x) = x$$

$$P_2(x) = \frac{1}{2}(3x^2 - 1)$$

$$P_3(x) = \frac{1}{2}(5x^3 - 3x)$$

$$P_4(x) = \frac{1}{8}(35x^4 - 30x^2 + 3).$$

(12)

Remember, $P_0(x)$, $P_1(x)$, $P_2(x)$, $P_3(x)$, and $P_4(x)$ are *particular solutions* of the differential equations

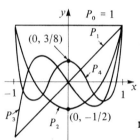

Figure 6.2

$n = 0,$	$(1 - x^2)y'' - 2xy' = 0$
$n = 1,$	$(1 - x^2)y'' - 2xy' + 2y = 0$
$n = 2,$	$(1 - x^2)y'' - 2xy' + 6y = 0$
$n = 3,$	$(1 - x^2)y'' - 2xy' + 12y = 0$
$n = 4,$	$(1 - x^2)y'' - 2xy' + 20y = 0,$

(13)

respectively.

The graphs of the first four Legendre polynomials on the interval $-1 \le x \le 1$ are given in Figure 6.2.

EXERCISES 6.4   *Answers to odd-numbered problems begin on page A-19.*

In Problems 1–6 find the general solution of the given differential equation on $0 < x < \infty$.

1. $x^2y'' + xy' + \left(x^2 - \frac{1}{9}\right)y = 0$ **2.** $x^2y'' + xy' + (x^2 - 1)y = 0$

3. $4x^2y'' + 4xy' + (4x^2 - 25)y = 0$

★ **4.** $16x^2y'' + 16xy' + (16x^2 - 1)y = 0$

5. $xy'' + y' + xy = 0$ **6.** $\frac{d}{dx}[xy'] + \left(x - \frac{4}{x}\right)y = 0$

7. Show that the general solution of the **parametric Bessel equation**
$$x^2y'' + xy' + (\lambda^2x^2 - \nu^2)y = 0$$
on $0 < x < \infty$ is given by
$$y = c_1 J_\nu(\lambda x) + c_2 J_{-\nu}(\lambda x), \qquad \nu \ne \text{integer}.$$

8. Find the general solution of

$$x^2y'' + xy' + (36x^2 - \tfrac{1}{4})y = 0$$

on $0 < x < \infty$.

9. Use the change of variables $y = x^{-1/2}v(x)$ to find the general solution of the equation

$$x^2y'' + 2xy' + \lambda^2x^2y = 0$$

on $0 < x < \infty$.

★10. Verify that the differential equation

$$xy'' + (1 - 2n)y' + xy = 0, \qquad x > 0,$$

possesses the particular solution $y = x^n J_n(x)$.

11. Verify that the differential equation

$$xy'' + (1 + 2n)y' + xy = 0, \qquad x > 0,$$

possesses the particular solution $y = x^{-n} J_n(x)$.

12. Verify that the differential equation

$$x^2y'' + (\lambda^2x^2 - \nu^2 + \tfrac{1}{4})y = 0, \qquad x > 0,$$

possesses the particular solution $y = \sqrt{x} J_\nu(\lambda x), \ \lambda > 0$.

In Problems 13–18 use the results of Problems 10, 11 and 12 to find a particular solution of the given differential equation on $0 < x < \infty$.

13. $y'' + y = 0$ **14.** $xy'' - y' + xy = 0$

15. $xy'' + 3y' + xy = 0$ **16.** $4x^2y'' + (16x^2 + 1)y = 0$

17. $x^2y'' + (x^2 - 2)y = 0$ **★18.** $xy'' - 5y' + xy = 0$

Recurrence formulas are very important in the study of Bessel functions. In Problems 19–23 derive the given formula.

EXAMPLE

$$xJ_\nu'(x) = \nu J_\nu(x) - xJ_{\nu+1}(x)$$

Solution: $\displaystyle J_\nu(x) = \sum_{n=0}^{\infty} \frac{(-1)^n}{n!\Gamma(1 + \nu + n)} \left(\frac{x}{2}\right)^{2n + \nu}$

$\displaystyle xJ_\nu'(x) = \sum_{n=0}^{\infty} \frac{(-1)^n(2n + \nu)}{n!\Gamma(1 + \nu + n)} \left(\frac{x}{2}\right)^{2n + \nu}$

$\displaystyle \qquad = \nu \sum_{n=0}^{\infty} \frac{(-1)^n}{n!\Gamma(1 + \nu + n)} \left(\frac{x}{2}\right)^{2n + \nu}$

$\displaystyle \qquad + 2 \sum_{n=0}^{\infty} \frac{(-1)^n n}{n!\Gamma(1 + \nu + n)} \left(\frac{x}{2}\right)^{2n + \nu}$

$$= \nu J_\nu(x) + x \underbrace{\sum_{n=1}^{\infty} \frac{(-1)^n}{(n-1)!\Gamma(1+\nu+n)} \left(\frac{x}{2}\right)^{2n+\nu-1}}_{k=n-1}$$

$$= \nu J_\nu(x) - x \sum_{k=0}^{\infty} \frac{(-1)^k}{k!\Gamma(2+\nu+k)} \left(\frac{x}{2}\right)^{2k+\nu+1}$$

$$= \nu J_\nu(x) - x J_{\nu+1}(x).$$

The expression $x J_\nu'(x) = \nu J_\nu(x) - x J_{\nu+1}(x)$ is called a differential recurrence relation.

---

19. $x J_\nu'(x) = -\nu J_\nu(x) + x J_{\nu-1}(x)$ [*Hint*: $2n + \nu = 2(n+\nu) - \nu$]

20. $2 J_\nu'(x) = J_{\nu-1}(x) - J_{\nu+1}(x)$

21. $2\nu J_\nu(x) = x J_{\nu+1}(x) + x J_{\nu-1}(x)$ ★**22.** $\frac{d}{dx}[x^\nu J_\nu(x)] = x^\nu J_{\nu-1}(x)$

23. $\frac{d}{dx}[x^{-\nu} J_\nu(x)] = -x^{-\nu} J_{\nu+1}(x)$

In Problems 24–26 use Problems 19–23 to obtain the given result.

24. $J_0'(x) = J_{-1}(x) = -J_1(x)$ **25.** $\displaystyle\int_0^x r J_0(r)\,dr = x J_1(x)$

★**26.** $\displaystyle\int x^3 J_0(x)\,dx = x^3 J_1(x) - 2x^2 J_2(x) + c$

EXAMPLE   Find an alternative expression for $J_{1/2}(x)$. Use the fact that $\Gamma(\frac{1}{2}) = \sqrt{\pi}$.

Solution: With $\nu = 1/2$, we have from (7)

$$J_{1/2}(x) = \sum_{n=0}^{\infty} \frac{(-1)^n}{n!\Gamma(1+\frac{1}{2}+n)} \left(\frac{x}{2}\right)^{2n+\frac{1}{2}}$$

Now observe that

$$n = 0, \quad \Gamma\left(1 + \frac{1}{2}\right) = \frac{1}{2}\Gamma\left(\frac{1}{2}\right) = \frac{1}{2}\sqrt{\pi}$$

$$n = 1, \quad \Gamma\left(1 + \frac{3}{2}\right) = \frac{3}{2}\Gamma\left(\frac{3}{2}\right) = \frac{3}{2^2}\sqrt{\pi}$$

$$n = 2, \quad \Gamma\left(1 + \frac{5}{2}\right) = \frac{5}{2}\Gamma\left(\frac{5}{2}\right) = \frac{5 \cdot 3}{2^3}\sqrt{\pi} = \frac{5 \cdot 4 \cdot 3 \cdot 2 \cdot 1}{2^3 4 \cdot 2}\sqrt{\pi} = \frac{5!}{2^5 2!}\sqrt{\pi}$$

$$n = 3, \quad \Gamma\left(1 + \frac{7}{2}\right) = \frac{7}{2}\Gamma\left(\frac{7}{2}\right) = \frac{7 \cdot 5 \cdot 3}{2^4}\sqrt{\pi} = \frac{7 \cdot 6 \cdot 5 \cdot 4 \cdot 3 \cdot 2 \cdot 1}{2^4 6 \cdot 4 \cdot 2}\sqrt{\pi}$$

$$= \frac{7!}{2^7 3!}\sqrt{\pi}.$$

In general, $\qquad \Gamma\left(1 + \dfrac{1}{2} + n\right) = \dfrac{(2n + 1)!}{2^{2n + 1} n!}\sqrt{\pi}.$

Hence we can write

$$J_{1/2}(x) = \sum_{n=0}^{\infty} \frac{(-1)^n}{n!\,\dfrac{(2n + 1)!\sqrt{\pi}}{2^{2n + 1} n!}} \left(\frac{x}{2}\right)^{2n + \frac{1}{2}}$$

$$= \sqrt{\frac{2}{\pi x}} \sum_{n=0}^{\infty} \frac{(-1)^n}{(2n + 1)!} x^{2n + 1}$$

$$= \sqrt{\frac{2}{\pi x}} \sin x.$$

27. Express $J_{-1/2}(x)$ in terms of cos x and a power of x.

When $\nu =$ half an odd integer, $J_\nu(x)$ can be expressed in terms of sin x, cos x, and powers of x. Such Bessel functions are usually called **spherical Bessel functions**. In Problems 28–33 use the results of Problems 21 and 27 and the preceding example to find an alternative expression for the given function.

28. $J_{3/2}(x)$ **29.** $J_{-3/2}(x)$

★**30.** $J_{5/2}(x)$ **31.** $J_{-5/2}(x)$

32. $J_{7/2}(x)$ **33.** $J_{-7/2}(x)$

34. Show that $i^{-\nu}J_\nu(ix)$, $i^2 = -1$, is a real function. The function defined by $I_\nu(x) = i^{-\nu}J_\nu(ix)$ is called a **modified Bessel function of the first kind** of order ν.

35. Find the general solution of the differential equation
$$x^2 y'' + xy' - (x^2 + \nu^2)y = 0, \quad x > 0, \qquad \nu \neq \text{integer}.$$
[*Hint*: $i^2 x^2 = -x^2$.]

36. If $y_1 = J_0(x)$ is one solution of the zero-order Bessel equation verify that another solution is
$$y_2 = J_0(x)\ln x + \frac{x^2}{4} - \frac{3x^4}{128} + \frac{11x^6}{13,824} - \cdots.$$

37. Use (8) with $\nu = m$, where m is a positive integer, and the fact that $1/\Gamma(N) = 0$ when N is a negative integer, to show that
$$J_{-m}(x) = (-1)^m J_m(x).$$

38. Show that between two consecutive positive roots of $J_1(x)$ there exists a root of $J_0(x)$. [*Hint:* Look up Rolle's theorem and then inspect Problem 24.]

[6.4.2]   **39. (a)** Use the explicit solutions $y_1(x)$ and $y_2(x)$ of Legendre's equation and the appropriate choices of c_0 and c_1 to find the Legendre polynomials $P_5(x)$ and $P_6(x)$.
   (b) Write the explicit differential equations for which $P_5(x)$ and $P_6(x)$ are particular solutions.

★**40.** Show that Legendre's equation has the alternative form

$$\frac{d}{dx}\left[(1-x^2)\frac{dy}{dx}\right] + n(n+1)y = 0.$$

41. Show that the equation

$$\sin\theta\frac{d^2y}{d\theta^2} + \cos\theta\frac{dy}{d\theta} + n(n+1)(\sin\theta)y = 0$$

can be transformed in Legendre's equation by means of the subsitution $x = \cos\theta$.

42. The general Legendre polynomial can be written as

$$P_n(x) = \sum_{k=0}^{[n/2]} \frac{(-1)^k(2n-2k)!}{2^n k!(n-k)!(n-2k)!}x^{n-2k}$$

where $[n/2]$ is the greatest integer not greater than $n/2$. Verify the results for $n = 0, 1, 2, 3, 4, 5$.

43. Use the binomial theorem to formally show that

$$(1-2xt+t^2)^{-1/2} = \sum_{n=0}^{\infty} P_n(x)t^n.$$

The expression $(1-2xt+t^2)^{-1/2}$ is called a **generating function** for the Legendre polynomials.

★**44.** Use Problem 43 to show that $P_n(1) = 1$ and $P_n(-1) = (-1)^n$.

EXAMPLE   Differentiating the generating function given in Problem 43 with respect to t gives

$$(1-2xt+t^2)^{-3/2}(x-t) = \sum_{n=0}^{\infty} nP_n(x)t^{n-1}$$

$$= \sum_{n=1}^{\infty} nP_n(x)t^{n-1}$$

so that after multiplying by $1-2xt+t^2$ we have

$$(x-t)(1-2xt+t^2)^{-1/2} = (1-2xt+t^2)\sum_{n=1}^{\infty} nP_n(x)t^{n-1}$$

or $$(x-t)\sum_{n=0}^{\infty} P_n(x)t^n = (1-2xt+t^2)\sum_{n=1}^{\infty} nP_n(x)t^{n-1}. \qquad (14)$$

Multiply out and rewrite (14) as

$$\sum_{n=0}^{\infty} xP_n(x)t^n - \sum_{n=0}^{\infty} P_n(x)t^{n+1} - \sum_{n=1}^{\infty} nP_n(x)t^{n-1} + 2x\sum_{n=1}^{\infty} nP_n(x)t^n - \sum_{n=1}^{\infty} nP_n(x)t^{n+1} = 0$$

or

$$x + x^2t + \sum_{n=2}^{\infty} xP_n(x)t^n - t - \sum_{n=1}^{\infty} P_n(x)t^{n+1} - x - 2\left(\frac{3x^2 - 1}{2}\right)t$$

$$- \sum_{n=3}^{\infty} nP_n(x)t^{n-1} + 2x^2t + 2x\sum_{n=2}^{\infty} nP_n(x)t^n - \sum_{n=1}^{\infty} nP_nt^{n+1} = 0$$

Observing the appropriate cancellations, simplifying, and changing the summation indices gives

$$\sum_{k=2}^{\infty} [-(k+1)P_{k+1}(x) + (2k+1)xP_k(x) - kP_{k-1}(x)]t^k = 0.$$

Equating the total coefficient of t^k to be zero gives the three term recurrence relation

$$(k+1)P_{k+1}(x) - (2k+1)xP_k(x) + kP_{k-1}(x) = 0, \quad k = 2, 3, 4 \ldots.$$

This formula is also valid when $k = 1$.

---

45. Use the results of the preceding example and the fact that $P_0(x) = 1$, $P_1(x) = x$ to generate the next three Legendre polynominals.

46. The Legendre polynominals are also generated by **Rodrigues' formula***

$$P_n(x) = \frac{1}{2^n n!}\frac{d^n}{dx^n}(x^2 - 1)^n.$$

Verify the results for $n = 0, 1, 2, 3$.

47. Use the explicit Legendre polynomials $P_0(x)$, $P_1(x)$, $P_2(x)$, and $P_3(x)$ to evaluate $\int_{-1}^{1} P_n^2(x)dx$ for $n = 0, 1, 2, 3$. Generalize the results.

48. Use the explicit Legendre polynomials $P_0(x)$, $P_1(x)$, $P_2(x)$, and $P_3(x)$ to evaluate $\int_{-1}^{1} P_n(x)P_m(x)dx$ for $n \neq m$. Generalize the results.

49. Find constants c_0, c_1, c_2, and c_3 so that

$$x^3 = c_0P_0(x) + c_1P_1(x) + c_2P_2(x) + c_3P_3(x)$$

on the interval $-1 \leq x \leq 1$.

★50. We know that $y_1 = x$ is a solution of Legendre's equation when $n = 1$, $(1 - x^2)y'' - 2xy' + 2y = 0$. Show that a second linearly independent solution on the interval $-1 < x < 1$ is

---

*Named after a minor French mathematician, O. Rodrigues (1794–1885).

$$y_2 = \frac{x}{2} \ln \left(\frac{1+x}{1-x} \right) - 1$$

51. In the study of applied partial differential equations, the equation

$$\frac{\partial}{\partial r} \left[r^2 \frac{\partial u}{\partial r} \right] = -\frac{\partial^2 u}{\partial \theta^2} - \frac{\cos \theta}{\sin \theta} \frac{\partial u}{\partial \theta}$$

where $u = u(r, \theta)$, occurs in certain problems involving temperatures in a sphere.

(a) If we let $u = R(r)\Theta(\theta)$ show that

$$r^2 \frac{R''}{R} + 2r \frac{R'}{R} = -\frac{\Theta''}{\Theta} - \frac{\cos \theta}{\sin \theta} \frac{\Theta'}{\Theta}$$

(b) If we equate both sides equal to the same constant $n(n + 1)$, show that particular product solutions of the partial differential equation can be found provided we can solve the ordinary differential equations

$$r^2 R'' + 2rR' - n(n + 1)R = 0$$

$$(\sin \theta)\Theta'' + (\cos \theta)\Theta' + n(n + 1)(\sin \theta)\Theta = 0.$$

(c) Solve both ordinary differential equations in part (b), and find particular product solutions $u = R(r)\Theta(\theta)$ of the original partial differential equation.

52. (a) By differentiating Legendre's equation (2) m times show that

$$v = \frac{d^m}{dx^m} P_n(x)$$

is a solution of

$$(1 - x^2) \frac{d^2 v}{dx^2} - 2x(m + 1) \frac{dv}{dx} + [n(n+1) - m(m+1)]v = 0.$$

(b) By letting $w = (1 - x^2)^{m/2} v$ show that the differential equation in (a) becomes

$$(1 - x^2) \frac{d^2 w}{dx^2} - 2x \frac{dw}{dx} + \left[n(n+1) - \frac{m^2}{1 - x^2} \right] w = 0.$$

The solution w of the last equation is denoted by $P_n^m(x)$. The functions defined by

$$P_n^m(x) = (1 - x^2)^{m/2} \frac{d^m}{dx^m} P_n(x)$$

are called **associated Legendre functions**.

(c) Explain why $P_n^m(x) = 0$ for $m > n$.

CHAPTER 6 SUMMARY

The remarkable characteristic of a Cauchy-Euler equation is the fact that even though it is a differential equation with variable coefficients it can be solved in terms of elementary functions. A **second-order Cauchy-Euler** equation is any differential equation of the form

$$ax^2 y'' + bxy' + cy = g(x)$$

**CHAPTER 6
SUMMARY**

where a, b, and c are constants. To solve the homogeneous equation we try a solution of the form $y = x^m$ and this in turn leads to an algebraic **auxiliary equation** $am(m - 1) + bm + c = 0$. Accordingly, when the roots are: real and distinct, real and equal, and complex conjugates, the general solution on the interval $0 < x < \infty$ is

$$y = c_1 x^{m_1} + c_2 x^{m_2}$$

$$y = c_1 x^{m_1} + c_2 x^{m_1} \ln x$$

$$y = x^\alpha [c_1 \cos (\beta \ln x) + c_2 \sin (\beta \ln x)],$$

respectively.

We say that $x = 0$ is an **ordinary point** of the linear second-order differential equation

$$a_2(x)y'' + a_1(x)y' + a_0(x)y = 0 \tag{1}$$

when $a_2(0) \neq 0$ and $a_2(x)$, $a_1(x)$, $a_0(x)$ are polynomials having no common factors. Every solution of (1) has the form of a power series in x;

$$y = \sum_{n=0}^{\infty} c_n x^n. \tag{2}$$

To find the coefficients c_n we substitute the basic assumption (2) into (1), and after appropriate algebraic manipulations we determine a **recurrence relation** by equating to zero the combined total coefficient of x^k. Iteration of the recurrence relation yields two distinct sets of coefficients, one set containing the arbitrary coefficient c_0 and the other containing c_1. Using each set of coefficients we form two linearly independent solutions $y_1(x)$ and $y_2(x)$. A solution is valid at least on an interval defined by $|x| < R_1$ where R_1 is the distance to the closest singular point of the equation.

If $a_2(0) = 0$ then $x = 0$ is said to be a **singular point** of (1). Singular points are classified as either **regular** or **irregular**. To determine whether $x = 0$ is a regular singular point, we examine the denominators of the rational functions P and Q that result when (1) is written as $y'' + P(x)y' + Q(x)y = 0$. It is understood that $a_1(x)/a_2(x)$ and $a_0(x)/a_2(x)$ are reduced to lowest terms. If x appears *at most* to the first power in the denominator of $P(x)$ and *at most* to the second power in the denominator of $Q(x)$, then $x = 0$ is a regular singular point. Around the regular singular point $x = 0$ the **method of Frobenius** guarantees that there exists *at least one* solution of the form

$$y = \sum_{n=0}^{\infty} c_n x^{n+r}. \tag{3}$$

The exponent r is a root of a quadratic **indicial equation**.

When the indicial roots r_1 and $r_2(r_1 > r_2)$ satisfy $r_1 - r_2 \neq$ an integer, then we can always find *two* linearly independent solutions of the assumed form (3). When $r_1 - r_2 =$ positive integer, then we could *possibly* find two solutions of form (3) by working with the smaller root r_2 and the recurrence relation which defines the c_n. Lastly, when $r_1 - r_2 = 0$ or $r_1 = r_2$, we can

**CHAPTER 6
SUMMARY**

find only one solution of form (3); the second solution *always* contains a logarithm.

Bessel's equation $x^2y'' + xy' + (x^2 - v^2)y = 0$ has a regular singular point at $x = 0$, whereas $x = 0$ is an ordinary point of **Legendre's equation** $(1 - x^2)y'' - 2xy' + n(n + 1)y = 0$. The latter equation possesses a polynomial solution when n is a nonnegative integer.

**CHAPTER 6
TEST**

Answers to odd-numbered problems begin on page A-21.

In Problems 1–4 solve the given Cauchy-Euler equation.

1. $6x^2y'' + 5xy' - y = 0$

2. $2x^3y''' + 19x^2y'' + 39xy' + 9y = 0$

3. $x^2y'' - 4xy' + 6y = 2x^4 + x^2$

4. $x^2y'' - xy' + y = x^3$

5. Specify the ordinary points of $(x^3 - 8)y'' - 2xy' + y = 0$._____

6. Specify the singular points of $(x^4 - 16)y'' + 2y = 0$.   _____

In Problems 7–10 specify the regular and irregular singular points of the given differential equation.

7. $(x^3 - 10x^2 + 25x)y'' + y' = 0$ RSP_____ ISP_____

8. $(x^3 - 10x^2 + 25x)y'' + y = 0$ RSP_____ ISP_____

9. $x^2(x^2 - 9)^2y'' - (x^2 - 9)y' + xy = 0$ RSP_____ ISP_____

10. $x(x^2 + 1)^3y'' + y' - 8xy = 0$ RSP_____ ISP_____

In Problems 11 and 12 specify an interval around $x = 0$ for which a power series solution of the given differential equation will converge.

11. $y'' - xy' + 6y = 0$ _____

12. $(x^2 - 4)y'' - 2xy' + 9y = 0$_____

In Problems 13–16 for each differential equation find two power series solutions about the ordinary point $x = 0$.

13. $y'' + xy = 0$ **14.** $y'' - 4y = 0$

15. $(x - 1)y'' + 3y = 0$ **16.** $y'' - x^2y' + xy = 0$

In Problems 17–22 find two linearly independent solutions of each equation.

17. $2x^2y'' + xy' - (x + 1)y = 0$ **18.** $2xy'' + y' + y = 0$

19. $x(1 - x)y'' - 2y' + y = 0$

continued

**CHAPTER 6
TEST**

20. $x^2y'' - xy' + (x^2 + 1)y = 0$

21. $xy'' - (2x - 1)y' + (x - 1)y = 0$

22. $x^2y'' - x^2y' + (x^2 - 2)y = 0$

23. Without referring to Section 6.4, use the method of Frobenius to obtain a solution of the Bessel equation for $v = 0$: $xy'' + y' + xy = 0$.

Chapter 6 Appendix   The Gamma Function

Euler's integral definition of the **Gamma function** is

$$\Gamma(x) = \int_0^\infty t^{x-1}e^{-t}\, dt. \tag{1}$$

Convergence of the integral requires that $x - 1 > -1$ or $x > 0$. The recurrence relation

$$\boxed{\Gamma(x + 1) = x\Gamma(x)} \tag{2}$$

which we have seen in Section 6.4 can be obtained from (1) by employing integration by parts. Now when $x = 1$

$$\Gamma(1) = \int_0^\infty e^{-t}\, dt = 1$$

and thus (2) gives

$$\Gamma(2) = 1\Gamma(1) = 1$$
$$\Gamma(3) = 2\Gamma(2) = 2 \cdot 1$$
$$\Gamma(4) = 3\Gamma(3) = 3 \cdot 2 \cdot 1,$$

and so on. In this manner it is seen that when n is a positive ingeger

$$\boxed{\Gamma(n + 1) = n!.}$$

For this reason the Gamma function is often called the **generalized factorial function**.

Although the integral form (1) does not converge for $x < 0$, it can be shown by means of alternative definitions that the Gamma function is defined for all real and complex number *except* $x = -n$, $n = 0, 1, 2, \ldots$. As a consequence (2) is actually valid for $x \neq -n$. Considered as a function of a real variable x, the graph of $\Gamma(x)$ is as indicated in Figure 6.3. Observe that the nonpositive integers correspond to vertical asymptotes of the graph.

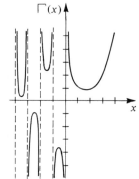
$\Gamma(x)$

Figure 6.3

In Problems 27–33 of Exercises 6.4 we utilized the fact that $\Gamma(\frac{1}{2}) = \sqrt{\pi}$. This result can be derived from (1) by setting $x = 1/2$

$$\Gamma(\tfrac{1}{2}) = \int_0^\infty t^{-1/2} e^{-t} \, dt. \tag{3}$$

By letting $t = u^2$, (3) can be written as

$$\Gamma(\tfrac{1}{2}) = 2 \int_0^\infty e^{-u^2} \, du.$$

But

$$\int_0^\infty e^{-u^2} \, du = \int_0^\infty e^{-v^2} \, dv$$

and so

$$[\Gamma(\tfrac{1}{2})]^2 = 2 \int_0^\infty e^{-u^2} \, du \cdot 2 \int_0^\infty e^{-v^2} \, dv$$

$$= 4 \int_0^\infty \int_0^\infty e^{-(u^2 + v^2)} \, du \, dv.$$

Switching to polar coordinates $x = r \cos \theta$, $y = r \sin \theta$ enables us to evaluate the double integral,

$$4 \int_0^\infty \int_0^\infty e^{-(u^2 + v^2)} \, du \, dv = 4 \int_0^{\pi/2} \int_0^\infty e^{-r^2} r \, dr \, d\theta = \pi.$$

Hence

$$[\Gamma(\tfrac{1}{2})]^2 = \pi \quad \text{or} \quad \Gamma(\tfrac{1}{2}) = \sqrt{\pi}.$$

In view of (2) it follows that with $x = -1/2$

$$\Gamma(\tfrac{1}{2}) = -\tfrac{1}{2}\Gamma(-\tfrac{1}{2})$$
$$\Gamma(-\tfrac{1}{2}) = -2\Gamma(\tfrac{1}{2})$$
$$= -2\sqrt{\pi}.$$

CHAPTER 6 APPENDIX EXERCISES

Answers to odd-numbered problems begin on page A-21.

1. Evaluate
   (a) $\Gamma(5)$ **(b)** $\Gamma(7)$ **(c)** $\Gamma(-\tfrac{3}{2})$ **(d)** $\Gamma(-\tfrac{5}{2})$

2. Use (1) and the fact that $\Gamma(6/5) = 0.92$ to evaluate

$$\int_0^\infty x^5 e^{-x^5} \, dx.$$

[*Hint:* Let $t = x^5$.]

3. Use (1) and the fact that $\Gamma(5/3) = 0.89$ to evaluate

$$\int_0^\infty x^4 e^{-x^3} \, dx.$$

★**4.** Evaluate $\displaystyle\int_0^1 x^3 (\ln \frac{1}{x})^3 \, dx$.

[*Hint*: Let $t = -\ln x$.]

5. Use the fact that $\Gamma(x) > \displaystyle\int_0^1 t^{x-1} e^{-t} \, dt$ to show that $\Gamma(x)$ is unbounded as $x \to 0^+$.

6. Use (1) to derive (2) for $x > 0$.

The Laplace Transform

7.1 The Laplace Transform

7.1.1 Basic Definition

In elementary calculus you have studied the operations of differentiation and integration; recall that

$$\frac{d}{dx}[\alpha f(x) + \beta g(x)] = \alpha \frac{d}{dx} f(x) + \beta \frac{d}{dx} g(x)$$

$$\int [\alpha f(x) + \beta g(x)]\, dx = \alpha \int f(x)\, dx + \beta \int g(x)\, dx. \tag{1}$$

for any real constants α and β. Any operation having the property illustrated in (1) is said to be a **linear operation**. A definite integral of a sum can, of course, be written

$$\int_a^b [\alpha f(x) + \beta g(x)]\, dx = \alpha \int_a^b f(x)\, dx + \beta \int_a^b g(x)\, dx$$

provided each integral exists. Hence, definite integration is a linear operation. In this section we are particularly interested in an improper integral for which the integrand contains a parameter s (a constant during integration). That is,

$$\int_0^\infty f(s, t)\, dt = \lim_{b \to \infty} \int_0^b f(s, t)\, dt. \tag{2}$$

If the limit in (2) exists, the integral is said to be convergent.

The Laplace transform

The foregoing linear operations of differentiation and integration *transform* a function into another expression. For example,

271

$$\frac{d}{dx}x^2 = 2x$$

$$\int x^2 \, dx = \frac{x^3}{3} + c$$

$$\int_0^3 x^2 \, dx = 9.$$

Specifically we are concerned with an improper integral which transforms a function $f(t)$ into a function of a parameter s. As we shall see in Section 7.3, the following transform is useful in solving certain differential equations connected with problems in physics and engineering.

DEFINITION 7.1   Let $f(t)$ be defined for $t \geq 0$, then the integral

$$\int_0^\infty e^{-st} f(t) \, dt = \lim_{b \to \infty} \int_0^b e^{-st} f(t) \, dt \qquad (3)$$

is said to be the **Laplace transform** of f provided the limit exists.* $\square$

Symbolically the Laplace transform of f is denoted by $\mathscr{L}\{f(t)\}$, and since the answer depends on s we write $\mathscr{L}\{f(t)\} = F(s)$.

EXAMPLE   Evaluate $\mathscr{L}\{1\}$.

Solution: $$\mathscr{L}\{1\} = \int_0^\infty e^{-st}(1) \, dt$$

$$= \lim_{b \to \infty} \int_0^b e^{-st} \, dt$$

$$= \lim_{b \to \infty} \left. \frac{-e^{-st}}{s} \right|_0^b$$

$$= \lim_{b \to \infty} \frac{-e^{-sb} + 1}{s}$$

$$= \frac{1}{s} \qquad \text{provided } s > 0.$$

The use of the limit sign becomes somewhat tedious, so we shall adopt the notation $\Big|_0^\infty$ as a shorthand to writing

$$\lim_{b \to \infty} (\ ) \Big|_0^b .$$

---

*Pierre Simon de Laplace (1749–1827) was a noted French mathematician and astronomer, called by some of his enthusiastic contemporaries the "Newton of France." Although Laplace made use of this particular integral transformation in his work in probability theory, it is likely that the integral was first discovered by Euler.

For example,
$$\mathscr{L}\{1\} = \int_0^\infty e^{-st}\, dt$$

$$= \left. \frac{-e^{-st}}{s} \right|_0^\infty$$

$$= \frac{1}{s}, \qquad s > 0$$

where it is understood that at the upper limit we mean $e^{-st} \to 0$ as $t \to \infty$ for $s > 0$.

$\mathscr{L}$ is a linear operation.

For a sum of functions we can write
$$\int_0^\infty e^{-st}[\alpha f(t) + \beta g(t)]\, dt = \alpha \int_0^\infty e^{-st} f(t)\, dt + \beta \int_0^\infty e^{-st} g(t)\, dt$$
whenever both integrals converge. Hence, it follows that
$$\mathscr{L}\{\alpha f(t) + \beta g(t)\} = \alpha \mathscr{L}\{f(t)\} + \beta \mathscr{L}\{g(t)\}$$
$$= \alpha F(s) + \beta G(s). \tag{4}$$

The integral which defines the Laplace transform does not have to converge. For example, neither $\mathscr{L}\{1/t\}$ nor $\mathscr{L}\{e^{t^2}\}$ exists. We state sufficient conditions which will guarantee the existence of $\mathscr{L}\{f(t)\}$.

THEOREM 7.1 Let $f(t)$ be piecewise continuous for $t \geq 0$.* If
$$|f(t)| \leq Me^{ct}, \qquad t > T \tag{5}$$
for some constant c and $T > 0$, then $\mathscr{L}\{f(t)\}$ exists for $s > c$.

Proof:
$$\mathscr{L}\{f(t)\} = \int_0^\infty e^{-st} f(t)\, dt$$

$$= \int_0^T e^{-st} f(t)\, dt + \int_T^\infty e^{-st} f(t)\, dt$$

$$= I_1 + I_2.$$

Now I_1 exists since it can be written as a sum of integrals over intervals for which f is continuous. I_2 exists since by (5)
$$|I_2| \leq \int_T^\infty \left| e^{-st} f(t) \right|\, dt$$

$$\leq M \int_T^\infty e^{-st} e^{ct}\, dt$$

$$= M \int_T^\infty e^{-(s-c)t}\, dt$$

---

*In any interval $0 \leq a \leq t \leq b$, the function f possesses at most a finite number of points t_k, $k = 1, 2, \ldots, n$ $(t_{k-1} < t_k)$ at which it has finite discontinuities; it is continuous on each open interval $t_{k-1} < t < t_k$.

$$= -M \frac{e^{-(s-c)t}}{s-c} \Big|_T^\infty$$

$$= M \frac{e^{-(s-c)T}}{s-c} \quad \text{for } s > c. \qquad \square$$

Exponential order

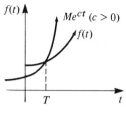

Figure 7.1

If, say, f is an increasing function, then the condition $|f(t)| \le Me^{ct}$, $t > T$, simply states that its graph on the interval (T, ∞) does not grow faster than the graph of Me^{ct} where c is a positive constant (see Figure 7.1). Functions with this exponential function as an upper bound for $t > T$ are said to be of **exponential order**. For example, $f(t) = t$, $f(t) = e^{-t}$, and $f(t) = 2\cos t$ are all of exponential order for $t > 0$ since on this interval we have, respectively,

$$|t| \le e^t$$

$$|e^{-t}| \le e^t$$

$$|2\cos t| \le 2e^t$$

(See Figure 7.2.)

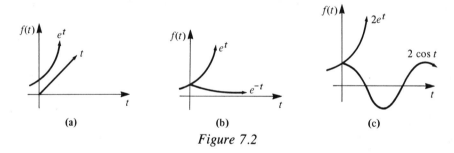

(a)　　　　　(b)　　　　　(c)

Figure 7.2

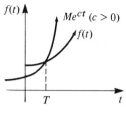

Figure 7.3

A function such as $f(t) = e^{t^2}$ is not of exponential order since, as shown in Figure 7.3, its graph grows faster than any positive linear power of e for $t > c > 0$.

A positive integral power of t is always of exponential order since for $c > 0$

$$|t^n| \le Me^{ct} \quad \text{or} \quad \left| \frac{t^n}{e^{ct}} \right| \le M \quad \text{for } t > T$$

is equivalent to showing that $\lim_{t \to \infty} t^n/e^{ct}$ is a finite limit for $n = 1, 2, 3 \ldots$. The result follows by n applications of L'Hôpital's rule.

Throughout this entire chapter we shall be concerned only with functions which are both piecewise continuous and of exponential order. We note, however, that these conditions are sufficient but not necessary for the existence of a Laplace transform. The function $f(t) = t^{-1/2}$ is not piecewise continuous on $t \ge 0$ but its Laplace transform exists (see Problem 38).

EXAMPLE

Evaluate $\mathcal{L}\{t\}$.

Solution: From Definition 7.1 we have

$$\mathcal{L}\{t\} = \int_0^\infty e^{-st}t\, dt.$$

Integrating by parts and using the fact that $\lim_{t\to\infty} te^{-st} = 0$ for $s > 0$, we obtain

$$\mathcal{L}\{t\} = \left.\frac{-te^{-st}}{s}\right|_0^\infty + \frac{1}{s}\int_0^\infty e^{-st}\, dt$$

$$= \frac{1}{s}\mathcal{L}\{1\} = \frac{1}{s}\left(\frac{1}{s}\right)$$

$$= \frac{1}{s^2}, \qquad s > 0.$$

EXAMPLE

Evaluate $\mathcal{L}\{e^{-3t}\}$.

Solution: From Definition 7.1

$$\mathcal{L}\{e^{-3t}\} = \int_0^\infty e^{-st}e^{-3t}\, dt$$

$$= \int_0^\infty e^{-(s+3)t}\, dt$$

$$= \left.\frac{-e^{-(s+3)t}}{s+3}\right|_0^\infty$$

$$= \frac{1}{s+3}, \qquad s > -3.$$

The last result follows from the fact that $\lim_{t\to\infty} e^{-(s+3)t} = 0$ for $s + 3 > 0$ or $s > -3$.

EXAMPLE

Evaluate $\mathcal{L}\{\sin 2t\}$.

Solution: From Definition 7.1 and integration by parts

$$\mathcal{L}\{\sin 2t\} = \int_0^\infty e^{-st}\sin 2t\, dt$$

$$= \left.\frac{-e^{-st}\sin 2t}{s}\right|_0^\infty + \frac{2}{s}\int_0^\infty e^{-st}\cos 2t\, dt$$

$$= \frac{2}{s} \int_0^\infty e^{-st} \cos 2t \, dt, \quad s > 0$$

$$= \frac{2}{s} \left[\frac{-e^{-st} \cos 2t}{s} \bigg|_0^\infty - \frac{2}{s} \int_0^\infty e^{-st} \sin 2t \, dt \right]$$

$$= \frac{2}{s^2} - \frac{4}{s^2} \mathcal{L}\{\sin 2t\}, \quad s > 0.$$

Now solve for $\mathcal{L}\{\sin 2t\}$,

$$\left[1 + \frac{4}{s^2} \right] \mathcal{L}\{\sin 2t\} = \frac{2}{s^2}$$

$$\mathcal{L}\{\sin 2t\} = \frac{2}{s^2 + 4}, \quad s > 0.$$

EXAMPLE   Evaluate $\mathcal{L}\{3t - 5 \sin 2t\}$.

Solution: From the preceding examples and the linearity property of the Laplace transform we can write

$$\mathcal{L}\{3t - 5 \sin 2t\} = 3\mathcal{L}\{t\} - 5\mathcal{L}\{\sin 2t\}$$

$$= 3 \cdot \frac{1}{s^2} - 5 \cdot \frac{2}{s^2 + 4}$$

$$= \frac{-7s^2 + 12}{s^2(s^2 + 4)}, \quad s > 0.$$

EXAMPLE   Evaluate **(a)** $\mathcal{L}\{te^{-2t}\}$, **(b)** $\mathcal{L}\{t^2 e^{-2t}\}$.

Solution: **(a)** From Definition 7.1 and integration by parts

$$\mathcal{L}\{te^{-2t}\} = \int_0^\infty e^{-st}(te^{-2t}) \, dt$$

$$= \int_0^\infty te^{-(s+2)t} \, dt$$

$$= \frac{-te^{-(s+2)t}}{s+2} \bigg|_0^\infty + \frac{1}{s+2} \int_0^\infty e^{-(s+2)t} \, dt$$

$$= \frac{-e^{-(s+2)t}}{(s+2)^2} \bigg|_0^\infty \quad (s > -2)$$

$$= \frac{1}{(s+2)^2}, \quad (s > -2).$$

(b) Again, integration by parts gives

$$\mathcal{L}\{t^2 e^{-2t}\} = \frac{-t^2 e^{-(s+2)t}}{s+2} \Big|_0^\infty + \frac{2}{s+2} \int_0^\infty t e^{-(s+2)t} \, dt$$

$$= \frac{2}{s+2} \int_0^\infty e^{-st}(t e^{-2t}) \, dt \qquad (s > -2)$$

$$= \frac{2}{s+2} \mathcal{L}\{t e^{-2t}\} = \frac{2}{s+2} \left[\frac{1}{(s+2)^2} \right]$$

$$= \frac{2}{(s+2)^3}, \qquad (s > -2).$$

We state the generalizations of some of the preceding examples by means of the next theorem. From this point on we shall also refrain from stating any restrictions on s; it is understood that s is sufficiently restricted to guarantee the convergence of the appropriate Laplace transform.

THEOREM 7.2   **(a)**
$$\mathcal{L}\{1\} = \frac{1}{s}$$

(b)
$$\mathcal{L}\{t^n\} = \frac{n!}{s^{n+1}}, \qquad n = 1, 2, 3, \ldots$$

(c)
$$\mathcal{L}\{e^{at}\} = \frac{1}{s-a}$$

(d)
$$\mathcal{L}\{\sin kt\} = \frac{k}{s^2 + k^2}$$

(e)
$$\mathcal{L}\{\cos kt\} = \frac{s}{s^2 + k^2}$$

(f)
$$\mathcal{L}\{\sinh kt\} = \frac{k}{s^2 - k^2}$$

(g)
$$\mathcal{L}\{\cosh kt\} = \frac{s}{s^2 - k^2}. \qquad \square$$

Part **(b)** can be justified in the following manner. Integration by parts yields

$$\mathcal{L}\{t^n\} = \int_0^\infty e^{-st} t^n \, dt$$

$$= -\frac{1}{s} e^{-st} t^n \Big|_0^\infty + \frac{n}{s} \int_0^\infty e^{-st} t^{n-1} \, dt$$

$$= \frac{n}{s} \int_0^\infty e^{-st} t^{n-1} \, dt$$

or
$$\mathscr{L}\{t^n\} = \frac{n}{s}\mathscr{L}\{t^{n-1}\}, \qquad n = 1, 2, 3 \ldots .$$

Now $\mathscr{L}\{1\} = 1/s$, so it follows by iteration that

$$\mathscr{L}\{t\} = \frac{1}{s}\mathscr{L}\{1\} = \frac{1}{s^2},$$

$$\mathscr{L}\{t^2\} = \frac{2}{s}\mathscr{L}\{t\} = \frac{2}{s}\left(\frac{1}{s^2}\right) = \frac{2}{s^3},$$

$$\mathscr{L}\{t^3\} = \frac{3}{s}\mathscr{L}\{t^2\} = \frac{3}{s}\left(\frac{2}{s^3}\right) = \frac{3!}{s^4}.$$

In general, it seems reasonable that

$$\mathscr{L}\{t^n\} = \frac{n}{s}\mathscr{L}\{t^{n-1}\} = \frac{n}{s}\left[\frac{(n-1)!}{s^n}\right] = \frac{n!}{s^{n+1}}.*$$

The justifications of parts **(f)** and **(g)** of Theorem 7.2 are left to the student (see Problems 27 and 28).

EXAMPLE

Evaluate $\mathscr{L}\{\sin^2 t\}$.

Solution: With the aid of a trigonometric identity and parts (a) and (e) of Theorem 7.2, we obtain

$$\mathscr{L}\{\sin^2 t\} = \mathscr{L}\left\{\frac{1 - \cos 2t}{2}\right\}$$

$$= \frac{1}{2}\mathscr{L}\{1\} - \frac{1}{2}\mathscr{L}\{\cos 2t\}$$

$$= \frac{1}{2}\cdot\frac{1}{s} - \frac{1}{2}\cdot\frac{s}{s^2 + 4}$$

$$= \frac{2}{s(s^2 + 4)}.$$

7.1.2   The Inverse Transform

By using the integral definition of the Laplace transform of a function f, we determine another function F, that is, a function of the transform parameter s. We have denoted this symbolically by $\mathscr{L}\{f(t)\} = F(s)$.

We now turn the problem around, namely, given $F(s)$ find the function $f(t)$ corresponding to this transform. We say $f(t)$ is the **inverse Laplace transform** of $F(s)$ and write $f(t) = \mathscr{L}^{-1}\{F(s)\}$. The analogue of Theorem 7.2 for the inverse transform is the following.

---

*A rigorous proof requires mathematical induction.

THEOREM 7.3

(a) $$1 = \mathcal{L}^{-1}\left\{\frac{1}{s}\right\}$$

(b) $$t^n = \mathcal{L}^{-1}\left\{\frac{n!}{s^{n+1}}\right\}, \quad n = 1, 2, 3, \ldots$$

(c) $$e^{at} = \mathcal{L}^{-1}\left\{\frac{1}{s-a}\right\}$$

(d) $$\sin kt = \mathcal{L}^{-1}\left\{\frac{k}{s^2+k^2}\right\}$$

(e) $$\cos kt = \mathcal{L}^{-1}\left\{\frac{s}{s^2+k^2}\right\}$$

(f) $$\sinh kt = \mathcal{L}^{-1}\left\{\frac{k}{s^2-k^2}\right\}$$

(g) $$\cosh kt = \mathcal{L}^{-1}\left\{\frac{s}{s^2-k^2}\right\}.$$

$\mathcal{L}^{-1}$ **is also a linear operation.**

We shall assume that the inverse Laplace transform is itself a linear transformation,* that is, for constants α and β

$$\mathcal{L}^{-1}\{\alpha F(s) + \beta G(s)\} = \alpha \mathcal{L}^{-1}\{F(s)\} + \beta \mathcal{L}^{-1}\{G(s)\},$$

where F and G are the transforms of some functions f and g.

It should also be noted that the inverse Laplace transform of a function $F(s)$ may *not* be unique (see Problems 65 and 66.) However, if $f_1(t)$ and $f_2(t)$ are continuous for $t \geq 0$ and $\mathcal{L}\{f_1(t)\} = \mathcal{L}\{f_2(t)\}$ then necessarily $f_1(t) = f_2(t)$.

EXAMPLE

Evaluate $\mathcal{L}^{-1}\left\{\dfrac{1}{s^5}\right\}$.

Solution: We multiply and divide by 4! and then use part (b) of Theorem 7.3. It follows that

$$\mathcal{L}^{-1}\left\{\frac{1}{s^5}\right\} = \frac{1}{4!}\mathcal{L}^{-1}\left\{\frac{4!}{s^5}\right\} = \frac{1}{24}t^4.$$

EXAMPLE

Evaluate $\mathcal{L}^{-1}\left\{\dfrac{3s+5}{s^2+7}\right\}$.

---

* The inverse Laplace transform is actually another integral. However, evaluation of this integral demands the use of complex variables which is beyond the scope of this text.

Solution: Use termwise division and the linearity property of the inverse transform. From parts (e) and (d) of Theorem 7.3 we have

$$\mathcal{L}^{-1}\left\{\frac{3s + 5}{s^2 + 7}\right\} = 3\mathcal{L}^{-1}\left\{\frac{s}{s^2 + 7}\right\} + \frac{5}{\sqrt{7}}\mathcal{L}^{-1}\left\{\frac{\sqrt{7}}{s^2 + 7}\right\}$$

$$= 3\cos\sqrt{7}t + \frac{5}{\sqrt{7}}\sin\sqrt{7}t.$$

Partial fractions

The use of **partial fractions** is very important in finding inverse Laplace transforms. Here we review three basic cases of that theory. For example, the denominators of

(a) $F(s) = \dfrac{1}{(s - 1)(s + 2)(s + 4)}$

(b) $F(s) = \dfrac{s + 1}{s^2(s + 2)^3}$

(c) $F(s) = \dfrac{3s - 2}{s^3(s^2 + 4)}$

contain, respectively,

(a)   only distinct linear factors,
(b)   repeated linear factors,
(c)   a quadratic factor.[*]

EXAMPLE

Evaluate $\mathcal{L}^{-1}\left\{\dfrac{1}{(s - 1)(s + 2)(s + 4)}\right\}.$

Solution: There exist constants A, B, and C so that

$$\frac{1}{(s - 1)(s + 2)(s + 4)} = \frac{A}{s - 1} + \frac{B}{s + 2} + \frac{C}{s + 4}$$

$$= \frac{A(s + 2)(s + 4) + B(s - 1)(s + 4) + C(s - 1)(s + 2)}{(s - 1)(s + 2)(s + 4)}.$$

Since the numerators are identical we must then have

$$1 = A(s + 2)(s + 4) + B(s - 1)(s + 4) + C(s - 1)(s + 2).$$

By comparing coefficients of powers of s on both sides of the equality we know that the last equation is equivalent to a system of three equations in the three unknowns A, B, and C. However, you might recall the following shortcut for determining these unknowns. If we set $s = 1$, $s = -2$, and $s = -4$ [the zeros of the common denominator $(s - 1)(s + 2)(s + 4)$] we obtain, respectively,

---

[*] Usually irreducible, that is, nonfactorable, over the set of real numbers.

$$1 = A(3)(5), \qquad A = 1/15,$$

$$1 = B(-3)(2), \qquad B = -1/6$$

$$1 = C(-5)(-2), \qquad C = 1/10.$$

Hence we can write

$$\frac{1}{(s-1)(s+2)(s+4)} = \frac{1/15}{s-1} - \frac{1/6}{s+2} + \frac{1/10}{s+4}$$

and thus from part (c) of Theorem 7.3

$$\mathscr{L}^{-1}\left\{\frac{1}{(s-1)(s+2)(s+4)}\right\} = \frac{1}{15}\mathscr{L}^{-1}\left\{\frac{1}{s-1}\right\} - \frac{1}{6}\mathscr{L}^{-1}\left\{\frac{1}{s+2}\right\}$$

$$+ \frac{1}{10}\mathscr{L}^{-1}\left\{\frac{1}{s+4}\right\}$$

$$= \frac{1}{15}e^{t} - \frac{1}{6}e^{-2t} + \frac{1}{10}e^{-4t}.$$

EXAMPLE

Evaluate $\mathscr{L}^{-1}\left\{\dfrac{s+1}{s^{2}(s+2)^{3}}\right\}$.

Solution: Assume

$$\frac{s+1}{s^{2}(s+2)^{3}} = \frac{A}{s} + \frac{B}{s^{2}} + \frac{C}{s+2} + \frac{D}{(s+2)^{2}} + \frac{E}{(s+2)^{3}}$$

so that

$$s + 1 = As(s+2)^{3} + B(s+2)^{3} + Cs^{2}(s+2)^{2} + Ds^{2}(s+2) + Es^{2}.$$

Setting $s = 0$ and $s = -2$ gives, respectively,

$$1 = B(2)^{3}, \qquad B = 1/8$$

$$-1 = E(-2)^{2}, \qquad E = -1/4.$$

By equating coefficients of s^{4}, s^{3}, and s, we obtain

$$0 = A + C$$

$$0 = 6A + B + 4C + D$$

$$1 = 8A + 12B$$

from which it follows that $A = -1/16, C = 1/16, D = 0$. Hence from parts (a), (b), and (c) of Theorem 7.3

$$\mathscr{L}^{-1}\left\{\frac{s+1}{s^{2}(s+2)^{3}}\right\} = \mathscr{L}^{-1}\left\{-\frac{1/16}{s} + \frac{1/8}{s^{2}} + \frac{1/16}{s+2} - \frac{1/4}{(s+2)^{3}}\right\}$$

$$= -\frac{1}{16}\mathscr{L}^{-1}\left\{\frac{1}{s}\right\} + \frac{1}{8}\mathscr{L}^{-1}\left\{\frac{1}{s^{2}}\right\} + \frac{1}{16}\mathscr{L}^{-1}\left\{\frac{1}{s+2}\right\} - \frac{1}{8}\mathscr{L}^{-1}\left\{\frac{2}{(s+2)^{3}}\right\}$$

$$= -\frac{1}{16} + \frac{1}{8}t + \frac{1}{16}e^{-2t} - \frac{1}{8}t^2 e^{-2t}.$$

Here we have also used $\mathcal{L}^{-1}\{2/(s + 2)^3\} = t^2 e^{-2t}$. (See page 277.)

EXAMPLE

Evaluate $\mathcal{L}^{-1}\left\{\dfrac{3s - 2}{s^3(s^2 + 4)}\right\}$.

Solution: Assume $\dfrac{3s - 2}{s^3(s^2 + 4)} = \dfrac{A}{s} + \dfrac{B}{s^2} + \dfrac{C}{s^3} + \dfrac{Ds + E}{s^2 + 4}$ so that

$$3s - 2 = As^2(s^2 + 4) + Bs(s^2 + 4) + C(s^2 + 4) + (Ds + E)s^3.$$

Setting $s = 0$ gives immediately $C = -1/2$. Now the coefficients of s^4, s^3, s^2, and s are

$$0 = A + D$$
$$0 = B + E$$
$$0 = 4A + C$$
$$3 = 4B$$

from which we obtain $B = 3/4$, $E = -3/4$, $A = 1/8$, $D = -1/8$. Therefore from parts (a), (b), (e) and (d) of Theorem 7.3

$$\mathcal{L}^{-1}\left\{\frac{3s - 2}{s^3(s^2 + 4)}\right\} = \mathcal{L}^{-1}\left\{\frac{1/8}{s} + \frac{3/4}{s^2} - \frac{1/2}{s^3} + \frac{-s/8 - 3/4}{s^2 + 4}\right\}$$

$$= \frac{1}{8}\mathcal{L}^{-1}\left\{\frac{1}{s}\right\} + \frac{3}{4}\mathcal{L}^{-1}\left\{\frac{1}{s^2}\right\} - \frac{1}{4}\mathcal{L}^{-1}\left\{\frac{2}{s^3}\right\}$$

$$- \frac{1}{8}\mathcal{L}^{-1}\left\{\frac{s}{s^2 + 4}\right\} - \frac{3}{8}\mathcal{L}^{-1}\left\{\frac{2}{s^2 + 4}\right\}$$

$$= \frac{1}{8} + \frac{3}{4}t - \frac{1}{4}t^2 - \frac{1}{8}\cos 2t - \frac{3}{8}\sin 2t.$$

Not every arbitrary function of s is a Laplace transform of a piecewise function of exponential order.

THEOREM 7.4   Let $f(t)$ be piecewise continuous for $t \geq 0$ and of exponential order for $t > T$. then

$$\lim_{s \to \infty} \mathcal{L}\{f(t)\} = 0.$$

Proof: Since $f(t)$ is piecewise continuous on $0 \le t \le T$ it is necessarily bounded on this interval:

$$|f(t)| < M_1 = M_1 e^{0t}.$$

Also

$$|f(t)| < M_2 e^{\gamma t}$$

for $t > T$. If M denotes the maximum of $\{M_1, M_2\}$ and c denotes the maximum of $\{0, \gamma\}$, then

$$|\mathcal{L}\{f(t)\}| \le \int_0^\infty e^{-st}|f(t)|\, dt$$

$$< M \int_0^\infty e^{-st} \cdot e^{ct}\, dt$$

$$= -M \frac{e^{-(s-c)t}}{s-c} \bigg|_0^\infty$$

$$= \frac{M}{s-c} \qquad \text{for } s > c.$$

As $s \to \infty$ we have $|\mathcal{L}\{f(t)\}| \to 0$ and so $\mathcal{L}\{f(t)\} \to 0$. $\square$

EXAMPLE $F(s) = s^2$ is not the Laplace transform of any piecewise continuous function of exponential order since $F(s) \not\to 0$ as $s \to \infty$. Hence we shall say that $\mathcal{L}^{-1}\{F(s)\}$ does not exist.

EXERCISES 7.1   *Answers to odd-numbered problems begin on page A–21.*

[**7.1.1**]   In Problems 1–12 use Definition 7.1 to find $\mathcal{L}\{f(t)\}$.

EXAMPLE   Find $\mathcal{L}\{f(t)\}$ for $f(t) = \begin{cases} 0, & 0 \le t < 3 \\ 2, & t \ge 3. \end{cases}$

Solution: $\mathcal{L}\{f(t)\} = \displaystyle\int_0^\infty e^{-st} f(t)\, dt$

$$= \int_0^3 e^{-st} f(t)\, dt + \int_3^\infty e^{-st} f(t)\, dt$$

$$= \int_0^3 e^{-st}(0)\, dt + \int_3^\infty e^{-st}(2)\, dt$$

$$= -\frac{2e^{-st}}{s} \bigg|_3^\infty$$

$$= \frac{2e^{-3s}}{s}.$$

1. $f(t) = \begin{cases} -1, & 0 < t < 1, \\ 1, & t \geq 1. \end{cases}$

2. $f(t) = \begin{cases} 4, & 0 < t < 2, \\ 0, & t \geq 2. \end{cases}$

3. $f(t) = \begin{cases} t, & 0 < t < 1, \\ 1, & t \geq 1. \end{cases}$

4. $f(t) = \begin{cases} 2t + 1, & 0 < t < 1 \\ 0, & t \geq 1 \end{cases}$

5. $f(t) = e^{t + 7}$

6. $f(t) = e^{-2t - 5}$

7. $f(t) = te^{4t}$

8. $f(t) = t^2 e^{3t}$

9. $f(t) = e^{-t}\sin t$

10. $f(t) = e^t\cos t$

11. $f(t) = t \cos t$

★12. $f(t) = t \sin t$

In Problems 13–36 use Theorem 7.2 to find $\mathcal{L}\{f(t)\}$.

13. $f(t) = 2t^4$

14. $f(t) = t^5$

15. $f(t) = 4t - 10$

16. $f(t) = 7t + 3$

17. $f(t) = t^2 + 6t - 3$

18. $f(t) = -4t^2 + 16t + 9$

19. $f(t) = (t + 1)^3$

20. $f(t) = (2t - 1)^3$

21. $f(t) = 1 + e^{4t}$

22. $f(t) = t^2 - e^{-9t} + 5$

23. $f(t) = (1 + e^{2t})^2$

24. $f(t) = (e^t - e^{-t})^2$

25. $f(t) = 4t^2 - 5 \sin 3t$

26. $f(t) = \cos 5t + \sin 2t$

27. $f(t) = \sinh kt$

28. $f(t) = \cosh kt$

29. $f(t) = e^t\sinh t$

30. $f(t) = e^{-t}\cosh t$

31. $f(t) = \sin 2t \cos 2t$

★32. $f(t) = \cos^2 t$

33. $f(t) = \cos t \cos 2t$ [*Hint*: Examine $\cos(t_1 \pm t_2)$.]

★34. $f(t) = \sin t \sin 2t$

35. $f(t) = \sin t \cos 2t$ [*Hint*: Examine $\sin(t_1 \pm t_2)$.]

36. $f(t) = \sin^3 t$ [*Hint*: $\sin^3 t = \sin t \sin^2 t$.]

37. Recall that the gamma function is defined by the integral

$$\Gamma(\alpha) = \int_0^\infty t^{\alpha - 1}e^{-t}\, dt, \qquad \alpha > 0.$$

Show that $\mathcal{L}\{t^\alpha\} = \dfrac{\Gamma(\alpha + 1)}{s^{\alpha + 1}}, \qquad \alpha > -1.$

In Problems 38–40 use the result of Problem 37 to find $\mathcal{L}\{f(t)\}$.

38. $f(t) = t^{-1/2}$

39. $f(t) = t^{1/2}$

★40. $f(t) = t^{3/2}$

[**7.1.2**]   In Problems 41–64 use Theorem 7.3 to find the given inverse transform.

41. $\mathcal{L}^{-1}\left\{\dfrac{1}{s^3}\right\}$

42. $\mathcal{L}^{-1}\left\{\dfrac{1}{s^4}\right\}$

43. $\mathcal{L}^{-1}\left\{\dfrac{(s+1)^3}{s^4}\right\}$

44. $\mathcal{L}^{-1}\left\{\dfrac{(s+2)^2}{s^3}\right\}$

45. $\mathcal{L}^{-1}\left\{\dfrac{1}{s^2}-\dfrac{1}{s}+\dfrac{1}{s-2}\right\}$

46. $\mathcal{L}^{-1}\left\{\dfrac{4}{s}+\dfrac{6}{s^5}-\dfrac{1}{s+8}\right\}$

47. $\mathcal{L}^{-1}\left\{\dfrac{1}{4s+1}\right\}$

★48. $\mathcal{L}^{-1}\left\{\dfrac{1}{5s-2}\right\}$

49. $\mathcal{L}^{-1}\left\{\dfrac{4s}{4s^2+1}\right\}$

50. $\mathcal{L}^{-1}\left\{\dfrac{1}{4s^2+1}\right\}$

51. $\mathcal{L}^{-1}\left\{\dfrac{1}{s^2-16}\right\}$

52. $\mathcal{L}^{-1}\left\{\dfrac{10s}{s^2-25}\right\}$

53. $\mathcal{L}^{-1}\left\{\dfrac{2s-6}{s^2+9}\right\}$

54. $\mathcal{L}^{-1}\left\{\dfrac{s+1}{s^2+2}\right\}$

55. $\mathcal{L}^{-1}\left\{\dfrac{1}{s^2+3s}\right\}$

56. $\mathcal{L}^{-1}\left\{\dfrac{s+1}{s^2-4s}\right\}$

57. $\mathcal{L}^{-1}\left\{\dfrac{s}{s^2+2s-3}\right\}$

58. $\mathcal{L}^{-1}\left\{\dfrac{1}{s^2+s-20}\right\}$

59. $\mathcal{L}^{-1}\left\{\dfrac{2s+4}{(s-2)(s^2+4s+3)}\right\}$

★60. $\mathcal{L}^{-1}\left\{\dfrac{s+1}{(s^2-4s)(s+5)}\right\}$

61. $\mathcal{L}^{-1}\left\{\dfrac{1}{s^2(s^2+4)}\right\}$

62. $\mathcal{L}^{-1}\left\{\dfrac{s-1}{s^2(s^2+1)}\right\}$

63. $\mathcal{L}^{-1}\left\{\dfrac{s}{(s^2+4)(s+2)}\right\}$

★64. $\mathcal{L}^{-1}\left\{\dfrac{1}{s^4-9}\right\}$

The inverse Laplace transform may not be unique. In Problems 65 and 66 evaluate $\mathcal{L}\{f(t)\}$.

65. $f(t) = \begin{cases} 1, & t \geq 0,\, t \neq 1,\, t \neq 2 \\ 3, & t = 1 \\ -3, & t = 2 \end{cases}$

66. $f(t) = \begin{cases} e^{3t} & 0 \leq t < 5,\, t > 5 \\ 1, & t = 5 \end{cases}$

Miscellaneous problems

67. Show that the function $f(t) = 1/t^2$ does not possess a Laplace transform. [*Hint*: Write

$$\mathcal{L}\{f(t)\} = \int_0^1 e^{-st}f(t)\, dt + \int_1^\infty e^{-st}f(t)\, dt.$$

Use the definition of an improper integral to show $\int_0^1 e^{-st}f(t)\,dt$ does not exist.]

7.2   Operational Properties

7.2.1   Translation Theorems and Derivatives of a Transform

It is not convenient to use Definition 7.1 each time we wish to find the Laplace transform of a function $f(t)$. For example, the integration by parts involved in evaluating, say, $\mathscr{L}\{e^t t^2 \sin 3t\}$ is formidable to say the least. In the discussion that follows, we present several labor saving theorems, and these in turn enable us to build up a more extensive list of transforms without the necessity of using the definition of the Laplace transform. Indeed, we shall see that evaluating transforms such as $\mathscr{L}\{e^{4t}\cos 6t\}$, $\mathscr{L}\{t^3\sin 2t\}$, and $\mathscr{L}\{t^{10}e^{-t}\}$ are fairly straightforward, provided we know $\mathscr{L}\{\cos 6t\}$, $\mathscr{L}\{\sin 2t\}$, and $\mathscr{L}\{t^{10}\}$, respectively. Though extensive tables can be constructed, and we have included three tables in the Chapter Summary, it is nonetheless a good idea to know the Laplace transforms of basic functions such as t^n, e^{at}, $\sin kt$, $\cos kt$, $\sinh kt$, and $\cosh kt$.

The first translation theorem

THEOREM 7.5   If a is any real number then
$$\mathscr{L}\{e^{at}f(t)\} = F(s - a)$$
where $F(s) = \mathscr{L}\{f(t)\}$.

Proof: The proof is immediate, since by Definition 7.1

$$\mathscr{L}\{e^{at}f(t)\} = \int_0^\infty e^{-st}e^{at}f(t)\,dt$$
$$= \int_0^\infty e^{-(s-a)t}f(t)\,dt$$
$$= F(s - a). \qquad \square$$

Thus if we already know $\mathscr{L}\{f(t)\} = F(s)$, we can compute $\mathscr{L}\{e^{at}f(t)\}$ with no additional effort other than translating, or shifting, $F(s)$ to $F(s - a)$. For emphasis it is also sometimes useful to employ the symbolism
$$\mathscr{L}\{e^{at}f(t)\} = \mathscr{L}\{f(t)\}_{s \to s-a}.$$
Theorem 7.5 is known as the **first translation theorem**.

EXAMPLES

Evaluate **(a)** $\mathscr{L}\{e^{5t}t^3\}$, **(b)** $\mathscr{L}\{e^{-2t}\cos 4t\}$.

Solutions: The results follow from Theorem 7.5.

(a) $\mathscr{L}\{e^{5t}t^3\} = \mathscr{L}\{t^3\}_{s \to s-5}$

$$= \left.\frac{3!}{s^4}\right|_{s \to s-5} \qquad = \frac{6}{(s - 5)^4}.$$

(b) $\mathcal{L}\{e^{-2t}\cos 4t\} = \mathcal{L}\{\cos 4t\}_{s\to s+2}$ [*Note*: $a = -2$, so

$$s - a = s - (-2) = s + 2.]$$

$$= \left.\frac{s}{s^2 + 16}\right|_{s\to s+2}$$

$$= \frac{s + 2}{(s + 2)^2 + 16}.$$

The inverse form of Theorem 7.5 can be written

$$e^{at}f(t) = \mathcal{L}^{-1}\{F(s - a)\}$$

$$= \mathcal{L}^{-1}\{F(s)|_{s\to s-a}\} \tag{1}$$

where $f(t) = \mathcal{L}^{-1}\{F(s)\}$.

EXAMPLE   Evaluate $\mathcal{L}^{-1}\left\{\dfrac{s}{s^2 + 6s + 11}\right\}$.

***Solution*:**

$$\mathcal{L}^{-1}\left\{\frac{s}{s^2 + 6s + 11}\right\} = \mathcal{L}^{-1}\left\{\frac{s}{(s + 3)^2 + 2}\right\} \quad \text{[completion of square]}$$

$$= \mathcal{L}^{-1}\left\{\frac{s + 3 - 3}{(s + 3)^2 + 2}\right\} \quad \text{[adding zero in the numerator]}$$

$$= \mathcal{L}^{-1}\left\{\frac{s + 3}{(s + 3)^2 + 2} - \frac{3}{(s + 3)^2 + 2}\right\} \quad \text{[termwise division]}$$

$$= \mathcal{L}^{-1}\left\{\frac{s + 3}{(s + 3)^2 + 2}\right\} - 3\mathcal{L}^{-1}\left\{\frac{1}{(s + 3)^2 + 2}\right\}$$

$$= \mathcal{L}^{-1}\left\{\left.\frac{s}{s^2 + 2}\right|_{s\to s+3}\right\} - \frac{3}{\sqrt{2}}\mathcal{L}^{-1}\left\{\left.\frac{\sqrt{2}}{s^2 + 2}\right|_{s\to s+3}\right\}$$

$$= e^{-3t}\cos \sqrt{2}t - \frac{3}{\sqrt{2}}e^{-3t}\sin \sqrt{2}t. \quad \text{[from (1) and Theorem 7.3]}$$

EXAMPLE   Evaluate $\mathcal{L}^{-1}\left\{\dfrac{1}{(s - 1)^3} + \dfrac{1}{s^2 + 2s - 8}\right\}$.

Solution:

$$\mathcal{L}^{-1}\left\{\frac{1}{(s-1)^3} + \frac{1}{s^2+2s-8}\right\}$$

$$= \mathcal{L}^{-1}\left\{\frac{1}{(s-1)^3} + \frac{1}{(s+1)^2-9}\right\}$$

$$= \frac{1}{2!}\mathcal{L}^{-1}\left\{\frac{2!}{(s-1)^3}\right\} + \frac{1}{3}\mathcal{L}^{-1}\left\{\frac{3}{(s+1)^2-9}\right\}$$

$$= \frac{1}{2!}\mathcal{L}^{-1}\left\{\frac{2!}{s^3}\Big|_{s\to s-1}\right\} + \frac{1}{3}\mathcal{L}^{-}\left\{\frac{3}{s^2-9}\Big|_{s\to s+1}\right\}$$

$$= \frac{1}{2}e^t t^2 + \frac{1}{3}e^{-t}\sinh 3t.$$

The unit step function

In engineering one frequently encounters functions which can be either "on" or "off." For example, an external force acting on a mechanical system or a voltage impressed on a circuit can be turned off after a period of time. It is thus convenient to define a special function called the **unit step function.**

DEFINITION 7.2   The function $\mathcal{U}(t-a)$ is defined to be

$$\mathcal{U}(t-a) = \begin{cases} 0, & 0 \le t < a \\ 1, & t \ge a. \end{cases} \qquad \square$$

Notice that we define $\mathcal{U}(t-a)$ only on the nonnegative t-axis since this is all that we are concerned with in the study of the Laplace transform. In a broader sense, $\mathcal{U}(t-a) = 0$ for $t < a$.

EXAMPLES

Graph **(a)** $\mathcal{U}(t)$, **(b)** $\mathcal{U}(t-2)$.

Solutions: **(a)** $\mathcal{U}(t) = 1, \qquad t \ge 0$

(b) $\mathcal{U}(t-2) = \begin{cases} 0, & 0 \le t < 2 \\ 1, & t \ge 2. \end{cases}$

The respective graphs are given in Figure 7.4.

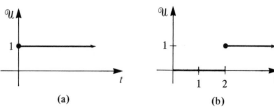

(a)   **(b)**

Figure 7.4

When combined with other functions defined for $t \geq 0$, the unit step function "turns off" a portion of their graphs. For example, Figure 7.5 illustrates the graph $y = f(t)$ where

$$f(t) = \sin t \, \mathcal{U}(t - 2\pi), \qquad t \geq 0.$$

$$= \begin{cases} 0, & 0 \leq t < 2\pi \\ \sin t, & t \geq 2\pi. \end{cases}$$

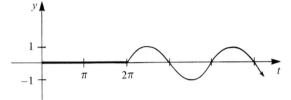

Figure 7.5

EXAMPLE

Consider the function $y = f(t)$ defined by $f(t) = t^3$. Compare the graphs of
(a) $f(t) = t^3$,   **(b)** $f(t) = t^3, t \geq 0$,
(c) $f(t - 2), t \geq 0$,   **(d)** $f(t - 2) \, \mathcal{U}(t - 2), t \geq 0$.

Solution: The respective graphs are given in Figure 7.6.

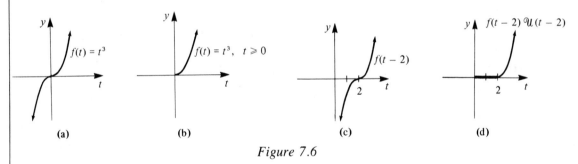

Figure 7.6

We have already seen in Theorem 7.5 that an exponential multiple of $f(t)$ results in a shift or translation of the transform $F(s)$. In the next theorem we see that whenever $F(s)$ is multiplied by an appropriate exponential function, the graph of $f(t)$ is not only translated but a portion of the graph is turned off as well.

Second translation theorem

THEOREM 7.6   If $a > 0$ then

$$\mathcal{L}\{f(t - a) \mathcal{U}(t - a)\} = e^{-as} \mathcal{L}\{f(t)\}$$

$$= e^{-as} F(s).$$

Proof: From Definition 7.1

$$\mathcal{L}\{f(t - a)\mathcal{U}(t - a)\} = \int_0^\infty e^{-st}f(t - a)\mathcal{U}(t - a)\, dt$$

$$= \int_0^a e^{-st}f(t - a)\mathcal{U}(t - a)\, dt$$

zero for $0 \le t < a$

$$+ \int_a^\infty e^{-st}f(t - a)\mathcal{U}(t - a)\, dt$$

one for $t \ge a$

$$= \int_a^\infty e^{-st}f(t - a)\, dt.$$

Now let $v = t - a$, $dv = dt$

$$\mathcal{L}\{f(t - a)\mathcal{U}(t - a)\} = \int_0^\infty e^{-s(v + a)}f(v)\, dv$$

$$= e^{-as}\int_0^\infty e^{-sv}f(v)\, dv$$

$$= e^{-as}\mathcal{L}\{f(t)\}. \qquad \square$$

Theorem 7.6 is known as the **second translation theorem**.

EXAMPLE

Evaluate $\mathcal{L}\{(t - 2)^3\mathcal{U}(t - 2)\}$.

Solution: With the identification $a = 2$ it follows from Theorem 7.6 that

$$\mathcal{L}\{(t - 2)^3\mathcal{U}(t - 2)\} = e^{-2s}\mathcal{L}\{t^3\}$$

$$= e^{-2s}\frac{3!}{s^4}$$

$$= \frac{6}{s^4}e^{-2s}.$$

Note: This result is equivalent to evaluating the integral

$$\int_2^\infty e^{-st}(t - 2)^3\, dt.$$

EXAMPLE

Evaluate $\mathcal{L}\{\mathcal{U}(t - 5)\}$.

Solution: Making the identifications $f(t) = 1$ and $a = 5$ in Theorem 7.6 we have

$$\mathcal{L}\{\mathcal{U}(t - 5)\} = e^{-5s}\mathcal{L}\{1\}$$

$$= \frac{e^{-5s}}{s}.$$

EXAMPLE

Find the Laplace transform of the function shown in Figure 7.7.

Solution: With the aid of the unit step function we can write

$$f(t) = 2 - 3\mathcal{U}(t - 2) + \mathcal{U}(t - 3).$$

Hence from Theorem 7.6 it follows that

$$\mathcal{L}\{f(t)\} = \mathcal{L}\{2\} - 3\mathcal{L}\{\mathcal{U}(t - 2)\} + \mathcal{L}\{\mathcal{U}(t - 3)\}$$

$$= \frac{2}{s} - 3\frac{e^{-2s}}{s} + \frac{e^{-3s}}{s}.$$

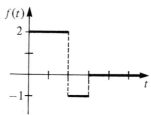

Figure 7.7

EXAMPLE

Evaluate $\mathcal{L}\{\sin t\,\mathcal{U}(t - 2\pi)\}$.

Solution: With $a = 2\pi$ we have from Theorem 7.6

$$\mathcal{L}\{\sin t\,\mathcal{U}(t - 2\pi)\} = \mathcal{L}\{\sin(t - 2\pi)\mathcal{U}(t - 2\pi)\} \quad [\sin t \text{ has period } 2\pi]$$

$$= e^{-2\pi s}\mathcal{L}\{\sin t\}$$

$$= \frac{e^{-2\pi s}}{s^2 + 1}.$$

The inverse form of Theorem 7.6 is

$$f(t - a)\mathcal{U}(t - a) = \mathcal{L}^{-1}\{e^{-as}F(s)\} \tag{2}$$

where $a > 0$ and $f(t) = \mathcal{L}^{-1}\{F(s)\}$.

EXAMPLE

Evaluate $\mathcal{L}^{-1}\left\{\dfrac{e^{-\pi s/2}}{s^2 + 9}\right\}$.

Solution: We identify $a = \dfrac{\pi}{2}$ and $f(t) = \mathscr{L}^{-1}\left\{\dfrac{1}{s^2 + 9}\right\} = \dfrac{1}{3}\sin 3t$. Thus from (2)

$$\mathscr{L}^{-1}\left\{\frac{e^{-\pi s/2}}{s^2 + 9}\right\} = \frac{1}{3}\mathscr{L}^{-1}\left\{\frac{3}{s^2 + 9}\right\}_{t \to t - \pi/2} \mathscr{U}\left(t - \frac{\pi}{2}\right)$$

$$= \frac{1}{3}\sin 3\left(t - \frac{\pi}{2}\right)\mathscr{U}\left(t - \frac{\pi}{2}\right)$$

$$= \frac{1}{3}\cos 3t\, \mathscr{U}\left(t - \frac{\pi}{2}\right).$$

If $F(s) = \mathscr{L}\{f(t)\}$, and if we assume that interchanging of differentiation and integration is possible, then

$$\frac{d}{ds}F(s) = \frac{d}{ds}\int_0^\infty e^{-st}f(t)\, dt$$

$$= \int_0^\infty \frac{\partial}{\partial s}\left[e^{-st}f(t)\, dt\right]$$

$$= -\int_0^\infty e^{-st}t f(t)\, dt$$

$$= -\mathscr{L}\{t f(t)\}.$$

That is $$\mathscr{L}\{t f(t)\} = -\frac{d}{ds}\mathscr{L}\{f(t)\}.$$

Similarly, $$\mathscr{L}\{t^2 f(t)\} = \mathscr{L}\{t \cdot t f(t)\}$$

$$= -\frac{d}{ds}\mathscr{L}\{t f(t)\}$$

$$= -\frac{d}{ds}\left(-\frac{d}{ds}\mathscr{L}\{f(t)\}\right)$$

$$= \frac{d^2}{ds^2}\mathscr{L}\{f(t)\}.$$

We have formally justified two cases of the following theorem.

Derivatives of a transform

THEOREM 7.7   For $n = 1, 2, 3, \ldots$

$$\mathscr{L}\{t^n f(t)\} = (-1)^n \frac{d^n}{ds^n}\mathscr{L}\{f(t)\}$$

$$= (-1)^n \frac{d^n}{ds^n}F(s)$$

where $F(s) = \mathscr{L}\{f(t)\}$.   □

EXAMPLES   Evaluate **(a)** $\mathcal{L}\{te^{3t}\}$, **(b)** $\mathcal{L}\{t \sin kt\}$, **(c)** $\mathcal{L}\{t^2\sin kt\}$, **(d)** $\mathcal{L}\{te^{-t}\cos t\}$.

Solutions: The results follow from Theorem 7.7.

(a) $\mathcal{L}\{te^{3t}\} = -\dfrac{d}{ds}\,\mathcal{L}\{e^{3t}\}$

$\qquad\qquad = -\dfrac{d}{ds}\left(\dfrac{1}{s-3}\right)$

$\qquad\qquad = \dfrac{1}{(s-3)^2}.$

(b) $\mathcal{L}\{t \sin kt\} = -\dfrac{d}{ds}\,\mathcal{L}\{\sin kt\}$

$\qquad\qquad = -\dfrac{d}{ds}\left(\dfrac{k}{s^2+k^2}\right)$

$\qquad\qquad = \dfrac{2ks}{(s^2+k^2)^2}$

(c) $\mathcal{L}\{t^2\sin kt\} = \dfrac{d^2}{ds^2}\,\mathcal{L}\{\sin kt\}$

$\qquad\qquad = -\dfrac{d}{ds}\,\mathcal{L}\{t \sin kt\}$

$\qquad\qquad = -\dfrac{d}{ds}\left[\dfrac{2ks}{(s^2+k^2)^2}\right]$ [from part **(b)**]

$\qquad\qquad = -\dfrac{(s^2+k^2)^2 2k - 8ks^2(s^2+k^2)}{(s^2+k^2)^4}$

$\qquad\qquad = \dfrac{6ks^2 - 2k^3}{(s^2+k^2)^3}$

(d) $\mathcal{L}\{te^{-t}\cos t\} = -\dfrac{d}{ds}\,\mathcal{L}\{e^{-t}\cos t\}$

$\qquad\qquad = -\dfrac{d}{ds}\,\mathcal{L}\{\cos t\}_{s\to s+1}$

$\qquad\qquad = -\dfrac{d}{ds}\left[\dfrac{s+1}{(s+1)^2+1}\right]$

$\qquad\qquad = \dfrac{(s+1)^2 - 1}{[(s+1)^2+1]^2}$

Of course, part **(a)** of the preceding example can also be obtained from Theorem 7.5.

7.2.2   Transforms of Derivatives and Integrals

Our goal is to use the Laplace transform to solve certain kinds of differential equations. To that end we need to evaluate quantities such as $\mathscr{L}\{dy/dt\}$ and $\mathscr{L}\{d^2y/dt^2\}$. For example, if $y = f(t)$ then

$$\mathscr{L}\{f'(t)\} = \int_0^\infty e^{-st} f'(t)\, dt$$

$$= e^{-st} f(t) \Big|_0^\infty + s \int_0^\infty e^{-st} f(t)\, dt$$

$$= -f(0) + s\mathscr{L}\{f(t)\}$$

$$= sF(s) - f(0).$$

Similarly, $$\mathscr{L}\{f''(t)\} = \int_0^\infty e^{-st} f''(t)\, dt$$

$$= e^{-st} f'(t) \Big|_0^\infty + s \int_0^\infty e^{-st} f'(t)\, dt$$

$$= -f'(0) + s\mathscr{L}\{f'(t)\}$$

$$= s[sF(s) - f(0)] - f'(0)$$

$$= s^2 F(s) - sf(0) - f'(0).$$

We state the general case in the following theorem.

Transform of a derivative

THEOREM 7.8   If $f(t), f'(t), \ldots, f^{(n-1)}(t)$ are continuous for $t \geq 0$ and are of exponential order, and if $f^{(n)}(t)$ is piecewise continuous for $t \geq 0$, then

$$\mathscr{L}\{f^{(n)}(t)\} = s^n F(s) - s^{n-1} f(0) - s^{n-2} f'(0) - \cdots - f^{(n-1)}(0)$$

where $F(s) = \mathscr{L}\{f(t)\}$.   □

EXAMPLE

If $\mathscr{L}\{1\} = 1/s$, compute $\mathscr{L}\{t\}$.

Solution: Let $f(t) = t$ so that $f'(t) = 1$ and $f(0) = 0$. By Theorem 7.8 with $n = 1$ we have

$$\mathscr{L}\{1\} = s\mathscr{L}\{t\} - f(0)$$

which implies $$\mathscr{L}\{t\} = \frac{1}{s}\mathscr{L}\{1\}$$

$$= \frac{1}{s^2}.$$

EXAMPLE

If $\mathscr{L}\{\cos t\} = s/(s^2 + 1)$, compute $\mathscr{L}\{\sin t\}$.

Solution: Let $f(t) = \cos t$, $f'(t) = -\sin t$, $f(0) = 1$. Therefore from Theorem 7.8

$$\mathscr{L}\{\sin t\} = -[s\mathscr{L}\{\cos t\} - \cos 0]$$

$$= \frac{-s^2}{s^2 + 1} + 1$$

$$= \frac{1}{s^2 + 1}.$$

EXAMPLE

$$\mathscr{L}\{kt \cos kt + \sin kt\} = \mathscr{L}\left\{\frac{d}{dt}(t \sin kt)\right\}$$

$$= s\mathscr{L}\{t \sin kt\} \qquad \text{[Theorem 7.8]}$$

$$= s\left[-\frac{d}{ds}\mathscr{L}\{\sin kt\}\right] \qquad \text{[Theorem 7.7]}$$

$$= s\left[\frac{2ks}{(s^2 + k^2)^2}\right]$$

$$= \frac{2ks^2}{(s^2 + k^2)^2}.$$

Transform of an integral

THEOREM 7.9   Let $f(t)$ be piecewise continuous for $t \geq 0$ and of exponential order.* Then

$$\mathscr{L}\left\{\int_0^t f(\tau)d\tau\right\} = \frac{1}{s}\mathscr{L}\{f(t)\}$$

$$= \frac{F(s)}{s}.$$

Proof: We make use of the facts that

$$\int_0^0 f(\tau)d\tau = 0 \qquad \text{and} \qquad \frac{d}{dt}\int_0^t f(\tau)d\tau = f(t).$$

Now by Definition 7.1 and integration by parts we have

---

*It can be shown that when $f(t)$ is piecewise continuous for $t \geq 0$ and of exponential order, then $\int_0^t f(\tau)\, d\tau$ is also of exponential order.

$$\mathcal{L}\left\{\int_0^t f(\tau)d\tau\right\} = \int_0^\infty e^{-st}\left\{\int_0^t f(\tau)d\tau\right\} dt$$

$$= -\frac{1}{s}e^{-st}\int_0^t f(\tau)d\tau\Big|_0^\infty + \frac{1}{s}\int_0^\infty e^{-st}f(t)\, dt$$

$$= \frac{1}{s}\int_0^\infty e^{-st}f(t)\, dt$$

$$= \frac{1}{s}\mathcal{L}\{f(t)\}. \qquad \qquad \square$$

The convolution theorem

Theorem 7.9 is a special case of the following result known as the **convolution theorem.**

THEOREM 7.10   Let $f(t)$ and $g(t)$ be piecewise continuous for $t \geq 0$ and of exponential order. Then

$$\mathcal{L}\left\{\int_0^t f(\tau)g(t-\tau)d\tau\right\} = \mathcal{L}\{f(t)\}\mathcal{L}\{g(t)\}$$

$$= F(s)G(s).$$

Proof: Let $\quad F(s) = \mathcal{L}\{f(t)\} = \displaystyle\int_0^\infty e^{-s\tau}f(\tau)\, d\tau$

$$G(s) = \mathcal{L}\{g(t)\} = \int_0^\infty e^{-s\beta}g(\beta)\, d\beta.$$

Proceeding formally we have

$$F(s)G(s) = \left(\int_0^\infty e^{-s\tau}f(\tau)\, d\tau\right)\left(\int_0^\infty e^{-s\beta}g(\beta)\, d\beta\right)$$

$$= \int_0^\infty \int_0^\infty e^{-s(\tau+\beta)}f(\tau)g(\beta)\, d\tau\, d\beta$$

$$= \int_0^\infty f(\tau)\, d\tau \int_0^\infty e^{-s(\tau+\beta)}g(\beta)\, d\beta.$$

Holding τ fixed, we let $t = \tau + \beta$, $dt = d\beta$ so that

$$F(s)G(s) = \int_0^\infty f(\tau)\, d\tau \int_\tau^\infty e^{-st}g(t-\tau)\, dt.$$

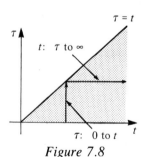

Figure 7.8

In the $t\tau$ plane we are integrating over the shaded region in Figure 7.8. Since f and g are piecewise continuous for $t \geq 0$ and of exponential order, it can be shown that it is possible to interchange the order of integration:

$$F(s)G(s) = \int_0^\infty e^{-st}\, dt \int_0^t f(\tau)g(t-\tau)\, d\tau$$

$$= \int_0^\infty e^{-st}\left\{\int_0^t f(\tau)g(t-\tau)\, d\tau\right\}dt$$

$$= \mathscr{L}\left\{\int_0^t f(\tau)g(t-\tau)\,d\tau\right\}. \qquad \square$$

The integral in Theorem 7.10 is called the **convolution** of $f(t)$ and $g(t)$, and it is commonly denoted by

$$f * g = \int_0^t f(\tau)g(t-\tau)\,d\tau.$$

We leave it as an exercise to show that $f * g = g * f$. Also, notice that Theorem 7.10 reduces to Theorem 7.9 when $g(t) = 1$ and $G(s) = 1/s$.

EXAMPLE   Evaluate $\mathscr{L}\left\{\displaystyle\int_0^t e^\tau \sin(t-\tau)\,d\tau\right\}$.

Solution: With the identifications $f(t) = e^t$ and $g(t) = \sin t$ we have from Theorem 7.10

$$\mathscr{L}\left\{\int_0^t e^\tau \sin(t-\tau)\,d\tau\right\} = \mathscr{L}\{e^t\}\cdot\mathscr{L}\{\sin t\}$$

$$= \frac{1}{s-1}\cdot\frac{1}{s^2+1}$$

$$= \frac{1}{(s-1)(s^2+1)}.$$

The convolution theorem is sometimes useful in finding the inverse Laplace transform of a product of two Laplace transforms. That is, if

$$\mathscr{L}\{f * g\} = F(s)G(s)$$

then

$$f * g = \mathscr{L}^{-1}\{F(s)G(s)\}.$$

EXAMPLE   Evaluate $\mathscr{L}^{-1}\left\{\dfrac{1}{(s-1)(s+4)}\right\}$.

Solution: Admittedly, we could use partial fractions, but if

$$F(s) = \frac{1}{s-1} \quad\text{and}\quad G(s) = \frac{1}{s+4}$$

then $\mathscr{L}^{-1}\{F(s)\} = f(t) = e^t$ and $\mathscr{L}^{-1}\{G(s)\} = g(t) = e^{-4t}$.

By the convolution theorem we can write

$$\mathscr{L}^{-1}\left\{\frac{1}{(s-1)(s+4)}\right\} = \int_0^t f(\tau)g(t-\tau)\,d\tau$$

$$= \int_0^t e^\tau e^{-4(t-\tau)}\,d\tau$$

$$= e^{-4t} \int_0^t e^{5\tau} \, d\tau$$

$$= e^{-4t} \frac{1}{5} e^{5\tau} \Big|_0^t$$

$$= \frac{e^{-4t}}{5} [e^{5t} - 1]$$

$$= \frac{1}{5} e^t - \frac{1}{5} e^{-4t}.$$

EXAMPLE Evaluate $\mathcal{L}^{-1}\left\{\dfrac{1}{(s^2 + k^2)^2}\right\}$.

Solution: Let

$$F(s) = G(s) = \frac{1}{s^2 + k^2}$$

so that

$$f(t) = g(t) = \frac{1}{k} \mathcal{L}^{-1}\left\{\frac{k}{s^2 + k^2}\right\}$$

$$= \frac{1}{k} \sin kt.$$

Thus

$$\mathcal{L}^{-1}\left\{\frac{1}{(s^2 + k^2)^2}\right\} = \frac{1}{k^2} \int_0^t \sin k\tau \, \sin k(t - \tau) \, d\tau.$$

Now recall from trigonometry that

$$\cos(A + B) = \cos A \cos B - \sin A \sin B$$

and

$$\cos(A - B) = \cos A \cos B + \sin A \sin B.$$

Subtracting the first from the second gives the identity

$$\sin A \sin B = \tfrac{1}{2}[\cos(A - B) - \cos(A + B)].$$

If we set $A = k\tau$ and $B = k(t - \tau)$, it follows that

$$\mathcal{L}^{-1}\left\{\frac{1}{(s^2 + k^2)^2}\right\} = \frac{1}{2k^2} \int_0^t [\cos k(2\tau - t) - \cos kt] \, d\tau$$

$$= \frac{1}{2k^2}\left[\frac{1}{2k} \sin k(2\tau - t) - \tau \cos kt\right]_0^t$$

$$= \frac{1}{2k^2}\left[\frac{1}{2k} \sin kt - \frac{1}{2k} \sin(-kt) - t \cos kt\right]$$

$$= \frac{\sin kt - kt \cos kt}{2k^3}.$$

7.2.3   Transform of a Periodic Function

If a periodic function has period T, $T > 0$, then $f(t + T) = f(t)$. The Laplace transform of a periodic function can be obtained by an integration over one period.

> **THEOREM 7.11**   Let $f(t)$ be piecewise continuous for $t \geq 0$ and of exponential order. If $f(t)$ is periodic with period T, then
>
> $$\mathscr{L}\{f(t)\} = \frac{1}{1 - e^{-sT}} \int_0^T e^{-st}f(t) \, dt. \tag{3}$$

Proof: Write the Laplace transform as

$$\mathscr{L}\{f(t)\} = \int_0^T e^{-st}f(t) \, dt + \int_T^\infty e^{-st}f(t) \, dt. \tag{4}$$

By letting $t = u + T$, the last integral in (4) becomes

$$\int_T^\infty e^{-st}f(t) \, dt = \int_0^\infty e^{-s(u + T)}f(u + T) \, du$$

$$= e^{-sT} \int_0^\infty e^{-su}f(u) \, du$$

$$= e^{-sT}\mathscr{L}\{f(t)\}.$$

Hence (4) is

$$\mathscr{L}\{f(t)\} = \int_0^T e^{-st}f(t) \, dt + e^{-sT}\mathscr{L}\{f(t)\}.$$

Solving for $\mathscr{L}\{f(t)\}$ yields

$$\mathscr{L}\{f(t)\} = \frac{1}{1 - e^{-sT}} \int_0^T e^{-st}f(t) \, dt. \qquad \square$$

EXAMPLE   Find the Laplace transform of the periodic function shown in Figure 7.9.

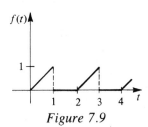

Figure 7.9

Solution: On the interval $0 \leq t < 2$ the function can be defined by

$$f(t) = \begin{cases} t, & 0 \leq t < 1, \\ 0, & 1 \leq t < 2 \end{cases}$$

and outside the interval by $f(t + 2) = f(t)$. Identifying $T = 2$, we use (3) and integration by parts,

$$\mathcal{L}\{f(t)\} = \frac{1}{1 - e^{-2s}} \int_0^2 e^{-st} f(t) \, dt$$

$$= \frac{1}{1 - e^{-2s}} \left[\int_0^1 e^{-st} t \, dt + \int_1^2 e^{-st} 0 \, dt \right]$$

$$= \frac{1}{1 - e^{-2s}} \left[-\frac{e^{-s}}{s} + \frac{1 - e^{-s}}{s^2} \right]$$

$$= \frac{1 - (s + 1)e^{-s}}{s^2(1 - e^{-2s})}.$$

Although we already know the results, $\mathcal{L}\{\sin t\}$ and $\mathcal{L}\{\cos t\}$ can also be obtained from (3).

EXERCISES 7.2   *Answers to odd-numbered problems begin on page A-22.*

[7.2.1]   In problems 1–40 find either $F(s)$ or $f(t)$ as indicated.

1. $\mathcal{L}\{te^{10t}\}$

2. $\mathcal{L}\{te^{-6t}\}$

3. $\mathcal{L}\{t^3 e^{-2t}\}$

4. $\mathcal{L}\{t^{10} e^{-7t}\}$

5. $\mathcal{L}\{e^t \sin 3t\}$

6. $\mathcal{L}\{e^{-2t} \cos 4t\}$

7. $\mathcal{L}\{e^{5t} \sinh 3t\}$

8. $\mathcal{L}\left\{\dfrac{\cosh t}{e^t}\right\}$

9. $\mathcal{L}\{t(e^t + e^{2t})^2\}$

10. $\mathcal{L}\{e^{2t}(t - 1)^2\}$

11. $\mathcal{L}\{e^{-t} \sin^2 t\}$

★12. $\mathcal{L}\{e^t \cos^2 3t\}$

13. $\mathcal{L}^{-1}\left\{\dfrac{1}{(s + 2)^3}\right\}$

14. $\mathcal{L}^{-1}\left\{\dfrac{1}{(s - 1)^4}\right\}$

15.. $\mathcal{L}^{-1}\left\{\dfrac{1}{s^2 - 6s + 10}\right\}$

16. $\mathcal{L}^{-1}\left\{\dfrac{1}{s^2 + 2s + 5}\right\}$

17. $\mathcal{L}^{-1}\left\{\dfrac{s}{s^2 + 4s + 5}\right\}$

★18. $\mathcal{L}^{-1}\left\{\dfrac{2s + 5}{s^2 + 6s + 34}\right\}$

19. $\mathcal{L}^{-1}\left\{\dfrac{s}{(s + 1)^2}\right\}$

20. $\mathcal{L}^{-1}\left\{\dfrac{5s}{(s - 2)^2}\right\}$

21. $\mathcal{L}^{-1}\left\{\dfrac{2s - 1}{s^2(s + 1)^3}\right\}$

22. $\mathcal{L}^{-1}\left\{\dfrac{(s + 1)^2}{(s + 2)^4}\right\}$

23. $\mathcal{L}\{(t - 1)\mathcal{U}(t - 1)\}$

★24. $\mathcal{L}\{e^{2-t}\mathcal{U}(t - 2)\}$

25. $\mathcal{L}\{t\,\mathcal{U}(t - 2)\}$

26. $\mathcal{L}\{\sin t\,\mathcal{U}(t - \frac{\pi}{2})\}$ [*Hint*: Examine $\cos (t - \frac{\pi}{2})$.]

27. $\mathcal{L}^{-1}\left\{\dfrac{e^{-2s}}{s^3}\right\}$

28. $\mathcal{L}^{-1}\left\{\dfrac{(1 + e^{-2s})^2}{s + 2}\right\}$

29. $\mathcal{L}^{-1}\left\{\dfrac{e^{-\pi s}}{s^2 + 1}\right\}$

30. $\mathcal{L}^{-1}\left\{\dfrac{se^{-\pi s/2}}{s^2 + 4}\right\}$

31. $\mathcal{L}^{-1}\left\{\dfrac{e^{-s}}{s(s + 1)}\right\}$

★32. $\mathcal{L}^{-1}\left\{\dfrac{e^{-2s}}{s^2(s - 1)}\right\}$

33. $\mathcal{L}\{t \cos 2t\}$

34. $\mathcal{L}\{t \sinh 3t\}$

35. $\mathcal{L}\{t^2\sinh t\}$

36. $\mathcal{L}\{t^2\cos t\}$

37. $\mathcal{L}\{te^{2t}\sin 6t\}$

★38. $\mathcal{L}\{te^{-3t}\cos 3t\}$

39. $\mathcal{L}^{-1}\left\{\dfrac{s}{(s^2 + 1)^2}\right\}$

40. $\mathcal{L}^{-1}\left\{\dfrac{s + 1}{(s^2 + 2s + 2)^2}\right\}$

In Problems 41–46 write the given function in terms of unit step functions. Find the Laplace transform of each function.

41. $f(t) = \begin{cases} 0, & 0 \le t < 1, \\ t^2, & t \ge 1. \end{cases}$

★42. $f(t) = \begin{cases} 0, & 0 \le t < \dfrac{3\pi}{2} \\ \sin t, & t \ge \dfrac{3\pi}{2} \end{cases}$

43. $f(t) = \begin{cases} t, & 0 \le t < 2, \\ 0, & t \ge 2 \end{cases}$

44. $f(t) = \begin{cases} \sin t, & 0 \le t < 2\pi \\ 0, & t \ge 2\pi \end{cases}$

45. $f(t)$

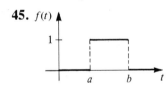

rectangular pulse

Figure 7.10

★46. $f(t)$

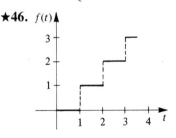

staircase function

Figure 7.11

In Problems 47 and 48 sketch the graph of the given function.

47. $f(t) = \mathcal{L}^{-1}\left\{\dfrac{1}{s^2} - \dfrac{e^{-s}}{s^2}\right\}$

48. $f(t) = \mathcal{L}^{-1}\left\{\dfrac{2}{s} - \dfrac{3e^{-s}}{s^2} + \dfrac{5e^{-2s}}{s^2}\right\}$

In Problems 49–52 use Theorem 7.7 in the form ($n = 1$)

$$f(t) = -\frac{1}{t}\mathcal{L}^{-1}\left\{\frac{d}{ds}F(s)\right\}$$

to evaluate the given inverse Laplace transformations.

EXAMPLE   Evaluate $\mathcal{L}^{-1}\left\{\tan^{-1}\dfrac{1}{s}\right\}$.

Solution: $f(t) = -\dfrac{1}{t}\mathcal{L}^{-1}\left\{\dfrac{d}{ds}\tan^{-1}\dfrac{1}{s}\right\}$

$= -\dfrac{1}{t}\mathcal{L}^{-1}\left\{\dfrac{1}{1 + (1/s)^2}\cdot(-s^{-2})\right\}$

$= -\dfrac{1}{t}\mathcal{L}^{-1}\left\{\dfrac{-1}{s^2 + 1}\right\}$

$= \dfrac{\sin t}{t}.$

---

49. $\mathcal{L}^{-1}\left\{\ln\dfrac{s - 3}{s + 1}\right\}$ **★50.** $\mathcal{L}^{-1}\left\{\ln\dfrac{s^2 + 1}{s^2 + 4}\right\}$

51. $\mathcal{L}^{-1}\left\{\dfrac{\pi}{2} - \tan^{-1}\dfrac{s}{2}\right\}$ **52.** $\mathcal{L}^{-1}\left\{\dfrac{1}{s} - \cot^{-1}\dfrac{4}{s}\right\}$

[7.2.2]   **53.** Use Theorem 7.8 to evaluate $\mathcal{L}\{e^t\}$.

54. Use Theorem 7.8 to evaluate $\mathcal{L}\{\cos^2 t\}$. [*Hint:* If $f(t) = \cos^2 t$ then $f'(t) = -2\sin t$.]

In Problems 55–66 evaluate the given Laplace transform.

55. $\mathcal{L}\left\{\displaystyle\int_0^t e^{-\tau}\cos\tau\,d\tau\right\}$ **56.** $\mathcal{L}\left\{\displaystyle\int_0^t \tau\sin\tau\,d\tau\right\}$

57. $\mathcal{L}\left\{\displaystyle\int_0^t \tau e^{t-\tau}\,d\tau\right\}$ **58.** $\mathcal{L}\left\{\displaystyle\int_0^t \sin\tau\cos(t - \tau)\,d\tau\right\}$

59. $\mathcal{L}\left\{t\displaystyle\int_0^t \sin\tau\,d\tau\right\}$ **★60.** $\mathcal{L}\left\{t\displaystyle\int_0^t \tau e^{-\tau}\,d\tau\right\}$

61. $\mathcal{L}\{1 * t^3\}$ **62.** $\mathcal{L}\{1 * e^{-2t}\}$

63. $\mathcal{L}\{t^2 * t^4\}$ **64.** $\mathcal{L}\{t^2 * te^t\}$

65. $\mathcal{L}\{e^{-t} * e^t\cos t\}$ **66.** $\mathcal{L}\{e^{2t} * \sin t\}$

In Problems 67–72 use Theorem 7.10 to find $f(t)$.

67. $\mathcal{L}^{-1}\left\{\dfrac{1}{s(s+1)}\right\}$

68. $\mathcal{L}^{-1}\left\{\dfrac{1}{s(s^2+1)}\right\}$

69. $\mathcal{L}^{-1}\left\{\dfrac{1}{(s+1)(s-2)}\right\}$

★70. $\mathcal{L}^{-1}\left\{\dfrac{1}{(s+1)^2}\right\}$

71. $\mathcal{L}^{-1}\left\{\dfrac{s}{(s^2+4)^2}\right\}$

72. $\mathcal{L}^{-1}\left\{\dfrac{1}{(s^2+1)(s^2+4)}\right\}$

[7.2.3]   In Problems 73–80 use Theorem 7.11 to find the Laplace transform of the given periodic function.

73.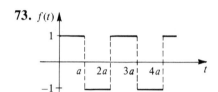

meander function

Figure 7.12

74.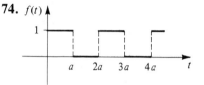

square wave

Figure 7.13

75.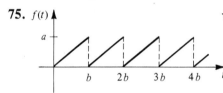

sawtooth function

Figure 7.14

★76.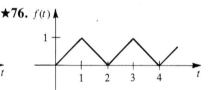

triangular wave

Figure 7.15

77.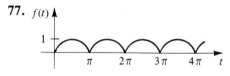

full-wave rectification of sin t

Figure 7.16

78.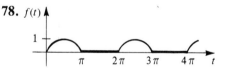

half-wave rectification of sin t

Figure 7.17

79. $f(t) = \sin t$
$f(t+2\pi) = f(t)$

80. $f(t) = \cos t$
$f(t+2\pi) = f(t)$

Miscellaneous problems

81. Show that $f*g = g*f$, that is,

$$\int_0^t f(\tau)g(t-\tau)\,d\tau = \int_0^t g(\tau)f(t-\tau)\,dt.$$

82. **(a)** Use Theorem 7.9 to show that

$$\mathcal{L}^{-1}\left\{\frac{F(s)}{s^2}\right\} = \int_0^t \int_0^\tau f(\lambda)\,d\lambda\,d\tau.$$

(b) Show that part (a) is the same as

$$\mathcal{L}^{-1}\left\{\frac{F(s)}{s^2}\right\} = t \int_0^t f(\tau) \, d\tau - \int_0^t \tau f(\tau) \, d\tau.$$

83. If $F(s) = \mathcal{L}\{f(t)\}$ show that

$$\mathcal{L}\{f(t)\cosh at\} = \frac{1}{2}[F(s-a) + F(s+a)].$$

84. Use the result of Problem 83 to find $\mathcal{L}\{\sin kt \cosh at\}$.

85. If $a > 0$ show that

$$\mathcal{L}\{f(at)\} = \frac{1}{a}F\left(\frac{s}{a}\right).$$

86. If $f(t)$ is piecewise continuous and of exponential order and $\lim_{t\to 0^+} \dfrac{f(t)}{t}$ exists, show that

$$\mathcal{L}\left\{\frac{f(t)}{t}\right\} = \int_s^\infty F(s) \, ds,$$

where $F(s) = \mathcal{L}\{f(t)\}$.

87. Use the results of Problem 86 to evaluate

$$\mathcal{L}\left\{\frac{e^t - e^{-t}}{t}\right\}.$$

7.3 Applications

Since $\mathcal{L}\{y^{(n)}(t)\}$, $n > 1$, depends on $y(t)$ and its $n - 1$ derivatives evaluated at $t = 0$, the Laplace transform is ideally suited to initial-value problems for linear differential equations with constant coefficients. This kind of differential equation can be reduced to an algebraic equation in the transformed function $Y(s)$. To see this, consider the initial-value problem

$$a_n\frac{d^n y}{dt^n} + a_{n-1}\frac{d^{n-1} y}{dt^{n-1}} + \cdots + a_1\frac{dy}{dt} + a_0 y = g(t)$$

$$y(0) = y_0$$
$$y'(0) = y_0'$$

$$\vdots$$

$$y^{(n-1)}(0) = y_0^{(n-1)}$$

where a_i, $i = 0, 1, \ldots, n$ and $y_0, y_0', \ldots, y_0^{(n-1)}$ are constants. By the linearity property of the Laplace transform we can write

$$a_n\mathcal{L}\left\{\frac{d^n y}{dt^n}\right\} + a_{n-1}\mathcal{L}\left\{\frac{d^{n-1} y}{dt^{n-1}}\right\} + \cdots + a_0\mathcal{L}\{y\} = \mathcal{L}\{g(t)\}. \qquad (1)$$

Using Theorem 7.8, (1) becomes

$$a_n[s^nY(s) - s^{n-1}y(0) - \cdots - y^{(n-1)}(0)] + a_{n-1}[s^{n-1}Y(s) - s^{n-2}y(0)$$
$$- \cdots - y^{(n-2)}(0)] + \cdots + a_0Y(s) = G(s)$$

or

$$[a_ns^n + a_{n-1}s^{n-1} + \cdots + a_0]Y(s) = a_n[s^{n-1}y_0 + \cdots + y_0^{(n-1)}]$$
$$+ a_{n-1}[s^{n-2}y_0 + \cdots + y_0^{(n-2)}] + \cdots + G(s), \qquad (2)$$

where $Y(s) = \mathcal{L}\{y(t)\}$ and $G(s) = \mathcal{L}\{g(t)\}$. By solving (2) for $Y(s)$ we find $y(t)$ by determining the inverse transform

$$y(t) = \mathcal{L}^{-1}\{Y(s)\}.$$

EXAMPLE

Solve

$$\frac{dy}{dt} - 3y = e^{2t}$$

subject to $y(0) = 1$.

Solution: We first take the transform of each member of the given differential equation,

$$\mathcal{L}\left\{\frac{dy}{dt}\right\} - 3\mathcal{L}\{y\} = \mathcal{L}\{e^{2t}\}.$$

We then use $\mathcal{L}\{dy/dt\} = sY(s) - y(0) = sY(s) - 1$ and $\mathcal{L}\{e^{2t}\} = 1/(s - 2)$. Therefore,

$$sY(s) - 1 - 3Y(s) = \frac{1}{s - 2}$$

or

$$(s - 3)Y(s) = 1 + \frac{1}{s - 2}$$

$$= \frac{s - 1}{s - 2},$$

$$Y(s) = \frac{s - 1}{(s - 2)(s - 3)}.$$

By partial fractions:

$$\frac{s - 1}{(s - 2)(s - 3)} = \frac{A}{s - 2} + \frac{B}{s - 3}$$

which yields $\qquad s - 1 = A(s - 3) + B(s - 2).$

Setting $s = 2$ and $s = 3$ in the last equation we obtain $A = -1$ and $B = 2$, respectively. Consequently,

$$Y(s) = \frac{-1}{s - 2} + \frac{2}{s - 3}$$

and

$$y(t) = -\mathcal{L}^{-1}\left\{\frac{1}{s - 2}\right\} + 2\mathcal{L}^{-1}\left\{\frac{1}{s - 3}\right\}.$$

From part (c) of Theorem 7.3 it follows that

$$y(t) = -e^{2t} + 2e^{3t}.$$

EXAMPLE   Solve $y'' - 6y' + 9y = t^2 e^{3t}$

subject to $y(0) = 2$ and $y'(0) = 6$.

Solution: $\mathcal{L}\{y''\} - 6\mathcal{L}\{y'\} + 9\mathcal{L}\{y\} = \mathcal{L}\{t^2 e^{3t}\}$

$$s^2 Y(s) - sy(0) - y'(0) - 6[sY(s) - y(0)] + 9Y(s) = \frac{2}{(s-3)^3}.$$

Using the initial conditions and simplifying gives

$$(s^2 - 6s + 9)Y(s) = 2s - 6 + \frac{2}{(s-3)^3}$$

$$(s-3)^2 Y(s) = 2(s-3) + \frac{2}{(s-3)^3}$$

$$Y(s) = \frac{2}{s-3} + \frac{2}{(s-3)^5}$$

and so $y(t) = 2\mathcal{L}^{-1}\left\{\frac{1}{s-3}\right\} + \frac{2}{4!}\mathcal{L}^{-1}\left\{\frac{4!}{(s-3)^5}\right\}.$

Recall from the first translation theorem that

$$\mathcal{L}^{-1}\left\{\frac{4!}{s^5}\bigg|_{s \to s-3}\right\} = t^4 e^{3t}.$$

Hence we have

$$y(t) = 2e^{3t} + \frac{1}{12}t^4 e^{3t}.$$

EXAMPLE   Solve $y'' + 4y' + 6y = 1 + e^{-t}$

subject to $y(0) = 0$ and $y'(0) = 0$.

Solution: $\mathcal{L}\{y''\} + 4\mathcal{L}\{y'\} + 6\mathcal{L}\{y\} = \mathcal{L}\{1\} + \mathcal{L}\{e^{-t}\}$

$$s^2 Y(s) - sy(0) - y'(0) + 4[sY(s) - y(0)] + 6Y(s) = \frac{1}{s} + \frac{1}{s+1}$$

$$(s^2 + 4s + 6)Y(s) = \frac{2s+1}{s(s+1)}$$

$$Y(s) = \frac{2s+1}{s(s+1)(s^2 + 4s + 6)}.$$

By partial fractions:

$$\frac{2s + 1}{s(s + 1)(s^2 + 4s + 6)} = \frac{A}{s} + \frac{B}{s + 1} + \frac{Cs + D}{s^2 + 4s + 6},$$

which implies

$$2s + 1 = A(s + 1)(s^2 + 4s + 6)$$
$$+ Bs(s^2 + 4s + 6) + (Cs + D)s(s + 1).$$

Setting $s = 0$ and $s = -1$ gives, respectively, $A = 1/6$ and $B = 1/3$. Equating the coefficients of s^3 and s gives

$$A + B + C = 0$$

$$10A + 6B + D = 2,$$

so it follows that $C = -1/2$ and $D = -5/3$. Thus,

$$Y(s) = \frac{1/6}{s} + \frac{1/3}{s + 1} + \frac{-s/2 - 5/3}{s^2 + 4s + 6}$$

$$= \frac{1/6}{s} + \frac{1/3}{s + 1} + \frac{-1/2(s + 2) - 2/3}{(s + 2)^2 + 2}$$

$$= \frac{1/6}{s} + \frac{1/3}{s + 1} - \frac{1}{2}\frac{s + 2}{(s + 2)^2 + 2} - \frac{2}{3}\frac{1}{(s + 2)^2 + 2}.$$

Finally, from parts (a) and (c) of Theorem 7.3 and the first translation theorem we obtain

$$y(t) = \frac{1}{6}\mathscr{L}^{-1}\left\{\frac{1}{s}\right\} + \frac{1}{3}\mathscr{L}^{-1}\left\{\frac{1}{s + 1}\right\} - \frac{1}{2}\mathscr{L}^{-1}\left\{\frac{s + 2}{(s + 2)^2 + 2}\right\} - \frac{2}{3\sqrt{2}}\mathscr{L}^{-1}\left\{\frac{\sqrt{2}}{(s + 2)^2 + 2}\right\}$$

$$= \frac{1}{6} + \frac{1}{3}e^{-t} - \frac{1}{2}e^{-2t}\cos\sqrt{2}t - \frac{\sqrt{2}}{3}e^{-2t}\sin\sqrt{2}t.$$

EXAMPLE

Solve
$$x'' + 16x = \cos 4t$$
subject to $x(0) = 0$ and $x'(0) = 1$.

Solution: Recall that this initial value problem could describe the forced, undamped, and resonant motion of a mass on a spring. The mass starts with an initial velocity of one foot per second in the downward direction from the equilibrium position. Now, one could readily solve this problem by, say variation of parameters, but the use of Laplace transforms obviates the necessity of determining the constants which would naturally occur in the general solution $x = x_c(t) + x_p(t)$.

Transforming the equation gives

$$(s^2 + 16)X(s) = 1 + \frac{s}{s^2 + 16}$$

$$X(s) = \frac{1}{s^2 + 16} + \frac{s}{(s^2 + 16)^2}.$$

With the aid of part (d) of Theorem 7.3 and Theorem 7.7 we find

$$x(t) = \frac{1}{4}\mathcal{L}^{-1}\left\{\frac{4}{s^2 + 16}\right\} + \frac{1}{8}\mathcal{L}^{-1}\left\{\frac{8s}{(s^2 + 16)^2}\right\}$$

$$= \frac{1}{4}\sin 4t + \frac{1}{8}t\sin 4t.$$

EXAMPLE   Solve $x'' + 16x = f(t)$

where
$$f(t) = \begin{cases} \cos 4t, & 0 \le t < \pi, \\ 0, & t \ge \pi, \end{cases}$$

and $x(0) = 0$, $x'(0) = 1$.

Solution: The function $f(t)$ can be interpreted as an external force which is acting on a mechanical system only for a short period of time and then is removed. While this problem could be solved by conventional means, the procedure is not at all convenient when $f(t)$ is defined in a piecewise manner.

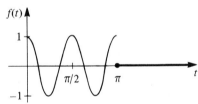

Figure 7.18

Since we can write

$$f(t) = \cos 4t - \cos 4t\, \mathcal{U}(t - \pi)$$

$$= \cos 4t - \cos 4(t - \pi)\mathcal{U}(t - \pi) \qquad \text{[by periodicity of the cosine]}$$

it follows that

$$\mathcal{L}\{x''\} + 16\mathcal{L}\{x\} = \mathcal{L}\{f(t)\}$$

$$s^2X(s) - sx(0) - x'(0) + 16X(s) = \frac{s}{s^2 + 16} - \frac{s}{s^2 + 16}e^{-\pi s}$$

$$(s^2 + 16)X(s) = 1 + \frac{s}{s + 16} - \frac{s}{s^2 + 16}e^{-\pi s}$$

$$X(s) = \frac{1}{s^2 + 16} + \frac{s}{(s^2 + 16)^2} - \frac{s}{(s^2 + 16)^2}e^{-\pi s}.$$

From the preceding example and (2) of Section 7.2 we obtain

$$x(t) = \frac{1}{4}\mathcal{L}^{-1}\left\{\frac{4}{s^2 + 16}\right\} + \frac{1}{8}\mathcal{L}^{-1}\left\{\frac{8s}{(s^2 + 16)^2}\right\} - \frac{1}{8}\mathcal{L}^{-1}\left\{\frac{8s}{(s^2 + 16)^2}e^{-\pi s}\right\}$$

$$= \frac{1}{4}\sin 4t + \frac{1}{8}t\sin 4t - \frac{1}{8}(t - \pi)\sin 4(t - \pi)\mathcal{U}(t - \pi).$$

The foregoing solution is the same as

$$x(t) = \begin{cases} \dfrac{1}{4}\sin 4t + \dfrac{1}{8}t\sin 4t, & 0 \le t < \pi \\ \dfrac{2 + \pi}{8}\sin 4t, & t \ge \pi. \end{cases}$$

Observe from the graph of $x(t)$ in Figure 7.19 that the amplitudes of vibration become steady as soon as the external force is turned off.

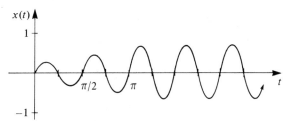

Figure 7.19

EXAMPLE

Solve
$$y'' + 2y' + y = f(t)$$
where
$$f(t) = \mathcal{U}(t - 1) - 2\mathcal{U}(t - 2) + \mathcal{U}(t - 3)$$
and $y(0) = 0$, $y'(0) = 0$.*

Solution: By the second translation theorem and simplification the transform of the differential equation is

$$(s + 1)^2Y(s) = \frac{e^{-s}}{s} - 2\frac{e^{-2s}}{s} + \frac{e^{-3s}}{s}$$

or

$$Y(s) = \frac{e^{-s}}{s(s + 1)^2} - 2\frac{e^{-2s}}{s(s + 1)^2} + \frac{e^{-3s}}{s(s + 1)^2}.$$

With the aid of partial fractions the last equation becomes

$$Y(s) = \left[\frac{1}{s} - \frac{1}{s + 1} - \frac{1}{(s + 1)^2}\right]e^{-s} - 2\left[\frac{1}{s} - \frac{1}{s + 1} - \frac{1}{(s + 1)^2}\right]e^{-2s}$$
$$+ \left[\frac{1}{s} - \frac{1}{s + 1} - \frac{1}{(s + 1)^2}\right]e^{-3s}.$$

---

*For practice you should graph $f(t)$.

Again employing the inverse form of the second translation theorem we find

$$y(t) = [1 - e^{-(t-1)} - (t - 1)e^{-(t-1)}]\mathcal{U}(t - 1)$$
$$- 2[1 - e^{-(t-2)} - (t - 2)e^{-(t-2)}]\mathcal{U}(t - 2)$$
$$+ [1 - e^{-(t-3)} - (t - 3)e^{-(t-3)}]\mathcal{U}(t - 3).$$

An integrodifferential equation

In a single loop or series circuit, Kirchoff's second law states that the sum of the voltage drops across an inductor, resistor, and capacitor is equal to the impressed voltage $E(t)$. Now it is known (see Section 1.2) that the

$$\text{voltage drop across the inductor} = L\frac{di}{dt},$$

$$\text{voltage drop across the resistor} = Ri(t),$$

$$\text{voltage drop across the capacitor} = \frac{1}{C}\int_0^t i(\tau)\,d\tau,$$

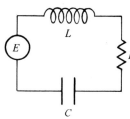

Figure 7.20

where $i(t)$ is the current and L, R, and C are constants. It follows that the current in a circuit, such as that shown in Figure 7.20, is governed by the **integrodifferential equation**

$$L\frac{di}{dt} + Ri + \frac{1}{C}\int_0^t i(\tau)\,d\tau = E(t). \tag{3}$$

EXAMPLE

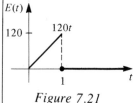

Figure 7.21

Determine the current $i(t)$ in a single loop L-R-C circuit when $L = 0.1$ henry, $R = 20$ ohms, $C = 10^{-3}$ farads, $i(0) = 0$, and if the impressed voltage $E(t)$ is as given in Figure 7.21.

Solution: Since the voltage is off for $t \geq 1$, we can write

$$E(t) = 120t - 120t\mathcal{U}(t - 1). \tag{4}$$

But, in order to use the second translation theorem, we must rewrite (4) as

$$E(t) = 120t - 120(t - 1)\mathcal{U}(t - 1) - 120\mathcal{U}(t - 1)$$

Equation (3) then becomes

$$0.1\frac{di}{dt} + 20i + 10^3\int_0^t i(\tau)\,d\tau = 120t - 120(t - 1)\mathcal{U}(t - 1) - 120\mathcal{U}(t - 1). \tag{5}$$

Now recall from Theorem 7.9 that

$$\mathcal{L}\left\{\int_0^t i(\tau)\,d\tau\right\} = I(s)/s$$

where $I(s) = \mathcal{L}\{i(t)\}$. Thus, the transform of equation (5) is

$$0.1sI(s) + 20I(s) + 10^3\frac{I(s)}{s} = 120\left[\frac{1}{s^2} - \frac{1}{s^2}e^{-s} - \frac{1}{s}e^{-s}\right],$$

or after multiplying by $10s$,

$$(s + 100)^2 I(s) = 1200\left[\frac{1}{s} - \frac{1}{s}e^{-s} - e^{-s}\right]$$

$$I(s) = 1200\left[\frac{1}{s(s + 100)^2} - \frac{1}{s(s + 100)^2}e^{-s} - \frac{1}{(s + 100)^2}e^{-s}\right].$$

By partial fractions we can write

$$I(s) = 1200\left[\frac{1/10,000}{s} - \frac{1/10,000}{s + 100} - \frac{1/100}{(s + 100)^2} - \frac{1/10,000}{s}e^{-s}\right.$$

$$\left. + \frac{1/10,000}{s + 100}e^{-s} + \frac{1/100}{(s + 100)^2}e^{-s} - \frac{1}{(s + 100)^2}e^{-s}\right].$$

Employing the second translation theorem, we obtain

$$i(t) = \tfrac{3}{25}[1 - \mathcal{U}(t - 1)] - \tfrac{3}{25}[e^{-100t} - e^{-100(t-1)}\mathcal{U}(t - 1)]$$

$$- 12te^{-100t} - 1188(t - 1)e^{-100(t - 1)}\mathcal{U}(t - 1).$$

In some circumstances the Laplace transform can be used to solve equations with variable coefficients. However, in this case the transform of the equation does not result in an algebraic expression, but rather in another differential equation in the transformed function $Y(s)$.

EXAMPLE

Solve
$$ty''(t) + 2(t - 1)y'(t) - 2y(t) = 0$$
subject to $y(0) = 0$.

Solution: Since $y'(0)$ is not specified, we expect the solution to contain an arbitrary constant. From Theorem 7.7 we know that

$$\mathcal{L}\{tf(t)\} = -\frac{d}{ds}\mathcal{L}\{f(t)\}$$

from which it follows

$$\mathcal{L}\{ty'(t)\} = -\frac{d}{ds}\mathcal{L}\{y'(t)\}$$

$$= -\frac{d}{ds}[sY(s) - y(0)]$$

$$= -s\frac{dY}{ds} - Y$$

and
$$\mathcal{L}\{ty''(t)\} = -\frac{d}{ds}\mathcal{L}\{y''(t)\}$$

$$= -\frac{d}{ds}[s^2Y(s) - sy(0) - y'(0)]$$

$$= -s^2\frac{dY}{ds} - 2sY + y(0).$$

Using $y(0) = 0$ we have

$$\mathcal{L}\{ty''\} + 2\mathcal{L}\{ty'\} - 2\mathcal{L}\{y'\} - 2\mathcal{L}\{y\} = \mathcal{L}\{0\}$$

$$-s^2\frac{dY}{ds} - 2sY - 2s\frac{dY}{ds} - 2Y - 2sY - 2Y = 0$$

$$(s^2 + 2s)\frac{dY}{ds} + (4s + 4)Y = 0.$$

This latter equation is separable, so we write

$$\frac{dY}{Y} + \frac{4s + 4}{s(s + 2)}\,ds = 0$$

$$\frac{dY}{Y} + \left[\frac{2}{s} + \frac{2}{s + 2}\right]ds = 0$$

and integrate

$$\ln|Y| + 2\ln|s| + 2\ln|s + 2| = \ln|c|$$

$$\ln[s^2(s + 2)^2|Y|] = \ln|c|$$

$$Y = \frac{c}{s^2(s + 2)^2}.$$

By partial fractions:

$$Y = c\left[\frac{-1/4}{s} + \frac{1/4}{s^2} + \frac{1/4}{s + 2} + \frac{1/4}{(s + 2)^2}\right].$$

The common factor of 1/4 can, of course, be absorbed in the arbitrary constant. Writing c_1 for $c/4$ we obtain

$$y(t) = c_1[-1 + t + e^{-2t} + te^{-2t}].$$

However, the Laplace transform does not provide a general method for solving differential equations with variable coefficients.

EXERCISES 7.3   *Answers to odd-numbered problems begin on page A–23.*

In Problems 1–26 use the Laplace transform to solve the given differential equation subject to the indicated initial conditions.

1. $\dfrac{dy}{dt} - y = 1, \quad y(0) = 0$

2. $\dfrac{dy}{dt} + 2y = t, \quad y(0) = -1$

3. $y' + 4y = e^{-4t}, \quad y(0) = 2$

4. $y' - y = \sin t, \quad y(0) = 0$

5. $y'' + 5y' + 4y = 0$, $y(0) = 1$, $y'(0) = 0$

6. $y'' - 6y' + 13y = 0$, $y(0) = 0$, $y'(0) = -3$

7. $y'' - 6y' + 9y = t$, $y(0) = 0$, $y'(0) = 1$

★ 8. $y'' - 4y' + 4y = t^3$, $y(0) = 1$, $y'(0) = 0$

9. $y'' - 4y' + 4y = t^3 e^{2t}$, $y(0) = 0$, $y'(0) = 0$

10. $y'' - 2y' + 5y = 1 + t$, $y(0) = 0$, $y'(0) = 4$

11. $y'' + y = \sin t$, $y(0) = 1$, $y'(0) = -1$

12. $y'' + 16y = 1$, $y(0) = 1$, $y'(0) = 2$

13. $y'' - y' = e^t \cos t$, $y(0) = 0, y'(0) = 0$

14. $y'' - 2y' = e^t \sinh t$, $y(0) = 0$, $y'(0) = 0$

15. $2y''' + 3y'' - 3y' - 2y = e^{-t}$, $y(0) = 0$, $y'(0) = 0$, $y''(0) = 1$

★16. $y''' + 2y'' - y' - 2y = \sin 3t$, $y(0) = 0$, $y'(0) = 0$, $y''(0) = 1$

17. $y^{(4)} - y = 0$, $y(0) = 1$, $y'(0) = 0$, $y''(0) = -1$, $y'''(0) = 0$

18. $y^{(4)} - y = t$, $y(0) = 0$, $y'(0) = 0$, $y''(0) = 0$, $y'''(0) = 0$

EXAMPLE

$y' + 2y = f(t)$ where $f(t) = \begin{cases} 2, & 0 \le t < 1, \\ 0, & t \ge 1, \end{cases}$ $y(0) = 1$.

Solution: By the usual operational properties, the transformation of the left-hand side of the equation is

$$sY(s) - y(0) + 2Y(s) = (s + 2)Y(s) - 1.$$

Whereas to transform the right-hand side, we use Definition 7.1*

$$\mathcal{L}\{f(t)\} = \int_0^\infty e^{-st} f(t)\, dt$$

$$= 2 \int_0^1 e^{-st}\, dt$$

$$= \frac{2}{s}[1 - e^{-s}].$$

Hence the differential equation becomes

$$(s + 2)Y(s) - 1 = \frac{2}{s}[1 - e^{-s}]$$

$$Y(s) = \frac{2}{s(s + 2)}[1 - e^{-s}] + \frac{1}{s + 2}$$

---

*Observe that $f(t) = 2 - 2\mathcal{U}(t - 1)$. Hence $\mathcal{L}\{f(t)\}$ can also be gotten from Theorem 7.6.

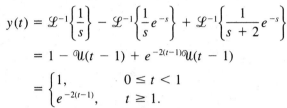

$$= \left[\frac{1}{s} - \frac{1}{s+2} \right][1 - e^{-s}] + \frac{1}{s+2}$$

$$= \frac{1}{s} - \frac{1}{s}e^{-s} + \frac{1}{s+2}e^{-s}.$$

Thus from the second translation theorem it follows that

$$y(t) = \mathcal{L}^{-1}\left\{ \frac{1}{s} \right\} - \mathcal{L}^{-1}\left\{ \frac{1}{s}e^{-s} \right\} + \mathcal{L}^{-1}\left\{ \frac{1}{s+2}e^{-s} \right\}$$

$$= 1 - \mathcal{U}(t - 1) + e^{-2(t-1)}\mathcal{U}(t - 1)$$

$$= \begin{cases} 1, & 0 \le t < 1 \\ e^{-2(t-1)}, & t \ge 1. \end{cases}$$

Figure 7.22   The graph of $y(t)$ is given in Figure 7.22.

19. $y' + y = f(t)$ where $f(t) = \begin{cases} 0, & 0 \le t < 1, \\ 5, & t \ge 1 \end{cases}$ $y(0) = 0$

20. $y' + y = f(t)$ where $f(t) = \begin{cases} 1, & 0 \le t < 1, \\ -1, & t \ge 1, \end{cases}$ $y(0) = 0$

21. $y' + 2y = f(t)$ where $f(t) = \begin{cases} t, & 0 \le t < 1, \\ 0, & t \ge 1, \end{cases}$ $y(0) = 0$

22. $y'' + 4y = f(t)$ where $f(t) = \begin{cases} 1, & 0 \le t < 1, \\ 0, & t \ge 1, \end{cases}$ $y(0) = 0, \ y'(0) = -1$

23. $y'' + 4y = f(t)$ where $f(t) = \sin t\,\mathcal{U}(t - 2\pi),$ $y(0) = 1, \ y'(0) = 0$

★**24.** $y'' - 5y' + 6y = \mathcal{U}(t - 1),$ $y(0) = 0, \ y'(0) = 1$

25. $y'' + y = f(t)$ where $f(t) = \begin{cases} 0, & 0 \le t < \pi \\ 1, & \pi \le t < 2\pi \\ 0, & t \ge 2\pi \end{cases}$ $y(0) = 0, \ y'(0) = 1$

26. $y'' + 4y' + 3y = 1 - \mathcal{U}(t - 2) - \mathcal{U}(t - 4)$
$\qquad\qquad + \mathcal{U}(t - 6), \quad y(0) = 0, \quad y'(0) = 0$

In Problems 27–30 use the Laplace transform and Theorem 7.7 to find a solution of the given equation.

27. $ty'' - y' = t^2, \quad y(0) = 0$

28. $ty'' + 2ty' + 2y = 0, \quad y(0) = 0$

29. $ty'' + y' + ty = 0$ [*Hint*: Expand $[1 + 1/s^2]^{-1/2}$ by the binomial theorem.]

30. $ty'' + (t + 2)y' + y = -1, \quad y(0) = 0$

In Problems 31–38 use Theorem 7.10 to solve the given integral equation for $f(t)$.

EXAMPLE

Solve for $f(t)$ $\qquad f(t) = 3t^2 - e^{-t} - \int_0^t f(\tau)e^{t-\tau}\,d\tau.$

Solution: It follows from the convolution theorem that

$$\mathcal{L}\{f(t)\} = 3\mathcal{L}\{t^2\} - \mathcal{L}\{e^{-t}\} - \mathcal{L}\{f(t)\}\mathcal{L}\{e^t\}$$

$$F(s) = 3 \cdot \frac{2}{s^3} - \frac{1}{s+1} - F(s) \cdot \frac{1}{s-1}$$

$$\left[1 + \frac{1}{s-1}\right]F(s) = \frac{6}{s^3} - \frac{1}{s+1}$$

$$\frac{s}{s-1}F(s) = \frac{6}{s^3} - \frac{1}{s+1}$$

$$F(s) = \frac{6(s-1)}{s^4} - \frac{s-1}{s(s+1)}$$

$$= \frac{6}{s^3} - \frac{6}{s^4} + \frac{1}{s} - \frac{2}{s+1} \qquad \begin{array}{l}\text{[termwise}\\\text{division}\\\text{and partial}\\\text{functions].}\end{array}$$

Therefore,

$$f(t) = 3\mathcal{L}^{-1}\left\{\frac{2!}{s^3}\right\} - \mathcal{L}^{-1}\left\{\frac{3!}{s^4}\right\} + \mathcal{L}^{-1}\left\{\frac{1}{s}\right\} - 2\mathcal{L}^{-1}\left\{\frac{1}{s+1}\right\}$$

$$= 3t^2 - t^3 + 1 - 2e^{-t}.$$

---

31. $f(t) + \displaystyle\int_0^t (t - \tau)f(\tau)\,d\tau = t$

32. $f(t) = 2t - 4\displaystyle\int_0^t \sin \tau f(t - \tau)\,d\tau$

33. $f(t) = te^t + \displaystyle\int_0^t \tau f(t - \tau)\,d\tau$

34. $f(t) + 2\displaystyle\int_0^t f(\tau)\cos(t - \tau)\,d\tau = 4e^{-t} + \sin t$

35. $f(t) + \displaystyle\int_0^t f(\tau)\,d\tau = 1$

36. $f(t) = \cos t + \displaystyle\int_0^t e^{-\tau}f(t - \tau)\,d\tau$

37. $f(t) = 1 + t - \dfrac{8}{3} \displaystyle\int_0^t (\tau - t)^3 f(\tau) \, d\tau$

★38. $t - 2f(t) = \displaystyle\int_0^t (e^\tau - e^{-\tau}) f(t - \tau) \, d\tau$

In Problems 39 and 40 solve the integrodifferential equation.

39. $y'(t) = 1 - \sin t - \displaystyle\int_0^t y(\tau) \, d\tau, \quad y(0) = 0.$

40. $\dfrac{dy}{dt} + 6y(t) + 9 \displaystyle\int_0^t y(\tau) \, d\tau = 1, \quad y(0) = 0$

41. Use equation (3) to determine the current $i(t)$ in a single loop L-R-C circuit when $L = 0.005$ henry, $R = 1$ ohms, $C = 0.02$ farads, $E(t) = 100[1 - \mathcal{U}(t - 1)]$ volts, and $i(0) = 0$.

★42. Solve Problem 41 when $E(t) = 100[t - (t - 1)\mathcal{U}(t - 1)]$.

43. Recall that the differential equation for the current $i(t)$ in a series circuit containing an inductor and a resistor is

$$L\frac{di}{dt} + Ri = E(t), \tag{6}$$

where $E(t)$ is the impressed voltage (see Section 3.2). Use the Laplace transform to determine the current $i(t)$ when $L = 1$ henry, $R = 10$ ohms,

$$E(t) = \begin{cases} \sin t, & 0 \le t < \dfrac{3\pi}{2} \\ 0, & t \ge \dfrac{3\pi}{2} \end{cases}$$

and $i(0) = 0$.

★44. Solve equation (6) subject to $i(0) = 0$ where $E(t)$ is given in Figure 7.23. [*Hint:* See Problem 73, Exercises 7.2. Also note that $1/(1 - X) = 1 + X + X^2 + \dots, \ |X| < 1$.]

45. Solve equation (6) subject to $i(0) = 0$ where $E(t)$ is given in Figure 7.24. Specify the solution for $0 \le t < 2$. [*Hint:* See Problem 75, Exercises 7.2.]

46. Recall that when a series circuit contains an inductor, resistor, and a capacitor, the differential equation for the instantaneous charge $q(t)$ on the capacitor is given by

$$L\frac{d^2q}{dt^2} + R\frac{dq}{dt} + \frac{1}{C}q = E(t),$$

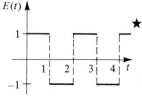

Figure 7.23

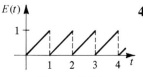

Figure 7.24

(see Section 5.4). Use the Laplace transform to determine $q(t)$ when $L = 1$ henry, $R = 20$ ohms, $C = 0.005$ farad, $E(t) = 150$ volts, $t > 0$, and $q(0) = 0$, $i(0) = 0$. What is the current $i(t)$? What is the charge $q(t)$ if the same constant voltage is turned off for $t \geq 2$?

47. Determine the charge $q(t)$ and current $i(t)$ for a series circuit in which $L = 1$ henry, $R = 20$ ohms, $C = 0.01$ farad, $E(t) = 120 \sin 10t$ volts, $q(0) = 0$, and $i(0) = 0$. What is the steady-state current?

★**48.** Suppose a 32-lb. weight stretches a spring 2 ft. If the weight is released from rest at the equilibrium position, determine the equation of a motion if an impressed force $f(t) = \sin t$ acts on the system for $0 \leq t < 2\pi$ and is then removed. Ignore any damping forces. [*Hint:* Write the impressed force in terms of the unit step function.]

49. A 4-lb. weight stretches a spring 2 ft. The weight is released from rest 18 in. above the equilibrium position, and the resulting motion takes place in a medium offering a damping force numerically equal to 7/8 times the instantaneous velocity. Use the Laplace transform to determine the equation of motion.

50. A 16-lb. weight is attached to a spring whose constant is $k = 4.5$ lb/ft. Beginning at $t = 0$, a force equal to $f(t) = 4 \sin 3t + 2 \cos 3t$ acts on the system. Assuming that no damping forces are present, use the Laplace transform to find the equation of motion if the weight is released from rest from the equilibrium position.

[O] 7.4 The Unit Impulse

Mechanical systems are often acted upon by an external force (or an emf in electrical circuits) of large magnitude that acts only for a very short period of time. For example, a vibrating airplane wing could be struck by lightning or a mass on a spring could be given a sharp blow by a ball-peen hammer. In an attempt to describe a brief but violent force scientists and engineers devised a "function" called the unit impulse. This formally defined function gave rise to an area of mathematics known as *generalized functions* or *distributions*.

Dirac delta function For $t_0 > 0$ consider the function defined by

$$\delta_a(t - t_0) = \begin{cases} \dfrac{1}{2a}, & t_0 - a < t < t_0 + a \\ 0, & t \leq t_0 - a, \quad \text{or} \quad t \geq t_0 + a \end{cases} \tag{1}$$

where a is a positive constant. For a small value of a, $\delta_a(t - t_0)$ is essentially a constant function of large magnitude that is "on" for only a short interval of time around t_0. The behavior of $\delta_a(t - t_0)$ as $a \to 0$ is illustrated in Figure 7.25 (b). The "function" defined by the limit

$$\delta(t - t_0) = \lim_{a \to 0} \delta_a(t - t_0)$$

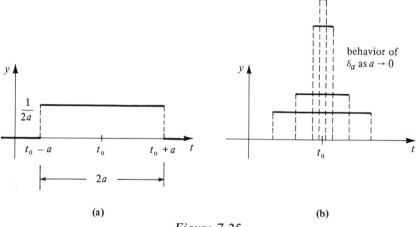

$$\text{Figure 7.25}$$

is called a **unit impulse** or **Dirac delta function**.* One might characterize the delta function by the two properties

(i) $\delta(t - t_0)$ is infinite at $t = t_0$ but zero everywhere else, and

(ii) $\displaystyle\int_{-\infty}^{\infty} \delta(t - t_0)\, dt = 1.$[†]

It is possible to obtain the Laplace transform of $\delta(t - t_0)$ by the formal assumption

$$\mathcal{L}\{\delta(t - t_0)\} = \lim_{a \to 0} \mathcal{L}\{\delta_a(t - t_0)\}.$$

We leave it as an exercise to show that

$$\mathcal{L}\{\delta_a(t - t_0)\} = e^{-st_0}\left(\frac{e^{sa} - e^{-sa}}{2sa}\right). \tag{2}$$

Since (2) is an indeterminate form as $a \to 0$ we apply L'Hôpital's rule:

$$\lim_{a \to 0} \mathcal{L}\{\delta_a(t - t_0)\} = e^{-st_0} \lim_{a \to 0}\left(\frac{e^{sa} + e^{-sa}}{2}\right)$$

$$= e^{-st_0}.$$

In other words,

$$\mathcal{L}\{\delta(t - t_0)\} = e^{-st_0}. \tag{3}$$

Now when $t_0 = 0$ it seems plausible to conclude from (3) that

$$\mathcal{L}\{\delta(t)\} = 1.$$

---

*Named after Paul A.M. Dirac (1902–), and English physicist who won the Nobel Prize in 1933.

[†]Note that the area under the graph of $\delta_a(t - t_0)$ on the interval $t_0 - a \leq t \leq t_0 + a$ is unity.

This last result emphasizes the fact that $\delta(t)$ is no ordinary function since we expect $\mathscr{L}\{f(t)\}\to 0$ as $s\to\infty$ (see Theorem 7.4).

EXAMPLE

The initial-value problem

$$y'' + y = \delta(t - 2\pi)$$
$$y(0) = 1 \qquad y'(0) = 0$$

could serve as a model for describing the motion of a mass on a spring. The mass is released from rest 1 unit below the equilibrium position in a medium in which damping is negligible. At $t = 2\pi$ sec the mass is given a sharp blow. In view of (3) the Laplace transform of the differential equation is

$$(s^2 + 1)Y(s) = s + e^{-2\pi s}$$

or

$$Y(s) = \frac{s}{s^2 + 1} + \frac{e^{-2\pi s}}{s^2 + 1}.$$

Utilizing the second translation theorem we find

$$y(t) = \cos t + \sin(t - 2\pi)\mathscr{U}(t - 2\pi) \tag{4}$$

Since $\sin(t - 2\pi) = \sin t$, (4) can be written

$$y(t) = \begin{cases} \cos t, & 0 \le t < 2\pi \\ \cos t + \sin t, & t \ge 2\pi. \end{cases}$$

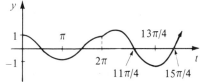

Figure 7.26

Figure 7.26 shows that the weight is exhibiting simple harmonic motion until it is struck at $t = 2\pi$. The influence of the unit impulse is to increase the amplitude of vibration to $\sqrt{2}$ for $t > 2\pi$.

EXERCISES 7.4

Answers to odd-numbered problems begin on page A–24.

In Problems 1–10 use the Laplace transform to solve the given differential equation subject to the indicated initial conditions.

1. $y' - 3y = \delta(t - 2), \quad y(0) = 0$

2. $y' + y = \delta(t - 1), \quad y(0) = 2$

3. $y'' + y = \delta(t - 2\pi), \quad y(0) = 0, \quad y'(0) = 1$

4. $y'' + 16y = \delta(t - 2\pi), \quad y(0) = 0, \quad y'(0) = 0$

5. $y'' + y = \delta\left(t - \frac{\pi}{2}\right) + \delta\left(t - \frac{3\pi}{2}\right), \quad y(0) = 0, \quad y'(0) = 0$

★ 6. $y'' + y = \delta(t - 2\pi) + \delta(t - 4\pi)$, $y(0) = 1$, $y'(0) = 0$

7. $y'' + 2y' = \delta(t - 1)$, $y(0) = 0$, $y'(0) = 1$

★ 8. $y'' - 2y' = 1 + \delta(t - 2)$, $y(0) = 0$, $y'(0) = 1$

9. $y'' + 4y' + 5y = \delta(t - 2\pi)$, $y(0) = 0$, $y'(0) = 0$

10. $y'' + 2y' + y = \delta(t - 1)$, $y(0) = 0$, $y'(0) = 0$

Miscellaneous problems

11. Use the definition of the Laplace transform and (1) to obtain equation (2).

12. Use (1) and the Mean Value Theorem for definite integrals to show formally that

$$\int_{-\infty}^{\infty} f(t)\, \delta(t - t_0)\, dt = f(t_0).$$

13. Use the result of Problem 12 to solve

$$y'' + 2y' + 2y = \cos t\, \delta(t - 3\pi)$$

subject to $y(0) = 1$ and $y'(0) = -1$.

★14. To emphasize the unusual nature of the delta function, show that the "solution" of the initial-value problem $y'' + \omega^2 y = \delta(t)$, $y(0) = 0$, $y'(0) = 0$ does *not* satisfy the initial condition $y'(0) = 0$.

15. Solve the initial-value problem

$$L\frac{di}{dt} + Ri = \delta(t), \qquad i(0) = 0,$$

where L and R are constants. Does the solution satisfy the condition at $t = 0$?

CHAPTER 7 SUMMARY

The **Laplace transform** of a function $f(t)$, $t \geq 0$, is defined by the integral

$$\mathscr{L}\{f(t)\} = \int_0^{\infty} e^{-st} f(t)\, dt = F(s).$$

The parameter s is usually restricted in such a manner to guarantee convergence of the integral. When applied to a linear equation with constant coefficients such as $ay'' + by' + cy = g(t)$ there results an algebraic equation

$$a[s^2 Y(s) - sy(0) - y'(0)] + b[sY(s) - y(0)] + cY(s) = G(s)$$

which depends on the initial conditions $y(0)$ and $y'(0)$. When these values are known, we determine $y(t)$ by evaluating $y(t) = \mathscr{L}^{-1}\{Y(s)\}$.

The following tables summarize all of the basic results obtained in this chapter.

Table I.   Transforms of Some Basic Functions

$f(t)$	$\mathscr{L}\{f(t)\} = F(s)$
1. 1	$\dfrac{1}{s}$
2. t^n, $n = 1, 2, 3, \ldots$	$\dfrac{n!}{s^{n+1}}$
3. e^{at}	$\dfrac{1}{s - a}$
4. $\sin kt$	$\dfrac{k}{s^2 + k^2}$
5. $\cos kt$	$\dfrac{s}{s^2 + k^2}$
6. $\sinh kt$	$\dfrac{k}{s^2 - k^2}$
7. $\cosh kt$	$\dfrac{s}{s^2 - k^2}$

Table II.   Operational Properties

8. $e^{at} f(t)$	$F(s - a)$
9. $f(t - a)\mathscr{U}(t - a), a > 0$	$e^{-as} F(s)$
10. $t^n f(t)$, $n = 1, 2, 3, \ldots$	$(-1)^n \dfrac{d^n}{ds^n} F(s)$
11. $f^{(n)}(t)$, $n = 1, 2, 3, \ldots$	$s^n F(s) - s^{n-1} f(0) - \cdots - f^{(n-1)}(0)$
12. $\displaystyle\int_0^t f(\tau)\, d\tau$	$\dfrac{F(s)}{s}$
13. $\displaystyle\int_0^t f(\tau) g(t - \tau)\, d\tau$	$F(s) G(s)$

Table III. Some Consequences of Tables I and II

$f(t)$	$\mathscr{L}\{f(t)\} = F(s)$
14. $t^n e^{at}, \quad n = 1, 2, 3, \ldots$	$\dfrac{n!}{(s - a)^{n+1}}$
15. $e^{at} \sin kt$	$\dfrac{k}{(s - a)^2 + k^2}$
16. $e^{at} \cos kt$	$\dfrac{s - a}{(s - a)^2 + k^2}$
17. $t \sin kt$	$\dfrac{2ks}{(s^2 + k^2)^2}$
18. $t \cos kt$	$\dfrac{s^2 - k^2}{(s^2 + k^2)^2}$
19. $\sin kt - kt \cos kt$	$\dfrac{2k^3}{(s^2 + k^2)^2}$
20. $\sin kt + kt \cos kt$	$\dfrac{2ks^2}{(s^2 + k^2)^2}$

CHAPTER 7
TEST

Answers to odd-numbered problems begin on page A–24.

In Problems 1 and 2 use the definition of the Laplace transform to find $\mathcal{L}\{f(t)\}$.

1. $f(t) = \begin{cases} t, & 0 \le t < 1 \\ 2 - t, & t \ge 1 \end{cases}$

2. $f(t) = \begin{cases} 0, & 0 \le t < 2 \\ 1, & 2 \le t < 4 \\ 0, & t \ge 4 \end{cases}$

In Problems 3-8, without referring back to the text, fill in the blanks.

3. $\mathcal{L}\{e^{-7t}\} = $ _____

4. $\mathcal{L}\{te^{-7t}\} = $ _____

5. $\mathcal{L}\{\sin 2t\} = $ _____

6. $\mathcal{L}\{e^{-3t}\sin 2t\} = $ _____

7. $\mathcal{L}\{t \sin 2t\} = $ _____

8. $\mathcal{L}\{\sin 2t\, \mathcal{U}(t - \pi)\} = $ _____

9. $\mathcal{L}^{-1}\left\{\dfrac{20}{s^6}\right\} = $ _____

10. $\mathcal{L}^{-1}\left\{\dfrac{1}{4s + 1}\right\} = $ _____

11. $\mathcal{L}^{-1}\left\{\dfrac{1}{(s - 5)^3}\right\} = $ _____

12. $\mathcal{L}^{-1}\left\{\dfrac{1}{s^2 - 5}\right\} = $ _____

13. $\mathcal{L}^{-1}\left\{\dfrac{s}{s^2 - 10s + 29}\right\} = $ _____

14. $\mathcal{L}^{-1}\left\{\dfrac{e^{-5s}}{s^2}\right\} = $ _____

In Problems 15–19 use the Laplace transform to solve the given equation.

15. $y'' - 2y' + y = e^t, \quad y(0) = 0, \quad y'(0) = 5$

16. $y'' - 8y' + 20y = te^t, \quad y(0) = 0, \quad y'(0) = 0$

17. $y' - 5y = f(t)$ where $f(t) = \begin{cases} t^2, & 0 \le t < 1, \\ 0, & t \ge 1, \end{cases} \quad y(0) = 1.$

18. $f(t) = 1 - 2\displaystyle\int_0^t e^{-3\tau}f(t - \tau)\, d\tau$

19. $y'(t) = \cos t + \displaystyle\int_0^t y(\tau)\cos(t - \tau)\, d\tau, \quad y(0) = 1$

20. A series circuit contains an inductor, resistor, and capacitor for which $L = 1/2$ henry, $R = 10$ ohms, and $C = 0.01$ farad, respectively. The voltage

$$E(t) = \begin{cases} 10, & 0 \le t < 5, \\ 0, & t \ge 5, \end{cases}$$

is applied to the circuit. Determine the instantaneous charge $q(t)$ on the capacitor for $t > 0$ if $q(0) = 0$ and $q'(0) = 0$.

Systems of Linear Differential Equations

8.1 The Operator Method

Until now we have been considering methods of solution for single ordinary differential equations. However, in practice, a physical situation may demand that more than one differential equation be used in its mathematical description. In Chapter 5, for example, we saw that the vibrations of a mass attached to a spring could be described by one relatively simple equation; were we to attach two such springs we would then need two coupled, or simultaneous, differential equations to represent the motion.

If we suppose x and y are functions of the independent variable t, then

$$x' - 3x + y' + y = 5 \qquad 4\frac{d^2x}{dt^2} = -5x + y$$

and

$$x + y' + 3y = t \qquad 2\frac{d^2y}{dt^2} = 3x - y,$$

are examples of simultaneous differential equations.

Solution

A **solution** of a system is a set of functions $x = f(t)$, $y = g(t)$, $z = h(t)$, and so on, which satisfies each equation of the system on some interval I.

Linear systems

Note that each differential equation in the foregoing two examples is linear in the dependent variables x and y. Throughout this chapter we shall confine our attention to the solution of systems of linear differential equations, or simply, **linear systems**, in which all coefficients are constants.

Systematic elimination

The first technique that we shall consider for solving such systems is based on the fundamental principle of systematic algebraic elimination of variables. We shall see that the analogue of multiplying an algebraic equation by a

324

constant is operating on a differential equation with some combination of derivatives. Recall, a linear differential equation

$$a_n y^{(n)} + a_{n-1} y^{(n-1)} + \cdots + a_1 y' + a_0 y = g(t)$$

where the a_i, $i = 0, 1, \ldots, n$ are constants, can be written as

$$(a_n D^n + a_{n-1} D^{n-1} + \cdots + a_1 D + a_0) y = g(t).$$

EXAMPLE

Write the system of differential equations

$$x'' + 2x' + y'' = x + 3y + \sin t$$

$$x' + y' = -4x + 2y + e^{-t}$$

in operator notation.

Solution: Rewrite the given system as

$$x'' + 2x' - x + y'' - 3y = \sin t$$

$$x' + 4x + y' - 2y = e^{-t}$$

so that

$$(D^2 + 2D - 1)x + (D^2 - 3)y = \sin t$$

$$(D + 4)x + (D - 2)y = e^{-t}.$$

Method of solution

Consider the simple system of linear first-order equations

$$\begin{aligned} Dy &= 2x \\ Dx &= 3y \end{aligned} \tag{1}$$

or equivalently

$$\begin{aligned} 2x - Dy &= 0 \\ Dx - 3y &= 0. \end{aligned} \tag{2}$$

Operating on the first equation in (2) by D while multiplying the second by 2 and then subtracting will eliminate x from the system. It follows that

$$-D^2 y + 6y = 0 \quad \text{or} \quad D^2 y - 6y = 0.$$

Since the roots of the auxiliary equation are $m_1 = \sqrt{6}$ and $m_2 = -\sqrt{6}$ we obtain

$$y(t) = c_1 e^{\sqrt{6}t} + c_2 e^{-\sqrt{6}t}. \tag{3}$$

Whereas multiplying the first equation by -3 while operating on the second by D and then adding gives the differential equation for x, $D^2 x - 6x = 0$. It follows immediately that

$$x(t) = c_3 e^{\sqrt{6}t} + c_4 e^{-\sqrt{6}t}. \tag{4}$$

Now (3) and (4) do not satisfy the system (1) for every choice of c_1, c_2, c_3, and c_4. Substituting $x(t)$ and $y(t)$ into the first equation of the original system (1) gives

$$c_1 \sqrt{6} e^{\sqrt{6}t} - c_2 \sqrt{6} e^{-\sqrt{6}t} = 2c_3 e^{\sqrt{6}t} + 2c_4 e^{-\sqrt{6}t}$$

or

$$(\sqrt{6}c_1 - 2c_3)e^{\sqrt{6}t} + (-\sqrt{6}c_2 - 2c_4)e^{-\sqrt{6}t} = 0.$$

Since the latter expression is to be zero for all values of t we must have

$$\sqrt{6}c_1 - 2c_3 = 0$$

$$-\sqrt{6}c_2 - 2c_4 = 0$$

or
$$c_3 = \frac{\sqrt{6}}{2}c_1, \qquad c_4 = -\frac{\sqrt{6}}{2}c_2. \tag{5}$$

Hence we conclude that a solution of the system must be

$$x(t) = \frac{\sqrt{6}}{2}c_1 e^{\sqrt{6}t} - \frac{\sqrt{6}}{2}c_2 e^{-\sqrt{6}t}$$

$$y(t) = c_1 e^{\sqrt{6}t} + c_2 e^{-\sqrt{6}t}.$$

The reader is urged to substitute (3) and (4) into the second equation of (1) and verify that the same relationship (5) between the constants obtains.

EXAMPLE

Solve
$$Dx + (D + 2)y = 0$$
$$(D - 3)x - \quad 2y = 0. \tag{6}$$

Solution: Operating on the first equation by $D - 3$ and on the second by D and subtracting eliminates x from the system. It follows that the differential equation for y is

$$[(D - 3)(D + 2) + 2D]y = 0 \quad \text{or} \quad (D^2 + D - 6)y = 0.$$

Since the characteristic equation of this last differential equation is $m^2 + m - 6 = (m - 2)(m + 3) = 0$, we obtain the solution

$$y(t) = c_1 e^{2t} + c_2 e^{-3t}. \tag{7}$$

Eliminating y in a similar manner yields $(D^2 + D - 6)x = 0$ from which we find

$$x(t) = c_3 e^{2t} + c_4 e^{-3t}. \tag{8}$$

As we noted in the foregoing discussion, a solution of (6) does not contain four independent constants since the system itself puts a constraint on the actual number which can be chosen arbitrarily. Substituting (7) and (8) into the first equation of (6) gives

$$2c_3 e^{2t} - 3c_4 e^{-3t} + 2c_1 e^{2t} - 3c_2 e^{-3t} + 2c_1 e^{2t} + 2c_2 e^{-3t}$$

$$= (4c_1 + 2c_3)e^{2t} + (-c_2 - 3c_4)e^{-3t}$$

$$= 0$$

and so
$$4c_1 + 2c_3 = 0$$
$$-c_2 - 3c_4 = 0.$$

Therefore
$$c_3 = -2c_1 \quad \text{and} \quad c_4 = -\tfrac{1}{3}c_2.$$

Accordingly, a solution of the system is

$$x(t) = -2c_1 e^{2t} - \tfrac{1}{3}c_2 e^{-3t}$$

$$y(t) = \quad c_1 e^{2t} + \quad c_2 e^{-3t}.$$

Since we could just as easily solve for c_3 and c_4 in terms of c_1 and c_2, the solution of the preceding example can be written in the alternative form

$$x(t) = c_3 e^{2t} + c_4 e^{-3t}$$

$$y(t) = -\tfrac{1}{2} c_3 e^{2t} - 3c_4 e^{-3t}.$$

Also, it sometimes pays to keep one's eyes open when solving systems. Had we solved for x first, then y, and the relationship between the constants could be determined by simply using the last equation in the form

$$y = \tfrac{1}{2}(D - 3)x$$

$$= \tfrac{1}{2}(Dx - 3x)$$

$$= \tfrac{1}{2}[2c_3 e^{2t} - 3c_4 e^{-3t} - 3c_3 e^{2t} - 3c_4 e^{-3t}]$$

$$= -\tfrac{1}{2} c_3 e^{2t} - 3c_4 e^{-3t}.$$

EXAMPLE   Solve
$$x' - 4x + y'' = t^2$$
$$x' + x + y' = 0. \tag{9}$$

Solution: First write the system in differential operator notation

$$(D - 4)x + D^2 y = t^2$$
$$(D + 1)x + Dy = 0. \tag{10}$$

Then by eliminating x we obtain

$$[(D + 1)D^2 - (D - 4)D]y = (D + 1)t^2 - (D - 4)0$$

or $(D^3 + 4D)y = t^2 + 2t.$

Since the roots of the auxiliary equation $m(m^2 + 4) = 0$ are $m_1 = 0$, $m_2 = 2i$, and $m_3 = -2i$, the complementary function is

$$y_c = c_1 + c_2 \cos 2t + c_3 \sin 2t.$$

To determine the particular solution y_p we use undetermined coefficients by assuming

$$y_p = At^3 + Bt^2 + Ct.$$

Therefore $y_p' = 3At^2 + 2Bt + C$

$$y_p'' = 6At + 2B$$

$$y_p''' = 6A,$$

$$y_p''' + 4y_p' = 12At^2 + 8Bt + 6A + 4C$$

$$= t^2 + 2t.$$

The last equality implies

$$12A = 1, \qquad 8B = 2, \qquad 6A + 4C = 0$$

and hence $A = \tfrac{1}{12}, \qquad B = \tfrac{1}{4}, \qquad\qquad C = -\tfrac{1}{8}.$

Thus $y = y_c + y_p = c_1 + c_2 \cos 2t + c_3 \sin 2t + \frac{1}{12}t^3 + \frac{1}{4}t^2 - \frac{1}{8}t.$

$$(11)$$

Eliminating y from the system (10) leads to

$$[(D - 4) - D(D + 1)]x = t^2 \quad \text{or} \quad (D^2 + 4)x = -t^2.$$

It should be obvious that

$$x_c = c_4 \cos 2t + c_5 \sin 2t$$

and that undetermined coefficients can be applied to obtain a particular solution of the form

$$x_p = At^2 + Bt + C.$$

In this case the usual differentiations and algebra yield

$$x_p = -\tfrac{1}{4}t^2 + \tfrac{1}{8}$$

and so $x = x_c + x_p$

$$= c_4 \cos 2t + c_5 \sin 2t - \tfrac{1}{4}t^2 + \tfrac{1}{8}. \qquad (12)$$

Now c_4 and c_5 can be expressed in terms of c_2 and c_3 by substituting (11) and (12) into either equation of (9). By using the second equation we find after combining terms

$$(c_5 - 2c_4 - 2c_2)\sin 2t + (2c_5 + c_4 + 2c_3)\cos 2t = 0$$

so that $c_5 - 2c_4 - 2c_2 = 0$

$$2c_5 + c_4 + 2c_3 = 0.$$

Solving for c_4 and c_5 in terms of c_2 and c_3 gives

$$c_4 = -\tfrac{1}{5}(4c_2 + 2c_3) \quad \text{and} \quad c_5 = \tfrac{1}{5}(2c_2 - 4c_3).$$

Finally, a solution of (9) is found to be

$$x(t) = -\tfrac{1}{5}(4c_2 + 2c_3)\cos 2t + \tfrac{1}{5}(2c_2 - 4c_3)\sin 2t - \tfrac{1}{4}t^2 + \tfrac{1}{8}$$

$$y(t) = c_1 + c_2 \cos 2t + c_3 \sin 2t + \tfrac{1}{12}t^3 + \tfrac{1}{4}t^2 - \tfrac{1}{8}t.$$

Use of determinants   Symbolically, if $L_1, L_2, L_3,$ and L_4 denote differential operators with constant coefficients, then a system of linear differential equations in two variables x and y can be written as

$$L_1x + L_2y = g_1(t)$$
$$L_3x + L_4y = g_2(t). \qquad (13)$$

Eliminating variables, as we would for algebraic equations, leads to

$$(L_1L_4 - L_2L_3)x = f_1(t) \quad \text{and} \quad (L_1L_4 - L_2L_3)y = f_2(t) \qquad (14)$$

where

$$f_1(t) = L_4g_1(t) - L_2g_2(t) \quad \text{and} \quad f_2(t) = L_1g_2(t) - L_3g_1(t).$$

Formally, the results in (14) can be written in terms of determinants similar to that used in Cramer's rule:

$$\begin{vmatrix} L_1 & L_2 \\ L_3 & L_4 \end{vmatrix} x = \begin{vmatrix} g_1 & L_2 \\ g_2 & L_4 \end{vmatrix} \quad \text{and} \quad \begin{vmatrix} L_1 & L_2 \\ L_3 & L_4 \end{vmatrix} y = \begin{vmatrix} L_1 & g_1 \\ L_3 & g_2 \end{vmatrix}. \tag{15}$$

The left-hand determinant in each equation in (15) can be expanded in the usual algebraic sense, the result then operating on the functions $x(t)$ and $y(t)$. However, some care should be excercised in the expansion of the the right-hand determinants in (15). We must expand these determinants in the sense of the internal differential operators actually operating upon the functions $g_1(t)$ and $g_2(t)$.

If $$\begin{vmatrix} L_1 & L_2 \\ L_3 & L_4 \end{vmatrix} \neq 0$$

in (15) and is a differential operator of order n, then

(a) The system (13) can be decoupled into two nth-order differential equations in x and y.

(b) The characteristic equation and hence the complementary function of each of these differential equations is the same.

(c) Since x and y both contain n constants, there is a total of $2n$ constants appearing.

(d) The total number of *independent* constants in the solution of the system is n.

If $$\begin{vmatrix} L_1 & L_2 \\ L_3 & L_4 \end{vmatrix} \equiv 0$$

in (13) then the system may have a solution containing any number of independent constants or may have no solution at all. Similar remarks hold for systems larger than indicated in (13).

EXAMPLE Solve

$$x' = 3x - y - 1$$
$$y' = x + y + 4e^t. \tag{16}$$

Solution: Write the system in terms of differential operators

$$(D - 3)x + \qquad y = -1$$
$$-x + (D - 1)y = 4e^t$$

and then use determinants

$$\begin{vmatrix} D - 3 & 1 \\ -1 & D - 1 \end{vmatrix} x = \begin{vmatrix} -1 & 1 \\ 4e^t & D - 1 \end{vmatrix}$$

$$\begin{vmatrix} D - 3 & 1 \\ -1 & D - 1 \end{vmatrix} y = \begin{vmatrix} D - 3 & -1 \\ -1 & 4e^t \end{vmatrix}.$$

After expanding we find that

$$(D - 2)^2 x = 1 - 4e^t$$
$$(D - 2)^2 y = -1 - 8e^t.$$

By the usual methods it follows that

$$x = x_c + x_p$$
$$= c_1 e^{2t} + c_2 t e^{2t} + \tfrac{1}{4} - 4e^t \tag{17}$$
$$y = y_c + y_p$$
$$= c_3 e^{2t} + c_4 t e^{2t} - \tfrac{1}{4} - 8e^t. \tag{18}$$

Substituting (17) and (18) into the second equation of (16) gives

$$(c_3 - c_1 + c_4)e^{2t} + (c_4 - c_2)t e^{2t} = 0$$

which then implies

$$c_4 = c_2 \qquad \text{and} \qquad c_3 = c_1 - c_4 = c_1 - c_2.$$

Thus a solution of (16) is

$$x(t) = c_1 e^{2t} + c_2 t e^{2t} + \tfrac{1}{4} - 4e^t$$
$$y(t) = (c_1 - c_2)e^{2t} + c_2 t e^{2t} - \tfrac{1}{4} - 8e^t.$$

EXAMPLE

Given the system

$$Dx + \qquad\qquad Dz = t^2$$
$$2x + D^2 y \qquad\qquad = e^t$$
$$-2Dx - 2y + (D + 1)z = 0$$

find the differential equation for the variable y.

Solution: By determinants we can write

$$\begin{vmatrix} D & 0 & D \\ 2 & D^2 & 0 \\ -2D & -2 & D+1 \end{vmatrix} y = \begin{vmatrix} D & t^2 & D \\ 2 & e^t & 0 \\ -2D & 0 & D+1 \end{vmatrix}.$$

In turn, expanding each determinant by cofactors of the first row gives

$$\left\{ D \begin{vmatrix} D^2 & 0 \\ -2 & D+1 \end{vmatrix} + D \begin{vmatrix} 2 & D^2 \\ -2D & -2 \end{vmatrix} \right\} y = D \begin{vmatrix} e^t & 0 \\ 0 & D+1 \end{vmatrix}$$
$$- \begin{vmatrix} 2 & 0 \\ -2D & D+1 \end{vmatrix} t^2 + D \begin{vmatrix} 2 & e^t \\ -2D & 0 \end{vmatrix}$$

or

$$D(3D^3 + D^2 - 4)y = 4e^t - 2t^2 - 4t.$$

Again, we remind the reader that the D symbol in the left-hand brace is to be treated as an algebraic quantity, but this is not the case on the right-hand side.

EXERCISES 8.1   *Answers to odd-numbered problems begin on page A-24.*

In Problems 1–22 solve, if possible, the given system of differential equations by either systematic elimination or determinants.

1. $\dfrac{dx}{dt} = 2x - y$

$\dfrac{dy}{dt} = x$

2. $\dfrac{dx}{dt} = 4x + 7y$

$\dfrac{dy}{dt} = x - 2y$

3. $\dfrac{dx}{dt} = -y + t$

$\dfrac{dy}{dt} = x - t$

4. $\dfrac{dx}{dt} - 4y = 1$

$x + \dfrac{dy}{dt} = 2$

5. $(D^2 + 5)x - \qquad 2y = 0$

$\qquad -2x + (D^2 + 2)y = 0$

6. $(D + 1)x + (D - 1)y = 2$

$\qquad 3x + (D + 2)y = -1$

7. $\dfrac{d^2x}{dt^2} = 4y + e^t$

$\dfrac{d^2y}{dt^2} = 4x - e^t$

★8. $\dfrac{d^2x}{dt^2} + \dfrac{dy}{dt} = -5x$

$\dfrac{dx}{dt} + \dfrac{dy}{dt} = -x + 4y$

9. $\qquad Dx + \qquad D^2y = e^{3t}$

$(D + 1)x + (D - 1)y = 4e^{3t}$

10. $\qquad D^2x - \qquad Dy = t$

$(D + 3)x + (D + 3)y = 2$

11. $(D^2 - 1)x - \quad y = 0$

$(D - 1)x + Dy = 0$

12. $(2D^2 - D - 1)x - (2D + 1)y = 1$

$(D - 1)x + \qquad Dy = -1$

13. $2\dfrac{dx}{dt} - 5x + \dfrac{dy}{dt} = e^t$

$\dfrac{dx}{dt} - x + \dfrac{dy}{dt} = 5e^t$

14. $\dfrac{dx}{dt} + \dfrac{dy}{dt} \qquad = e^t$

$-\dfrac{d^2x}{dt^2} + \dfrac{dx}{dt} + x + y = 0$

15. $(D - 1)x + (D^2 + 1)y = 1$

$(D^2 - 1)x + (D + 1)y = 2$

16. $D^2x - 2(D^2 + D)y = \sin t$

$x + \qquad Dy = 0$

17. $Dx = y$

$Dy = z$

$Dz = x$

18. $\qquad Dx + \qquad z = e^t$

$(D - 1)x + Dy + Dz = 0$

$\qquad x + 2y + Dz = e^t$

19. $\dfrac{dx}{dt} - 6y \quad\quad = 0$

$x - \dfrac{dy}{dt} + z = 0$

$x + y - \dfrac{dz}{dt} = 0$

★20. $\dfrac{dx}{dt} = -x + z$

$\dfrac{dy}{dt} = -y + z$

$\dfrac{dz}{dt} = -x + y$

21. $2Dx + (D - 1)y = t$

$Dx + \quad\quad Dy = t^2$

★22. $\quad\quad Dx - \quad\quad 2Dy = t^2$

$(D + 1)x - 2(D + 1)y = 1$

23. Determine, if possible, a system of differential equations having

$$x(t) = c_1 + c_2 e^{2t}$$

$$y(t) = -c_1 + c_2 e^{2t}$$

as its solution.

8.2   The Laplace Transform Method

When initial conditions are specified, the Laplace transform will reduce a system of linear differential equations with constant coefficients to a set of simultaneous algebraic equations in the transformed functions.

EXAMPLE

Solve

$$2x' + y' - y = t$$
$$x' + y' \quad\quad = t^2 \tag{1}$$

subject to $x(0) = 1$, $y(0) = 0$.

Solution: If $X(s) = \mathcal{L}\{x(t)\}$ and $Y(s) = \mathcal{L}\{y(t)\}$, then after transforming each equation we obtain

$$2[sX(s) - x(0)] + sY(s) - y(0) - Y(s) = \frac{1}{s^2}$$

$$sX(s) - x(0) + sY(s) - y(0) \quad\quad = \frac{2}{s^3}$$

or

$$2sX(s) + (s - 1)Y(s) = 2 + \frac{1}{s^2}$$

$$sX(s) + \quad\quad sY(s) = 1 + \frac{2}{s^3}.$$

Multiplying the second equation of (2) by 2 and subtracting yields

$$(-s - 1)Y(s) = \frac{1}{s^2} - \frac{4}{s^3}$$

$$Y(s) = \frac{4-s}{s^3(s+1)}. \tag{3}$$

Now by partial fractions

$$\frac{4-s}{s^3(s+1)} = \frac{A}{s} + \frac{B}{s^2} + \frac{C}{s^3} + \frac{D}{s+1}$$

so that

$$4-s = As^2(s+1) + Bs(s+1) + C(s+1) + Ds^3.$$

Setting $s = 0$ and $s = -1$ in the last line gives $C = 4$ and $D = -5$, respectively; whereas equating the coefficients of s^3 and s^2 on each side of the equality yields

$$A + D = 0 \qquad \text{and} \qquad A + B = 0.$$

It follows that $A = 5$, $B = -5$. Thus (3) becomes

$$Y(s) = \frac{5}{s} - \frac{5}{s^2} + \frac{4}{s^3} - \frac{5}{s+1}$$

and so

$$y(t) = 5\mathcal{L}^{-1}\left\{\frac{1}{s}\right\} - 5\mathcal{L}^{-1}\left\{\frac{1}{s^2}\right\} + 2\mathcal{L}^{-1}\left\{\frac{2!}{s^3}\right\} - 5\mathcal{L}^{-1}\left\{\frac{1}{s+1}\right\}$$

$$= 5 - 5t + 2t^2 - 5e^{-t}.$$

By the second equation of (2)

$$X(s) = -Y(s) + \frac{1}{s} + \frac{2}{s^4}$$

from which it follows that

$$x(t) = -\mathcal{L}^{-1}\{Y(s)\} + \mathcal{L}^{-1}\left\{\frac{1}{s}\right\} + \frac{2}{3!}\mathcal{L}^{-1}\left\{\frac{3!}{s^4}\right\}$$

$$= -4 + 5t - 2t^2 + \tfrac{1}{3}t^3 + 5e^{-t}.$$

Hence we conclude that the solution of the given system (1) is

$$x(t) = -4 + 5t - 2t^2 + \tfrac{1}{3}t^3 + 5e^{-t}$$

$$y(t) = \quad 5 - 5t + 2t^2 - 5e^{-t}. \tag{4}$$

Applications

Let us turn now to some elementary applications involving systems of differential equations. The solutions of the problems that we shall consider can be obtained either by the method of the preceding section or through the use of the Laplace transformation. We shall confine our attention to the latter method.

Coupled springs

Suppose two masses m_1 and m_2 are attached to two springs A and B having spring constants k_1 and k_2, respectively. In turn the springs are connected and the resulting system is set in motion. Let $x_1(t)$ and $x_2(t)$ represent the vertical displacements of the two masses beyond their equilibrium positions as shown in Figure 8.1. We saw in Section 1.2 that the vertical motion in a straight line through the centers of mass is described by the system of linear second-order equations.

$$m_1 x_1'' = -k_1 x_1 + k_2(x_2 - x_1)$$
$$m_2 x_2'' = -k_2(x_2 - x_1). \tag{5}$$

Recall that the right side of the first equation is the net force acting on m_1 due to the elongation of spring A and the net elongation of spring B; whereas the right side of the second equation in (5) is the force acting on m_2 due only to net elongation of spring B.

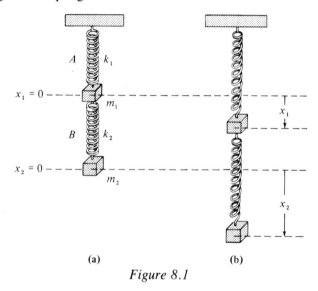

(a)   (b)

Figure 8.1

In the next example we shall solve the system (5) under the assumption
$$k_1 = 6, \qquad k_2 = 4, \qquad m_1 = 1, \qquad m_2 = 1$$
and that the masses start from their equilibrium positions with opposite unit velocities.

EXAMPLE   Solve
$$x_1'' + 10x_1 \qquad - 4x_2 = 0$$
$$- 4x_1 + x_2'' + 4x_2 = 0. \tag{6}$$

subject to
$$x_1(0) = 0 \qquad x_1'(0) = 1 \qquad x_2(0) = 0 \qquad x_2'(0) = -1.$$

Solution: The Laplace transform of each equation is
$$s^2 X_1(s) - sx_1(0) - x_1'(0) + 10X_1(s) - 4X_2(s) = 0$$
$$-4X_1(s) + s^2 X_2(s) - sx_2(0) - x_2'(0) + 4X_2(s) = 0$$

where $X_1(s) = \mathcal{L}\{x_1(t)\}$ and $X_2(s) = \mathcal{L}\{x_2(t)\}$. The preceding system is the same as

$$(s^2 + 10)X_1(s) - \qquad 4X_2(s) = 1$$
$$-4X_1(s) + (s^2 + 4)X_2(s) = -1. \qquad (7)$$

Eliminating X_2 gives

$$X_1(s) = \frac{s^2}{(s^2 + 2)(s^2 + 12)}.$$

By partial fractions we can write

$$\frac{s^2}{(s^2 + 2)(s^2 + 12)} = \frac{As + B}{s^2 + 2} + \frac{Cs + D}{s^2 + 12}$$

and $\qquad s^2 = (As + B)(s^2 + 12) + (Cs + D)(s^2 + 2).$

Comparing coefficients of s on each side of the last equality gives

$$A + C = 0$$
$$B + D = 1$$
$$12A + 2C = 0$$
$$12B + 2D = 0$$

so that $\qquad A = 0, \qquad C = 0, \qquad B = -\tfrac{1}{5}, \qquad D = \tfrac{6}{5}.$

Hence $\qquad X_1(s) = -\dfrac{1/5}{s^2 + 2} + \dfrac{6/5}{s^2 + 12}$

and therefore

$$x_1(t) = -\frac{1}{5\sqrt{2}}\mathcal{L}^{-1}\left\{\frac{\sqrt{2}}{s^2 + 2}\right\} + \frac{6}{5\sqrt{12}}\mathcal{L}^{-1}\left\{\frac{\sqrt{12}}{s^2 + 12}\right\}$$

$$= -\frac{\sqrt{2}}{10}\sin\sqrt{2}t + \frac{\sqrt{3}}{5}\sin 2\sqrt{3}t.$$

From the first equation of (7) it follows that

$$X_2(s) = -\frac{s^2 + 6}{(s^2 + 2)(s^2 + 12)}.$$

Proceeding as before with partial fractions we find

$$X_2(s) = -\frac{2/5}{s^2 + 2} - \frac{3/5}{s^2 + 12}$$

and so

$$x_2(t) = -\frac{2}{5\sqrt{2}}\mathcal{L}^{-1}\left\{\frac{\sqrt{2}}{s^2 + 2}\right\} - \frac{3}{5\sqrt{12}}\mathcal{L}^{-1}\left\{\frac{\sqrt{12}}{s^2 + 12}\right\}$$

$$= -\frac{\sqrt{2}}{5}\sin\sqrt{2}t - \frac{\sqrt{3}}{10}\sin 2\sqrt{3}t.$$

Finally, the solution to the given system (6) is

$$x_1(t) = -\frac{\sqrt{2}}{10} \sin \sqrt{2}t + \frac{\sqrt{3}}{5} \sin 2\sqrt{3}t$$

$$x_2(t) = -\frac{\sqrt{2}}{5} \sin \sqrt{2}t - \frac{\sqrt{3}}{10} \sin \sqrt{3}t \tag{8}$$

Networks

An electrical network having more than one loop also gives rise to simultaneous differential equations. As shown in Figure 8.2, the current $i_1(t)$ splits in the directions shown at point B_1 called a *branch point* of the network. By Kirchoff's first law we can write

$$i_1(t) = i_2(t) + i_3(t). \tag{9}$$

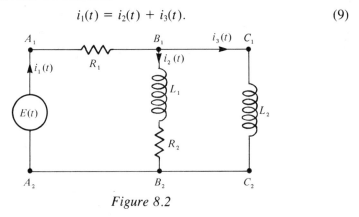

Figure 8.2

In addition, we can also apply *Kirchoff's second law* to each loop. For loop $A_1B_1B_2A_2A_1$, summing the voltage drops across each part of the loop gives

$$E(t) = i_1R_1 + L_1\frac{di_2}{dt} + i_2R_2. \tag{10}$$

Similarly, for loop $A_1B_1C_1C_2B_2A_2A_1$, we find

$$E(t) = i_1R_1 + L_2\frac{di_3}{dt}. \tag{11}$$

Using (9) to eliminate i_1 in (10) and (11) yields two first-order equations for the currents $i_2(t)$ and $i_3(t)$

$$L_1\frac{di_2}{dt} + (R_1 + R_2)i_2 + R_1i_3 = E(t)$$

$$L_2\frac{di_3}{dt} + \qquad R_1i_2 + R_1i_3 = E(t). \tag{12}$$

Given the natural initial conditions $i_2(0) = 0$, $i_3(0) = 0$, the system (12) is amenable to solution by the Laplace transform.

We leave it as an exercise (see Problem 16) to show that the system of differential equations describing the currents $i_1(t)$ and $i_2(t)$ in the network containing a resistor, inductor, and capacitor shown in Figure 8.3 is

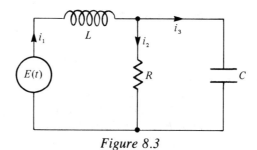

Figure 8.3

$$L\frac{di_1}{dt} + Ri_2 = E(t)$$

$$RC\frac{di_2}{dt} + i_2 - i_1 = 0.$$

(13)

EXAMPLE   Solve the system (13) under the conditions $E = 60$ volts, $L = 1$ henry, $R = 50$ ohms, $C = 10^{-4}$ farads, and i_1 and i_2 are initially zero.

Solution: We must solve

$$\frac{di_1}{dt} + 50i_2 = 60$$

$$50(10^{-4})\frac{di_2}{dt} + i_2 - i_1 = 0$$

subject to $i_1(0) = 0$, $i_2(0) = 0$.

Applying the Laplace transform to each equation of the system and simplifying gives

$$sI_1(s) + 50I_2(s) = \frac{60}{s}$$

$$-200I_1(s) + (s + 200)I_2(s) = 0$$

where $I_1(s) = \mathcal{L}\{i_1(t)\}$ and $I_2(s) = \mathcal{L}\{i_2(t)\}$. Solving the system for I_1 and I_2 yields

$$I_1(s) = \frac{60s + 12{,}000}{s(s + 100)^2}$$

$$I_2(s) = \frac{12{,}000}{s(s + 100)^2}.$$

By partial fractions we can write

$$I_1(s) = \frac{6/5}{s} - \frac{6/5}{s + 100} - \frac{60}{(s + 100)^2}$$

$$I_2(s) = \frac{6/5}{s} - \frac{6/5}{s + 100} - \frac{120}{(s + 100)^2}$$

from which it follows that

$$i_1(t) = \tfrac{6}{5} - \tfrac{6}{5}e^{-100t} - 60te^{-100t}$$

$$i_2(t) = \tfrac{6}{5} - \tfrac{6}{5}e^{-100t} - 120te^{-100t}.$$

Note that both $i_1(t)$ and $i_2(t)$ in the preceding example tend toward the value $E/R = 6/5$ as $t\to\infty$. Furthermore, since the current through the capacitor is $i_3(t) = i_1(t) - i_2(t) = 60te^{-100t}$ we observe $i_3(t)\to 0$ as $t\to\infty$.

EXERCISES 8.2   *Answers to odd-numbered problems begin on page A–25.*

In Problems 1–12 use the Laplace transform to solve the given system of differential equations.

1. $\dfrac{dx}{dt} = -x + y$

$\dfrac{dy}{dt} = 2x,$

$x(0) = 0, \quad y(0) = 1$

★2. $\dfrac{dx}{dt} = 2y + e^t$

$\dfrac{dy}{dt} = 8x - t,$

$x(0) = 1, \quad y(0) = 1$

3. $\dfrac{dx}{dt} = x - 2y$

$\dfrac{dy}{dt} = 5x - y,$

$x(0) = -1, \quad y(0) = 2$

4. $\dfrac{dx}{dt} + 3x + \dfrac{dy}{dt} = 1$

$\dfrac{dx}{dt} - x + \dfrac{dy}{dt} - y = e^t,$

$x(0) = 0, \quad y(0) = 0$

5. $2\dfrac{dx}{dt} + \dfrac{dy}{dt} - 2x = 1$

$\dfrac{dx}{dt} + \dfrac{dy}{dt} - 3x - 3y = 2,$

$x(0) = 0, \quad y(0) = 0$

★6. $\dfrac{dx}{dt} + x - \dfrac{dy}{dt} + y = 0$

$\dfrac{dx}{dt} + \dfrac{dy}{dt} + 2y = 0,$

$x(0) = 0, \quad y(0) = 1$

7. $\dfrac{d^2x}{dt^2} + x - y = 0$

$\dfrac{d^2y}{dt^2} + y - x = 0,$

$x(0) = 0, \quad x'(0) = -2,$

$y(0) = 0, \quad y'(0) = 1$

8. $\dfrac{d^2x}{dt^2} + \dfrac{dx}{dt} + \dfrac{dy}{dt} = 0$

$\dfrac{d^2y}{dt^2} + \dfrac{dy}{dt} - 4\dfrac{dx}{dt} = 0,$

$x(0) = 1, \quad x'(0) = 0,$

$y(0) = -1, \quad y'(0) = 5$

9. $\dfrac{d^2x}{dt^2} + \dfrac{d^2y}{dt^2} = t^2$

$\dfrac{d^2x}{dt^2} - \dfrac{d^2y}{dt^2} = 4t,$

$x(0) = 8, \quad x'(0) = 0,$

$y(0) = 0, \quad y'(0) = 0$

10. $\dfrac{dx}{dt} - 4x + \dfrac{d^3y}{dt^3} = 6 \sin t$

$\dfrac{dx}{dt} + 2x - 2\dfrac{d^3y}{dt^3} = 0,$

$x(0) = 0, \quad y(0) = 0,$

$y'(0) = 0, \quad y''(0) = 0$

11. $\dfrac{d^2x}{dt^2} + 3\dfrac{dy}{dt} + 3y = 0$

$\dfrac{d^2x}{dt^2} + 3y = te^{-t},$

$x(0) = 0, \quad x'(0) = 2, \quad y(0) = 0$

★12. $\dfrac{dx}{dt} = 4x - 2y + 2\mathcal{U}(t - 1)$

$\dfrac{dy}{dt} = 3x - y + \mathcal{U}(t - 1),$

$x(0) = 0, \quad y(0) = \tfrac{1}{2}$

13. Solve system (5) when

$$k_1 = 3, \quad k_2 = 2, \quad m_1 = 1, \quad m_2 = 1$$

and

$$x_1(0) = 0, \qquad x_1'(0) = 1$$
$$x_2(0) = 1, \qquad x_2'(0) = 0.$$

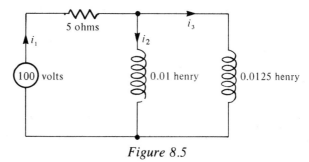

Figure 8.4

14. Derive the system of differential equations describing the straight line vertical motion of the coupled springs shown in Figure 8.4. Use the Laplace transform to solve the system when $k_1 = 1$, $k_2 = 1$, $k_3 = 1$, $m_1 = 1$, $m_2 = 1$, and $x_1(0) = 0$, $x_1'(0) = -1$, $x_2(0) = 0$, $x_2'(0) = 1$.

15. Find the current $i_1(t)$ at any time in the network shown in Figure 8.5. Assume $i_2(0) = 0$, $i_3(0) = 0$.

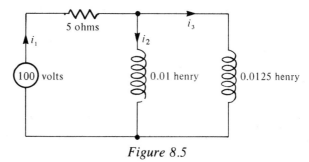

Figure 8.5

★16. Derive the system of equations (13).

17. Solve (13) when $E = 60$ volts, $L = 1/2$ henry, $R = 50$ ohms, $C = 10^{-4}$ farads, $i_1(0) = 0$, $i_2(0) = 0$.

18. Solve (13) when $E = 60$ volts, $L = 2$ henry, $R = 50$ ohms, $C = 10^{-4}$ farads, $i_1(0) = 0$, $i_2(0) = 0$.

19. Solve the system given in (12) when $R_1 = 6$ ohms, $R_2 = 5$ ohms, $L_1 = 1$ henry, $L_2 = 1$ henry, $E(t) = 50 \sin t$ volts.

20. The system
$$\frac{dx_1}{dt} = -\frac{2}{25}x_1 + \frac{1}{50}x_2 \qquad \frac{dx_2}{dt} = \frac{2}{25}x_1 - \frac{2}{25}x_2$$

results from the analysis of a particular problem that involves the pumping of a well-mixed salt solution between two tanks (see Section 8.3). Use the Laplace transform to solve the system when $x_1(0) = 25$ and $x_2(0) = 0$.

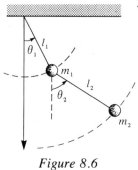

21. A double pendulum oscillates in a vertical plane under the influence of gravity (see Figure 8.6). For small displacements $\theta_1(t)$ and $\theta_2(t)$, it can be shown* that the differential equations of motion are
$$(m_1 + m_2)l_1^2\theta_1'' + m_2l_1l_2\theta_2'' + (m_1 + m_2)l_1g\theta_1 = 0$$
$$m_2l_2^2\theta_2'' + m_2l_1l_2\theta_1'' + m_2l_2g\theta_2 = 0.$$

Figure 8.6

Use the Laplace transform to solve the system when $m_1 = 3$, $m_2 = 1$, $l_i = l_2 = 16$, $\theta_1(0) = 1$, $\theta_2(0) = -1$, $\theta_1'(0) = 0$, $\theta_2'(0) = 0$.

Use the Laplace transform to solve the system when $m_1 = 3$, $m_2 = 1$, $l_i = l_2 = 16$, $\theta_1(0) = 1$, $\theta_2(0) = -1$, $\theta_1'(0) = 0$, $\theta_2'(0) = 0$.

8.3 Systems of Linear First-Order Equations

In the preceding two sections we have dealt with linear systems which were of the form

$$P_{11}(D)x_1 + P_{12}(D)x_2 + \cdots + P_{1n}(D)x_n = b_1(t)$$
$$P_{21}(D)x_1 + P_{22}(D)x_2 + \cdots + P_{2n}(D)x_n = b_2(t)$$
$$\vdots \qquad \vdots \qquad \vdots \qquad \vdots \qquad (1)$$
$$P_{n1}(D)x_1 + P_{n2}(D)x_2 + \cdots + P_{nn}(D)x_n = b_n(t)$$

where the P_{ij} are polynomials in the differential operator D. However, the study of systems of *first-order* differential equations

$$\frac{dx_1}{dt} = g_1(t, x_1, x_2, \ldots, x_n)$$

$$\frac{dx_2}{dt} = g_2(t, x_1, x_2, \ldots, x_n)$$

$$\vdots \qquad (2)$$

$$\frac{dx_n}{dt} = g_n(t, x_1, x_2, \ldots, x_n)$$

*See W. Hauser, *Introduction to the Principles of Mechancis* (Reading, MA: Addison-Wesley, 1965), pp. 268–70.

is particularly important in advanced mathematics since every nth-order differential equation

$$y^{(n)} = F(t, y, y', \ldots, y^{(n-1)})$$

as well as most systems of differential equations can be reduced to form (2).

Linear normal form

Of course, a system such as (2) need not be linear and need not have constant coefficients. Consequently, the system may not be readily solvable, if at all. In the remaining sections of this chapter we shall be interested only in a particular, but important, case of (2), namely, those systems having the linear **normal**, or **canonical**, form

$$\frac{dx_1}{dt} = a_{11}(t)x_1 + a_{12}(t)x_2 + \cdots + a_{1n}(t)x_n + f_1(t)$$

$$\frac{dx_2}{dt} = a_{21}(t)x_1 + a_{22}(t)x_2 + \cdots + a_{2n}(t)x_n + f_2(t)$$

$$\vdots \qquad\qquad \vdots \qquad\quad \vdots \qquad\qquad (3)$$

$$\frac{dx_n}{dt} = a_{n1}(t)x_1 + a_{n2}(t)x_2 + \cdots + a_{nn}(t)x_n + f_n(t).$$

where the coefficients a_{ij} and the f_i are functions continuous on a common interval I. When $f_i(t) = 0$, $i = 1, 2, \ldots, n$, the system (3) is said to be **homogeneous**, otherwise it is called **nonhomogeneous**.

We shall now show that every linear nth-order differential equation with constant coefficients and most linear systems of form (1) can be reduced to a linear system having the normal form (3).

Equation to a system

Suppose a linear nth-order differential equation with constant coefficients is first written as

$$\frac{d^n y}{dt^n} = -\frac{a_0}{a_n}y - \frac{a_1}{a_n}y' - \cdots - \frac{a_{n-1}}{a_n}y^{(n-1)} + f(t). \qquad (4)$$

If we then introduce the variables

$$y = x_1$$

$$y' = x_2$$

$$y'' = x_3 \qquad\qquad (5)$$

$$\vdots$$

$$y^{(n-1)} = x_n$$

it follows that $y' = x_1' = x_2$, $y'' = x_2' = x_3$, and so on. Hence, with the help of (4), the system (5) becomes

$$x_1' = x_2$$
$$x_2' = x_3$$
$$x_3' = x_4$$
$$\vdots \qquad\qquad\qquad (6)$$
$$x_{n-1}' = x_n$$

$$x_n' = -\frac{a_0}{a_n}x_1 - \frac{a_1}{a_n}x_2 - \cdots - \frac{a_{n-1}}{a_n}x_n + f(t).$$

Inspection of (6) reveals that it has the same form as (3).

EXAMPLE

Reduce the third-order equation

$$2y''' - 6y'' + 4y' + y = \sin t$$

to the normal form (3).

Solution: Write the differential equation as

$$y''' = -\tfrac{1}{2}y - 2y' + 3y'' + \tfrac{1}{2}\sin t$$

and then let $y = x_1$, $y' = x_2$, $y'' = x_3$.

Since

$$x_1' = y' = x_2$$
$$x_2' = y'' = x_3$$
$$x_3' = y'''$$

we find that

$$x_1' = x_2$$
$$x_2' = x_3$$
$$x_3' = -\tfrac{1}{2}x_1 - 2x_2 + 3x_3 + \tfrac{1}{2}\sin t.$$

The requirement that the differential equation have constant coefficients is not necessary; a general linear equation can be reduced to a normal form identical with (6) except that some of the a_{ij} would be functions of t (specifically, the coefficients in the last line of the system).

System to an equation

A linear system in normal form with constant coefficients and n dependent variables can be reduced to a single linear nth-order differential equation. Of course, in this case we must make the additional assumption that the $f_i(t)$, $i = 1, 2, \ldots, n$ of (3) are sufficiently differentiable. For example, if the homogeneous system

$$\frac{dx}{dt} = a_{11}x + a_{12}y \qquad \frac{dy}{dt} = a_{21}x + a_{22}y \qquad (7)$$

is written in operator form

$$(D - a_{11})x - \qquad a_{12}y = 0$$

$$- a_{12}x + (D - a_{22})y = 0$$

We can then use the elimination technique of Section 8.1 to obtain one differential equation in either $x(t)$ or $y(t)$.

EXAMPLE   Reduce $$\frac{dx}{dt} = x - y$$

$$\frac{dy}{dt} = 2x - y + t$$

to a single second-order differential equation.

Solution: We write the original system as

$$(D - 1)x + \qquad y = 0$$

$$- 2x + (D + 1)y = t$$

and operate on the first equation by $D + 1$ and subtract. It follows that

$$[(D^2 - 1) + 2]x = -t \qquad \text{or} \qquad (D^2 + 1)x = -t.$$

Systems reduced to normal form

Using a procedure similar to that just outlined, we can reduce *most* systems of the linear form (1) to the linear normal form (3). To accomplish this it is necessary to first solve the system for the highest order derivative of each dependent variable. As we shall see, this may not always be possible.

EXAMPLE   Reduce $$(D^2 - D + 5)x + \qquad 2D^2y = e^t$$

$$- 2x + (D^2 + 2)y = 3t^2$$

to the normal form (3)

Solution: Write the system as

$$D^2x + 2D^2y = e^t - 5x + Dx$$

$$D^2y = 3t^2 + 2x - 2y$$

and then eliminate D^2y by multiplying the second equation by 2 and subtracting. We have

$$D^2x = e^t - 6t^2 - 9x + 4y + Dx.$$

Since the second equation of the system already expresses the highest order derivative of y in terms of the remaining functions, we are now in a position to introduce new variables. If we let

$$Dx = u \quad \text{and} \quad Dy = v$$

the expressions for D^2x and D^2y become, respectively,

$$Du = e^t - 6t^2 - 9x + 4y + u$$

$$Dv = 3t^2 + 2x - 2y.$$

Thus the original system can be written in the normal form

$$Dx = u$$

$$Dy = v$$

$$Du = -9x + 4y + u + e^t - 6t^2$$

$$Dv = 2x - 2y + 3t^2.$$

Degenerate systems

Those systems of differential equations of form (1) which cannot be reduced to a linear system in normal form are said to be **degenerate**. For example, the contradictory system

$$x'' + y' = 1$$

$$x'' + y' = -1$$

is degenerate because it is impossible to solve the system for the highest order derivative of each variable. It should be obvious that this system possesses no solution. However, as the next example shows, a system may be degenerate but yet possess a solution.

EXAMPLE

It is a straightforward matter to show that the system

$$(D + 1)x + (D + 1)y = 0$$

$$2Dx + (2D + 1)y = 0$$

cannot be reduced to the normal form (3), and hence is degenerate. However, by the elimination procedure of Section 8.1, it can also be shown that the system possesses the solution

$$x(t) = c_1 e^{-t}$$

$$y(t) = -2c_1 e^{-t}.$$

By this time the reader may be wondering why anyone would want to convert a single differential equation to a system of equations, or for that matter, a system of differential equations to an even larger system. While we are not in a position to completely justify their importance, suffice it to say that these procedures are more than a theoretical exercise. There are times where it is actually desirable to work with a system rather than with one equation. In the numerical analysis of differential equations, almost all computational algorithms are established for first-order equations. Since these algorithms can be generalized directly to systems, to compute numerically, say, a second-order equation, we could reduce it to a system of two first-order equations (see Chapter 9).

A linear system such as (3) also arises naturally in some physical applications. The following example illustrates a homogeneous system in two dependent variables.

EXAMPLE

Tank A contains 50 gallons of water in which 25 pounds of salt are dissolved. A second tank, B, contains 50 gallons of pure water. Liquid is pumped in and out of the tanks at rates shown in Figure 8.7. Derive the differential equations which describe the number of pounds $x_1(t)$ and $x_2(t)$ of salt at any time in tanks A and B, respectively.

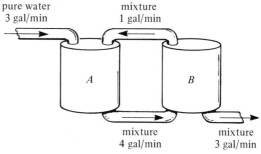

pure water
3 gal/min

mixture
1 gal/min

A B

mixture
4 gal/min

mixture
3 gal/min

Figure 8.7

Solution: By an analysis similar to that used in Section 3.2 we see that the net rate of change in $x_1(t)$ in lb/min is

$$\overbrace{\hspace{5cm}}^{\text{input}} \qquad \overbrace{\hspace{3cm}}^{\text{output}}$$

$$\frac{dx_1}{dt} = (3 \text{ gal/min}) \cdot (0 \text{ lb/gal}) + (1 \text{ gal/min}) \cdot \left(\frac{x_2}{50} \text{ lb/gal}\right) - (4 \text{ gal/min}) \cdot \left(\frac{x_1}{50} \text{ lb/gal}\right)$$

$$= -\frac{2}{25}x_1 + \frac{1}{50}x_2.$$

In addition, we find the net rate of change in $x_2(t)$ is

$$\frac{dx_2}{dt} = 4 \cdot \frac{x_1}{50} - 3 \cdot \frac{x_2}{50} - 1 \cdot \frac{x_2}{50}$$

$$= \frac{2}{25}x_1 - \frac{2}{25}x_2.$$

Thus we obtain the first-order system

$$\frac{dx_1}{dt} = -\frac{2}{25}x_1 + \frac{1}{50}x_2$$

$$\frac{dx_2}{dt} = \frac{2}{25}x_1 - \frac{2}{25}x_2.$$

Observe that the foregoing system is accompanied by the initial conditions $x_1(0) = 25$, $x_2(0) = 0$.

EXERCISES 8.3   *Answers to odd-numbered problems begin on page A–25.*

In Problems 1–6 rewrite the given differential equations as a system in normal form (3).

1. $y'' - 3y' + 4y = \sin 3t$ **2.** $2\dfrac{d^2y}{dt^2} + 4\dfrac{dy}{dt} - 5y = 0$

3. $y''' - 3y'' + 6y' - 10y = t^2 + 1$ **4.** $4y''' + y = e^t$

5. $\dfrac{d^4y}{dt^4} - 2\dfrac{d^2y}{dt^2} + 4\dfrac{dy}{dt} + y = t$ **★6.** $t^2y'' + ty' + (t^2 - 4)y = 0$

7. Reduce the nonlinear equation

$$y''' - yy'' + y^2 = 2t - 1$$

to a system of first-order differential equations.

8. Reduce the system $\begin{aligned} \dfrac{dx}{dt} &= x + 2y \\[2mm] \dfrac{dy}{dt} &= -x + y \end{aligned}$

to a single, second-order differential equation in $x(t)$.

9. Reduce the system $\begin{aligned} \dfrac{dx}{dt} &= 3x - y + 1 \\[2mm] \dfrac{dy}{dt} &= x + 4y - t \end{aligned}$

to a single, second-order differential equation in $y(t)$.

★10. Reduce the system $\begin{aligned} \dfrac{dx}{dt} &= x - y \\[2mm] \dfrac{dy}{dt} &= y - z \\[2mm] \dfrac{dz}{dt} &= -x + z \end{aligned}$

to a single, third-order differential equation in $z(t)$.

In Problems 11–16 rewrite, if possible, the given systems in the normal form (3).

11. $(D - 1)x - Dy = t^2$ **★12.** $x'' - 2y'' = \sin t$

$\qquad x + Dy = 5t - 2$ $\qquad x'' + y'' = \cos t$

13. $(2D + 1)x - 2Dy = 4$

$$Dx - Dy = e^t$$

14. $m_1 x_1'' = -k_1 x_1 + k_2(x_2 - x_1)$

$$m_2 x_2'' = -k_2(x_2 - x_1)$$

15. $\dfrac{d^3 y}{dt^3} = 4x - 3\dfrac{d^2 x}{dt^2} + 4\dfrac{dy}{dt}$

$$\dfrac{d^2 y}{dt^2} = 10t^2 - 4\dfrac{dx}{dt} + 3\dfrac{dy}{dt}$$

16. $D^2 x + Dy = 4t$

$$-D^2 x + (D + 1)y = 6t^2 + 10$$

17. Consider two tanks A and B with liquid being pumped in and out at the same rates as given in the last example of this section. What is the system of differential equations if, instead of pure water, a brine solution containing 2 lb of salt per gallon is pumped into tank A?

18. Using the information given in Figure 8.8, derive the system of differential equations describing the number of pounds of salt x_1, x_2, and x_3 at any time in tanks A, B, and C, respectively.

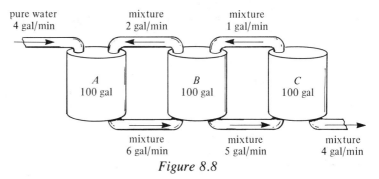

Figure 8.8

Miscellaneous problems

19. Consider the first-order system

$$(a_1 D - b_1)x + (a_2 D - b_2)y = 0$$

$$(a_3 D - b_3)x + (a_4 D - b_4)y = 0$$

where the a_i are nonzero constants. Determine a condition on the a_i such that the system is degenerate.

8.4   Matrices: A Brief Introduction

Matrix

Before examining the systematic procedure for solving linear first-order systems in normal form we need the new and useful concept of a **matrix**.

DEFINITION 8.1   A **matrix A** is any rectangular array of numbers or functions:

$$
\mathbf{A} = \begin{pmatrix} a_{11} & a_{12} & \cdots & a_{1n} \\ a_{21} & a_{22} & \cdots & a_{2n} \\ \vdots & & & \vdots \\ a_{m1} & a_{m2} & \cdots & a_{mn} \end{pmatrix}. \tag{1}
$$

A matrix with m rows and n columns is usually referred to as an $m \times n$ (m by n) matrix. An $n \times n$ matrix is called a *square* matrix.

The element, or entry, in the ith row and jth column of an $m \times n$ matrix $\mathbf{A}$ is written a_{ij}. An $m \times n$ matrix $\mathbf{A}$ is then abbreviated as $\mathbf{A} = (a_{ij})_{m \times n}$ or simply $\mathbf{A} = (a_{ij})$. A 1×1 matrix is simply one constant or function.

Equality

> **DEFINITION 8.2** Two $m \times n$ matrices $\mathbf{A}$ and $\mathbf{B}$ are **equal** if $a_{ij} = b_{ij}$ for each i and j. ☐

Vectors

> **DEFINITION 8.3** A **column matrix** $\mathbf{X}$ is any matrix having n rows and one column:
>
> $$
> \mathbf{X} = \begin{pmatrix} b_{11} \\ b_{21} \\ \vdots \\ b_{n1} \end{pmatrix} = (b_{i1})_{n \times 1}. \qquad ☐
> $$

A column matrix is also called a **column vector** or simply a **vector**.

Multiples of a matrix

> **DEFINITION 8.4** A **multiple** of a matrix $\mathbf{A}$ is defined to be
>
> $$
> k\mathbf{A} = \begin{pmatrix} ka_{11} & ka_{12} & \cdots & ka_{1n} \\ ka_{21} & ka_{22} & \cdots & ka_{2n} \\ \vdots & & & \vdots \\ ka_{m1} & ka_{m2} & \cdots & ka_{mn} \end{pmatrix} = (ka_{ij})_{m \times n}
> $$
>
> where k is a constant or a function. ☐

EXAMPLES

(a)
$$
5\begin{pmatrix} 2 & -3 \\ 4 & -1 \\ \frac{1}{5} & 6 \end{pmatrix} = \begin{pmatrix} 10 & -15 \\ 20 & -5 \\ 1 & 30 \end{pmatrix}.
$$

(b)
$$
e^t\begin{pmatrix} 1 \\ -2 \\ 4 \end{pmatrix} = \begin{pmatrix} e^t \\ -2e^t \\ 4e^t \end{pmatrix}.
$$

We note in passing that for any matrix **A** the product $k\mathbf{A}$ is the same as $\mathbf{A}k$. For example,

$$e^{-3t}\begin{pmatrix} 2 \\ 5 \end{pmatrix} = \begin{pmatrix} 2e^{-3t} \\ 5e^{-3t} \end{pmatrix} = \begin{pmatrix} 2 \\ 5 \end{pmatrix} e^{-3t}.$$

Addition of matrices

DEFINITION 8.5   The **sum** of two $m \times n$ matrices **A** and **B** is defined to be the matrix

$$\mathbf{A} + \mathbf{B} = (a_{ij} + b_{ij})_{m \times n}. \qquad \square$$

In other words, when adding two matrices of the same size we add the corresponding elements.

EXAMPLE   The sum of

$$\mathbf{A} = \begin{pmatrix} 2 & -1 & 3 \\ 0 & 4 & 6 \\ -7 & 10 & -5 \end{pmatrix} \quad \text{and} \quad \mathbf{B} = \begin{pmatrix} 4 & 7 & -8 \\ 9 & 3 & 5 \\ 1 & -1 & 2 \end{pmatrix}$$

is

$$\mathbf{A} + \mathbf{B} = \begin{pmatrix} 2+4 & -1+7 & 3+(-8) \\ 0+9 & 4+3 & 6+5 \\ -7+1 & 10+(-1) & -5+2 \end{pmatrix}$$

$$= \begin{pmatrix} 6 & 6 & -5 \\ 9 & 7 & 11 \\ -5 & 9 & -3 \end{pmatrix}.$$

EXAMPLE   The single matrix

$$\begin{pmatrix} 3t^2 - 2e^t \\ t^2 + 7t \\ 5t \end{pmatrix}$$

can be written as the sum of three column vectors

$$\begin{pmatrix} 3t^2 - 2e^t \\ t^2 + 7t \\ 5t \end{pmatrix} = \begin{pmatrix} 3t^2 \\ t^2 \\ 0 \end{pmatrix} + \begin{pmatrix} 0 \\ 7t \\ 5t \end{pmatrix} + \begin{pmatrix} -2e^t \\ 0 \\ 0 \end{pmatrix}$$

$$= \begin{pmatrix} 3 \\ 1 \\ 0 \end{pmatrix} t^2 + \begin{pmatrix} 0 \\ 7 \\ 5 \end{pmatrix} t + \begin{pmatrix} -2 \\ 0 \\ 0 \end{pmatrix} e^t.$$

The **difference** of two $m \times n$ matrices is defined in the usual manner: $\mathbf{A} - \mathbf{B} = \mathbf{A} + (-\mathbf{B})$ where $-\mathbf{B} = (-1)\mathbf{B}$.

DEFINITION 8.6   Let **A** be a matrix having m rows and n columns, and **B** be a matrix having n rows and p columns. We define the **product AB** to be the $m \times p$ matrix

$$\mathbf{AB} = \begin{pmatrix} a_{11} & a_{12} & \cdots & a_{1n} \\ a_{21} & a_{22} & \cdots & a_{2n} \\ \vdots & & & \vdots \\ a_{m1} & a_{m2} & \cdots & a_{mn} \end{pmatrix} \begin{pmatrix} b_{11} & b_{12} & \cdots & b_{1p} \\ b_{21} & b_{22} & \cdots & b_{2p} \\ \vdots & & & \vdots \\ b_{n1} & b_{n2} & \cdots & b_{np} \end{pmatrix}$$

$$= \begin{pmatrix} a_{11}b_{11} + a_{12}b_{21} + \cdots + a_{1n}b_{n1} & \cdots & a_{11}b_{1p} + a_{12}b_{2p} + \cdots + a_{1n}b_{np} \\ a_{21}b_{11} + a_{22}b_{21} + \cdots + a_{2n}b_{n1} & \cdots & a_{21}b_{1p} + a_{22}b_{2p} + \cdots + a_{2n}b_{np} \\ \vdots & & \vdots \\ a_{m1}b_{11} + a_{m2}b_{21} + \cdots + a_{mn}b_{n1} & \cdots & a_{m1}b_{1p} + a_{m2}b_{2p} + \cdots + a_{mn}b_{np} \end{pmatrix}$$

$$= \left(\sum_{k=1}^{n} a_{ik}b_{kj} \right)_{m \times p}. \qquad \square$$

The reader might recognize that the entries in, say, the ith row of the final matrix **AB** are formed by using the component definition of the inner or dot product of the ith row of **A** with each of the columns of **B**.

EXAMPLES

(a) For $\mathbf{A} = \begin{pmatrix} 4 & 7 \\ 3 & 5 \end{pmatrix}$ and $\mathbf{B} = \begin{pmatrix} 9 & -2 \\ 6 & 8 \end{pmatrix}$

$$\mathbf{AB} = \begin{pmatrix} 4 \cdot 9 + 7 \cdot 6 & 4 \cdot (-2) + 7 \cdot 8 \\ 3 \cdot 9 + 5 \cdot 6 & 3 \cdot (-2) + 5 \cdot 8 \end{pmatrix} = \begin{pmatrix} 78 & 48 \\ 57 & 34 \end{pmatrix}.$$

(b) For $\mathbf{A} = \begin{pmatrix} 5 & 8 \\ 1 & 0 \\ 2 & 7 \end{pmatrix}$ and $\mathbf{B} = \begin{pmatrix} -4 & -3 \\ 2 & 0 \end{pmatrix}$

$$\mathbf{AB} = \begin{pmatrix} 5 \cdot (-4) + 8 \cdot 2 & 5 \cdot (-3) + 8 \cdot 0 \\ 1 \cdot (-4) + 0 \cdot 2 & 1 \cdot (-3) + 0 \cdot 0 \\ 2 \cdot (-4) + 7 \cdot 2 & 2 \cdot (-3) + 7 \cdot 0 \end{pmatrix} = \begin{pmatrix} -4 & -15 \\ -4 & -3 \\ 6 & -6 \end{pmatrix}.$$

In general matrix multiplication is not commutative. That is, $\mathbf{AB} \neq \mathbf{BA}$. Observe that in part (a) of the preceding example.

$$\mathbf{BA} = \begin{pmatrix} 30 & 53 \\ 48 & 82 \end{pmatrix}$$

whereas in part (b) the product **BA** is not defined since Definition 8.6 requires that the first matrix (in this case **B**) have the same number of columns as the second matrix has rows.

We are particularly interested in the product of a square matrix and a column vector.

EXAMPLES

(a)
$$\begin{pmatrix} 2 & -1 & 3 \\ 0 & 4 & 5 \\ 1 & -7 & 9 \end{pmatrix} \begin{pmatrix} -3 \\ 6 \\ 4 \end{pmatrix} = \begin{pmatrix} 2 \cdot (-3) + (-1) \cdot 6 + 3 \cdot 4 \\ 0 \cdot (-3) + 4 \cdot 6 \quad + 5 \cdot 4 \\ 1 \cdot (-3) + (-7) \cdot 6 + 9 \cdot 4 \end{pmatrix}$$

$$= \begin{pmatrix} 0 \\ 44 \\ -9 \end{pmatrix}$$

(b)
$$\begin{pmatrix} -4 & 2 \\ 3 & 8 \end{pmatrix} \begin{pmatrix} x \\ y \end{pmatrix} = \begin{pmatrix} -4x + 2y \\ 3x + 8y \end{pmatrix}$$

Identities

For a given positive integer n, the $n \times n$ matrix

$$I = \begin{pmatrix} 1 & 0 & 0 & \cdots & 0 \\ 0 & 1 & 0 & \cdots & 0 \\ \vdots & & & & \vdots \\ 0 & 0 & 0 & \cdots & 1 \end{pmatrix}$$

is called the **multiplicative identity matrix**.* It follows from Definition 8.6 that for any $n \times n$ matrix **A**.

$$\mathbf{AI} = \mathbf{IA} = \mathbf{A}.$$

Also, it is readily verified that if **X** is an $n \times 1$ column matrix $\mathbf{IX} = \mathbf{X}$. Now the $n \times n$ matrix **0** consisting of all zero entries is called the **additive identity matrix** and has the property

$$\mathbf{A} + \mathbf{0} = \mathbf{A}.$$

Associative law

Although we shall not prove it, matrix multiplication is **associative**. If **A** is an $m \times p$ matrix, **B** a $p \times r$ matrix, and **C** an $r \times n$ matrix then

$$\mathbf{A}(\mathbf{BC}) = (\mathbf{AB})\mathbf{C}$$

is an $m \times n$ matrix.

Distributive law

If **B** and **C** are $r \times n$ matrices then the **distributive law** is **A** is an $m \times r$ matrix and

$$\mathbf{A}(\mathbf{B} + \mathbf{C}) = \mathbf{AB} + \mathbf{AC}.$$

Furthermore, if the product $(\mathbf{B} + \mathbf{C})\mathbf{A}$ is defined then

$$(\mathbf{B} + \mathbf{C})\mathbf{A} = \mathbf{BA} + \mathbf{CA}.$$

Determinant of a matrix

Associated with every *square* matrix **A** there is a number called the **determinant of the matrix**, which is denoted by det **A** or $|\mathbf{A}|$.

---

*Do not confuse the symbol I, denoting an interval, with the multiplicative identity matrix **I**.

EXAMPLE   For

$$\mathbf{A} = \begin{pmatrix} 3 & 6 & 2 \\ 2 & 5 & 1 \\ -1 & 2 & 4 \end{pmatrix}$$

we expand det $\mathbf{A}$ by cofactors of the first row:

$$\begin{vmatrix} 3 & 6 & 2 \\ 2 & 5 & 1 \\ -1 & 2 & 4 \end{vmatrix} = 3 \begin{vmatrix} 5 & 1 \\ 2 & 4 \end{vmatrix} - 6 \begin{vmatrix} 2 & 1 \\ -1 & 4 \end{vmatrix} + 2 \begin{vmatrix} 2 & 5 \\ -1 & 2 \end{vmatrix}$$

$$= 3(20 - 2) - 6(8 + 1) + 2(4 + 5)$$

$$= 18.$$

DEFINITION 8.7   The **transpose** of the $m \times n$ matrix (1) is the $n \times m$ matrix $\mathbf{A}^T$ given by

$$\mathbf{A}^T = \begin{pmatrix} a_{11} & a_{21} & \cdots & a_{m1} \\ a_{12} & a_{22} & \cdots & a_{m2} \\ \vdots & & & \vdots \\ a_{1n} & a_{2n} & \cdots & a_{mn} \end{pmatrix}.$$ □

In other words the rows of a matrix A become the columns of its transpose $\mathbf{A}^T$.

EXAMPLES   **(a)** The transpose of the matrix $\mathbf{A}$ in the preceding example is

$$\mathbf{A}^T = \begin{pmatrix} 3 & 2 & -1 \\ 6 & 5 & 2 \\ 2 & 1 & - 4 \end{pmatrix}.$$

(b) If $\mathbf{X} = \begin{pmatrix} 5 \\ 0 \\ 3 \end{pmatrix}$ then $\mathbf{X}^T = (5 \quad 0 \quad 3).$

Multiplicative inverse

DEFINITION 8.8   Let $\mathbf{A}$ be an $n \times n$ matrix. If there exists an $n \times n$ matrix $\mathbf{B}$ such that

$$\mathbf{AB} = \mathbf{BA} = \mathbf{I},$$

where $\mathbf{I}$ is the multiplicative identity, then $\mathbf{B}$ is said to be the **multiplicative inverse of $\mathbf{A}$** and is denoted by $\mathbf{B} = \mathbf{A}^{-1}$.   □

DEFINITION 8.9   Let $\mathbf{A}$ be an $n \times n$ matrix. If det $\mathbf{A} \neq 0$ then $\mathbf{A}$ is said to be **nonsingular**. If det $\mathbf{A} = 0$ then $\mathbf{A}$ is said to be **singular**.□

The following gives a necessary and sufficient condition for a square matrix to have a multiplicative inverse.

THEOREM 8.1 An $n \times n$ matrix $\mathbf{A}$ has a multiplicative inverse $\mathbf{A}^{-1}$ if and only if $\mathbf{A}$ is nonsingular. □

The following theorem gives one way of finding the multiplicative inverse for a nonsingular matrix.

THEOREM 8.2 Let $\mathbf{A}$ be an $n \times n$ nonsingular matrix and let $A_{ij} = (-1)^{i+j}M_{ij}$, where M_{ij} is the $(n-1) \times (n-1)$ determinant obtained by deleting the ith row and jth column from det $\mathbf{A}$. Then

$$\mathbf{A}^{-1} = \frac{1}{\det \mathbf{A}} (A_{ij})^T. \qquad \square(2)$$

Each A_{ij} in Theorem 8.2 is simply the *cofactor* (signed minor) of the corresponding entry a_{ij} in det $\mathbf{A}$. Note that the transpose is utilized in formula (2).

For future reference we observe in the case of a 2×2 nonsingular matrix

$$\mathbf{A} = \begin{pmatrix} a_{11} & a_{12} \\ a_{21} & a_{22} \end{pmatrix}$$

that $A_{11} = a_{22}$, $A_{12} = -a_{21}$, $A_{21} = -a_{12}$, and $A_{22} = a_{11}$.

Thus $\qquad \mathbf{A}^{-1} = \frac{1}{\det \mathbf{A}} \begin{pmatrix} a_{22} & -a_{12} \\ -a_{12} & a_{11} \end{pmatrix}^T = \frac{1}{\det \mathbf{A}} \begin{pmatrix} a_{22} & -a_{12} \\ -a_{21} & a_{11} \end{pmatrix}.$ \qquad (3)

For a 3×3 nonsingular matrix

$$\mathbf{A} = \begin{pmatrix} a_{11} & a_{12} & a_{13} \\ a_{21} & a_{22} & a_{23} \\ a_{31} & a_{32} & a_{33} \end{pmatrix}$$

$$A_{11} = \begin{vmatrix} a_{22} & a_{23} \\ a_{32} & a_{33} \end{vmatrix}, \quad A_{12} = - \begin{vmatrix} a_{21} & a_{23} \\ a_{31} & a_{33} \end{vmatrix},$$

and so on. Carrying out the transposition gives

$$\mathbf{A}^{-1} = \frac{1}{\det \mathbf{A}} \begin{pmatrix} A_{11} & A_{21} & A_{31} \\ A_{12} & A_{22} & A_{32} \\ A_{13} & A_{23} & A_{33} \end{pmatrix} \qquad (4)$$

EXAMPLE

Find the multiplicative inverse for

$$\mathbf{A} = \begin{pmatrix} 1 & 4 \\ 2 & 10 \end{pmatrix}.$$

Solution: Since det $\mathbf{A} = 10 - 8 = 2 \neq 0$, $\mathbf{A}$ is nonsingular. It follows from Theorem 8.1 $\mathbf{A}^{-1}$ exists. From (3) we find

$$\mathbf{A}^{-1} = \frac{1}{2}\begin{pmatrix} 10 & -4 \\ -2 & 1 \end{pmatrix}$$

$$= \begin{pmatrix} 5 & -2 \\ -1 & \frac{1}{2} \end{pmatrix}.$$

Check:

$$\mathbf{A}\mathbf{A}^{-1} = \begin{pmatrix} 1 & 4 \\ 2 & 10 \end{pmatrix}\begin{pmatrix} 5 & -2 \\ -1 & \frac{1}{2} \end{pmatrix} = \begin{pmatrix} 5 - 4 & -2 + 2 \\ 10 - 10 & -4 + 5 \end{pmatrix} = \begin{pmatrix} 1 & 0 \\ 0 & 1 \end{pmatrix}.$$

$$\mathbf{A}^{-1}\mathbf{A} = \begin{pmatrix} 5 & -2 \\ -1 & \frac{1}{2} \end{pmatrix}\begin{pmatrix} 1 & 4 \\ 2 & 10 \end{pmatrix} = \begin{pmatrix} 5 - 4 & 20 - 20 \\ -1 + 1 & -4 + 5 \end{pmatrix} = \begin{pmatrix} 1 & 0 \\ 0 & 1 \end{pmatrix}.$$

EXAMPLE

The matrix $\mathbf{A} = \begin{pmatrix} 2 & 2 \\ 3 & 3 \end{pmatrix}$ is singular since det $\mathbf{A} = 2(3) - 2(3) = 0$. We conclude that $\mathbf{A}^{-1}$ does not exist.

EXAMPLE

Find the multiplicative inverse for

$$\mathbf{A} = \begin{pmatrix} 2 & 2 & 0 \\ -2 & 1 & 1 \\ 3 & 0 & 1 \end{pmatrix}.$$

Solution: Since det $\mathbf{A} = 12 \neq 0$ the given matrix is nonsingular.

The cofactors of the entries in each row of det $\mathbf{A}$ are

$$A_{11} = \begin{vmatrix} 1 & 1 \\ 0 & 1 \end{vmatrix} \qquad A_{12} = -\begin{vmatrix} -2 & 1 \\ 3 & 1 \end{vmatrix} \qquad A_{13} = \begin{vmatrix} -2 & 1 \\ 3 & 0 \end{vmatrix}$$

$$= 1 \qquad\qquad = 5 \qquad\qquad = -3$$

$$A_{21} = -\begin{vmatrix} 2 & 0 \\ 0 & 1 \end{vmatrix} \qquad A_{22} = \begin{vmatrix} 2 & 0 \\ 3 & 1 \end{vmatrix} \qquad A_{23} = -\begin{vmatrix} 2 & 2 \\ 3 & 0 \end{vmatrix}$$

$$= -2 \qquad\qquad = 2 \qquad\qquad = 6$$

$$A_{31} = \begin{vmatrix} 2 & 0 \\ 1 & 1 \end{vmatrix} \qquad A_{32} = -\begin{vmatrix} 2 & 0 \\ -2 & 1 \end{vmatrix} \qquad A_{33} = \begin{vmatrix} 2 & 2 \\ -2 & 1 \end{vmatrix}$$

$$= 2 \qquad\qquad = -2 \qquad\qquad = 6.$$

It follows from (4) that

$$\mathbf{A}^{-1} = \frac{1}{12}\begin{pmatrix} 1 & -2 & 2 \\ 5 & 2 & -2 \\ -3 & 6 & 6 \end{pmatrix} = \begin{pmatrix} 1/12 & -1/6 & 1/6 \\ 5/12 & 1/6 & -1/6 \\ -1/4 & 1/2 & 1/2 \end{pmatrix}.$$

The reader is urged to verify that $\mathbf{A}^{-1}\mathbf{A} = \mathbf{A}\mathbf{A}^{-1} = \mathbf{I}$.

Formula (2) presents obvious difficulties for nonsingular matrices larger than 3×3. For example, to apply (2) to a 4×4 matrix we would have to calculate *sixteen* 3×3 determinants! In the case of a large matrix there are more efficient ways of finding $\mathbf{A}^{-1}$. The curious reader is referred to any text in linear algebra.

Since our goal is to apply the concept of a matrix to systems of linear differential equations in normal form we need two final definitions.

Derivative of a matrix of functions

DEFINITION 8.10   If $\mathbf{A}(t) = (a_{ij}(t))_{m \times n}$ is a matrix whose entries are functions differentiable on a common interval then

$$\frac{d\mathbf{A}}{dt} = \left(\frac{d}{dt}a_{ij}\right)_{m \times n}. \qquad \square$$

Integral of a matrix of functions

DEFINITION 8.11   If $\mathbf{A}(t) = a_{ij}(t))_{m \times n}$ matrix whose entries are functions continuous on a common interval containing t and t_0 then

$$\int_{t_0}^{t} \mathbf{A}(s)\, ds = \left(\int_{t_0}^{t} a_{ij}(s)\, ds\right)_{m \times n}. \qquad \square$$

To differentiate (integrate) a matrix of functions we simply differentiate (integrate) each entry. The derivative of a matrix is also denoted by $\mathbf{A}'(t)$.

EXAMPLE

If

$$\mathbf{X}(t) = \begin{pmatrix} \sin 2t \\ e^{3t} \\ 8t - 1 \end{pmatrix}$$

then

$$\mathbf{X}'(t) = \begin{pmatrix} \dfrac{d}{dt}\sin 2t \\ \dfrac{d}{dt}e^{3t} \\ \dfrac{d}{dt}(8t - 1) \end{pmatrix} = \begin{pmatrix} 2\cos 2t \\ 3e^{3t} \\ 8 \end{pmatrix}$$

and

$$\int_0^t \mathbf{X}(s)\, ds = \begin{pmatrix} \displaystyle\int_0^t \sin 2s\, ds \\ \displaystyle\int_0^t e^{3s}\, ds \\ \displaystyle\int_0^t (8s - 1)\, ds \end{pmatrix} = \begin{pmatrix} -\dfrac{1}{2}\cos 2t + \dfrac{1}{2} \\ \dfrac{1}{3}e^{3t} - \dfrac{1}{3} \\ 4t^2 - t \end{pmatrix}.$$

EXERCISES 8.4   *Answers to odd-numbered problems begin on page A–25.*

1. If $\qquad \mathbf{A} = \begin{pmatrix} 4 & 5 \\ -6 & 9 \end{pmatrix}$ and $\mathbf{B} = \begin{pmatrix} -2 & 6 \\ 8 & -10 \end{pmatrix}$,

   find **(a)** $\mathbf{A} + \mathbf{B}$,   **(b)** $\mathbf{B} - \mathbf{A}$,   **(c)** $2\mathbf{A} + 3\mathbf{B}$.

2. If $\qquad \mathbf{A} = \begin{pmatrix} -2 & 0 \\ 4 & 1 \\ 7 & 3 \end{pmatrix}$ and $\mathbf{B} = \begin{pmatrix} 3 & -1 \\ 0 & 2 \\ -4 & -2 \end{pmatrix}$,

   find **(a)** $\mathbf{A} - \mathbf{B}$,   **(b)** $\mathbf{B} - \mathbf{A}$,   **(c)** $2(\mathbf{A} + \mathbf{B})$.

3. If $\qquad \mathbf{A} = \begin{pmatrix} 2 & -3 \\ -5 & 4 \end{pmatrix}$ and $\mathbf{B} = \begin{pmatrix} -1 & 6 \\ 3 & 2 \end{pmatrix}$,

   find **(a)** $\mathbf{AB}$,   **(b)** $\mathbf{BA}$,   **(c)** $\mathbf{A}^2 = \mathbf{AA}$,   **(d)** $\mathbf{B}^2 = \mathbf{BB}$

4. If $\qquad \mathbf{A} = \begin{pmatrix} 1 & 4 \\ 5 & 10 \\ 8 & 12 \end{pmatrix}$ and $\mathbf{B} = \begin{pmatrix} -4 & 6 & -3 \\ 1 & -3 & 2 \end{pmatrix}$,

   find **(a)** $\mathbf{AB}$,   **(b)** $\mathbf{BA}$.

5. If

   $$\mathbf{A} = \begin{pmatrix} 1 & -2 \\ -2 & 4 \end{pmatrix}, \qquad \mathbf{B} = \begin{pmatrix} 6 & 3 \\ 2 & 1 \end{pmatrix}, \qquad \text{and} \qquad \mathbf{C} = \begin{pmatrix} 0 & 2 \\ 3 & 4 \end{pmatrix},$$

   find **(a)** $\mathbf{BC}$,   **(b)** $\mathbf{A}(\mathbf{BC})$,   **(c)** $\mathbf{C}(\mathbf{BA})$,   **(d)** $\mathbf{A}(\mathbf{B} + \mathbf{C})$

★6. If

   $$\mathbf{A} = (5 \quad -6 \quad 7), \qquad \mathbf{B} = \begin{pmatrix} 3 \\ 4 \\ -1 \end{pmatrix}, \qquad \text{and} \qquad \mathbf{C} = \begin{pmatrix} 1 & 2 & 4 \\ 0 & 1 & -1 \\ 3 & 2 & 1 \end{pmatrix},$$

   find **(a)** $\mathbf{AB}$,   **(b)** $\mathbf{BA}$,   **(c)** $(\mathbf{BA})\mathbf{C}$,   **(d)** $(\mathbf{AB})\mathbf{C}$

7. If $\qquad \mathbf{A} = \begin{pmatrix} 4 \\ 8 \\ -10 \end{pmatrix}$ and $\mathbf{B} = (2 \quad 4 \quad 5)$,

   find **(a)** $\mathbf{A}^T\mathbf{A}$,   **(b)** $\mathbf{B}^T\mathbf{B}$,   **(c)** $\mathbf{A} + \mathbf{B}^T$

★8. If $\qquad \mathbf{A} = \begin{pmatrix} 1 & 2 \\ 2 & 4 \end{pmatrix}$ and $\mathbf{B} = \begin{pmatrix} -2 & 3 \\ 5 & 7 \end{pmatrix}$,

   find **(a)** $\mathbf{A} + \mathbf{B}^T$,   **(b)** $2\mathbf{A}^T - \mathbf{B}^T$,   **(c)** $\mathbf{A}^T(\mathbf{A} - \mathbf{B})$

9. If $\qquad \mathbf{A} = \begin{pmatrix} 3 & 4 \\ 8 & 1 \end{pmatrix}$ and $\mathbf{B} = \begin{pmatrix} 5 & 10 \\ -2 & -5 \end{pmatrix}$,

   (a) $(\mathbf{AB})^T$,   **(b)** $\mathbf{B}^T\mathbf{A}^T$

10. If $\qquad \mathbf{A} = \begin{pmatrix} 5 & 9 \\ -4 & 6 \end{pmatrix}$ and $\mathbf{B} = \begin{pmatrix} -3 & 11 \\ -7 & 2 \end{pmatrix}$,

    find **(a)** $\mathbf{A}^T + \mathbf{B}^T$,   **(b)** $(\mathbf{A} + \mathbf{B})^T$

In Problems 11–14 write the indicated sum as a one column matrix.

11. $4\begin{pmatrix} -1 \\ 2 \end{pmatrix} - 2\begin{pmatrix} 2 \\ 8 \end{pmatrix} + 3\begin{pmatrix} -2 \\ 3 \end{pmatrix}$

★12. $3t\begin{pmatrix} 2 \\ t \\ -1 \end{pmatrix} + (t-1)\begin{pmatrix} -1 \\ -t \\ 3 \end{pmatrix} - 2\begin{pmatrix} 3t \\ 4 \\ -5t \end{pmatrix}$

13. $\begin{pmatrix} 2 & -3 \\ 1 & 4 \end{pmatrix}\begin{pmatrix} -2 \\ 5 \end{pmatrix} - \begin{pmatrix} -1 & 6 \\ -2 & 3 \end{pmatrix}\begin{pmatrix} -7 \\ 2 \end{pmatrix}$

14. $\begin{pmatrix} 1 & -3 & 4 \\ 2 & 5 & -1 \\ 0 & -4 & -2 \end{pmatrix}\begin{pmatrix} t \\ 2t-1 \\ -t \end{pmatrix} + \begin{pmatrix} -t \\ 1 \\ 4 \end{pmatrix} - \begin{pmatrix} 2 \\ 8 \\ -6 \end{pmatrix}$

In Problems 15–22 determine whether the given matrix is singular or non-singular. If nonsingular, find A^{-1}.

15. $A = \begin{pmatrix} -3 & 6 \\ -2 & 4 \end{pmatrix}$

16. $A = \begin{pmatrix} 2 & 5 \\ 1 & 4 \end{pmatrix}$

17. $A = \begin{pmatrix} 4 & 8 \\ -3 & -5 \end{pmatrix}$

18. $A = \begin{pmatrix} 7 & 10 \\ 2 & 2 \end{pmatrix}$

19. $A = \begin{pmatrix} 2 & 1 & 0 \\ -1 & 2 & 1 \\ 1 & 2 & 1 \end{pmatrix}$

★20. $A = \begin{pmatrix} 3 & 2 & 1 \\ 4 & 1 & 0 \\ -2 & 5 & -1 \end{pmatrix}$

21. $A = \begin{pmatrix} 2 & 1 & 1 \\ 1 & -2 & -3 \\ 3 & 2 & 4 \end{pmatrix}$

22. $A = \begin{pmatrix} 4 & 1 & -1 \\ 6 & 2 & -3 \\ -2 & -1 & 2 \end{pmatrix}$

In Problems 23 and 24 show that the given matrix is nonsingular for every real value of t. Find $A^{-1}(t)$.

23. $A(t) = \begin{pmatrix} 2e^{-t} & e^{4t} \\ 4e^{-t} & 3e^{4t} \end{pmatrix}$

★24. $A(t)\begin{pmatrix} 2e^{t}\sin t & -2e^{t}\cos t \\ e^{t}\cos t & e^{t}\sin t \end{pmatrix}$

In Problems 25 and 26 find the value of λ for which $\det(A - \lambda I) = 0$. I is the multiplicative identity.

25. $A = \begin{pmatrix} 2 & 3 \\ 5 & 4 \end{pmatrix}$

26. $A = \begin{pmatrix} 9 & 0 & 1 \\ 0 & -3 & 0 \\ 0 & 8 & 4 \end{pmatrix}$

In Problems 27–30 find dX/dt.

27. $X = \begin{pmatrix} 5e^{-t} \\ 2e^{-t} \\ -7e^{-t} \end{pmatrix}$

28. $X = \begin{pmatrix} \dfrac{1}{2}\sin 2t - 4\cos 2t \\ -3\sin 2t + 5\cos 2t \end{pmatrix}$

29. $\mathbf{X} = 2\begin{pmatrix} 1 \\ -1 \end{pmatrix} e^{2t} + 4\begin{pmatrix} 2 \\ 1 \end{pmatrix} e^{-3t}$ ★**30.** $\mathbf{X} = \begin{pmatrix} 5te^{2t} \\ t\,\sin 3t \end{pmatrix}$

31. Let $\mathbf{A}(t) = \begin{pmatrix} e^{4t} & \cos \pi t \\ 2t & 3t^2 - 1 \end{pmatrix}.$

Find **(a)** $\dfrac{d\mathbf{A}}{dt}$,   **(b)** $\displaystyle\int_0^2 \mathbf{A}(t)\,dt$,   **(c)** $\displaystyle\int_0^t \mathbf{A}(s)\,ds$.

32. Let $\mathbf{A}(t) = \begin{pmatrix} \dfrac{1}{t^2 + 1} & 3t \\ t^2 & t \end{pmatrix}$ and $\mathbf{B}(t) = \begin{pmatrix} 6t & 2 \\ \dfrac{1}{t} & 4t \end{pmatrix}.$

Find **(a)** $\dfrac{d\mathbf{A}}{dt}$,   **(b)** $\dfrac{d\mathbf{B}}{dt}$,   **(c)** $\displaystyle\int_0^1 \mathbf{A}(t)\,dt$,   **(d)** $\displaystyle\int_1^2 \mathbf{B}(t)\,dt$,

(e) $\mathbf{A}(t)\mathbf{B}(t)$,   **(f)** $\dfrac{d}{dt}\mathbf{A}(t)\mathbf{B}(t)$,   **(g)** $\displaystyle\int_1^t \mathbf{A}(s)\mathbf{B}(s)\,ds$.

Miscellaneous problems

33. If $\mathbf{A}(t)$ is a 2×2 matrix of differentiable functions and $\mathbf{X}(t)$ is a 2×1 column matrix of differentiable functions, prove the product rule

$$\frac{d}{dt}[\mathbf{A}(t)\mathbf{X}(t)] = \mathbf{A}(t)\mathbf{X}'(t) + \mathbf{A}'(t)\mathbf{X}(t).$$

34. Derive formula (3). [*Hint:* Find a matrix

$$\mathbf{B} = \begin{pmatrix} b_{11} & b_{12} \\ b_{21} & b_{22} \end{pmatrix}$$

for which $\mathbf{AB} = \mathbf{I}$. Solve for b_{11}, b_{12}, b_{21} and b_{22}. Then show that $\mathbf{BA} = \mathbf{I}$.]

35. If $\mathbf{A}$ is nonsingular and $\mathbf{AB} = \mathbf{AC}$, show that $\mathbf{B} = \mathbf{C}$.

36. If $\mathbf{A}$ and $\mathbf{B}$ are nonsingular show that $(\mathbf{AB})^{-1} = \mathbf{B}^{-1}\mathbf{A}^{-1}$

37. Let $\mathbf{A}$ and $\mathbf{B}$ be $n \times n$ matrices. In general, is $(\mathbf{A} + \mathbf{B})^2 = \mathbf{A}^2 + 2\mathbf{AB} + \mathbf{B}^2$?

8.5   Matrices and Systems of Linear First-Order Equations

8.5.1   Preliminary Theory

Matrix form of a system

If $\mathbf{X}$ and $\mathbf{A}(t)$ denote the respective matrices

$$\mathbf{X} = \begin{pmatrix} x_1 \\ x_2 \\ \vdots \\ x_n \end{pmatrix} = \begin{pmatrix} x_1(t) \\ x_2(t) \\ \vdots \\ x_n(t) \end{pmatrix}$$

$$
\mathbf{A}(t) = \begin{pmatrix}
a_{11}(t) & a_{12}(t) & \cdots & a_{1n}(t) \\
a_{21}(t) & a_{22}(t) & \cdots & a_{2n}(t) \\
\vdots & & & \vdots \\
a_{n1}(t) & a_{n2}(t) & \cdots & a_{nn}(t)
\end{pmatrix}
$$

then the homogeneous system of linear first-order differential equations

$$
\begin{aligned}
\frac{dx_1}{dt} &= a_{11}(t)x_1 + a_{12}(t)x_2 + \cdots + a_{1n}(t)x_n \\[2mm]
\frac{dx_2}{dt} &= a_{21}(t)x_1 + a_{22}(t)x_2 + \cdots + a_{2n}(t)x_n \\[2mm]
\vdots & \qquad\qquad \vdots \\[2mm]
\frac{dx_n}{dt} &= a_{n1}(t)x_1 + a_{n2}(t)x_2 + \cdots + a_{nn}(t)x_n
\end{aligned}
\tag{1}
$$

can be written as

$$
\frac{d}{dt}\begin{pmatrix} x_1 \\ x_2 \\ \vdots \\ x_n \end{pmatrix} = \begin{pmatrix}
a_{11}(t) & a_{12}(t) & \cdots & a_{1n}(t) \\
a_{21}(t) & a_{22}(t) & \cdots & a_{2n}(t) \\
\vdots & & & \vdots \\
a_{n1}(t) & a_{n2}(t) & \cdots & a_{nn}(t)
\end{pmatrix}\begin{pmatrix} x_1 \\ x_2 \\ \vdots \\ x_n \end{pmatrix}
$$

or simply

$$
\frac{d\mathbf{X}}{dt} = \mathbf{A}(t)\mathbf{X}.
\tag{2}
$$

Also, if

$$
\mathbf{F}(t) = \begin{pmatrix} f_1(t) \\ f_2(t) \\ \vdots \\ f_n(t) \end{pmatrix}
$$

then the nonhomogeneous system

$$
\begin{aligned}
\frac{dx_1}{dt} &= a_{11}(t)x_1 + a_{12}(t)x_2 + \cdots + a_{1n}(t)x_n + f_1(t) \\[2mm]
\frac{dx_2}{dt} &= a_{21}(t)x_1 + a_{22}(t)x_2 + \cdots + a_{2n}(t)x_n + f_2(t) \\[2mm]
\vdots & \qquad\qquad \vdots \\[2mm]
\frac{dx_n}{dt} &= a_{n1}(t)x_1 + a_{n2}(t)x_2 + \cdots + a_{nn}(t)x_n + f_n(t)
\end{aligned}
\tag{3}
$$

can be written as

$$
\frac{d\mathbf{X}}{dt} = \mathbf{A}(t)\mathbf{X} + \mathbf{F}(t).
\tag{4}
$$

EXAMPLE

The matrix form of the homogeneous system

$$\frac{dx}{dt} = 2x - 3y$$

$$\frac{dy}{dt} = 6x + 5y$$

is

$$\frac{d\mathbf{X}}{dt} = \begin{pmatrix} 2 & -3 \\ 6 & 5 \end{pmatrix} \mathbf{X}$$

where $\mathbf{X} = \begin{pmatrix} x \\ y \end{pmatrix}$.

EXAMPLE

The nonhomogeneous system

$$\frac{dx}{dt} = -2x + 5y + e^t - 2t$$

$$\frac{dy}{dt} = \quad 4x - 3y + 10t$$

can be written as

$$\mathbf{X}' = \begin{pmatrix} -2 & 5 \\ 4 & -3 \end{pmatrix} \mathbf{X} + \begin{pmatrix} e^t - 2t \\ 10t \end{pmatrix}$$

or

$$\mathbf{X}' = \begin{pmatrix} -2 & 5 \\ 4 & -3 \end{pmatrix} \mathbf{X} + \begin{pmatrix} e^t \\ 0 \end{pmatrix} + \begin{pmatrix} -2t \\ 10t \end{pmatrix}$$

$$= \begin{pmatrix} -2 & 5 \\ 4 & -3 \end{pmatrix} \mathbf{X} + \begin{pmatrix} 1 \\ 0 \end{pmatrix} e^t + \begin{pmatrix} -2 \\ 10 \end{pmatrix} t$$

where $\mathbf{X} = \begin{pmatrix} x \\ y \end{pmatrix}$.

DEFINITION 8.12   A **solution vector** on an interval I is any column matrix

$$\mathbf{X} = \begin{pmatrix} x_1(t) \\ x_2(t) \\ \vdots \\ x_n(t) \end{pmatrix}$$

whose entries are differentiable functions satisfying the system (4) on the interval.   □

EXAMPLE

Verify that

$$\mathbf{X}_1 = \begin{pmatrix} 1 \\ -1 \end{pmatrix} e^{-2t} = \begin{pmatrix} e^{-2t} \\ -e^{-2t} \end{pmatrix} \quad \text{and} \quad \mathbf{X}_2 = \begin{pmatrix} 3 \\ 5 \end{pmatrix} e^{6t} = \begin{pmatrix} 3e^{6t} \\ 5e^{6t} \end{pmatrix}$$

are solutions of

$$\mathbf{X}' = \begin{pmatrix} 1 & 3 \\ 5 & 3 \end{pmatrix} \mathbf{X} \tag{5}$$

on $-\infty < t < \infty$.

Solution: We have

$$\frac{d\mathbf{X}_1}{dt} = \begin{pmatrix} -2e^{-2t} \\ 2e^{-2t} \end{pmatrix}$$

and $\mathbf{AX}_1 = \begin{pmatrix} 1 & 3 \\ 5 & 3 \end{pmatrix} \begin{pmatrix} e^{-2t} \\ -e^{-2t} \end{pmatrix} = \begin{pmatrix} e^{-2t} - 3e^{-2t} \\ 5e^{-2t} - 3e^{-2t} \end{pmatrix} = \begin{pmatrix} -2e^{-2t} \\ 2e^{-2t} \end{pmatrix} = \mathbf{X}_1'.$

Now

$$\frac{d\mathbf{X}_2}{dt} = \begin{pmatrix} 18e^{6t} \\ 30e^{5t} \end{pmatrix}$$

and $\mathbf{AX}_2 = \begin{pmatrix} 1 & 3 \\ 5 & 3 \end{pmatrix} \begin{pmatrix} 3e^{6t} \\ 5e^{6t} \end{pmatrix} = \begin{pmatrix} 3e^{6t} + 15e^{6t} \\ 15e^{6t} + 15e^{6t} \end{pmatrix} = \begin{pmatrix} 18e^{6t} \\ 30e^{6t} \end{pmatrix} = \mathbf{X}_2'.$

Much of the theory of systems of n linear first-order differential equations is similar to that of linear nth-order differential equations (See Section 4.1). This should not be too surprising in view of the discussion in Section 8.3.

Initial-value problem

Let t_0 denote a point in an interval I and

$$\mathbf{X}(t_0) = \begin{pmatrix} x_1(t_0) \\ x_2(t_0) \\ \vdots \\ x_n(t_0) \end{pmatrix} \quad \text{and} \quad \mathbf{X}_0 = \begin{pmatrix} \gamma_1 \\ \gamma_2 \\ \vdots \\ \gamma_n \end{pmatrix}$$

where the γ_i, $i = 1, 2, \ldots, n$ are given constants. Then the problem

$$Solve: \frac{d\mathbf{X}}{dt} = \mathbf{A}(t)\mathbf{X} + \mathbf{F}(t)$$

$$Subject \ to: \ \mathbf{X}(t_0) = \mathbf{X}_0 \tag{6}$$

is an **initial-value problem** on the interval.

THEOREM 8.3   Let the entries of the matrices $\mathbf{A}(t)$ and $\mathbf{F}(t)$ be functions continuous on a common interval I that contains the point t_0. Then there exists a unique solution of the initial-value problem (6) on the interval.   □

Homogeneous systems

In the next several definitions and theorems we are concerned only with homogeneous systems. Without stating it, we shall always assume that the a_{ij} and the f_i are continuous functions of t on some common interval I.

The superposition principle

The following result is a **superposition principle** for solutions of linear systems.

> **THEOREM 8.4**   Let $X_1, X_2 \ldots \ldots , X_k$ be a set of solution vectors of the homogeneous system (2) on an interval I. Then the linear combination
>
> $$X = c_1 X_1 + c_2 X_2 + \cdots + c_k X_k,$$
>
> where the c_i, $i = 1, 2, \ldots , k$, are arbitrary constants, is also a solution on the interval.   □

It follows from Theorem 8.4 that a constant multiple of any solution vector of the homogeneous system (2) is a solution.

EXAMPLE

One solution of the system

$$X' = \begin{pmatrix} 1 & 0 & 1 \\ 1 & 1 & 0 \\ -2 & 0 & -1 \end{pmatrix} X \tag{7}$$

is

$$X_1 = \begin{pmatrix} \cos t \\ -\tfrac{1}{2}\cos t + \tfrac{1}{2}\sin t \\ -\cos t - \sin t \end{pmatrix}$$

For any constant c_1 the vector $X = c_1 X_1$ is also a solution since

$$\frac{dX}{dt} = \begin{pmatrix} -c_1 \sin t \\ \tfrac{1}{2}c_1 \sin t + \tfrac{1}{2}c_1 \cos t \\ c_1 \sin t - c_1 \cos t \end{pmatrix}$$

and

$$AX = \begin{pmatrix} 1 & 0 & 1 \\ 1 & 1 & 0 \\ -2 & 0 & -1 \end{pmatrix} \begin{pmatrix} c_1 \cos t \\ -\tfrac{1}{2}c_1 \cos t + \tfrac{1}{2}c_1 \sin t \\ -c_1 \cos t - c_1 \sin t \end{pmatrix}$$

$$= \begin{pmatrix} -c_1 \sin t \\ \tfrac{1}{2}c_1 \cos t + \tfrac{1}{2}c_1 \sin t \\ -c_1 \cos t + c_1 \sin t \end{pmatrix}.$$

Inspection of the resulting matrices shows that $X' = AX$.

EXAMPLE

Consider the system (7) of the preceding example.

If
$$\mathbf{X}_2 = \begin{pmatrix} 0 \\ 1 \\ 0 \end{pmatrix} e^t = \begin{pmatrix} 0 \\ e^t \\ 0 \end{pmatrix}$$

then
$$\mathbf{X}_2' = \begin{pmatrix} 0 \\ e^t \\ 0 \end{pmatrix}$$

and
$$\mathbf{A}\mathbf{X}_2 = \begin{pmatrix} 1 & 0 & 1 \\ 1 & 1 & 0 \\ -2 & 0 & -1 \end{pmatrix} \begin{pmatrix} 0 \\ e^t \\ 0 \end{pmatrix} = \begin{pmatrix} 0 \\ e^t \\ 0 \end{pmatrix} = \mathbf{X}_2'.$$

Thus we see that $\mathbf{X}_2$ is also a solution vector of the system. By the superposition principle the linear combination

$$\mathbf{X} = c_1\mathbf{X}_1 + c_2\mathbf{X}_2$$

$$= c_1 \begin{pmatrix} \cos t \\ -\tfrac{1}{2}\cos t + \tfrac{1}{2}\sin t \\ -\cos t - \sin t \end{pmatrix} + c_2 \begin{pmatrix} 0 \\ e^t \\ 0 \end{pmatrix}$$

is yet another solution.

Linear independence   We are primarily interested in linearly independent solutions of the homogeneous system (2).

> **DEFINITION 8.13**   Let $\mathbf{X}_1, \mathbf{X}_2, \ldots, \mathbf{X}_k$ be a set of solution vectors of the homogeneous system (2) on an interval I. We say that the set is **linearly dependent** on the interval if there exist constants c_1, $c_2, \ldots, c_k$, not all zero, such that
>
> $$c_1\mathbf{X}_1 + c_2\mathbf{X}_2 + \cdots + c_k\mathbf{X}_k = \mathbf{0}$$
>
> for every t in the interval. If the set of vectors is not linearly dependent on the interval, it is said to be **linearly independent**.   □

The case when $k = 2$ should be clear; two solution vectors $\mathbf{X}_1$ and $\mathbf{X}_2$ are linearly dependent if one is a constant multiple of the other, and conversely. For $k > 2$, a set of solution vectors is linearly dependent if we can express at least one solution vector as a nontrivial linear combination of the remaining vectors.

EXAMPLE   It can be verified that

$$\mathbf{X}_1 = \begin{pmatrix} 3 \\ 1 \end{pmatrix} e^t \quad \text{and} \quad \mathbf{X}_2 = \begin{pmatrix} 1 \\ 1 \end{pmatrix} e^{-t}$$

are solution vectors of the system

$$\mathbf{X}' = \begin{pmatrix} 2 & -3 \\ 1 & -2 \end{pmatrix} \mathbf{X}. \tag{8}$$

Now X_1 and X_2 are linearly independent on the interval $-\infty < t < \infty$ since

$$c_1 X_1 + c_2 X_2 = 0 \quad \text{or} \quad c_1 \binom{3}{1} e^t + c_2 \binom{1}{1} e^{-t} = \binom{0}{0}$$

is equivalent to

$$3c_1 e^t + c_2 e^{-t} = 0$$
$$c_1 e^t + c_2 e^{-t} = 0.$$

Solving this system for c_1 and c_2 immediately yields $c_1 = 0$ and $c_2 = 0$.

EXAMPLE

The vector

$$X_3 = \binom{e^t + \cosh t}{\cosh t}$$

is also a solution of the system (8) given in the preceding example. However, X_1, X_2 and X_3 are linearly dependent since

$$X_3 = \tfrac{1}{2} X_1 + \tfrac{1}{2} X_2.$$

The Wronskian

As in our earlier consideration of the theory of a single ordinary differential equation we can introduce the concept of the **Wronskian** determinant as a test for linear independence. We state the following theorem without proof.

THEOREM 8.5 Let

$$X_1 = \begin{pmatrix} x_{11} \\ x_{21} \\ \vdots \\ x_{n1} \end{pmatrix}, X_2 = \begin{pmatrix} x_{12} \\ x_{22} \\ \vdots \\ x_{n2} \end{pmatrix}, \ldots, X_n = \begin{pmatrix} x_{1n} \\ x_{2n} \\ \vdots \\ x_{nn} \end{pmatrix}$$

be n solution vectors of the homogeneous system (2) on an interval I. A necessary and sufficient condition that the set of solutions be linearly independent is that the Wronskian

$$W(X_1, X_2, \ldots, X_n) = \begin{vmatrix} x_{11} & x_{12} & \cdots & x_{1n} \\ x_{21} & x_{22} & \cdots & x_{2n} \\ \vdots & & & \vdots \\ x_{n1} & x_{n2} & \cdots & x_{nn} \end{vmatrix} \neq 0 \qquad (9)$$

for every t in I. ☐

In fact, it can be shown that if X_1, X_2, $\ldots$, X_n are solution vectors of (2), then either

$$W(X_1, X_2, \ldots, X_n) \neq 0$$

for every t in I, or

$$W(X_1, X_2, \ldots, X_n) = 0$$

for every t in the interval. Thus if we can show that $W \neq 0$ for some t_0 in I, then $W \neq 0$ for every t and hence the solutions are linearly independent on the interval.

Notice that, unlike our previous definition of the Wronskian, the determinant (9) does not involve differentiation.

EXAMPLE   We have already seen that

$$\mathbf{X}_1 = \begin{pmatrix} 1 \\ -1 \end{pmatrix} e^{-2t} \quad \text{and} \quad \mathbf{X}_2 = \begin{pmatrix} 3 \\ 5 \end{pmatrix} e^{-6t}$$

are solutions of the system (5). Clearly, $\mathbf{X}_1$ and $\mathbf{X}_2$ are linearly independent on $-\infty < t < \infty$ since neither vector is a constant multiple of the other. In addition, we have

$$W(\mathbf{X}_1, \mathbf{X}_2) = \begin{vmatrix} e^{-2t} & 3e^{6t} \\ -e^{-2t} & 5e^{6t} \end{vmatrix}$$

$$= 8e^{4t}$$

$$\neq 0$$

for all real values of t.

Fundamental set of solutions

DEFINITION 8.14   Any set $\mathbf{X}_1, \mathbf{X}_2, \ldots, \mathbf{X}_n$ of n linearly independent solution vectors of the homogeneous system (2) on an interval I is said to be a **fundamental set of solutions** on the interval.   ☐

THEOREM 8.6   There exists a fundamental set of solutions for the homogeneous system (2) on an interval I.   ☐

DEFINITION 8.15   Let $\mathbf{X}_1, \mathbf{X}_2, \ldots, \mathbf{X}_n$ be a fundamental set of solutions of the homogeneous system (2) on an interval I. The **general solution** of the system on the interval is defined to be

$$\mathbf{X} = c_1\mathbf{X}_1 + c_2\mathbf{X}_2 + \cdots + c_n\mathbf{X}_n,$$

where the c_i, $i = 1, 2, \ldots, n$ are arbitrary constants.   ☐

Although we shall not give the proof, it can be shown that, for appropriate choices of the constants $c_1, c_2, \ldots, c_n$, *any* solution of (2) on the interval I can be obtained from the general solution.

EXAMPLE   In the last example we saw that

$$\mathbf{X}_1 = \begin{pmatrix} 1 \\ -1 \end{pmatrix} e^{-2t} \quad \text{and} \quad \mathbf{X}_2 = \begin{pmatrix} 3 \\ 5 \end{pmatrix} e^{6t}$$

are linearly independent solutions of (5) on $-\infty < t < \infty$. Hence $\mathbf{X}_1$ and $\mathbf{X}_2$ form a fundamental set of solutions on the interval. The general solution of the system on the interval is then

$$\mathbf{X} = c_1\mathbf{X}_1 + c_2\mathbf{X}_2$$

$$= c_1\begin{pmatrix} 1 \\ -1 \end{pmatrix} e^{-2t} + c_2\begin{pmatrix} 3 \\ 5 \end{pmatrix} e^{6t}. \tag{10}$$

EXAMPLE

The vectors

$$\mathbf{X}_1 = \begin{pmatrix} \cos t \\ -\tfrac{1}{2}\cos t + \tfrac{1}{2}\sin t \\ -\cos t - \sin t \end{pmatrix}, \qquad \mathbf{X}_2 = \begin{pmatrix} 0 \\ 1 \\ 0 \end{pmatrix} e^t,$$

and

$$\mathbf{X}_3 = \begin{pmatrix} \sin t \\ -\tfrac{1}{2}\sin t - \tfrac{1}{2}\cos t \\ -\sin t + \cos t \end{pmatrix}$$

are solutions of the system (7).*

Now

$$W(\mathbf{X}_1, \mathbf{X}_2, \mathbf{X}_3) = \begin{vmatrix} \cos t & 0 & \sin t \\ -\tfrac{1}{2}\cos t + \tfrac{1}{2}\sin t & e^t & -\tfrac{1}{2}\sin t - \tfrac{1}{2}\cos t \\ -\cos t - \sin t & 0 & -\sin t + \cos t \end{vmatrix}$$

$$= e^t \begin{vmatrix} \cos t & \sin t \\ -\cos t - \sin t & -\sin t + \cos t \end{vmatrix}$$

$$= e^t$$

$$\neq 0$$

for all real values of t. We conclude that $\mathbf{X}_1$, $\mathbf{X}_2$ and $\mathbf{X}_3$ form a fundamental set of solutions on $-\infty < t < \infty$. Thus the general solution of the system on the interval is

$$\mathbf{X} = c_1\mathbf{X}_1 + c_2\mathbf{X}_2 + c_3\mathbf{X}_3$$

$$= c_1\begin{pmatrix} \cos t \\ -\tfrac{1}{2}\cos t + \tfrac{1}{2}\sin t \\ -\cos t - \sin t \end{pmatrix} + c_2\begin{pmatrix} 0 \\ 1 \\ 0 \end{pmatrix} e^t + c_3\begin{pmatrix} \sin t \\ -\tfrac{1}{2}\sin t - \tfrac{1}{2}\cos t \\ -\sin t + \cos t \end{pmatrix}.$$

Nonhomogeneous systems

For nonhomogeneous systems a **particular solution $\mathbf{X}_p$** on an interval I is any vector, free of arbitrary parameters, whose entries are functions satisfying the system (4).

---

*On pages 362 and 363 it was verified that $\mathbf{X}_1$ and $\mathbf{X}_2$ are solutions; we leave it as an exercise to demonstrate that $\mathbf{X}_3$ is also a solution.

EXAMPLE

Verify that the vector

$$\mathbf{X}_p = \begin{pmatrix} 3t - 4 \\ -5t + 6 \end{pmatrix}$$

is a particular solution of the system

$$\mathbf{X}' = \begin{pmatrix} 1 & 3 \\ 5 & 3 \end{pmatrix} \mathbf{X} + \begin{pmatrix} 12t - 11 \\ -3 \end{pmatrix}$$

on $-\infty < t < \infty$.

Solution: We have $\mathbf{X}_p' = \begin{pmatrix} 3 \\ -5 \end{pmatrix}$ and

$$\begin{pmatrix} 1 & 3 \\ 5 & 3 \end{pmatrix} \mathbf{X}_p + \begin{pmatrix} 12t - 11 \\ -3 \end{pmatrix} = \begin{pmatrix} 1 & 3 \\ 5 & 3 \end{pmatrix} \begin{pmatrix} 3t - 4 \\ -5t + 6 \end{pmatrix} + \begin{pmatrix} 12t - 11 \\ -3 \end{pmatrix}$$

$$= \begin{pmatrix} (3t - 4) + 3(-5t + 6) \\ 5(3t - 4) + 3(-5t + 6) \end{pmatrix} + \begin{pmatrix} 12t - 11 \\ -3 \end{pmatrix}$$

$$= \begin{pmatrix} -12t + 14 \\ -2 \end{pmatrix} + \begin{pmatrix} 12t - 11 \\ -3 \end{pmatrix}$$

$$= \begin{pmatrix} 3 \\ -5 \end{pmatrix} = \mathbf{X}_p'.$$

THEOREM 8.7   Let $\mathbf{X}_1, \mathbf{X}_2, \ldots, \mathbf{X}_k$ be a set of solution vectors of the homogeneous system (2) on an interval I and let $\mathbf{X}_p$ be any solution vector of the nonhomogeneous system (4) on the same interval. Then

$$\mathbf{X} = c_1\mathbf{X}_1 + c_2\mathbf{X}_2 + \cdots + c_k\mathbf{X}_k + \mathbf{X}_p$$

is also a solution of the nonhomogeneous system on the interval for any constants $c_1, c_2, \ldots, c_k$. $\square$

DEFINITION 8.16   Let $\mathbf{X}_p$ be a given solution of the nonhomogeneous system (4) on an interval I, and let

$$\mathbf{X}_c = c_1\mathbf{X}_1 + c_2\mathbf{X}_2 + \cdots + c_n\mathbf{X}_n$$

denote the general solution on the same interval of the corresponding homogeneous system (2). The **general solution** of the nonhomogeneous system on the interval is defined to be

$$\mathbf{X} = \mathbf{X}_c + \mathbf{X}_p.$$ $\square$

The general solution $\mathbf{X}_c$ of the homogeneous system (2) is called the **complementary function** of the nonhomogeneous system (4).

EXAMPLE

In the preceding example it was verified that a particular solution of

$$X' = \begin{pmatrix} 1 & 3 \\ 5 & 3 \end{pmatrix} X + \begin{pmatrix} 12t - 11 \\ -3 \end{pmatrix} \tag{11}$$

on $-\infty < t < \infty$ is

$$X_p = \begin{pmatrix} 3t - 4 \\ -5t + 6 \end{pmatrix}.$$

The complementary function of (11) on the same interval, or general solution of

$$X' = \begin{pmatrix} 1 & 3 \\ 5 & 3 \end{pmatrix} X,$$

has already been seen to be

$$X_c = c_1 \begin{pmatrix} 1 \\ -1 \end{pmatrix} e^{-2t} + c_2 \begin{pmatrix} 3 \\ 5 \end{pmatrix} e^{6t}.$$

Hence by Definition 8.16

$$X = X_c + X_p$$

$$= c_1 \begin{pmatrix} 1 \\ -1 \end{pmatrix} e^{-2t} + c_2 \begin{pmatrix} 3 \\ 5 \end{pmatrix} e^{6t} + \begin{pmatrix} 3t - 4 \\ -5t + 6 \end{pmatrix}.$$

is the general solution of (11) on $-\infty < t < \infty$.

As one might expect, if X is *any* solution of the nonhomogeneous system (4) on an interval I, then it is always possible to find appropriate constants $c_1, c_2, \ldots, c_n$ so that X can be obtained from the general solution.

8.5.2   A Fundamental Matrix

If $X_1, X_2, \ldots, X_n$ is a fundamental set of solutions of the homogeneous system (2) on an interval I then its general solution on the interval is

$$X = c_1 X_1 + c_2 X_2 + \cdots + c_n X_n$$

$$= c_1 \begin{pmatrix} x_{11} \\ x_{21} \\ \vdots \\ x_{n1} \end{pmatrix} + c_2 \begin{pmatrix} x_{12} \\ x_{22} \\ \vdots \\ x_{n2} \end{pmatrix} + \cdots + c_n \begin{pmatrix} x_{1n} \\ x_{2n} \\ \vdots \\ x_{nn} \end{pmatrix}$$

$$= \begin{pmatrix} c_1 x_{11} + c_2 x_{12} + \cdots + c_n x_{1n} \\ c_1 x_{21} + c_2 x_{22} + \cdots + c_n x_{2n} \\ \vdots \\ c_1 x_{n1} + c_2 x_{n2} + \cdots + c_n x_{nn} \end{pmatrix}. \tag{12}$$

Observe that (12) can be written as the matrix product

$$\mathbf{X} = \begin{pmatrix} x_{11} & x_{12} & \cdots & x_{1n} \\ x_{21} & x_{22} & \cdots & x_{2n} \\ \vdots & & & \vdots \\ x_{n1} & x_{n2} & \cdots & x_{nn} \end{pmatrix} \begin{pmatrix} c_1 \\ c_2 \\ \vdots \\ c_n \end{pmatrix} \tag{13}$$

We are led to the following definition.

DEFINITION 8.17   Let

$$\mathbf{X}_1 = \begin{pmatrix} x_{11} \\ x_{21} \\ \vdots \\ x_{n1} \end{pmatrix}, \; \mathbf{X}_2 = \begin{pmatrix} x_{12} \\ x_{22} \\ \vdots \\ x_{n2} \end{pmatrix}, \; \ldots, \; \mathbf{X}_n = \begin{pmatrix} x_{1n} \\ x_{2n} \\ \vdots \\ x_{nn} \end{pmatrix}$$

be a fundamdental set of n solution vectors of the homogeneous system (2) on an interval I. The matrix

$$\mathbf{\Phi}(t) = \begin{pmatrix} x_{11} & x_{12} & \cdots & x_{1n} \\ x_{21} & x_{22} & \cdots & x_{2n} \\ \vdots & & & \vdots \\ x_{n1} & x_{n2} & \cdots & x_{nn} \end{pmatrix}$$

is said to be a **fundamental matrix** of the system on the interval.   ☐

EXAMPLE   The vectors

$$\mathbf{X}_1 = \begin{pmatrix} 1 \\ -1 \end{pmatrix} e^{-2t} = \begin{pmatrix} e^{-2t} \\ -e^{-2t} \end{pmatrix} \quad \text{and} \quad \mathbf{X}_2 = \begin{pmatrix} 3 \\ 5 \end{pmatrix} e^{6t} = \begin{pmatrix} 3e^{6t} \\ 5e^{6t} \end{pmatrix}$$

have been shown to form a fundamental set of solutions of the system (5) on $-\infty < t < \infty$. A fundamental matrix of the system on the interval is then

$$\mathbf{\Phi}(t) = \begin{pmatrix} e^{-2t} & 3e^{6t} \\ -e^{-2t} & 5e^{6t} \end{pmatrix}. \tag{14}$$

The result given in (13) states that the general solution of any homogeneous system $\mathbf{X}' = \mathbf{A}(t)\mathbf{X}$ can always be written in terms of a fundamental matrix of the system: $\mathbf{X} = \mathbf{\Phi}(t)\mathbf{C}$ where $\mathbf{C}$ is an $n \times 1$ column vector of arbitrary constants.

EXAMPLE   The general solution given in (10) can be written

$$\mathbf{X} = \begin{pmatrix} e^{-2t} & 3e^{6t} \\ -e^{-2t} & 5e^{6t} \end{pmatrix} \begin{pmatrix} c_1 \\ c_2 \end{pmatrix}.$$

Furthermore, to say that $\mathbf{X} = \mathbf{\Phi}(t)\mathbf{C}$ is a solution of $\mathbf{X}' = \mathbf{A}(t)\mathbf{X}$ we mean

$$\mathbf{\Phi}'(t)\mathbf{C} = \mathbf{A}(t)\mathbf{\Phi}(t)\mathbf{C}$$

or

$$(\mathbf{\Phi}'(t) - \mathbf{A}(t)\mathbf{\Phi}(t))\mathbf{C} = \mathbf{0}.$$

Since the last equation is to hold for every t in the interval I and for every possible column matrix of constants $\mathbf{C}$, we must have

$$\mathbf{\Phi}'(t) - \mathbf{A}(t)\mathbf{\Phi}(t) = \mathbf{0}$$

or

$$\mathbf{\Phi}'(t) = \mathbf{A}(t)\mathbf{\Phi}(t). \tag{15}$$

This result will be useful in the discussion on the solution of nonhomogeneous systems.

A fundamental matrix is nonsingular.

Comparison of Theorem 8.5 and Definition 8.17 shows that det $\mathbf{\Phi}(t)$ is the same as the Wronskian $W(\mathbf{X}_1, \mathbf{X}_2, \ldots, \mathbf{X}_n)$*. Hence the linear independence of the columns of $\mathbf{\Phi}(t)$ on an interval I guarantees that det $\mathbf{\Phi}(t) \neq 0$ for every t in the interval. That is, $\mathbf{\Phi}(t)$ is nonsingular on the interval.

> **THEOREM 8.8**   Let $\mathbf{\Phi}(t)$ be a fundamental matrix of the homogeneous system (2) on an interval I. Then $\mathbf{\Phi}^{-1}(t)$ exists for every value of t in the interval. $\square$

EXAMPLE

For the fundamental matrix given in (14) we see that det $\mathbf{\Phi}(t) = 8e^{4t}$. It then follows from (4) of Section 8.4 that

$$\mathbf{\Phi}^{-1}(t) = \frac{1}{8e^{4t}} \begin{pmatrix} 5e^{6t} & -3e^{6t} \\ e^{-2t} & e^{-2t} \end{pmatrix}$$

$$= \begin{pmatrix} \frac{5}{8}e^{2t} & -\frac{3}{8}e^{2t} \\ \frac{1}{8}e^{-6t} & \frac{1}{8}e^{-6t} \end{pmatrix}.$$

A special matrix

In some instances it is convenient to form another special $n \times n$ matrix; a matrix in which the column vectors $\mathbf{V}_i$ are solutions of $\mathbf{X}' = \mathbf{A}(t)\mathbf{X}$ that satisfy the conditions

$$\mathbf{V}_1(t_0) = \begin{pmatrix} 1 \\ 0 \\ \vdots \\ 0 \end{pmatrix}, \quad \mathbf{V}_2(t_0) = \begin{pmatrix} 0 \\ 1 \\ \vdots \\ 0 \end{pmatrix}, \quad \ldots, \quad \mathbf{V}_n(t_0) = \begin{pmatrix} 0 \\ 0 \\ \vdots \\ 1 \end{pmatrix}. \tag{16}$$

Here t_0 is an arbitrarily chosen point in the interval on which the general solution of the system is defined. We shall denote this special matrix by the symbol $\mathbf{\Psi}(t)$. Observe that $\mathbf{\Psi}(t)$ has the property

---

*For this reason some texts will call $\mathbf{\Phi}(t)$ a *Wronski matrix*.

$$\Psi(t_0) = \begin{pmatrix} 1 & 0 & 0 & \cdots & 0 \\ 0 & 1 & 0 & \cdots & 0 \\ \vdots & & & & \vdots \\ 0 & 0 & 0 & \cdots & 1 \end{pmatrix} = I \qquad (17)$$

where I is the $n \times n$ multiplicative identity.

EXAMPLE

Find the matrix $\Psi(t)$ satisfying $\Psi(0) = I$ for the system given in (5).

Solution: From (10) we know that the general solution of (5) is given by

$$X = c_1 \begin{pmatrix} 1 \\ -1 \end{pmatrix} e^{-2t} + c_2 \begin{pmatrix} 3 \\ 5 \end{pmatrix} e^{6t}.$$

When $t = 0$ we first solve for constants c_1 and c_2 such that

$$c_1 \begin{pmatrix} 1 \\ -1 \end{pmatrix} + c_2 \begin{pmatrix} 3 \\ 5 \end{pmatrix} = \begin{pmatrix} 1 \\ 0 \end{pmatrix}$$

or
$$c_1 + 3c_2 = 1$$
$$-c_1 + 5c_2 = 0.$$

We find that $c_1 = 5/8$ and $c_2 = 1/8$. Hence we define the vector V_1 to be the linear combination

$$V_1 = \frac{5}{8} \begin{pmatrix} 1 \\ -1 \end{pmatrix} e^{-2t} + \frac{1}{8} \begin{pmatrix} 3 \\ 5 \end{pmatrix} e^{6t}.$$

Again when $t = 0$ we wish to find another pair of constants c_1 and c_2 for which

$$c_1 \begin{pmatrix} 1 \\ -1 \end{pmatrix} + c_2 \begin{pmatrix} 3 \\ 5 \end{pmatrix} = \begin{pmatrix} 0 \\ 1 \end{pmatrix}$$

or
$$c_1 + 3c_2 = 0$$
$$-c_1 + 5c_2 = 1.$$

In this case we find $c_1 = -3/8$ and $c_2 = 1/8$. We then define

$$V_2 = -\frac{3}{8} \begin{pmatrix} 1 \\ -1 \end{pmatrix} e^{-2t} + \frac{1}{8} \begin{pmatrix} 3 \\ 5 \end{pmatrix} e^{6t}.$$

Hence

$$\Psi(t) = \begin{pmatrix} \frac{5}{8}e^{-2t} + \frac{3}{8}e^{6t} & -\frac{3}{8}e^{-2t} + \frac{3}{8}e^{6t} \\ -\frac{5}{8}e^{-2t} + \frac{5}{8}e^{6t} & \frac{3}{8}e^{-2t} + \frac{5}{8}e^{6t} \end{pmatrix} \qquad (18)$$

Observe that $\Psi(0) = \begin{pmatrix} 1 & 0 \\ 0 & 1 \end{pmatrix} = I$.

Note in the preceding example that since the columns of $\Psi(t)$ are linear combinations of the solutions of $X' = A(t)X$ we know from the superposition principle that each column is a solution of the system.

Ψ(t) is a
fundamental matrix

From (17) it is seen that det $\Psi(t_0) \neq 0$ and hence we conclude from Theorem 8.5 that the columns of $\Psi(t)$ are linearly independent on the interval under consideration. Therefore $\Psi(t)$ is a fundamental matrix. Also, it follows from Theorem 8.3 that $\Psi(t)$ is the unique matrix satisfying the condition $\Psi(t_0) = \mathbf{I}$. Lastly, the fundamental matrices $\Phi(t)$ and $\Psi(t)$ are related by

$$\Psi(t) = \Phi(t)\Phi^{-1}(t_0). \tag{19}$$

Equation (19) provides an alternative method for determining $\Psi(t)$. (See Problem 37.)

The answer to why anyone would want to form an obviously complicated looking fundamental matrix such as (18) will be answered in Sections 8.8 and 8.9.

EXERCISES 8.5   *Answers to odd-numbered problems begin on page A–26.*

[8.5.1]   In Problems 1–6 write the given system in matrix form.

1. $\dfrac{dx}{dt} = 3x - 5y$

$\dfrac{dy}{dt} = 4x + 8y$

2. $\dfrac{dx}{dt} = 4x - 7y$

$\dfrac{dy}{dt} = 5x$

3. $\dfrac{dx}{dt} = -3x + 4y - 9z$

$\dfrac{dy}{dt} = 6x - y$

$\dfrac{dz}{dt} = 10x + 4y + 3z$

4. $\dfrac{dx}{dt} = x - y$

$\dfrac{dy}{dt} = x + 2z$

$\dfrac{dz}{dt} = -x + z$

5. $\dfrac{dx}{dt} = x - y + z + t - 1$

$\dfrac{dy}{dt} = 2x + y - z - 3t^2$

$\dfrac{dz}{dt} = x + y + z + t^2 - t + 2$

★6. $\dfrac{dx}{dt} = -3x + 4y + e^{-t}\sin 2t$

$\dfrac{dy}{dt} = 5x + 9y + 4e^{-t}\cos 2t$

In Problems 7–10 write the given system without the use of matrices.

7. $\mathbf{X}' = \begin{pmatrix} 4 & 2 \\ -1 & 3 \end{pmatrix} \mathbf{X} + \begin{pmatrix} 1 \\ -1 \end{pmatrix} e^t$

8. $\mathbf{X'} = \begin{pmatrix} 7 & 5 & -9 \\ 4 & 1 & 1 \\ 0 & -2 & 3 \end{pmatrix} \mathbf{X} + \begin{pmatrix} 0 \\ 2 \\ 1 \end{pmatrix} e^{5t} - \begin{pmatrix} 8 \\ 0 \\ 3 \end{pmatrix} e^{-2t}$

9. $\dfrac{d}{dt}\begin{pmatrix} x \\ y \\ z \end{pmatrix} = \begin{pmatrix} 1 & -1 & 2 \\ 3 & -4 & 1 \\ -2 & 5 & 6 \end{pmatrix}\begin{pmatrix} x \\ y \\ z \end{pmatrix} + \begin{pmatrix} 1 \\ 2 \\ 2 \end{pmatrix} e^{-t} - \begin{pmatrix} 3 \\ -1 \\ 1 \end{pmatrix} t$

★10. $\dfrac{d}{dt}\begin{pmatrix} x \\ y \end{pmatrix} = \begin{pmatrix} 3 & -7 \\ 1 & 1 \end{pmatrix}\begin{pmatrix} x \\ y \end{pmatrix} + \begin{pmatrix} 4 \\ 8 \end{pmatrix} \sin t + \begin{pmatrix} t - 4 \\ 2t + 1 \end{pmatrix} e^{4t}$

In Problems 11–16 verify that the vector **X** is a solution of the given system.

11. $\dfrac{dx}{dt} = 3x - 4y$

$\dfrac{dy}{dt} = 4x - 7y; \quad \mathbf{X} = \begin{pmatrix} 1 \\ 2 \end{pmatrix} e^{-5t}$

12. $\dfrac{dx}{dt} = -2x + 5y$

$\dfrac{dy}{dt} = -2x + 4y; \quad \mathbf{X} = \begin{pmatrix} 5 \cos t \\ 3 \cos t - \sin t \end{pmatrix} e^{t}$

13. $\mathbf{X'} = \begin{pmatrix} -1 & \frac{1}{4} \\ 1 & -1 \end{pmatrix} \mathbf{X}; \quad \mathbf{X} = \begin{pmatrix} -1 \\ 2 \end{pmatrix} e^{-3t/2}$

14. $\mathbf{X'} = \begin{pmatrix} 2 & 0 \\ -1 & 0 \end{pmatrix} \mathbf{X}; \quad \mathbf{X} = \begin{pmatrix} 1 \\ 3 \end{pmatrix} e^{t} + \begin{pmatrix} 4 \\ -4 \end{pmatrix} te^{t}$

15. $\dfrac{d\mathbf{X}}{dt} = \begin{pmatrix} 1 & 2 & 1 \\ 6 & -1 & 0 \\ -1 & -2 & -1 \end{pmatrix} \mathbf{X}; \quad \mathbf{X} = \begin{pmatrix} 1 \\ 6 \\ -13 \end{pmatrix}$

16. $\mathbf{X'} = \begin{pmatrix} 1 & 0 & 1 \\ 1 & 1 & 0 \\ -2 & 0 & -1 \end{pmatrix} \mathbf{X}; \quad \mathbf{X} = \begin{pmatrix} \sin t \\ -\frac{1}{2}\sin t - \frac{1}{2}\cos t \\ -\sin t + \cos t \end{pmatrix}$

In Problems 17–20 the given vectors are solutions of a system $\mathbf{X'} = \mathbf{AX}$. Determine whether the vectors form a fundamental set on $-\infty < t < \infty$.

17. $\mathbf{X_1} = \begin{pmatrix} 1 \\ 1 \end{pmatrix} e^{-2t}, \quad \mathbf{X_2} = \begin{pmatrix} 1 \\ -1 \end{pmatrix} e^{-6t}$

★18. $\mathbf{X_1} = \begin{pmatrix} 1 \\ -1 \end{pmatrix} e^{t}, \quad \mathbf{X_2} = \begin{pmatrix} 2 \\ 6 \end{pmatrix} e^{t} + \begin{pmatrix} 8 \\ -8 \end{pmatrix} te^{t}$

19. $\mathbf{X_1} = \begin{pmatrix} 1 \\ -2 \\ 4 \end{pmatrix} + t\begin{pmatrix} 1 \\ 2 \\ 2 \end{pmatrix}, \quad \mathbf{X_2} = \begin{pmatrix} 1 \\ -2 \\ 4 \end{pmatrix}, \quad \mathbf{X_3} = \begin{pmatrix} 3 \\ -6 \\ 12 \end{pmatrix} + t\begin{pmatrix} 2 \\ 4 \\ 4 \end{pmatrix}$

20. $X_1 = \begin{pmatrix} 1 \\ 6 \\ -13 \end{pmatrix}$, $X_2 = \begin{pmatrix} 1 \\ -2 \\ -1 \end{pmatrix} e^{-4t}$, $X_3 = \begin{pmatrix} 2 \\ 3 \\ -2 \end{pmatrix} e^{3t}$

In Problems 21–24 verify that the vector X_p is a particular solution of the given system.

21. $\dfrac{dx}{dt} = x + 4y + 2t - 7$

$\dfrac{dy}{dt} = 3x + 2y - 4t - 18$; $X_p = \begin{pmatrix} 2 \\ -1 \end{pmatrix} t + \begin{pmatrix} 5 \\ 1 \end{pmatrix}$

22. $X' = \begin{pmatrix} 2 & 1 \\ 1 & -1 \end{pmatrix} X + \begin{pmatrix} -5 \\ 2 \end{pmatrix}$; $X_p = \begin{pmatrix} 1 \\ 3 \end{pmatrix}$

23. $X' = \begin{pmatrix} 2 & 1 \\ 3 & 4 \end{pmatrix} X - \begin{pmatrix} 1 \\ 7 \end{pmatrix} e^t$; $X_p = \begin{pmatrix} 1 \\ 1 \end{pmatrix} e^t + \begin{pmatrix} 1 \\ -1 \end{pmatrix} te^t$

24. $X' = \begin{pmatrix} 1 & 2 & 3 \\ -4 & 2 & 0 \\ -6 & 1 & 0 \end{pmatrix} X + \begin{pmatrix} -1 \\ 4 \\ 3 \end{pmatrix} \sin 3t$; $X_p = \begin{pmatrix} \sin 3t \\ 0 \\ \cos 3t \end{pmatrix}$

25. Prove that the general solution of

$$X' = \begin{pmatrix} 0 & 6 & 0 \\ 1 & 0 & 1 \\ 1 & 1 & 0 \end{pmatrix} X$$

on $-\infty < t < \infty$ is

$$X = c_1 \begin{pmatrix} 6 \\ -1 \\ -5 \end{pmatrix} e^{-t} + c_2 \begin{pmatrix} -3 \\ 1 \\ 1 \end{pmatrix} e^{-2t} + c_3 \begin{pmatrix} 2 \\ 1 \\ 1 \end{pmatrix} e^{3t}.$$

26. Prove that the general solution of

$$X' = \begin{pmatrix} -1 & -1 \\ -1 & 1 \end{pmatrix} X + \begin{pmatrix} 1 \\ 1 \end{pmatrix} t^2 + \begin{pmatrix} 4 \\ -6 \end{pmatrix} t + \begin{pmatrix} -1 \\ 5 \end{pmatrix}$$

on $-\infty < t < \infty$ is

$$X = c_1 \begin{pmatrix} 1 \\ -1 - \sqrt{2} \end{pmatrix} e^{\sqrt{2}t} + c_2 \begin{pmatrix} 1 \\ -1 + \sqrt{2} \end{pmatrix} e^{-\sqrt{2}t}$$

$$+ \begin{pmatrix} 1 \\ 0 \end{pmatrix} t^2 + \begin{pmatrix} -2 \\ 4 \end{pmatrix} t + \begin{pmatrix} 1 \\ 0 \end{pmatrix}.$$

[8.5.2]   In Problems 27–30 the indicated column vectors form a fundamental set of solutions for the given system on $-\infty < t < \infty$. Form a fundamental matrix $\Phi(t)$ and compute $\Phi^{-1}(t)$.

27. $\mathbf{X}' = \begin{pmatrix} 4 & 1 \\ 6 & 5 \end{pmatrix} \mathbf{X}; \quad \mathbf{X}_1 = \begin{pmatrix} 1 \\ -2 \end{pmatrix} e^{2t}, \quad \mathbf{X}_2 = \begin{pmatrix} 1 \\ 3 \end{pmatrix} e^{7t}$

★28. $\mathbf{X}' = \begin{pmatrix} 2 & 3 \\ 3 & 2 \end{pmatrix} \mathbf{X}; \quad \mathbf{X}_1 = \begin{pmatrix} -1 \\ 1 \end{pmatrix} e^{-t}, \quad \mathbf{X}_2 = \begin{pmatrix} 1 \\ 1 \end{pmatrix} e^{5t}$

29. $\mathbf{X}' = \begin{pmatrix} 4 & 1 \\ -9 & -2 \end{pmatrix} \mathbf{X}; \quad \mathbf{X}_1 = \begin{pmatrix} -1 \\ 3 \end{pmatrix} e^{t},$

$\mathbf{X}_2 = \begin{pmatrix} -1 \\ 3 \end{pmatrix} te^{t} + \begin{pmatrix} 0 \\ -1 \end{pmatrix} e^{t}$

30. $\mathbf{X}' = \begin{pmatrix} 3 & -2 \\ 5 & -3 \end{pmatrix} \mathbf{X}; \quad \mathbf{X}_1 = \begin{pmatrix} 2 \cos t \\ 3 \cos t + \sin t \end{pmatrix},$

$\mathbf{X}_2 = \begin{pmatrix} -2 \sin t \\ \cos t - 3 \sin t \end{pmatrix}$

31. Find the fundamental matrix $\mathbf{\Psi}(t)$ satisfying $\mathbf{\Psi}(0) = \mathbf{I}$ for the system given in Problem 27.

★32. Find the fundamental matrix $\mathbf{\Psi}(t)$ satisfying $\mathbf{\Psi}(0) = \mathbf{I}$ for the system given in Problem 28.

33. Find the fundamental matrix $\mathbf{\Psi}(t)$ satisfying $\mathbf{\Psi}(0) = \mathbf{I}$ for the system given in Problem 29.

34. Find the fundamental matrix $\mathbf{\Psi}(t)$ satisfying $\mathbf{\Psi}(\pi/2) = \mathbf{I}$ for the system given in Problem 30.

Miscellaneous problems

35. If $\mathbf{X} = \mathbf{\Phi}(t)\mathbf{C}$ is the general solution of $\mathbf{X}' = \mathbf{A}\mathbf{X}$ show that the solution of the initial-value problem $\mathbf{X}' = \mathbf{A}\mathbf{X}$, $\mathbf{X}(t_0) = \mathbf{X}_0$ is $\mathbf{X} = \mathbf{\Phi}(t)\mathbf{\Phi}^{-1}(t_0)\mathbf{X}_0$.

36. Show that the solution of the initial-value problem given in problem 35 is also given by $\mathbf{X} = \mathbf{\Psi}(t)\mathbf{X}_0$.

37. Show that $\mathbf{\Psi}(t) = \mathbf{\Phi}(t)\mathbf{\Phi}^{-1}(t_0)$. [*Hint:* Compare Problems 35 and 36.]

8.6   Homogeneous Linear Systems

8.6.1   Distinct Real Eigenvalues

For the remainder of this chapter we shall be concerned only with linear systems with real constant coefficients.

We saw in the preceding section that the general solution of the system

$$\frac{dx}{dt} = x + 3y$$

$$\frac{dy}{dt} = 5x + 3y$$

is
$$\mathbf{X} = c_1 \begin{pmatrix} 1 \\ -1 \end{pmatrix} e^{-2t} + c_2 \begin{pmatrix} 3 \\ 5 \end{pmatrix} e^{6t}.$$

Since both solution vectors have the basic form

$$\mathbf{X}_i = \begin{pmatrix} k_1 \\ k_2 \end{pmatrix} e^{\lambda_i t} \qquad i = 1, 2,$$

k_1 and k_2 constants, we are prompted to ask whether we can always find a solution of the form

$$\mathbf{X} = \begin{pmatrix} k_1 \\ k_2 \\ \vdots \\ k_n \end{pmatrix} e^{\lambda t} = \mathbf{K} e^{\lambda t} \tag{1}$$

for the general homogeneous linear first-order system

$$\mathbf{X}' = \mathbf{A}\mathbf{X} \tag{2}$$

where $\mathbf{A}$ is an $n \times n$ matrix of constants. Given the discussion of Section 8.4 and the fact that a system such as (2) with constant coefficients can be reduced to a single linear nth-order differential equation, it should not come as any surprise that there do indeed exist solutions of form (1).

Eigenvalues and eigenvectors

If we make the assumption that (1) is a solution vector of (2) then

$$\mathbf{X}' = \mathbf{K}\lambda e^{\lambda t}$$

so that the system becomes

$$\mathbf{K}\lambda e^{\lambda t} = \mathbf{A}\mathbf{K} e^{\lambda t}.$$

After dividing out $e^{\lambda t}$ and rearranging we obtain

$$\mathbf{A}\mathbf{K} = \lambda \mathbf{K}$$

or
$$(\mathbf{A} - \lambda \mathbf{I})\mathbf{K} = \mathbf{0}. \tag{3}$$

Equation (3) is equivalent to the simultaneous algebraic equations

$$\begin{aligned} (a_{11} - \lambda)k_1 + \quad a_{12}k_2 + \cdots + \quad a_{1n}k_n &= 0 \\ a_{21}k_1 + (a_{22} - \lambda)k_2 + \cdots + \quad a_{2n}k_n &= 0 \\ \vdots \\ a_{n1}k_1 + \quad a_{n2}k_2 + \cdots + (a_{nn} - \lambda)k_n &= 0 \end{aligned} \tag{4}$$

Although an immediate solution to this homogeneous system is $k_1 = k_2 = \cdots = k_n = 0$, we are naturally seeking only nonzero solutions. Recall from algebra, a necessary and sufficient condition for the existence of a nonzero solution for systems such as (4) is that the determinant of the coefficients must be zero. That is,

$$\mathbf{K} = \begin{pmatrix} k_1 \\ k_2 \\ \vdots \\ k_n \end{pmatrix}$$

is a nonzero solution vector of (3) if and only if λ is a solution of

$$\det(\mathbf{A} - \lambda\mathbf{I}) = 0. \tag{5}$$

Expanding the determinant shows that (5) is an nth-degree polynomial equation

Eigenvalues and eigenvectors

The values of λ satisfying (5) are said to be **eigenvalues*** of the matrix $\mathbf{A}$. A solution vector $\mathbf{K}$ corresponding to a specific value of λ is called an **eigenvector.** Equation (5) is also called the **characteristic equation** of the matrix $\mathbf{A}$.

When the $n \times n$ matrix $\mathbf{A}$ possesses n distinct real eigenvalues λ_1, $\lambda_2, \ldots, \lambda_n$ then a set of n linearly independent eigenvectors $\mathbf{K}_1, \mathbf{K}_2, \ldots, \mathbf{K}_n$ can always be found and

$$\mathbf{X}_1 = \mathbf{K}_1 e^{\lambda_1 t}, \ \mathbf{X}_2 = \mathbf{K}_2 e^{\lambda_2 t}, \ \ldots, \ \mathbf{X}_n = \mathbf{K}_n e^{\lambda_n t}$$

is a fundamental set of solutions of (2) on $-\infty < t < \infty$.

THEOREM 8.9   Let $\lambda_1, \lambda_2, \ldots, \lambda_n$ be n distinct real eigenvalues of the coefficient matrix $\mathbf{A}$ of the homogeneous system (2) and let $\mathbf{K}_1$, $\mathbf{K}_2, \ldots, \mathbf{K}_n$ be the corresponding eigenvectors. Then the general solution of (2) on the interval $-\infty < t < \infty$ is given by

$$\mathbf{X} = c_1\mathbf{K}_1 e^{\lambda_1 t} + c_2\mathbf{K}_2 e^{\lambda_2 t} + \cdots + c_n\mathbf{K}_n e^{\lambda_n t}. \qquad \square$$

EXAMPLE

Solve

$$\frac{dx}{dt} = 2x + 3y$$

$$\frac{dy}{dt} = 2x + y. \tag{6}$$

Solution: The assumed solution

$$\mathbf{X} = \begin{pmatrix} x(t) \\ y(t) \end{pmatrix} = \begin{pmatrix} k_1 \\ k_2 \end{pmatrix} e^{\lambda t}$$

leads to the system

$$\begin{aligned} (2 - \lambda)k_1 + \quad\quad 3k_2 &= 0 \\ 2k_1 + (1 - \lambda)k_2 &= 0. \end{aligned} \tag{7}$$

And so in order to have nonzero solutions, λ must satisfy

$$\begin{vmatrix} 2 - \lambda & 3 \\ 2 & 1 - \lambda \end{vmatrix} = 0$$

or

$$(2 - \lambda)(1 - \lambda) - 6 = 0$$
$$\lambda^2 - 3\lambda - 4 = 0$$
$$(\lambda + 1)(\lambda - 4) = 0.$$

Thus the eigenvalues are $\lambda_1 = -1$ and $\lambda_2 = 4$. For $\lambda_1 = -1$ the system (7) becomes

---

*From the German word *eigenwert* meaning characteristic or particular value.

$$3k_1 + 3k_2 = 0$$
$$2k_1 + 2k_2 = 0$$

which implies that $k_2 = -k_1$. The related eigenvector is then

$$\begin{pmatrix} k_1 \\ k_2 \end{pmatrix} = k_1 \begin{pmatrix} 1 \\ -1 \end{pmatrix} \tag{8}$$

For $\lambda_2 = 4$ we find

$$-2k_1 + 3k_2 = 0$$
$$2k_1 - 3k_2 = 0$$

so that $k_2 = 2k_1/3$ and, therefore, the corresponding eigenvector is

$$\begin{pmatrix} k_1 \\ k_2 \end{pmatrix} = k_1 \begin{pmatrix} 1 \\ 2/3 \end{pmatrix} \tag{9}$$

We are really interested only in the basic eigenvectors in (8) and (9), hence to avoid confusion and to emphasize the fact that the values of k_1 in (8) and (9) *are not related*, we can choose $k_1 = 1$ in each case. Now

$$\mathbf{X}_1 = \begin{pmatrix} 1 \\ -1 \end{pmatrix} e^{-t} \quad \text{and} \quad \mathbf{X}_2 = \begin{pmatrix} 1 \\ 2/3 \end{pmatrix} e^{4t}$$

are solutions of (6). Hence the general solution of the system is

$$\mathbf{X} = c_1 \mathbf{X}_1 + c_2 \mathbf{X}_2$$
$$= c_1 \begin{pmatrix} 1 \\ -1 \end{pmatrix} e^{-t} + c_2 \begin{pmatrix} 1 \\ 2/3 \end{pmatrix} e^{4t}. \tag{10}$$

For the sake of review, the reader should keep firmly in mind that a solution of a system of first-order differential equations, when written in terms of matrices, is simply an alternative to the method that we employed in Section 8.1, namely, listing the individual functions and the relationships between the constants. By adding the vectors given in (10) we obtain

$$\begin{pmatrix} x(t) \\ y(t) \end{pmatrix} = \begin{pmatrix} c_1 e^{-t} + c_2 e^{4t} \\ -c_1 e^{-t} + \frac{2}{3} c_2 e^{4t} \end{pmatrix}$$

and this in turn yields the more familiar statement

$$x(t) = c_1 e^{-t} + c_2 e^{4t}$$
$$y(t) = -c_1 e^{-t} + \tfrac{2}{3} c_2 e^{4t}.$$

Also, it should be noted that we could just as well choose $k_1 = 1$ in (8) and $k_1 = 3$ in (9) so that an alternative, but equivalent, form of (10) is

$$\mathbf{X} = c_1 \begin{pmatrix} 1 \\ -1 \end{pmatrix} e^{-t} + c_2 \begin{pmatrix} 3 \\ 2 \end{pmatrix} e^{4t}. \tag{11}$$

EXAMPLE　　　　Solve　　　　　$$\dfrac{dx}{dt} = x + 2y + z$$

$$\frac{dy}{dt} = 6x - y \tag{12}$$

$$\frac{dz}{dt} = -x - 2y - z.$$

Solution: To expand the determinant in the characteristic equation

$$\begin{vmatrix} 1 - \lambda & 2 & 1 \\ 6 & -1 - \lambda & 0 \\ -1 & -2 & -1 - \lambda \end{vmatrix} = 0$$

we use the cofactors of the second row. It follows that

$$-\lambda^3 - \lambda^2 + 12\lambda = 0 \qquad \text{or} \qquad \lambda(\lambda + 4)(\lambda - 3) = 0.$$

Hence the eigenvalues are $\lambda_1 = 0$, $\lambda_2 = -4$, $\lambda_3 = 3$. We must now determine three eigenvectors of the form

$$\mathbf{K} = \begin{pmatrix} k_1 \\ k_2 \\ k_3 \end{pmatrix}$$

by solving the algebraic system

$$\begin{aligned}
(1 - \lambda)k_1 + \quad 2k_2 + \quad k_3 &= 0 \\
6k_1 - (1 + \lambda)k_2 \qquad\qquad &= 0 \\
-k_1 - \quad 2k_2 - (1 + \lambda)k_3 &= 0
\end{aligned} \tag{13}$$

three times corresponding to the three distinct values of λ.

For $\lambda_1 = 0$, (13) becomes

$$\begin{aligned}
k_1 + 2k_2 + k_3 &= 0 \\
6k_1 - \quad k_2 \quad\;\; &= 0 \\
-k_1 - 2k_2 - k_3 &= 0.
\end{aligned}$$

Choosing $k_1 = 1$ immediately gives $k_2 = 6$ and $k_3 = -13$. Hence the first eigenvector is

$$\mathbf{K}_1 = \begin{pmatrix} 1 \\ 6 \\ -13 \end{pmatrix}. \tag{14}$$

For $\lambda_2 = -4$, (13) becomes

$$\begin{aligned}
5k_1 + 2k_2 + \quad k_3 &= 0 \\
6k_1 + 3k_2 \qquad\;\; &= 0 \\
-k_1 - 2k_2 + 3k_3 &= 0.
\end{aligned}$$

Choosing $k_1 = 1$ gives $k_2 = -2$ and $k_3 = -1$. The second eigenvector is then

$$\mathbf{K}_2 = \begin{pmatrix} 1 \\ -2 \\ -1 \end{pmatrix}. \tag{15}$$

Finally, when $\lambda_3 = 3$, the system (13) becomes

$$-2k_1 + 2k_2 + \ k_3 = 0$$
$$6k_1 - 4k_2 \qquad = 0$$
$$-k_1 - 2k_2 - 4k_3 = 0.$$

The usual choice of $k_1 = 1$ leads to $k_2 = 3/2$ and $k_3 = -1$ so that the third eigenvector is

$$\mathbf{K}_3 = \begin{pmatrix} 1 \\ 3/2 \\ -1 \end{pmatrix}. \tag{16}$$

Multiplying the vectors (14), (15), and (16) by e^{0t}, e^{-4t}, and e^{3t}, respectively, yields three solutions of (12):

$$\mathbf{X}_1 = \begin{pmatrix} 1 \\ 6 \\ -13 \end{pmatrix}, \ \mathbf{X}_2 = \begin{pmatrix} 1 \\ -2 \\ -1 \end{pmatrix} e^{-4t}, \ \mathbf{X}_3 = \begin{pmatrix} 1 \\ 3/2 \\ -1 \end{pmatrix} e^{3t}.$$

The general solution of the system is

$$\mathbf{X} = c_1 \begin{pmatrix} 1 \\ 6 \\ -13 \end{pmatrix} + c_2 \begin{pmatrix} 1 \\ -2 \\ -1 \end{pmatrix} e^{-4t} + c_3 \begin{pmatrix} 1 \\ 3/2 \\ -1 \end{pmatrix} e^{3t}.$$

8.6.2   Complex Eigenvalues

If $\lambda_1 = \alpha + i\beta$ and $\lambda_2 = \alpha - i\beta$, $i^2 = -1$

are complex eigenvalues of the coefficient matrix $\mathbf{A}$, we can then certainly expect their corresponding eigenvectors to also have complex entries.*

For example, the characteristic equation of the system

$$\frac{dx}{dt} = 6x - \ y$$
$$\frac{dy}{dt} = 5x + 4y \tag{17}$$

is $\begin{vmatrix} 6 - \lambda & -1 \\ 5 & 4 - \lambda \end{vmatrix} = (6 - \lambda)(4 - \lambda) + 5$

$$= \lambda^2 - 10\lambda + 29$$

$$= 0.$$

From the quadratic formula we find

$$\lambda_1 = 5 + 2i, \qquad \lambda_2 = 5 - 2i.$$

---

*When the characteristic equation has real coefficients complex eigenvalues will always appear in conjugate pairs.

Now for $\lambda_1 = 5 + 2i$ we must solve

$$(1 - 2i)k_1 - \qquad k_2 = 0$$

$$5k_1 - (1 + 2i)k_2 = 0.$$

Since $k_2 = (1 - 2i)k_1$ it follows, after choosing $k_1 = 1$, that one eigenvector is

$$\mathbf{K}_1 = \begin{pmatrix} 1 \\ 1 - 2i \end{pmatrix}.$$

Similarly, for $\lambda_2 = 5 - 2i$, we find the other eigenvector to be

$$\mathbf{K}_2 = \begin{pmatrix} 1 \\ 1 + 2i \end{pmatrix}.$$

Consequently two solutions of (17) are

$$\mathbf{X}_1 = \begin{pmatrix} 1 \\ 1 - 2i \end{pmatrix} e^{(5 + 2i)t} \qquad \text{and} \qquad \mathbf{X}_2 = \begin{pmatrix} 1 \\ 1 + 2i \end{pmatrix} e^{(5 - 2i)t}.$$

By the superposition principle another solution is

$$\mathbf{X} = c_1 \begin{pmatrix} 1 \\ 1 - 2i \end{pmatrix} e^{(5 + 2i)t} + c_2 \begin{pmatrix} 1 \\ 1 + 2i \end{pmatrix} e^{(5 - 2i)t}. \tag{18}$$

Note that the entries in $\mathbf{K}_2$ corresponding to λ_2 are the complex conjugates of the entries in $\mathbf{K}_1$ corresponding to λ_1. The complex conjugate of λ_1 is, of course, λ_2. We write this as $\lambda_2 = \overline{\lambda}_1$ and $\mathbf{K}_2 = \overline{\mathbf{K}}_1$. We have illustrated the following general result.

THEOREM 8.10   Let $\mathbf{A}$ be the coefficient matrix having real entries of the homogeneous system (2), and let $\mathbf{K}$ be an eigenvector corresponding to the complex eigenvalue $\lambda_1 = \alpha + i\beta$, α and β real. Then

$$\mathbf{X}_1 = \mathbf{K}_1 e^{\lambda_1 t} \qquad \text{and} \qquad \mathbf{X}_2 = \overline{\mathbf{K}}_1 e^{\overline{\lambda}_1 t}$$

are solutions of (2).

It is desirable, and relatively easy to rewrite a solution such as (18) in terms of real functions. Since

$$x = c_1 e^{(5 + 2i)t} + c_2 e^{(5 - 2i)t}$$

$$y = c_1(1 - 2i)e^{(5 + 2i)t} + c_2(1 + 2i)e^{(5 - 2i)t}$$

it follows from Euler's formula that

$$x = e^{5t}[c_1 e^{2it} + c_2 e^{-2it}]$$

$$= e^{5t}[(c_1 + c_2)\cos 2t + (c_1 i - c_2 i)\sin 2t]$$

$$y = e^{5t}[(c_1(1 - 2i) + c_2(1 + 2i))\cos 2t + (c_1 i(1 - 2i) - c_2 i(1 + 2i))\sin 2t]$$

$$= e^{5t}[(c_1 + c_2) - 2(c_1 i - c_2 i)]\cos 2t + e^{5t}[2(c_1 + c_2) + (c_1 i - c_2 i)]\sin 2t.$$

If we replace $c_1 + c_2$ by C_1 and $c_1 i - c_2 i$ by C_2, then

$$x = e^{5t}[C_1 \cos 2t + C_2 \sin 2t]$$
$$y = e^{5t}[C_1 - 2C_2]\cos 2t + e^{5t}[2C_1 + C_2]\sin 2t,$$

or, in terms of vectors,

$$\mathbf{X} = \begin{pmatrix} x(t) \\ y(t) \end{pmatrix} = C_1 \begin{pmatrix} \cos 2t \\ \cos 2t + 2 \sin 2t \end{pmatrix} e^{5t} + C_2 \begin{pmatrix} \sin 2t \\ -2 \cos 2t + \sin 2t \end{pmatrix} e^{5t}. \tag{19}$$

Here, of course, it can be verified that each vector in (19) is a solution of (17). In addition, the solutions are linearly independent on $-\infty < t < \infty$. We may further assume that C_1 and C_2 are completely arbitrary and real. Thus (19) is the general solution of (17).

The foregoing process can be generalized. Let $\mathbf{K}_1$ be an eigenvector of the matrix $\mathbf{A}$ corresponding to the complex eigenvalue $\lambda_1 = \alpha + i\beta$. Then $\mathbf{X}_1$ and $\mathbf{X}_2$ in Theorem 8.10 can be written as

$$\mathbf{K}_1 e^{\lambda_1 t} = \mathbf{K}_1 e^{\alpha t} e^{i\beta t} = \mathbf{K}_1 e^{\alpha t}(\cos \beta t + i \sin \beta t)$$
$$\overline{\mathbf{K}}_1 e^{\overline{\lambda}_1 t} = \overline{\mathbf{K}}_1 e^{\alpha t} e^{-i\beta t} = \overline{\mathbf{K}}_1 e^{\alpha t}(\cos \beta t - i \sin \beta t). \tag{20}$$

The equations in (20) then yield

$$(\mathbf{K}_1 e^{\lambda_1 t} + \overline{\mathbf{K}}_1 e^{\overline{\lambda}_1 t}) = (\mathbf{K}_1 + \overline{\mathbf{K}}_1)e^{\alpha t} \cos \beta t + i(\mathbf{K}_1 - \overline{\mathbf{K}}_1)e^{\alpha t} \sin \beta t$$
$$i(\mathbf{K}_1 e^{\lambda_1 t} - \overline{\mathbf{K}}_1 e^{\overline{\lambda}_1 t}) = i(\mathbf{K}_1 - \overline{\mathbf{K}}_1)e^{\alpha t} \cos \beta t - (\mathbf{K}_1 + \overline{\mathbf{K}}_1)e^{\alpha t} \sin \beta t.$$

For *any* complex number $z = a + ib$, we note that $z + \overline{z} = 2a$ and $i(z - \overline{z}) = -2b$ are *real* numbers. Therefore, the entries in the column vectors $\mathbf{K}_1 + \overline{\mathbf{K}}_1$ and $i(\mathbf{K}_1 - \overline{\mathbf{K}}_1)$ are real numbers. By defining

$$\mathbf{B}_1 = \tfrac{1}{2}[\mathbf{K}_1 + \overline{\mathbf{K}}_1] \quad \text{and} \quad \mathbf{B}_2 = \tfrac{i}{2}[\mathbf{K}_1 - \overline{\mathbf{K}}_1] \tag{21}$$

we are led to the following theorem.

THEOREM 8.11 Let $\lambda_1 = \alpha + i\beta$ be a complex eigenvalue of the coefficient matrix $\mathbf{A}$ in the homogeneous system (2) and let $\mathbf{B}_1$ and $\mathbf{B}_2$ denote the column vectors defined in (21). Then

$$\mathbf{X}_1 = (\mathbf{B}_1 \cos \beta t + \mathbf{B}_2 \sin \beta t)e^{\alpha t}$$
$$\mathbf{X}_2 = (\mathbf{B}_2 \cos \beta t - \mathbf{B}_1 \sin \beta t)e^{\alpha t} \tag{22}$$

are linearly independent solutions of (2) on $-\infty < t < \infty$. □

The reader should verify that (19) can be obtained from (18) with

$$\mathbf{B}_1 = \frac{1}{2}\left[\begin{pmatrix} 1 \\ 1 - 2i \end{pmatrix} + \begin{pmatrix} 1 \\ 1 + 2i \end{pmatrix}\right] = \begin{pmatrix} 1 \\ 1 \end{pmatrix}$$

$$\mathbf{B}_2 = \frac{i}{2}\left[\begin{pmatrix} 1 \\ 1 - 2i \end{pmatrix} - \begin{pmatrix} 1 \\ 1 + 2i \end{pmatrix}\right] = \begin{pmatrix} 0 \\ 2 \end{pmatrix}.$$

EXAMPLE

Solve
$$\mathbf{X}' = \begin{pmatrix} 2 & 8 \\ -1 & -2 \end{pmatrix} \mathbf{X}.$$

Solution: First we obtain the eigenvalues from

$$\begin{vmatrix} 2 - \lambda & 8 \\ -1 & -2 - \lambda \end{vmatrix} = 0 \quad \text{or} \quad (2 - \lambda)(-2 - \lambda) + 8 = 0$$

$$\lambda^2 + 4 = 0.$$

Thus the eigenvalues are $\lambda_1 = 2i$ and $\lambda_2 = \overline{\lambda_1} = -2i$. For λ_1 we see that a solution of

$$(2 - \lambda)k_1 + 8k_2 = 0$$
$$-k_1 + (-2 - \lambda)k_2 = 0$$

is
$$\mathbf{K}_1 = \begin{pmatrix} 8 \\ -2 + 2i \end{pmatrix}.$$

Now from (21) we form

$$\mathbf{B}_1 = \frac{1}{2} \left[\begin{pmatrix} 8 \\ -2 + 2i \end{pmatrix} + \begin{pmatrix} 8 \\ -2 - 2i \end{pmatrix} \right] = \begin{pmatrix} 8 \\ -2 \end{pmatrix}$$

$$\mathbf{B}_2 = \frac{i}{2} \left[\begin{pmatrix} 8 \\ -2 + 2i \end{pmatrix} - \begin{pmatrix} 8 \\ -2 - 2i \end{pmatrix} \right] = \begin{pmatrix} 0 \\ -2 \end{pmatrix}.$$

Since $\alpha = 0$, it follows from (22) that the general solution of the system is

$$\mathbf{X} = c_1 \left[\begin{pmatrix} 8 \\ -2 \end{pmatrix} \cos 2t + \begin{pmatrix} 0 \\ -2 \end{pmatrix} \sin 2t \right] + c_2 \left[\begin{pmatrix} 0 \\ -2 \end{pmatrix} \cos 2t - \begin{pmatrix} 8 \\ -2 \end{pmatrix} \sin 2t \right]$$

$$= c_1 \begin{pmatrix} 8 \cos 2t \\ -2 \cos 2t - 2 \sin 2t \end{pmatrix} + c_2 \begin{pmatrix} -8 \sin 2t \\ -2 \cos 2t + 2 \sin 2t \end{pmatrix}.$$

EXAMPLE

Solve
$$\mathbf{X}' = \begin{pmatrix} 1 & 2 \\ \frac{1}{2} & 1 \end{pmatrix} \mathbf{X}.$$

Solution: The solutions of the characteristic equation

$$\begin{vmatrix} 1 - \lambda & 2 \\ -\frac{1}{2} & 1 - \lambda \end{vmatrix} = (1 - \lambda)^2 + 1$$

$$= \lambda^2 - 2\lambda + 2$$

$$= 0$$

are
$$\lambda_1 = 1 + i \quad \text{and} \quad \lambda_2 = \overline{\lambda_1} = 1 - i.$$

Now an eigenvector associated with λ_1 is seen to be

$$\mathbf{K}_1 = \begin{pmatrix} 2 \\ i \end{pmatrix}.$$

From (21) we find

$$\mathbf{B}_1 = \begin{pmatrix} 2 \\ 0 \end{pmatrix} \quad \text{and} \quad \mathbf{B}_2 = \begin{pmatrix} 0 \\ -1 \end{pmatrix}.$$

Thus

$$\mathbf{X} = c_1 \left[\begin{pmatrix} 2 \\ 0 \end{pmatrix} \cos t + \begin{pmatrix} 0 \\ -1 \end{pmatrix} \sin t \right] e^t$$

$$+ c_2 \left[\begin{pmatrix} 0 \\ -1 \end{pmatrix} \cos t - \begin{pmatrix} 2 \\ 0 \end{pmatrix} \sin t \right] e^t$$

$$= c_1 \begin{pmatrix} 2 \cos t \\ -\sin t \end{pmatrix} e^t + c_2 \begin{pmatrix} -2 \sin t \\ -\cos t \end{pmatrix} e^t.$$

Alternative method   When $\mathbf{A}$ is a 2×2 matrix having a complex eigenvalue $\lambda = \alpha + i\beta$ the general solution of the system can also be obtained from the assumption

$$\mathbf{X} = \begin{pmatrix} c_1 \\ c_2 \end{pmatrix} e^{\alpha t} \sin \beta t + \begin{pmatrix} c_3 \\ c_4 \end{pmatrix} e^{\alpha t} \cos \beta t$$

and then substituting $x(t)$ and $y(t)$ into one of the equations of the original system. This procedure is basically that of Section 8.1.

8.6.3   Repeated Eigenvalues

Up to this point we have not considered the case where some of the n eigenvalues $\lambda_1, \lambda_2, \ldots, \lambda_n$ of an $n \times n$ matrix are repeated. For example, the characteristic equation of the coefficient matrix in

$$\mathbf{X}' = \begin{pmatrix} 3 & -18 \\ 2 & -9 \end{pmatrix} \mathbf{X} \tag{23}$$

is readily shown to be $(\lambda + 3)^2 = 0$ and therefore $\lambda_1 = \lambda_2 = -3$. We say that $\lambda_1 = -3$ is a *double root* or a root of *multiplicity two*. Now for this value we find the single eigenvector

$$\mathbf{K}_1 = \begin{pmatrix} 3 \\ 1 \end{pmatrix}$$

and so one solution of (23) is

$$\mathbf{X}_1 = \begin{pmatrix} 3 \\ 1 \end{pmatrix} e^{-3t}. \tag{24}$$

But since we are obviously interested in forming the general solution of the system we need to pursue the question of finding a second solution.

In general, if m is a positive integer and $(\lambda - \lambda_1)^m$ is a factor of the characteristic equation, while $(\lambda - \lambda_1)^{m+1}$ is not a factor, then λ_1 is said to be an **eigenvalue of multiplicity** m. We distinguish two possibilities.

(a) For some $n \times n$ matrices $\mathbf{A}$ it may be possible to find m linearly independent eigenvectors $\mathbf{K}_1, \mathbf{K}_2, \ldots, \mathbf{K}_m$ corresponding to an eigenvalue λ_1 of multiplicity $m \leq n$. In this case the general solution of the system contains the linear combination

$$c_1 \mathbf{K}_1 e^{\lambda_1 t} + c_2 \mathbf{K}_2 e^{\lambda_1 t} + \cdots + c_m \mathbf{K}_m e^{\lambda_1 t}.$$

(b) If there is only one eigenvector corresponding to to the eigenvalue λ_1 of multiplicity m then m linearly independent solutions of the form

$$\mathbf{X}_1 = \mathbf{K}_{11} e^{\lambda_1 t}$$
$$\mathbf{X}_2 = \mathbf{K}_{21} t e^{\lambda_1 t} + \mathbf{K}_{22} e^{\lambda_1 t}$$
$$\vdots$$
$$\mathbf{X}_m = \mathbf{K}_{m1} \frac{t^{m-1}}{(m-1)!} e^{\lambda_1 t} + \mathbf{K}_{m2} \frac{t^{m-2}}{(m-2)!} e^{\lambda_1 t} + \cdots + \mathbf{K}_{mm} e^{\lambda_1 t},$$

where $\mathbf{K}_{ij}$ are column vectors, can always be found.

Eigenvalues of multiplicity two

We begin by considering eigenvalues of multiplicity two. In the first example we illustrate a matrix for which we can find two distinct eigenvectors corresponding to a double eigenvalue.

EXAMPLE

Solve
$$\mathbf{X}' = \begin{pmatrix} 1 & -2 & 2 \\ -2 & 1 & -2 \\ 2 & -2 & 1 \end{pmatrix} \mathbf{X}.$$

Solution: Expanding the determinant in the characteristic equation

$$\begin{vmatrix} 1 - \lambda & -2 & 2 \\ -2 & 1 - \lambda & -2 \\ 2 & -2 & 1 - \lambda \end{vmatrix} = 0$$

yields $-(\lambda + 1)^2(\lambda - 5) = 0$. We see that $\lambda_1 = \lambda_2 = -1$ and $\lambda_3 = 5$.

Now for $\lambda_1 = -1$ all three equations in the system

$$(1 - \lambda)k_1 - 2k_2 + 2k_3 = 0$$
$$-2k_1 + (1 - \lambda)k_2 - 2k_3 = 0 \qquad (25)$$
$$2k_1 - 2k_2 + (1 - \lambda)k_3 = 0$$

reduce to $k_1 - k_2 + k_3 = 0$. We can then express, say, k_2 in terms of k_1 and k_3. By choosing $k_1 = 1$ and $k_3 = 0$ in $k_2 = k_1 + k_3$ we obtain $k_2 = 1$ and so one eigenvector is

$$\mathbf{K}_1 = \begin{pmatrix} 1 \\ 1 \\ 0 \end{pmatrix}.$$

But the choice $k_1 = 0$, $k_3 = 1$ implies $k_2 = 1$. Hence a second eigenvector is

$$\mathbf{K}_2 = \begin{pmatrix} 0 \\ 1 \\ 1 \end{pmatrix}.$$

Since neither eigenvector is a constant multiple of the other, we have found, corresponding to the same eigenvalue, two linearly independent solutions

$$\mathbf{X}_1 = \begin{pmatrix} 1 \\ 1 \\ 0 \end{pmatrix} e^{-t} \quad \text{and} \quad \mathbf{X}_2 = \begin{pmatrix} 0 \\ 1 \\ 1 \end{pmatrix} e^{-t}.$$

Lastly, for $\lambda_3 = 5$, the system (25) becomes

$$-4k_1 - 2k_2 + 2k_3 = 0$$
$$-2k_1 - 4k_2 - 2k_3 = 0$$
$$2k_1 - 2k_2 - 4k_3 = 0.$$

In turn we find $k_2 = -k_1$ and $k_3 = k_1$ where k_1 is arbitrary. Picking $k_1 = 1$ implies $k_2 = -1$, $k_3 = 1$ and thus a third eigenvector is

$$\mathbf{K}_3 = \begin{pmatrix} 1 \\ -1 \\ 1 \end{pmatrix}$$

We conclude that the general solution of the system is

$$\mathbf{X} = c_1 \begin{pmatrix} 1 \\ 1 \\ 0 \end{pmatrix} e^{-t} + c_2 \begin{pmatrix} 0 \\ 1 \\ 1 \end{pmatrix} e^{-t} + c_3 \begin{pmatrix} 1 \\ -1 \\ 1 \end{pmatrix} e^{5t}.$$

A second solution

Now suppose λ_1 is an eigenvalue of multiplicity two but that there is only one eigenvector associated with this value. A second solution can be found of the form

$$\mathbf{X}_2 = \mathbf{K} t e^{\lambda_1 t} + \mathbf{P} e^{\lambda_1 t} \tag{26}$$

where

$$\mathbf{K} = \begin{pmatrix} k_1 \\ k_2 \\ \vdots \\ k_n \end{pmatrix} \quad \text{and} \quad \mathbf{P} = \begin{pmatrix} p_1 \\ p_2 \\ \vdots \\ p_n \end{pmatrix}.$$

To see this we substitute (26) into the system $\mathbf{X}' = \mathbf{AX}$ and simplify:

$$(\mathbf{AK} - \lambda_1 \mathbf{K}) t e^{\lambda_1 t} + (\mathbf{AP} - \lambda_1 \mathbf{P} - \mathbf{K}) e^{\lambda_1 t} = \mathbf{0}.$$

Since this last equation is to hold for all values of t we must have

$$(\mathbf{A} - \lambda_1 \mathbf{I})\mathbf{K} = \mathbf{0} \tag{27}$$

and

$$(\mathbf{A} - \lambda_1 \mathbf{I})\mathbf{P} = \mathbf{K}. \tag{28}$$

The first equation (27) simply states that $\mathbf{K}$ must be an eigenvector of $\mathbf{A}$ associated with λ_1. By solving (27) we find one solution $\mathbf{X}_1 = \mathbf{K}e^{\lambda_1 t}$. To find the second solution $\mathbf{X}_2$ we need only solve the additional system (28) for the vector $\mathbf{P}$.

EXAMPLE	Find the general solution of the system given in (23).

Solution: From (24) we know $\lambda_1 = -3$ and that one solution is

$$\mathbf{X}_1 = \begin{pmatrix} 3 \\ 1 \end{pmatrix} e^{-3t}.$$

Identifying $\mathbf{K} = \begin{pmatrix} 3 \\ 1 \end{pmatrix}$ and $\mathbf{P} = \begin{pmatrix} p_1 \\ p_2 \end{pmatrix}$ it follows from (28) that we must now solve

$$(\mathbf{A} + 3\mathbf{I})\mathbf{P} = \mathbf{K} \qquad \text{or} \qquad \begin{pmatrix} 6 & -18 \\ 2 & -6 \end{pmatrix} \begin{pmatrix} p_1 \\ p_2 \end{pmatrix} = \begin{pmatrix} 3 \\ 1 \end{pmatrix}.$$

Multiplying out this last expression gives

$$6p_1 - 18p_2 = 3$$
$$2p_1 - 6p_2 = 1.$$

Since this system is obviously equivalent to one equation we have an infinite number of choices for p_1 and p_2. For example, by choosing $p_1 = 1$, we find $p_2 = 1/6$. However, for simplicity we shall choose $p_1 = 1/2$ so that $p_2 = 0$. Hence $\mathbf{P} = \begin{pmatrix} 1/2 \\ 0 \end{pmatrix}$. Thus from (26) we find

$$\mathbf{X}_2 = \begin{pmatrix} 3 \\ 1 \end{pmatrix} te^{-3t} + \begin{pmatrix} 1/2 \\ 0 \end{pmatrix} e^{-3t}.$$

The general solution of (23) is then

$$\mathbf{X} = c_1 \begin{pmatrix} 3 \\ 1 \end{pmatrix} e^{-3t} + c_2 \left[\begin{pmatrix} 3 \\ 1 \end{pmatrix} te^{-3t} + \begin{pmatrix} 1/2 \\ 0 \end{pmatrix} e^{-3t} \right].$$

Eigenvalues of multiplicity three

When a matrix $\mathbf{A}$ has only one eigenvector associated with an eigenvalue λ_1 of multiplicity three we can find a second solution of form (26) and a third solution of the form

$$\mathbf{X}_3 = \mathbf{K}\frac{t^2}{2}e^{\lambda_1 t} + \mathbf{P}te^{\lambda_1 t} + \mathbf{Q}e^{\lambda_1 t} \qquad (29)$$

where

$$\mathbf{K} = \begin{pmatrix} k_1 \\ k_2 \\ \vdots \\ k_n \end{pmatrix}, \qquad \mathbf{P} = \begin{pmatrix} p_1 \\ p_2 \\ \vdots \\ p_n \end{pmatrix}, \qquad \text{and} \qquad \mathbf{Q} = \begin{pmatrix} q_1 \\ q_2 \\ \vdots \\ q_n \end{pmatrix}.$$

By substituting (29) into the system $\mathbf{X}' = \mathbf{A}\mathbf{X}$ we find the column vectors $\mathbf{K}$, $\mathbf{P}$, and $\mathbf{Q}$ must satisfy

$$(\mathbf{A} - \lambda_1\mathbf{I})\mathbf{K} = \mathbf{0} \tag{30}$$

$$(\mathbf{A} - \lambda_1\mathbf{I})\mathbf{P} = \mathbf{K} \tag{31}$$

$$(\mathbf{A} - \lambda_1\mathbf{I})\mathbf{Q} = \mathbf{P}. \tag{32}$$

Of course the solutions of (30) and (31) can be utilized in the formulation of the solutions $\mathbf{X}_1$ and $\mathbf{X}_2$.

EXAMPLE

Solve
$$\mathbf{X}' = \begin{pmatrix} 2 & 1 & 6 \\ 0 & 2 & 5 \\ 0 & 0 & 2 \end{pmatrix}\mathbf{X}.$$

Solution: The characteristic equation $(\lambda - 2)^3 = 0$ shows that $\lambda_1 = 2$ is an eigenvalue of multiplicity three. In succession we find that a solution of

$$(\mathbf{A} - 2\mathbf{I})\mathbf{K} = \mathbf{0} \quad \text{is} \quad \mathbf{K} = \begin{pmatrix} 1 \\ 0 \\ 0 \end{pmatrix};$$

a solution of

$$(\mathbf{A} - 2\mathbf{I})\mathbf{P} = \mathbf{K} \quad \text{is} \quad \mathbf{P} = \begin{pmatrix} 0 \\ 1 \\ 0 \end{pmatrix};$$

and finally a solution of

$$(\mathbf{A} - 2\mathbf{I})\mathbf{Q} = \mathbf{P} \quad \text{is} \quad \mathbf{Q} = \begin{pmatrix} 0 \\ -6/5 \\ 1/5 \end{pmatrix}.$$

We see from (26) and (29) that the general solution of the system is

$$\mathbf{X} = c_1\begin{pmatrix} 1 \\ 0 \\ 0 \end{pmatrix}e^{2t} + c_2\left[\begin{pmatrix} 1 \\ 0 \\ 0 \end{pmatrix}te^{2t} + \begin{pmatrix} 0 \\ 1 \\ 0 \end{pmatrix}e^{2t}\right]$$

$$+ c_3\left[\begin{pmatrix} 1 \\ 0 \\ 0 \end{pmatrix}\frac{t^2}{2}e^{2t} + \begin{pmatrix} 0 \\ 1 \\ 0 \end{pmatrix}te^{2t} + \begin{pmatrix} 0 \\ -6/5 \\ 1/5 \end{pmatrix}e^{2t}\right].$$

EXERCISES 8.6   *Answers to odd-numbered problems begin on page A-28.*

[8.6.1]   In Problems 1–10 find the general solution of the given system.

1. $\dfrac{dx}{dt} = x + 2y$

   $\dfrac{dy}{dt} = 4x + 3y$

2. $\dfrac{dx}{dt} = 2y$

   $\dfrac{dy}{dt} = 8x$

3. $\dfrac{dx}{dt} = -4x + 2y$

   $\dfrac{dy}{dt} = -\dfrac{5}{2}x + 2y$

4. $\dfrac{dx}{dt} = \dfrac{1}{2}x + 9y$

   $\dfrac{dy}{dt} = \dfrac{1}{2}x + 2y$

5. $\mathbf{X}' = \begin{pmatrix} 10 & -5 \\ 8 & -12 \end{pmatrix}\mathbf{X}$

★6. $\mathbf{X}' = \begin{pmatrix} -6 & 2 \\ -3 & 1 \end{pmatrix}\mathbf{X}$

7. $\dfrac{dx}{dt} = x + y - z$

   $\dfrac{dy}{dt} = 2y$

   $\dfrac{dz}{dt} = y - z$

8. $\dfrac{dx}{dt} = 2x - 7y$

   $\dfrac{dy}{dt} = 5x + 10y + 4z$

   $\dfrac{dz}{dt} = 5y + 2z$

9. $\mathbf{X}' = \begin{pmatrix} -1 & 1 & 0 \\ 1 & 2 & 1 \\ 0 & 3 & -1 \end{pmatrix}\mathbf{X}$

★10. $\mathbf{X}' = \begin{pmatrix} 1 & 0 & 1 \\ 0 & 1 & 0 \\ 1 & 0 & 1 \end{pmatrix}\mathbf{X}$

In Problems 11 and 12 solve the given system subject to the indicated initial condition.

11. $\mathbf{X}' = \begin{pmatrix} 1/2 & 0 \\ 1 & -1/2 \end{pmatrix}\mathbf{X}'$ $\mathbf{X}(0) = \begin{pmatrix} 3 \\ 5 \end{pmatrix}$

12. $\mathbf{X}' = \begin{pmatrix} 1 & 1 & 4 \\ 0 & 2 & 0 \\ 1 & 1 & 1 \end{pmatrix}\mathbf{X},$ $\mathbf{X}(0) = \begin{pmatrix} 1 \\ 3 \\ 0 \end{pmatrix}$

[8.6.2]   In Problems 13–22 find the general solution of the given system.

13. $\dfrac{dx}{dt} = 6x - y$

    $\dfrac{dy}{dt} = 5x + 2y$

14. $\dfrac{dx}{dt} = x + y$

    $\dfrac{dy}{dt} = -2x - y$

15. $\dfrac{dx}{dt} = 5x + y$

$\dfrac{dy}{dt} = -2x + 3y$

16. $\dfrac{dx}{dt} = 4x + 5y$

$\dfrac{dy}{dt} = -2x + 6y$

17. $\mathbf{X'} = \begin{pmatrix} 4 & -5 \\ 5 & -4 \end{pmatrix} \mathbf{X}$

18. $\mathbf{X'} = \begin{pmatrix} 1 & -8 \\ 1 & -3 \end{pmatrix} \mathbf{X}$

19. $\dfrac{dx}{dt} = z$

$\dfrac{dy}{dt} = -z$

$\dfrac{dz}{dt} = y$

★20. $\dfrac{dx}{dt} = 2x + y + 2z$

$\dfrac{dy}{dt} = 3x + 6z$

$\dfrac{dz}{dt} = -4x - 3z$

21. $\mathbf{X'} = \begin{pmatrix} 2 & 5 & 1 \\ -5 & -6 & 4 \\ 0 & 0 & 2 \end{pmatrix} \mathbf{X}$

22. $\mathbf{X'} = \begin{pmatrix} 2 & 4 & 4 \\ -1 & -2 & 0 \\ -1 & 0 & -2 \end{pmatrix} \mathbf{X}$

In Problems 23 and 24 solve the given system subject to the indicated initial condition.

23. $\mathbf{X'} = \begin{pmatrix} 1 & -12 & -14 \\ 1 & 2 & -3 \\ 1 & 1 & -2 \end{pmatrix} \mathbf{X}, \qquad \mathbf{X}(0) = \begin{pmatrix} 4 \\ 6 \\ -7 \end{pmatrix}$

★24. $\mathbf{X'} = \begin{pmatrix} 6 & -1 \\ 5 & 4 \end{pmatrix} \mathbf{X}, \qquad \mathbf{X}(0) = \begin{pmatrix} -2 \\ 8 \end{pmatrix}$

[8.6.3]   In Problems 25–34 find the general solution of the given system.

25. $\dfrac{dx}{dt} = 3x - y$

$\dfrac{dy}{dt} = 9x - 3y$

26. $\dfrac{dx}{dt} = -6x + 5y$

$\dfrac{dy}{dt} = -5x + 4y$

27. $\dfrac{dx}{dt} = -x + 3y$

$\dfrac{dy}{dt} = -3x + 5y$

28. $\dfrac{dx}{dt} = 12x - 9y$

$\dfrac{dy}{dt} = 4x$

29. $\dfrac{dx}{dt} = 3x - y - z$

$\dfrac{dy}{dt} = x + y - z$

$\dfrac{dz}{dt} = x - y + z$

★30. $\dfrac{dx}{dt} = 3x + 2y + 4z$

$\dfrac{dy}{dt} = 2x + 2z$

$\dfrac{dz}{dt} = 4x + 2y + 3z$

31. $\mathbf{X}' = \begin{pmatrix} 5 & -4 & 0 \\ 1 & 0 & 2 \\ 0 & 2 & 5 \end{pmatrix}\mathbf{X}$ **32.** $\mathbf{X}' = \begin{pmatrix} 1 & 0 & 0 \\ 0 & 3 & 1 \\ 0 & -1 & 1 \end{pmatrix}\mathbf{X}$

33. $\mathbf{X}' = \begin{pmatrix} 1 & 0 & 0 \\ 2 & 2 & -1 \\ 0 & 1 & 0 \end{pmatrix}\mathbf{X}$ **34.** $\mathbf{X}' = \begin{pmatrix} 4 & 1 & 0 \\ 0 & 4 & 1 \\ 0 & 0 & 4 \end{pmatrix}\mathbf{X}.$

In Problems 35 and 36 solve the given system subject to the indicated initial condition.

35. $\mathbf{X}' = \begin{pmatrix} 2 & 4 \\ -1 & 6 \end{pmatrix}\mathbf{X}, \qquad \mathbf{X}(0) = \begin{pmatrix} -1 \\ 6 \end{pmatrix}$

36. $\mathbf{X}' = \begin{pmatrix} 0 & 0 & 1 \\ 0 & 1 & 0 \\ 1 & 0 & 0 \end{pmatrix}\mathbf{X}, \qquad \mathbf{X}(0) = \begin{pmatrix} 1 \\ 2 \\ 5 \end{pmatrix}$

Miscellaneous problems

If $\mathbf{\Phi}(t)$ is a fundamental matrix of the system, the initial-value problem $\mathbf{X}' = \mathbf{AX}, \mathbf{X}(t_0) = \mathbf{X}_0$ has the solution $\mathbf{X} = \mathbf{\Phi}(t)\mathbf{\Phi}^{-1}(t_0)\mathbf{X}_0$. (See Problem 35 of Section 8.5.). In Problems 37 and 38 use this result to solve the given system subject to the indicated initial condition.

37. $\mathbf{X}' = \begin{pmatrix} 4 & 3 \\ 3 & -4 \end{pmatrix}\mathbf{X}, \qquad \mathbf{X}(0) = \begin{pmatrix} 1 \\ 1 \end{pmatrix}$

38. $\mathbf{X}' = \begin{pmatrix} -2/25 & 1/50 \\ 2/25 & -2/25 \end{pmatrix}\mathbf{X}, \qquad \mathbf{X}(0) = \begin{pmatrix} 25 \\ 0 \end{pmatrix}$

In Problems 39 and 40 find a solution of the given system of the form $\mathbf{X} = t^\lambda \mathbf{K}, t > 0$, where $\mathbf{K}$ is a column vector of constants.

39. $t\mathbf{X}' = \begin{pmatrix} 1 & 3 \\ -1 & 5 \end{pmatrix}\mathbf{X}$ **★40.** $t\mathbf{X}' = \begin{pmatrix} 2 & -2 \\ 2 & 7 \end{pmatrix}\mathbf{X}$

8.7   Undetermined Coefficients

When the entries of $\mathbf{F}(t)$ are constants, polynomials, exponential functions, sines and cosines, or finite sums and products of the functions it is possible to adapt the method of undetermined coefficients to find a particular solution of the nonhomogeneous linear system $\mathbf{X}' = \mathbf{AX} + \mathbf{F}(t)$.

EXAMPLE   Solve the system

$$\frac{dx}{dt} = 6x + y + 6t$$

$$\frac{dy}{dt} = 4x + 3y - 10t + 4$$

(1)

on $-\infty < t < \infty$.

Solution: We first solve the homogeneous system

$$\frac{dx}{dt} = 6x + y$$

$$\frac{dy}{dt} = 4x + 3y$$

by the method of Section 8.6. The eigenvalues are determined from

$$\begin{vmatrix} 6 - \lambda & 1 \\ 4 & 3 - \lambda \end{vmatrix} = (6 - \lambda)(3 - \lambda) - 4$$

$$= \lambda^2 - 9\lambda + 14$$

$$= (\lambda - 2)(\lambda - 7)$$

$$= 0.$$

Thus $\lambda_1 = 2$ and $\lambda_2 = 7$. It is then easily verified that the respective eigenvectors of the coefficient matrix are

$$\begin{pmatrix} 1 \\ -4 \end{pmatrix} \quad \text{and} \quad \begin{pmatrix} 1 \\ 1 \end{pmatrix}$$

Consequently, the complementary function of (1) is

$$\mathbf{X}_c = c_1 \begin{pmatrix} 1 \\ -4 \end{pmatrix} e^{2t} + c_2 \begin{pmatrix} 1 \\ 1 \end{pmatrix} e^{7t}.$$

Now observe that the given system (1) can be written as

$$\mathbf{X}' = \begin{pmatrix} 6 & 1 \\ 4 & 3 \end{pmatrix} \mathbf{X} + \begin{pmatrix} 6 \\ -10 \end{pmatrix} t + \begin{pmatrix} 0 \\ 4 \end{pmatrix} \tag{2}$$

Proceeding in a manner similar to that in Section 4.4, we expect a particular solution of the system to possess the form

$$\mathbf{X}_p = \begin{pmatrix} a_2 \\ b_2 \end{pmatrix} t + \begin{pmatrix} a_1 \\ b_1 \end{pmatrix} \tag{3}$$

were we must find specific values of the coefficients a_2, b_2, a_1, b_1. Substituting (3) in (2) gives

$$\mathbf{X}_p' = \begin{pmatrix} 6 & 1 \\ 4 & 3 \end{pmatrix} \mathbf{X}_p + \begin{pmatrix} 6 \\ -10 \end{pmatrix} t + \begin{pmatrix} 0 \\ 4 \end{pmatrix}$$

or

$$\begin{pmatrix} a_2 \\ b_2 \end{pmatrix} = \begin{pmatrix} 6 & 1 \\ 4 & 3 \end{pmatrix} \left[\begin{pmatrix} a_2 \\ b_2 \end{pmatrix} t + \begin{pmatrix} a_1 \\ b_1 \end{pmatrix} \right] + \begin{pmatrix} 6 \\ -10 \end{pmatrix} t + \begin{pmatrix} 0 \\ 4 \end{pmatrix}$$

$$\begin{pmatrix} 0 \\ 0 \end{pmatrix} = \begin{pmatrix} (6a_2 + b_2 + 6)t + 6a_1 + b_1 - a_2 \\ (4a_2 + 3b_2 - 10)t + 4a_1 + 3b_1 - b_2 + 4 \end{pmatrix}.$$

Hence we must have

$$6a_2 + b_2 + 6 = 0$$

$$4a_2 + 3b_2 - 10 = 0$$

$$\tag{4}$$

$$6a_1 + b_1 - a_2 = 0$$

$$4a_1 + 3b_1 - b_2 + 4 = 0.$$

Solving the first two equations of (4) simultaneously yields $a_2 = -2$ and $b_2 = 6$. Substituting these values into the last two equations and solving for a_1 and b_1 gives $a_1 = -4/7$, $b_1 = 10/7$. It follows therefore that a particular solution vector is

$$\mathbf{X}_p = \begin{pmatrix} -2 \\ 6 \end{pmatrix} t + \begin{pmatrix} -4/7 \\ 10/7 \end{pmatrix}$$

and so the general solution of (1) on $-\infty < t < \infty$ is given by

$$\mathbf{X} = \mathbf{X}_c + \mathbf{X}_p$$

$$= c_1 \begin{pmatrix} 1 \\ -4 \end{pmatrix} e^{2t} + c_2 \begin{pmatrix} 1 \\ 1 \end{pmatrix} e^{7t} + \begin{pmatrix} -2 \\ 6 \end{pmatrix} t + \begin{pmatrix} -4/7 \\ 10/7 \end{pmatrix}.$$

EXAMPLE

To find a particular solution vector for

$$\frac{dx}{dt} = -x + 4y - 8$$

$$\frac{dy}{dt} = 5x + 2y + 3$$

or

$$\mathbf{X}' = \begin{pmatrix} -1 & 4 \\ 5 & 2 \end{pmatrix} \mathbf{X} + \begin{pmatrix} -8 \\ 3 \end{pmatrix}$$

we assume a constant vector $\mathbf{X}_p = \begin{pmatrix} a_1 \\ b_1 \end{pmatrix}$. The reader should verify that this assumption leads to

$$-a_1 + 4b_1 - 8 = 0$$

$$5a_1 + 2b_1 + 3 = 0.$$

Solving the system gives $a_1 = -14/11$, $b_1 = 37/22$. Thus a particular solution vector is

$$\mathbf{X}_p = \begin{pmatrix} -14/11 \\ 37/22 \end{pmatrix}.$$

A word of caution is in order here, the assumption of a constant vector $\mathbf{X}_p$ in the preceding example is actually predicated on prior knowledge that the complementary function $\mathbf{X}_c$ does not contain a constant vector (that is, $\lambda = 0$ is not an eigenvalue). For example, it is readily shown that the complementary function of the system

$$\mathbf{X}' = \begin{pmatrix} 1 & -1 \\ -1 & 1 \end{pmatrix} \mathbf{X} + \begin{pmatrix} 2 \\ -5 \end{pmatrix}$$

is
$$\mathbf{X}_c = c_1 \begin{pmatrix} 1 \\ 1 \end{pmatrix} + c_2 \begin{pmatrix} 1 \\ -1 \end{pmatrix} e^{2t}$$

But, because of the constant vector present in $\mathbf{X}_c$, it is not unexpected that we are unable to find a constant particular solution vector

$$\begin{pmatrix} a_1 \\ b_1 \end{pmatrix}. \tag{5}$$

However, we leave it as an exercise to show that the natural alternative form

$$\begin{pmatrix} a_1 \\ b_1 \end{pmatrix} t \tag{6}$$

still does *not* yield a particular solution.* We need *both* (5) and (6) (that is, the sum) to find $\mathbf{X}_p$.

EXAMPLE

Find the general solution of the system

$$\mathbf{X}' = \begin{pmatrix} 1 & -1 \\ -1 & 1 \end{pmatrix} \mathbf{X} + \begin{pmatrix} 2 \\ -5 \end{pmatrix} \tag{7}$$

on $-\infty < t < \infty$.

Solution: From the foregoing discussion we shall seek a particular solution with four undetermined coefficients

$$\mathbf{X}_p = \begin{pmatrix} a_2 \\ b_2 \end{pmatrix} t + \begin{pmatrix} a_1 \\ b_1 \end{pmatrix} \tag{8}$$

Differentiating (8), substituting in (7), and gathering terms gives

$$(a_2 - b_2)t + a_1 - b_1 - a_2 + 2 = 0$$
$$(-a_2 + b_2)t - a_1 + b_1 - b_2 - 5 = 0$$

or
$$a_2 - b_2 = 0$$
$$a_1 - b_1 - a_2 + 2 = 0$$
$$-a_1 + b_1 - b_2 - 5 = 0.$$

Substituting $a_2 = b_2$ into the last two equations implies that we must solve the following system of two equations in three unknowns

$$a_1 - a_2 - b_1 = -2$$
$$-a_1 - a_2 + b_1 = 5.$$

This system yields a specific value of a_2 whereas a_1 can be expressed in terms of b_1:

---

*The situation is similar to finding particular solutions for, say, $y'' + 9y' = 7$ and $y'' + 9y = 7$. In the first instance we assume $y_p = Ax$ and in the second $y_p = A$. The analogy breaks down somewhat in this discussion since both forms (5) and (6) are needed in the form of the particular solution. See Equation (8).

$$a_1 = b_1 - \tfrac{7}{2}$$

$$a_2 = -\tfrac{3}{2}.$$

(9)

By choosing b_1 arbitrarily*, say $b_1 = 1$, we find $a_1 = -5/2$. Hence we conclude that a particular solution vector is

$$\mathbf{X}_p = \begin{pmatrix} -3/2 \\ -3/2 \end{pmatrix} t + \begin{pmatrix} -5/2 \\ 1 \end{pmatrix}$$

It follows that the general solution of (7) on $-\infty < t < \infty$ is

$$\mathbf{X} = \mathbf{X}_c + \mathbf{X}_p$$

$$= c_1 \begin{pmatrix} 1 \\ 1 \end{pmatrix} + c_2 \begin{pmatrix} 1 \\ -1 \end{pmatrix} e^{2t} + \begin{pmatrix} -3/2 \\ -3/2 \end{pmatrix} t + \begin{pmatrix} -5/2 \\ 1 \end{pmatrix}$$

It is of interest to observe that had we chosen b_1 in (9) to be some value other than $b_1 = 1$, we naturally would have found a *different* particular solution $\mathbf{X}_p$ of (7). For example, by picking $b_1 = 0$, we then find $a_2 = b_2 = -3/2$, $a_1 = -7/2$, and, therefore,

$$\mathbf{X}_p = \begin{pmatrix} -3/2 \\ -3/2 \end{pmatrix} t + \begin{pmatrix} -7/2 \\ 0 \end{pmatrix}.$$

If we designate the arbitrary parameter b_1 by α, we can obtain particular solution vectors from

$$\begin{pmatrix} -3/2 \\ -3/2 \end{pmatrix} t + \begin{pmatrix} \alpha - 7/2 \\ \alpha \end{pmatrix}$$

or

$$\begin{pmatrix} -3/2 \\ -3/2 \end{pmatrix} t + \begin{pmatrix} -7/2 \\ 0 \end{pmatrix} + \alpha \begin{pmatrix} 1 \\ 1 \end{pmatrix}.$$

In the sum $\mathbf{X}_c + \mathbf{X}_p$ we can obviously combine

$$c_1 \begin{pmatrix} 1 \\ 1 \end{pmatrix} \quad \text{and} \quad \alpha \begin{pmatrix} 1 \\ 1 \end{pmatrix}$$

as one vector. Recall, a particular solution vector should be free of arbitrary parameters.

EXAMPLE   Determine the form of the particular solution vector $\mathbf{X}_p$ for

$$\frac{dx}{dt} = 5x - 3y - 2e^{2t} + 1$$

$$\frac{dy}{dt} = -x + y + e^{2t} - 5t + 7.$$

---

*It might help to think of two planes

$$x - y - z = -2$$
$$-x - y + z = 5$$

intersecting in a line in the plane $y = -3/2$. Any point on this line will satisfy the system.

Solution: Write the system as

$$\mathbf{X}' = \begin{pmatrix} 5 & -3 \\ -1 & 1 \end{pmatrix} \mathbf{X} + \begin{pmatrix} -2 \\ 1 \end{pmatrix} e^{2t} + \begin{pmatrix} 0 \\ -5 \end{pmatrix} t + \begin{pmatrix} 1 \\ 7 \end{pmatrix}.$$

Now find the eigenvalues:

$$\begin{vmatrix} 5 - \lambda & -3 \\ -1 & 1 - \lambda \end{vmatrix} = (5 - \lambda)(1 - \lambda) + 3$$

$$= \lambda^2 - 6\lambda + 8$$

$$= (\lambda - 2)(\lambda - 4)$$

$$= 0$$

which implies $\lambda_1 = 2$ and $\lambda_2 = 4$. Because $\lambda_1 = 2$ is an eigenvalue, we know that both $\mathbf{X}_c$ and $\mathbf{F}(t)$ contain vectors with the multiple e^{2t}. Hence we must assume

$$\mathbf{X}_p = \begin{pmatrix} a_4 \\ b_4 \end{pmatrix} t e^{2t} + \begin{pmatrix} a_3 \\ b_3 \end{pmatrix} e^{2t} + \begin{pmatrix} a_2 \\ b_2 \end{pmatrix} t + \begin{pmatrix} a_1 \\ b_1 \end{pmatrix}. *$$

EXERCISES 8.7 *Answers to odd-numbered problems begin on page A-29.*

In Problems 1–8 use the method of undetermined coefficients to solve the given system.

1. $\dfrac{dx}{dt} = 2x + 3y - 7$

   $\dfrac{dy}{dt} = -x - 2y + 5$

2. $\dfrac{dx}{dt} = 4x + 5y - 3e^t$

   $\dfrac{dy}{dt} = 9x + 6y + 10e^t$

3. $\dfrac{dx}{dt} = x + 3y - 2t^2$

   $\dfrac{dy}{dt} = 3x + y + t + 5$

★4. $\dfrac{dx}{dt} = 6x + 8y - 3$

   $\dfrac{dy}{dt} = 3x + 4y + 10t$

5. $\mathbf{X}' = \begin{pmatrix} 5 & 1 \\ -8 & -4 \end{pmatrix} \mathbf{X} + \begin{pmatrix} -2 \\ 3 \end{pmatrix} e^{4t}$

6. $\mathbf{X}' = \begin{pmatrix} -1 & 5 \\ -1 & 1 \end{pmatrix} \mathbf{X} + \begin{pmatrix} \sin t \\ -2 \cos t \end{pmatrix}$

7. $\mathbf{X}' = \begin{pmatrix} 1 & 1 & 1 \\ 0 & 2 & 3 \\ 0 & 0 & 5 \end{pmatrix} \mathbf{X} + \begin{pmatrix} 1 \\ -1 \\ 2 \end{pmatrix} e^{4t}$

8. $\mathbf{X}' = \begin{pmatrix} 6 & 2 & 0 \\ 3 & 1 & 0 \\ 0 & 0 & 3 \end{pmatrix} \mathbf{X} + \begin{pmatrix} 1 \\ 1 \\ 3 \end{pmatrix}$

---

*Once again compare this with the procedure of Section 4.4. For example, the particular solution for $y'' - 4y = e^{2x} + 3x - 1$ is of the form $y_p = Axe^{2x} + Bx + C$ whereas for $y'' + 4y = e^{2x} + 3x - 1$, $y_p = Ae^{2x} + Bx + C$. The term Axe^{2x} appears in y_p for the former equation since e^{2x} is part of the complementary function of the equation. The reader should ponder why this is so.

In Problems 9 and 10 solve the given system subject to the indicated initial condition.

9. $X' = \begin{pmatrix} -1 & -2 \\ 3 & 4 \end{pmatrix} X + \begin{pmatrix} 3 \\ 3 \end{pmatrix}, \quad X(0) = \begin{pmatrix} -4 \\ 5 \end{pmatrix}$

★10. $X' = \begin{pmatrix} 1 & 1 \\ 1 & 1 \end{pmatrix} X + \begin{pmatrix} 2 \\ 1 \end{pmatrix} e^{2t}, \quad X(0) = \begin{pmatrix} 1 \\ 0 \end{pmatrix}$

8.8   Variation of Parameters

We saw in Section 8.5 that the general solution of a homogeneous system $X' = AX$ can be written as the product

$$X = \Phi(t)C$$

where $\Phi(t)$ is a fundamental matrix of the system and C is an $n \times 1$ column vector of constants. Analogous to the procedure of Section 4.5 we ask whether it is possible to replace C by a column matrix of functions

$$U(t) = \begin{pmatrix} u_1(t) \\ u_2(t) \\ \vdots \\ u_n(t) \end{pmatrix}$$

so that
$$X_p = \Phi(t)U(t) \tag{1}$$

is a particular solution of the nonhomogeneous system

$$X' = AX + F(t). \tag{2}$$

By the product rule* the derivative of (1) is

$$X_p' = \Phi(t)U'(t) + \Phi'(t)U(t). \tag{3}$$

Substituting (3) and (1) into (2) gives

$$\Phi(t)U'(t) + \Phi'(t)U(t) = A\Phi(t)U(t) + F(t). \tag{4}$$

Now recall from (13) of Section 8.5 that $\Phi'(t) = A\Phi(t)$. Thus (4) becomes

$$\Phi(t)U'(t) + A\Phi(t)U(t) = A\Phi(t)U(t) + F(t).$$

or
$$\Phi(t)U'(t) = F(t). \tag{5}$$

Multiplying both sides of equation (5) by $\Phi^{-1}(t)$ gives

$$U'(t) = \Phi^{-1}(t)F(t)$$

or
$$U(t) = \int \Phi^{-1}(t)F(t)\, dt.$$

---

*See Problem 33 of Section 8.4. Note that the order of the products is very important. Since $U(t)$ is a column matrix, the products $U'(t)\Phi(t)$ and $U(t)\Phi'(t)$ are not defined.

Hence, by the assumption (1), we conclude that a particular solution of (2) is given by

$$\mathbf{X}_p = \mathbf{\Phi}(t) \int \mathbf{\Phi}^{-1}(t)\mathbf{F}(t) \, dt. \tag{6}$$

To calculate the indefinite integral of the column matrix $\mathbf{\Phi}^{-1}(t)\mathbf{F}(t)$ in (6), we integrate each entry. Also, the general solution of the system (2) can now be formally written as

$$\mathbf{X} = \mathbf{X}_c + \mathbf{X}_p$$

or

$$\mathbf{X} = \mathbf{\Phi}(t)\mathbf{C} + \mathbf{\Phi}(t) \int \mathbf{\Phi}^{-1}(t)\mathbf{F}(t) \, dt. \tag{7}$$

EXAMPLE

Find the general solution of the nonhomogeneous system

$$\mathbf{X}' = \begin{pmatrix} -3 & 1 \\ 2 & -4 \end{pmatrix}\mathbf{X} + \begin{pmatrix} 3t \\ e^{-t} \end{pmatrix}. \tag{8}$$

on the interval $-\infty < t < \infty$.

Solution: We first solve the homogeneous system

$$\mathbf{X}' = \begin{pmatrix} -3 & 1 \\ 2 & -4 \end{pmatrix}\mathbf{X}. \tag{9}$$

The characteristic equation of the coefficient matrix is

$$\begin{vmatrix} -3 - \lambda & 1 \\ 2 & -4 - \lambda \end{vmatrix} = (3 + \lambda)(4 + \lambda) - 2$$

$$= \lambda^2 + 7\lambda + 10$$

$$= (\lambda + 2)(\lambda + 5)$$

$$= 0$$

and so the eigenvalues are $\lambda_1 = -2$ and $\lambda_2 = -5$. By the usual method we find the eigenvectors corresponding to λ_1 and λ_2 are, respectively,

$$\begin{pmatrix} 1 \\ 1 \end{pmatrix} \quad \text{and} \quad \begin{pmatrix} 1 \\ -2 \end{pmatrix}.$$

The solution vectors of the system (9) are then

$$\mathbf{X}_1 = \begin{pmatrix} 1 \\ 1 \end{pmatrix} e^{-2t} \quad \text{and} \quad \mathbf{X}_2 = \begin{pmatrix} 1 \\ -2 \end{pmatrix} e^{-5t}.$$

Next we form

$$\boldsymbol{\Phi}(t) = \begin{pmatrix} e^{-2t} & e^{-5t} \\ e^{-2t} & -2e^{-5t} \end{pmatrix} \quad \text{and} \quad \boldsymbol{\Phi}^{-1}(t) = \begin{pmatrix} \frac{2}{3}e^{2t} & \frac{1}{3}e^{2t} \\ \frac{1}{3}e^{5t} & -\frac{1}{3}e^{5t} \end{pmatrix}.$$

From (6) we then obtain

$$\mathbf{X}_p = \boldsymbol{\Phi}(t) \int \boldsymbol{\Phi}^{-1}(t)\mathbf{F}(t)\, dt$$

$$= \begin{pmatrix} e^{-2t} & e^{-5t} \\ e^{-2t} & -2e^{-5t} \end{pmatrix} \int \begin{pmatrix} \frac{2}{3}e^{2t} & \frac{1}{3}e^{2t} \\ \frac{1}{3}e^{5t} & -\frac{1}{3}e^{5t} \end{pmatrix} \begin{pmatrix} 3t \\ e^{-t} \end{pmatrix} dt$$

$$= \begin{pmatrix} e^{-2t} & e^{-5t} \\ e^{-2t} & -2e^{-5t} \end{pmatrix} \int \begin{pmatrix} 2te^{2t} + \frac{1}{3}e^{t} \\ te^{5t} - \frac{1}{3}e^{4t} \end{pmatrix} dt$$

$$= \begin{pmatrix} e^{-2t} & e^{-5t} \\ e^{-2t} & -2e^{-5t} \end{pmatrix} \begin{pmatrix} te^{2t} - \frac{1}{2}e^{2t} + \frac{1}{3}e^{t} \\ \frac{1}{5}te^{5t} - \frac{1}{25}e^{5t} - \frac{1}{12}e^{4t} \end{pmatrix}$$

$$= \begin{pmatrix} \frac{6}{5}t - \frac{27}{50} + \frac{1}{4}e^{-t} \\ \frac{3}{5}t - \frac{21}{50} + \frac{1}{2}e^{-t} \end{pmatrix}.$$

Hence from (7) the general solution of (8) on the interval is

$$\mathbf{X} = \begin{pmatrix} e^{-2t} & e^{-5t} \\ e^{-2t} & -2e^{-5t} \end{pmatrix} \begin{pmatrix} c_1 \\ c_2 \end{pmatrix} + \begin{pmatrix} \frac{6}{5}t - \frac{27}{50} + \frac{1}{4}e^{-t} \\ \frac{3}{5}t - \frac{21}{50} + \frac{1}{2}e^{-t} \end{pmatrix}$$

$$= c_1 \begin{pmatrix} 1 \\ 1 \end{pmatrix} e^{-2t} + c_2 \begin{pmatrix} 1 \\ -2 \end{pmatrix} e^{-5t} + \begin{pmatrix} \frac{6}{5} \\ \frac{3}{5} \end{pmatrix} t - \begin{pmatrix} \frac{27}{50} \\ \frac{21}{50} \end{pmatrix} + \begin{pmatrix} \frac{1}{4} \\ \frac{1}{2} \end{pmatrix} e^{-t}.$$

The general solution of (2) on an interval can be written in the alternative manner

$$\mathbf{X} = \boldsymbol{\Phi}(t)\mathbf{C} + \boldsymbol{\Phi}(t) \int_{t_0}^{t} \boldsymbol{\Phi}^{-1}(s)\mathbf{F}(s)\, ds \tag{10}$$

where t and t_0 are points in the interval. This last form is useful in solving (2) subject to an initial condition $\mathbf{X}(t_0) = \mathbf{X}_0$. Substituting $t = t_0$ in (10) yields

$$\mathbf{X}_0 = \boldsymbol{\Phi}(t_0)\mathbf{C}$$

from which we see immediately that

$$\mathbf{C} = \boldsymbol{\Phi}^{-1}(t_0)\mathbf{X}_0.$$

We conclude that the solution of the initial-value problem is given by

$$\mathbf{X} = \boldsymbol{\Phi}(t)\boldsymbol{\Phi}^{-1}(t_0)\mathbf{X}_0 + \boldsymbol{\Phi}(t) \int_{t_0}^{t} \boldsymbol{\Phi}^{-1}(s)\mathbf{F}(s)\, ds. \tag{11}$$

Recall from Section 8.5 that an alternative way of forming a fundamental matrix is to choose its column vectors $\mathbf{V}_i$ in such a manner that

$$\mathbf{V}_1(t_0) = \begin{pmatrix} 1 \\ 0 \\ \vdots \\ 0 \end{pmatrix}, \quad \mathbf{V}_2(t_0) = \begin{pmatrix} 0 \\ 1 \\ \vdots \\ 0 \end{pmatrix}, \quad \dots, \quad \mathbf{V}_n(t_0) = \begin{pmatrix} 0 \\ 0 \\ \vdots \\ 1 \end{pmatrix}. \tag{12}$$

This fundamental matrix is denoted by $\mathbf{\Psi}(t)$. As a consequence of (12) we know $\mathbf{\Psi}(t)$ has the property

$$\mathbf{\Psi}(t_0) = \mathbf{I}. \qquad (13)$$

But since $\mathbf{\Psi}(t)$ is nonsingular for all values of t in an interval, (13) implies

$$\mathbf{\Psi}^{-1}(t_0) = \mathbf{I}. \qquad (14)$$

Thus when $\mathbf{\Psi}(t)$ is used rather than $\mathbf{\Phi}(t)$, it follows from (14) that (11) can be written

$$\mathbf{X} = \mathbf{\Psi}(t)\mathbf{X}_0 + \mathbf{\Psi}(t) \int_{t_0}^{t} \mathbf{\Psi}^{-1}(s)\mathbf{F}(s) \, ds. \qquad (15)$$

EXERCISES 8.8   *Answers to odd-numbered problems begin on page A-29.*

In Problems 1–20 use variation of parameters to solve the given system.

1. $\dfrac{dx}{dt} = 3x - 3y + 4$

$\dfrac{dy}{dt} = 2x - 2y - 1$

2. $\dfrac{dx}{dt} = 2x - y$

$\dfrac{dy}{dt} = 3x - 2y + 4t$

3. $\mathbf{X}' = \begin{pmatrix} 3 & -5 \\ 3/4 & -1 \end{pmatrix} \mathbf{X} + \begin{pmatrix} 1 \\ -1 \end{pmatrix} e^{t/2}$

4. $\mathbf{X}' = \begin{pmatrix} 2 & -1 \\ 4 & 2 \end{pmatrix} \mathbf{X} + \begin{pmatrix} \sin 2t \\ 2 \cos 2t \end{pmatrix} e^{2t}$

5. $\mathbf{X}' = \begin{pmatrix} 0 & 2 \\ -1 & 3 \end{pmatrix} \mathbf{X} + \begin{pmatrix} 1 \\ -1 \end{pmatrix} e^{t}$

★6. $\mathbf{X}' = \begin{pmatrix} 0 & 2 \\ -1 & 3 \end{pmatrix} \mathbf{X} + \begin{pmatrix} 2 \\ e^{-3t} \end{pmatrix}$

7. $\mathbf{X}' = \begin{pmatrix} 1 & 8 \\ 1 & -1 \end{pmatrix} \mathbf{X} + \begin{pmatrix} 12 \\ 12 \end{pmatrix} t$

8. $\mathbf{X}' = \begin{pmatrix} 1 & 8 \\ 1 & -1 \end{pmatrix} \mathbf{X} + \begin{pmatrix} e^{-t} \\ te^{t} \end{pmatrix}$

9. $\mathbf{X}' = \begin{pmatrix} 3 & 2 \\ -2 & -1 \end{pmatrix} \mathbf{X} + \begin{pmatrix} 2e^{-t} \\ e^{-t} \end{pmatrix}$

★10. $\mathbf{X}' = \begin{pmatrix} 3 & 2 \\ -2 & -1 \end{pmatrix} \mathbf{X} + \begin{pmatrix} 1 \\ 1 \end{pmatrix}$

11. $\mathbf{X}' = \begin{pmatrix} 0 & -1 \\ 1 & 0 \end{pmatrix} \mathbf{X} + \begin{pmatrix} \sec t \\ 0 \end{pmatrix}$

12. $\mathbf{X}' = \begin{pmatrix} 1 & -1 \\ 1 & 1 \end{pmatrix} \mathbf{X} + \begin{pmatrix} 3 \\ 3 \end{pmatrix} e^{t}$

13. $\mathbf{X}' = \begin{pmatrix} 1 & -1 \\ 1 & 1 \end{pmatrix} \mathbf{X} + \begin{pmatrix} \cos t \\ \sin t \end{pmatrix} e^{t}$

14. $\mathbf{X}' = \begin{pmatrix} 2 & -2 \\ 8 & -6 \end{pmatrix} \mathbf{X} + \begin{pmatrix} 1 \\ 3 \end{pmatrix} \dfrac{e^{-2t}}{t}$

15. $\mathbf{X}' = \begin{pmatrix} 0 & 1 \\ -1 & 0 \end{pmatrix} \mathbf{X} + \begin{pmatrix} 0 \\ \sec t \tan t \end{pmatrix}$

16. $\mathbf{X}' = \begin{pmatrix} 0 & 1 \\ -1 & 0 \end{pmatrix} \mathbf{X} + \begin{pmatrix} 1 \\ \cot t \end{pmatrix}$

17. $\mathbf{X}' = \begin{pmatrix} 1 & 2 \\ -\frac{1}{2} & 1 \end{pmatrix} \mathbf{X} + \begin{pmatrix} \csc t \\ \sec t \end{pmatrix} e^{t}$

18. $\mathbf{X}' = \begin{pmatrix} 1 & -2 \\ 1 & -1 \end{pmatrix} \mathbf{X} + \begin{pmatrix} \tan t \\ 1 \end{pmatrix}$

19. $\mathbf{X}' = \begin{pmatrix} 1 & 1 & 0 \\ 1 & 1 & 0 \\ 0 & 0 & 3 \end{pmatrix} \mathbf{X} + \begin{pmatrix} e^{t} \\ e^{2t} \\ te^{3t} \end{pmatrix}$

20. $\mathbf{X}' = \begin{pmatrix} 3 & -1 & -1 \\ 1 & 1 & -1 \\ 1 & -1 & 1 \end{pmatrix} \mathbf{X} + \begin{pmatrix} 0 \\ t \\ 2e^t \end{pmatrix}$

In Problems 21 and 22 use (11) to solve the given system subject to the indicated initial condition.

21. $\mathbf{X}' = \begin{pmatrix} 3 & -1 \\ -1 & 3 \end{pmatrix} \mathbf{X} + \begin{pmatrix} 4e^{2t} \\ 4e^{4t} \end{pmatrix}, \quad \mathbf{X}(0) = \begin{pmatrix} 1 \\ 1 \end{pmatrix}$

★22. $\mathbf{X}' = \begin{pmatrix} 1 & -1 \\ 1 & -1 \end{pmatrix} \mathbf{X} + \begin{pmatrix} 1/t \\ 1/t \end{pmatrix}, \quad \mathbf{X}(1) = \begin{pmatrix} 2 \\ -1 \end{pmatrix}$

In Problems 23 and 24 use (15) to solve the given system subject to the indicated initial condition. Use the results of Problems 31 and 34 of Section 8.5.

23. $\mathbf{X}' = \begin{pmatrix} 4 & 1 \\ 6 & 5 \end{pmatrix} \mathbf{X} + \begin{pmatrix} 50e^{7t} \\ 0 \end{pmatrix}, \quad \mathbf{X}(0) = \begin{pmatrix} 5 \\ -5 \end{pmatrix}$

24. $\mathbf{X}' = \begin{pmatrix} 3 & -2 \\ 5 & -3 \end{pmatrix} \mathbf{X} + \begin{pmatrix} 2 \\ 3 \end{pmatrix}, \quad \mathbf{X}(\pi/2) = \begin{pmatrix} 0 \\ 0 \end{pmatrix}$

[O]   8.9   The Matrix Exponential

Matrices can be utilized in an entirely different manner to solve a homogeneous system of linear first-order differential equations.

Recall that the simple linear first-order differential equation

$$x' = ax,$$

where a is a constant, has the general solution

$$x = ce^{at}.$$

It seems natural then to ask whether we can define a matrix exponential $e^{t\mathbf{A}}$ so that the homogeneous system

$$\mathbf{X}' = \mathbf{AX},$$

where $\mathbf{A}$ is an $n \times n$ matrix of constants, has a solution

$$\mathbf{X} = e^{t\mathbf{A}}\mathbf{C}. \tag{1}$$

Since $\mathbf{C}$ is to be an $n \times 1$ column vector of arbitrary constants we want $e^{t\mathbf{A}}$ to be an $n \times n$ matrix. While the complete development of the meaning of the matrix exponential would necessitate a more thorough investigation of matrix algebra, one means of computing $e^{t\mathbf{A}}$ is given in the following definition.

DEFINITION 8.18   For any $n \times n$ matrix $\mathbf{A}$,

$$e^{t\mathbf{A}} = \sum_{n=0}^{\infty} \frac{(t\mathbf{A})^n}{n!}$$

$$= \mathbf{I} + t\mathbf{A} + \frac{t^2}{2!}\mathbf{A}^2 + \frac{t^3}{3!}\mathbf{A}^3 + \cdots \qquad \square(2)$$

It can be shown that the series given in (2) converges to an $n \times n$ matrix for every value of t. Also, $\mathbf{A}^2 = \mathbf{AA}$, $\mathbf{A}^3 = \mathbf{A}(\mathbf{A}^2)$ and so on.

Now the general solution of the single differential equation

$$x' = ax + f(t),$$

where a is a constant, can be expressed as

$$x = x_c + x_p$$

$$= ce^{at} + e^{at} \int_{t_0}^{t} e^{-as} f(s)\, ds.$$

For systems of linear first-order differential equations, it can be shown that the general solution of

$$\mathbf{X}' = \mathbf{AX} + \mathbf{F}(t),$$

where $\mathbf{A}$ is an $n \times n$ matrix of constants, is

$$\mathbf{X} = \mathbf{X}_c + \mathbf{X}_p = e^{t\mathbf{A}}\mathbf{C} + e^{t\mathbf{A}} \int_{t_0}^{t} e^{-s\mathbf{A}}\mathbf{F}(s)\, ds. \qquad (3)$$

The matrix exponential $e^{t\mathbf{A}}$ is always nonsingular and $e^{-s\mathbf{A}} = (e^{s\mathbf{A}})^{-1}$. In practice, $e^{-s\mathbf{A}}$ can be obtained from $e^{t\mathbf{A}}$ by replacing t by $-s$.

Additional properties

From (2) it is seen that

$$e^{0} = \mathbf{I}. \qquad (4)$$

Also, formal termwise differentiation of (2) shows that

$$\frac{d}{dt} e^{t\mathbf{A}} = \mathbf{A}e^{t\mathbf{A}}. \qquad (5)$$

If we denote the matrix exponential by $\mathbf{\Psi}(t)$ then (5) and (4) are equivalent to

$$\mathbf{\Psi}'(t) = \mathbf{A}\mathbf{\Psi}(t) \qquad (6)$$

$$\mathbf{\Psi}(0) = \mathbf{I}, \qquad (7)$$

respectively. The notation here is chosen deliberately. Comparing (6) with (15) of Section 8.5 reveals that $e^{t\mathbf{A}}$ *is a fundamental matrix* of the system $\mathbf{X}' = \mathbf{AX}$. It is precisely this formulation of the fundamental matrix that was discussed on page 372 of Section 8.5.

By multiplying the series defining $e^{t\mathbf{A}}$ and $e^{-s\mathbf{A}}$ it can also be

$$e^{t\mathbf{A}}e^{-s\mathbf{A}} = e^{(t-s)\mathbf{A}} \qquad \text{or equivalently} \qquad \mathbf{\Psi}(t)\mathbf{\Psi}^{-1}(s) = \mathbf{\Psi}(t-s).*$$

---

*Although $e^{t\mathbf{A}}e^{-s\mathbf{A}} = e^{(t-s)\mathbf{A}}$ it is interesting to note that $e^{\mathbf{A}}e^{\mathbf{B}}$ is, in general, not the same as $e^{\mathbf{A}+\mathbf{B}}$ for $n \times n$ matrices $\mathbf{A}$ and $\mathbf{B}$.

This last result enables us to relate (3) with (10) of the preceding section

$$\mathbf{X} = e^{t\mathbf{A}}\mathbf{C} + \int_{t_0}^{t} e^{t\mathbf{A}}e^{-s\mathbf{A}}\mathbf{F}(s)\, ds$$

$$= e^{t\mathbf{A}}\mathbf{C} + \int_{t_0}^{t} e^{(t-s)\mathbf{A}}\mathbf{F}(s)\, ds \tag{8}$$

$$\mathbf{X} = \boldsymbol{\Psi}(t)\mathbf{C} + \int_{t_0}^{t} \boldsymbol{\Psi}(t-s)\mathbf{F}(s)\, ds. \tag{9}$$

Equation (9) possesses a form simpler than (10) of Section 8.8. In other words, there is no need to compute $\boldsymbol{\Psi}^{-1}$, we need only replace t by $t - s$ in $\boldsymbol{\Psi}(t)$.

EXERCISES 8.9   *Answers to odd-numbered problems begin on page A-30.*

In Problems 1 and 2 use (2) to compute $e^{t\mathbf{A}}$ and $e^{-t\mathbf{A}}$.

1. $\mathbf{A} = \begin{pmatrix} 0 & 1 \\ 1 & 0 \end{pmatrix}$ ⋆**2.** $\mathbf{A} = \begin{pmatrix} 1 & 0 \\ 0 & 2 \end{pmatrix}$

In Problems 3 and 4 use (1) to find the general solution of each system.

3. $\mathbf{X}' = \begin{pmatrix} 0 & 1 \\ 1 & 0 \end{pmatrix}\mathbf{X}$ **4.** $\mathbf{X}' = \begin{pmatrix} 1 & 0 \\ 0 & 2 \end{pmatrix}\mathbf{X}$

In Problems 5–8 use (3) to find the general solution of each system.

5. $\mathbf{X}' = \begin{pmatrix} 0 & 1 \\ 1 & 0 \end{pmatrix}\mathbf{X} + \begin{pmatrix} 1 \\ 1 \end{pmatrix}$ **6.** $\mathbf{X}' = \begin{pmatrix} 0 & 1 \\ 1 & 0 \end{pmatrix}\mathbf{X} + \begin{pmatrix} \cosh t \\ \sinh t \end{pmatrix}$

7. $\mathbf{X}' = \begin{pmatrix} 1 & 0 \\ 0 & 2 \end{pmatrix}\mathbf{X} + \begin{pmatrix} t \\ e^{4t} \end{pmatrix}$ ⋆**8.** $\mathbf{X}' = \begin{pmatrix} 1 & 0 \\ 0 & 2 \end{pmatrix}\mathbf{X} + \begin{pmatrix} 3 \\ -1 \end{pmatrix}$

Miscellaneous problems

Let $\mathbf{P}$ denote a matrix whose columns are eigenvectors $\mathbf{K}_1, \mathbf{K}_2, \ldots, \mathbf{K}_n$ corresponding to *distinct* eigenvalues $\lambda_1, \lambda_2, \ldots, \lambda_n$ of an $n \times n$ matrix $\mathbf{A}$. Then it can be shown that $\mathbf{A} = \mathbf{P}\mathbf{D}\mathbf{P}^{-1}$ where $\mathbf{D}$ is defined by

$$\mathbf{D} = \begin{pmatrix} \lambda_1 & 0 & \ldots & 0 \\ 0 & \lambda_2 & \ldots & 0 \\ \vdots & & & \vdots \\ 0 & 0 & \ldots & \lambda_n \end{pmatrix} \tag{10}$$

In Problems 9 and 10 verify the above result for the given matrix.

9. $\mathbf{A} = \begin{pmatrix} 2 & 1 \\ -3 & 6 \end{pmatrix}$ **10.** $\mathbf{A} = \begin{pmatrix} 2 & 1 \\ 1 & 2 \end{pmatrix}$

11. Suppose $\mathbf{A} = \mathbf{P}\mathbf{D}\mathbf{P}^{-1}$ where $\mathbf{D}$ is defined in (10). Use (2) to show that $e^{t\mathbf{A}} = \mathbf{P}e^{t\mathbf{D}}\mathbf{P}^{-1}$.

12. Use (2) to show that

$$e^{tD} = \begin{pmatrix} e^{\lambda_1 t} & 0 & \ldots 0 \\ 0 & e^{\lambda_2 t} & \ldots 0 \\ \vdots & & \\ 0 & 0 & \ldots e^{\lambda_n t} \end{pmatrix}$$

where **D** is defined in (10).

In Problems 13 and 14 use the results of Problems 9–12 to solve the given system.

13. $\mathbf{X}' = \begin{pmatrix} 2 & 1 \\ -3 & 6 \end{pmatrix}\mathbf{X}$

14. $\mathbf{X}' = \begin{pmatrix} 2 & 1 \\ 1 & 2 \end{pmatrix}\mathbf{X}$

CHAPTER 8 SUMMARY

Throughout this chapter we have considered **systems of linear** differential equations.

For linear systems, probably the most basic technique of solution consists of rewriting the entire system in **operator notation** and then using **systematic elimination** to obtain single differential equations in one dependent variable which can be solved by the usual procedures. Also, **determinants** can usually be utilized to accomplish the same result. Once all dependent variables have been determined, it is necessary to use the system itself to find various relationships between the parameters.

When initial conditions are specified, the **Laplace transform** can be used to reduce the system to simultaneous algebraic equations in the transformed functions. However, we must again use an elimination procedure to solve for these transformed variables.

A first-order system in **normal form** in *two* dependent variables is any system

$$\frac{dx}{dt} = a_{11}(t)x + a_{12}(t)y + f_1(t)$$
$$\frac{dy}{dt} = a_{21}(t)x + a_{22}(t)y + f_2(t) \tag{1}$$

where the coefficients $a_{ij}(t)$, $f_2(t)$ and $f_2(t)$ are continuous on some common interval I. When $f_1(t) = 0, f_2(t) = 0$, the system is said to be **homogeneous**, otherwise it is said to be **nonhomogeneous**. Any linear second-order differential equation can be expressed in this form.

Using matrices, the system (1) can be written compactly as

$$\frac{d\mathbf{X}}{dt} = \mathbf{A}(t)\mathbf{X} + \mathbf{F}(t) \tag{2}$$

where

$$\mathbf{X} = \begin{pmatrix} x \\ y \end{pmatrix}, \qquad \mathbf{A}(t) = \begin{pmatrix} a_{11}(t) & a_{12}(t) \\ a_{21}(t) & a_{22}(t) \end{pmatrix}, \qquad \text{and} \qquad \mathbf{F}(t) = \begin{pmatrix} f_1(t) \\ f_2(t) \end{pmatrix}.$$

The **general solution of the homogeneous system** in two dependent variables

$$\frac{d\mathbf{X}}{dt} = \mathbf{A}(t)\mathbf{X} \tag{3}$$

is defined to be the linear combination

$$\mathbf{X} = c_1\mathbf{X}_1 + c_2\mathbf{X}_2 \tag{4}$$

where $\mathbf{X}_1$ and $\mathbf{X}_2$ form a **fundamental set of solutions** of (3) on I. The **general solution of the nonhomogeneous system** (2) is defined to be

$$\mathbf{X} = \mathbf{X}_c + \mathbf{X}_p$$

where $\mathbf{X}_c$ is defined by (4) and $\mathbf{X}_p$ is *any* solution vector of (2).

To solve a homogeneous system (3) we determine the **eigenvalues** of the coefficient matrix $\mathbf{A}$ and then find the corresponding **eigenvectors**.

To solve a nonhomogeneous system we first solve the associated homogeneous system. A particular solution vector $\mathbf{X}_p$ of the nonhomogeneous system is found by either **undetermined coefficients** or **variation of parameters**.

A **fundamental matrix** of a homogeneous system (3) in two dependent variables is defined to be

$$\boldsymbol{\Phi}(t) = \begin{pmatrix} x_1 & x_2 \\ y_1 & y_2 \end{pmatrix}. \tag{5}$$

The columns in (5) are obtained from two linearly independent solution vectors $\mathbf{X}_1$ and $\mathbf{X}_2$ of (3). In terms of matrices, the method of variation of parameters leads to a particular solution given by:

$$\mathbf{X}_p = \boldsymbol{\Phi}(t) \int \boldsymbol{\Phi}^{-1}(t)\mathbf{F}(t) \, dt.$$

The general solution of (2) on an interval is

$$\mathbf{X} = \boldsymbol{\Phi}(t)\mathbf{C} + \boldsymbol{\Phi}(t) \int \boldsymbol{\Phi}^{-1}(t)\mathbf{F}(t) \, dt,$$

where $\mathbf{C}$ is a column matrix containing two arbitrary constants. The matrix, $\boldsymbol{\Phi}^{-1}(t)$ is called the **multiplicative inverse** of $\boldsymbol{\Phi}(t)$; the multiplicative inverse satisfies $\boldsymbol{\Phi}(t)\boldsymbol{\Phi}^{-1}(t) = \boldsymbol{\Phi}^{-1}(t)\boldsymbol{\Phi}(t) = \mathbf{I}$, where $\mathbf{I}$ is the 2×2 **multiplicative identity**.

CHAPTER 8
TEST

Answers to odd-numbered problems begin on page A-31.

In Problems 1 and 2 use systematic elimination to solve each system.

1. $x' + y' = 2x + 2y + 1$

$x' + 2y' = y + 3$

2. $\dfrac{dx}{dt} = 2x + y + t - 2$

$\dfrac{dy}{dt} = 3x + 4y - 4t$

In Problems 3 and 4 use determinants to solve each system.

3. $(D - 2)x - \quad y = -e^t$

$\quad - 3x + (D - 4)y = -7e^t$

4. $(D + 2)x + (D + 1)y = \sin 2t$

$\quad 5x + (D + 3)y = \cos 2t$

In Problems 5 and 6 use the Laplace transform to solve each system.

5. $x' + y = t$

$4x + y' = 0$

$x(0) = 1 \quad y(0) = 2$

6. $x'' + y'' = e^{2t}$

$2x' + y'' = -e^{2t}$

$x(0) = 0 \quad y(0) = 0$

7.(a) Write as one column matrix $\mathbf{X}$:

$$\begin{pmatrix} 3 & 1 & 1 \\ -1 & 2 & -1 \\ 0 & -2 & 4 \end{pmatrix}\begin{pmatrix} t \\ t^2 \\ t^3 \end{pmatrix} - \begin{pmatrix} 2 \\ -2 \\ -1 \end{pmatrix} + 2t\begin{pmatrix} 1 \\ 0 \\ 4 \end{pmatrix} + 2t^2\begin{pmatrix} 1 \\ -1 \\ 7 \end{pmatrix}.$$

(b) Find $d\mathbf{X}/dt$.

8. Write the differential equation

$$3y^{(4)} - 5y'' + 9y = 6e^t - 2t$$

as a system of first-order equations in linear normal form.

9. Write the system

$$(2D^2 + D)y - D^2x = \ln t$$

$$D^2y + (D + 1)x = 5t - 2$$

as a system of first-order equations in linear normal form.

CHAPTER 8 TEST

10. Verify that the general solution of the system

$$\frac{dx}{dt} = y$$

$$\frac{dy}{dt} = -x + 2y - 2\cos t$$

on $-\infty < t < \infty$ is

$$\mathbf{X} = c_1\begin{pmatrix}1\\1\end{pmatrix}e^t + c_2\left\{\begin{pmatrix}1\\1\end{pmatrix}te^t + \begin{pmatrix}0\\1\end{pmatrix}e^t\right\} + \begin{pmatrix}\sin t\\\cos t\end{pmatrix}.$$

In Problems 11–16 use the concept of eigenvalues and eigenvectors to solve each system.

11. $\dfrac{dx}{dt} = 2x + y$

$\dfrac{dy}{dt} = -x$

12. $\dfrac{dx}{dt} = -4x + 2y$

$\dfrac{dy}{dt} = 2x - 4y$

13. $\mathbf{X}' = \begin{pmatrix}1 & 2\\-2 & 1\end{pmatrix}\mathbf{X}$

14. $\mathbf{X}' = \begin{pmatrix}-2 & 5\\-2 & 4\end{pmatrix}\mathbf{X}$

15. $\mathbf{X}' = \begin{pmatrix}1 & 1 & 1\\1 & 1 & 1\\1 & 1 & 1\end{pmatrix}\mathbf{X}$

16. $\mathbf{X}' = \begin{pmatrix}1 & -1 & 1\\0 & 1 & 3\\4 & 3 & 1\end{pmatrix}\mathbf{X}$

In Problems 17–20 use either undetermined coefficients or variation of parameters to solve the given system.

17. $\mathbf{X}' = \begin{pmatrix}2 & 8\\1 & 4\end{pmatrix}\mathbf{X} + \begin{pmatrix}0\\3te^{6t}\end{pmatrix}$

18. $\dfrac{dx}{dt} = x + 2y$

$\dfrac{dy}{dt} = -\tfrac{1}{2}x + y + e^t\tan t$

19. $\mathbf{X}' = \begin{pmatrix}-1 & 1\\-2 & 1\end{pmatrix}\mathbf{X} + \begin{pmatrix}1\\\cot t\end{pmatrix}$

20. $\mathbf{X}' = \begin{pmatrix}3 & 1\\-1 & 1\end{pmatrix}\mathbf{X} + \begin{pmatrix}-2\\1\end{pmatrix}e^{2t}$

ND EDITION A FIRST COURSE IN DIFFERENTIAL EQUATIONS SECOND EDITION A FIRS
SE IN DIFFERENTIAL EQUATIONS SECOND EDITION A FIRST COURSE IN DIFFERENTIA
IONS SECOND EDITION A FIRST COURSE IN DIFFE   ND ED
FIRST COURSE IN DIFFERENTIAL EQUATIONS SEC   SE IN
NTIAL EQUATIONS SECOND EDITION A FIRST COURSE IN DIFFERENTIAL EQUATIONS
ND EDITION A FIRST COURSE IN DIFFERENTIAL EQUATIONS SECOND EDITION A FIRST
SE IN DIFFERENTIAL EQUATIONS SECOND EDITION A FIRST COURSE IN DIFFERENTIAL
IONS S   IN DIFFERENTIAL EQUATIONS SECOND EDIT
FIRST   ONS SECOND EDITION A FIRST COURSE IN D
NTIAL EQUATIONS SECOND EDITION A FIRST COURSE IN DIFFERENTIAL EQUATIONS
ND EDITION A FIRST COURSE IN DIFFERENTIAL EQUATIONS SECOND EDITION A FIRST
SE IN DIFFERENTIAL EQUATIONS SECOND EDITION A FIRST COURSE IN DIFFERENTIAL
IONS SECOND EDITION A FIRST COURSE IN DIFFERENTIAL EQUATIONS SECOND EDIT
FIRST COURSE IN DIFFERENTIAL EQUATIONS SECOND EDITION A FIRST COURSE IN D
NTIAL EQUATIONS SECOND EDITION A FIRST COURSE IN DIFFERENTIAL EQUATIONS

CHAPTER 9

Numerical Methods

Introduction

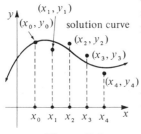

Figure 9.1

A differential equation does not have to possess a solution, or even if a solution exists, we need not always be able to find an explicit or implicit formula, containing elementary functions, satisfying the equation. In many instances, particularly in the study of nonlinear equations, we may have to be satisfied with obtaining only approximations to a solution. Assuming that a solution to a differential equation exists, it represents a locus of points (points connected by a smooth curve) in the Cartesian plane. Beginning in Section 9.2 we shall study methods that utilize the differential equation to obtain a sequence of discrete points which numerically approximate the points on the actual solution curve. See Figure 9.1.

Throughout this chapter we shall confine our attention to first-order equations $dy/dx = f(x, y)$. As we saw in the preceding chapter, higher order differential equations can always be reduced to a system of first-order equations. The numerical procedures that we shall study are then easily adaptable to this system of equations (see Section 9.5).

9.1 Direction Fields

Lineal elements

Suppose for the moment that we do not know the general solution of the simple equation $y' = y$. Specifically, the differential equation implies that the slope of the tangent line to a solution curve is given by the function $f(x, y) = y$. When $f(x, y)$ is held constant, that is, when

$$y = c \qquad (1)$$

where c is any constant, we are in effect stating that the slope of the tangents to the solution curves is the same constant value along a horizontal line. For

408

example, for $y = 2$ let us draw a sequence of short line segments, or **lineal elements**, each having slope 2 and its midpoint on the line. As shown in Figure 9.2 the solution curves pass through this horizontal line at every point tangent to the lineal elements.

Isoclines and direction fields

Equation (1) represents a one-parameter family of horizontal lines. In general, any member of the family $f(x, y) = c$ is called an **isocline** which literally means a curve along which the inclination (of the tangents) is the same. As the parameter c is varied, we obtain a collection of isoclines on which the lineal elements are judiciously constructed. The totality of these lineal elements is called a **direction field, slope field,** or **lineal element field** of the differential equation $y' = f(x, y)$. As we see in Figure 9.3(a) the direction field suggests the "flow pattern" for the family of solution curves of the differential equation $y' = y$. In particular, if we want the one solution passing through the point $(0, 1)$, then, as indicated in Figure 9.3(b), we construct a curve through this point and passing through the isoclines with the appropriate slopes.

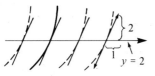

solution curves

Figure 9.2

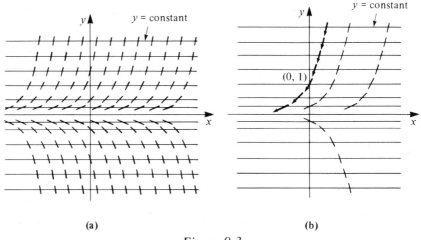

(a)   (b)

Figure 9.3

EXAMPLE

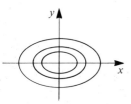

Figure 9.4

Determine the isoclines for the equation

$$\frac{dy}{dx} = 4x^2 + 9y^2.$$

Solution: For $c > 0$ the isoclines are the curves

$$4x^2 + 9y^2 = c.$$

As Figure 9.4 shows, the curves are a concentric family of ellipses with major axis along the x-axis.

EXAMPLE   Sketch the direction field and indicate several possible members of the family of solution curves for

$$\frac{dy}{dx} = \frac{x}{y}.$$

Solution: Before sketching the direction field corresponding to the isoclines $x/y = c$ or $y = x/c$ we note that the differential equation gives the following information:

   (a) if a solution curve crosses the x-axis ($y = 0$), it does so tangent to a vertical lineal element at every point except possibly $(0, 0)$.

   (b) If a solution curve crosses the y-axis ($x = 0$), it does so tangent to a horizontal lineal element at every point except possibly $(0,0)$.

   (c) The lineal elements corresponding to the isoclines $c = 1$ and $c = -1$ are collinear with the lines $y = x$ and $y = -x$, respectively. Indeed, it is easily verified that these isoclines are both particular solutions of the given differential equation. However, it should be noted that *in general* isoclines are themselves not solutions to a differential equation.*

Figure 9.5 shows the direction field and several possible solution curves. Remember, on any particular isocline all the lineal elements are parallel. Also, the lineal elements may be drawn in such a manner as to suggest the flow of a particular curve.† In other words, imagine the isoclines so close together that if the lineal elements were connected, we would have a polygonal curve suggestive of the shape of a smooth curve.

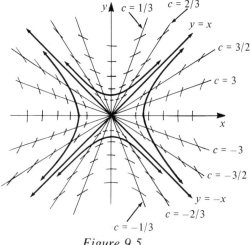

Figure 9.5

---

*When the isoclines are straight lines it is easy to determine which, if any, of these isoclines are also particular solutions of the differential equation. See Problems 21–26.
†Alternatively, the lineal elements can be drawn uniformly spaced on the isocline.

EXAMPLE

In Section 2.1 we indicated that the differential equation

$$\frac{dy}{dx} = x^2 + y^2$$

cannot be solved in terms of elementary functions. Use a direction field to locate an approximate solution satisfying

$$y(0) = 1.$$

Solution: The isoclines are concentric circles defined by

$$x^2 + y^2 = c, \qquad c > 0.$$

By choosing $c = 1/4$, $c = 1$, $c = 9/4$, and $c = 4$, we obtain the circles with radii $1/2$, 1, $3/2$, and 2 shown in Figure 9.6(a). The lineal elements superimposed on each circle have slope corresponding to the particular value of c. It seems plausible from inspection of Figure 9.6(a) that a solution curve of the given initial-value problem might have the shape given in Figure 9.6(b). Unfortunately, we are not able to obtain any formula that describes this curve.

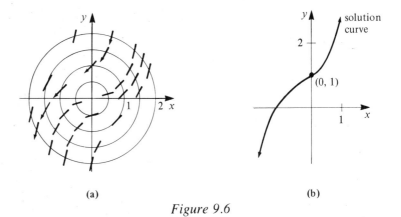

(a)   (b)

Figure 9.6

The concept of the direction field is used primarily to establish the existence and to possibly locate an approximate solution curve for a first-order differential equation which cannot be solved by the usual standard techniques. However, the preceding discussion is of little value in determining specific values of a solution $y(x)$ at given points. For example, if we want to know the approximate value of $y(0.5)$ for the solution of

$$\frac{dy}{dx} = x^2 + y^2$$

$$y(0) = 1,$$

then Figure 9.6(b) can do nothing more for us than to indicate that $y(0.5)$ may be in the same "ball park" as $y = 2$.

EXERCISES 9.1   *Answers to odd-numbered problems begin on page A-31.*

In Problems 1–10 identify the isoclines for the given differential equation.

1. $\dfrac{dy}{dx} = x + 4$

2. $\dfrac{dy}{dx} = 2x + y$

3. $\dfrac{dy}{dx} = x^2 - y^2$

4. $\dfrac{dy}{dx} = y - x^2$

5. $y' = \sqrt{x^2 + y^2 + 2y + 1}$

★6. $y' = (x^2 + y^2)^{-1}$

7. $\dfrac{dy}{dx} = y(x + y)$

8. $\dfrac{dy}{dx} = y + e^x$

9. $\dfrac{dy}{dx} = \dfrac{y - 1}{x - 2}$

10. $\dfrac{dy}{dx} = \dfrac{x - y}{x + y}$

In Problems 11–18 sketch the direction field for the given differential equation and indicate several possible solution curves.

11. $y' = x$

★12. $y' = x + y$

13. $y\dfrac{dy}{dx} = -x$

14. $\dfrac{dy}{dx} = \dfrac{1}{y}$

15. $\dfrac{dy}{dx} = xy$

16. $\dfrac{dy}{dx} = 1 - xy$

17. $y' = y - \cos\dfrac{\pi}{2}x$

18. $y' = 1 - \dfrac{y}{x}$

Miscellaneous problems

19. Formally show that the isoclines for the differential equation
$$\frac{dy}{dx} = \frac{\alpha x + \beta y}{\gamma x + \delta y}$$
are straight lines through the origin.

20. Show that $y = cx$ is a solution of the differential equation in Problem 19 if and only if $(\beta - \gamma)^2 + 4\alpha\delta \geq 0$.

In Problems 21–26 find those isoclines which are also solutions of the given differential equation. See Problems 19 and 20.

EXAMPLE   The isoclines of the differential equation
$$y' = 2x + y \tag{2}$$
are the straight lines
$$2x + y = c. \tag{3}$$
A line in this latter family will be a solution of the differential equation whenever its slope is the same as c. In other words, both the original equation

and the line will satisfy $y' = c$. Since the slope of (3) is -2 if we choose $c = -2$, then $2x + y = -2$ is a solution of (2).

Check: Write the solution as

$$y = -2x - 2$$

and so

$$y' = -2$$

$$= 2x + (-2x - 2)$$

$$= 2x + y$$

---

21. $y' = 3x + 2y$

22. $y' = 6x - 2y$

23. $y' = \dfrac{2x}{y}$

★24. $y' = \dfrac{2y}{x + y}$

25. $\dfrac{dy}{dx} = \dfrac{4x + 3y}{y}$

26. $\dfrac{dy}{dx} = \dfrac{5x + 10y}{x + 2y}$

9.2   The Euler Methods

9.2.1   Euler's Method

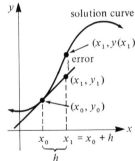

Figure 9.7

One of the simplest techniques for approximating solutions of differential equations is known as **Euler's method** or the method of **tangent lines**. Suppose we wish to approximate the solution of the equation

$$y' = f(x, y)$$

satisfying the initial condition

$$y(x_0) = y_0.$$

If h is a positive increment on the x-axis, then as Figure 9.7 shows, we can find a point $(x_1, y_1) = (x_0 + h, y_1)$ on the line tangent to the unknown solution curve at (x_0, y_0).

By the point-slope form of the equation of a line, we have

$$\frac{y_1 - y_0}{(x_0 + h) - x_0} = y_0' \qquad \text{or} \qquad y_1 = y_0 + h y_0'$$

where $y_0' = f(x_0, y_0)$. If we label $x_0 + h$ by x_1, then the point (x_1, y_1) on the tangent line is an approximation to the point $(x_1, y(x_1))$ on the solution curve. That is, $y_1 \approx y(x_1)$. Of course, the accuracy of the approximation depends heavily on the size of the increment h. Usually we must choose this step size to be "reasonably small."

Assuming a uniform (constant) value of h, we can obtain a succession of points $(x_1, y_1), (x_2, y_2), \ldots, (x_n, y_n)$ which we hope are proximate to the points $(x_1, y(x_1)), (x_2, y(x_2)), \ldots, (x_n, y(x_n))$. See Figure 9.8. Now using

(x_1, y_1), we can obtain the value of y_2 which is the ordinate of a point on a new "tangent" line. We have

$$\frac{y_2 - y_1}{h} = y_1' \quad \text{or} \quad y_2 = y_1 + hy_1'$$
$$= y_1 + hf(x_1, y_1).$$

In general it follows that

$$y_{n+1} = y_n + hy_n'$$
$$= y_n + hf(x_n, y_n) \qquad (1)$$

where $x_n = x_0 + nh$.

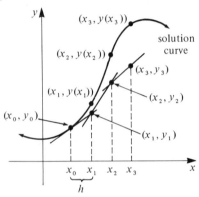

Figure 9.8

As an example, suppose we try the iteration scheme (1) on a differential equation for which we know the explicit solution; in this way we can compare the estimated values y_n and the true values $y(x_n)$.

EXAMPLE   Consider the initial-value problem

$$\frac{dy}{dx} = 2xy$$

$$y(1) = 1.$$

Use the Euler method to obtain an approximation to $y(1.5)$ using first $h = 0.1$ and then $h = 0.05$.

Solution: We first identify $f(x, y) = 2xy$ so that (1) becomes

$$y_{n+1} = y_n + h(2x_n y_n).$$

Then for $h = 0.1$ we find

$$y_1 = y_0 + (0.1)(2x_0 y_0)$$
$$= 1 + (0.1)[2(1)(1)]$$
$$= 1.2$$

which is an estimate to the value of $y(1.1)$. However, if we use $h = 0.05$ it takes *two* iterations to reach $x = 1.1$. We have

$$y_1 = 1 + (0.05)[2(1)(1)]$$

$$= 1.1,$$

$$y_2 = 1.1 + (0.05)[2(1.05)(1.1)]$$

$$= 1.2155.$$

Here we note that $y_1 \approx y(1.05)$ and $y_2 \approx y(1.1)$. The remainder of the calculations are summarized in Tables 9.1 and 9.2. Each entry is rounded to four decimal places.*

Table 9.1 Euler's Method with $h = 0.1$

x_n	y_n	True Value	Error	% Rel Error
1.00	1.0000	1.0000	0.0000	0.00
1.10	1.2000	1.2337	0.0337	2.73
1.20	1.4640	1.5527	0.0887	5.71
1.30	1.8154	1.9937	0.1784	8.95
1.40	2.2874	2.6117	0.3244	12.42
1.50	2.9278	3.4904	0.5625	16.12

Table 9.2 Euler's Method with $h = 0.05$

x_n	y_n	True Value	Error	% Rel Error
1.00	1.0000	1.0000	0.0000	0.00
1.05	1.1000	1.1079	0.0079	0.72
1.10	1.2155	1.2337	0.0182	1.47
1.15	1.3492	1.3806	0.0314	2.27
1.20	1.5044	1.5527	0.0483	3.11
1.25	1.6849	1.7551	0.0702	4.00
1.30	1.8955	1.9937	0.0982	4.93
1.35	2.1419	2.2762	0.1343	5.90
1.40	2.4311	2.6117	0.1806	6.92
1.45	2.7714	3.0117	0.2403	7.98
1.50	3.1733	3.4904	0.3171	9.08

---

*It is fairly easy to perform most of the calculations in this chapter on any scientific hand calculator. However, in each case a *Data General Nova 1200* computer was used.

In the preceding example the true values were calculated from the known solution $y = e^{x^2-1}$. Also, the percentage relative error is defined to be

$$\frac{|\text{true value} - \text{approximation}|}{\text{true value}} \times 100 = \frac{|\text{error}|}{\text{true value}} \times 100.$$

It should be apparent that in the case of the step size $h = 0.1$ a 16% relative error in the calculation of the approximation to $y(1.5)$ is totally unacceptable. At the expense of doubling the number of calculations a slight improvement in accuracy is obtained by halving the step size to $h = 0.05$.

Of course, in many instances we may not know the solution of a particular differential equation, or for that matter, whether a solution of an intitial-value problem actually exists. The following nonlinear equation does possess a solution in closed form, but we leave it as an exercise for the reader to find it (see Problem 1).

EXAMPLE

Use the Euler method to obtain the approximate value of $y(0.5)$ for the solution of

$$y' = (x + y - 1)^2$$

$$y(0) = 2.$$

Solution: For $n = 0$ and $h = 0.1$ we have

$$y_1 = y_0 + (0.1)(x_0 + y_0 - 1)^2$$

$$= 2 + (0.1)(1)^2$$

$$= 2.1.$$

The remaining calculations are summarized in Tables 9.3 and 9.4 for $h = 0.1$ and $h = 0.05$, respectively.

Table 9.3 Euler's Method with $h = 0.1$

x_n	y_n
0.00	2.0000
0.10	2.1000
0.20	2.2440
0.30	2.4525
0.40	2.7596
0.50	3.2261

Table 9.4 Euler's Method with $h = 0.05$

x_n	y_n
0.00	2.0000
0.05	2.0500
0.10	2.1105
0.15	2.1838
0.20	2.2727
0.25	2.3812
0.30	2.5142
0.35	2.6788
0.40	2.8845
0.45	3.1455
0.50	3.4823

We may want greater accuracy than that displayed, say, in Table 9.2, and so we could try a step size even smaller than $h = 0.05$. However, rather than resorting to this extra labor, it probably would be more advantageous to employ an alternative numerical procedure. The Euler formula by itself, through attractive in its simplicity, is seldom used in serious calculations.

9.2.2   The Improved Euler Method

The formula

$$y_{n+1} = y_n + h\frac{f(x_n, y_n) + f(x_{n+1}, y^*_{n+1})}{2}$$

where $\quad y^*_{n+1} = y_n + hf(x_n, y_n)$

(2)

is known as the **improved Euler formula** or **Heun's formula**. The values $f(x_n, y_n)$ and $f(x_{n+1}, y^*_{n+1})$ are approximations to the slope of the curve at $(x_n, y(x_n))$ and $(x_{n+1}, y(x_{n+1}))$ and consequently the ratio

$$\frac{f(x_n, y_n) + f(x_{n+1}, y^*_{n+1})}{2}$$

can be interpreted as an average slope on the interval between x_n and x_{n+1}.

The equations in (2) can be readily visualized. In Figure 9.9 we have shown the case when $n = 0$. Note that

$$f(x_0, y_0) \qquad \text{and} \qquad f(x_1, y_1^*)$$

are slopes of the indicated straight lines passing through the points (x_0, y_0) and (x_1, y_1^*), respectively. By taking an average of these slopes, we obtain the slope of the dotted skew lines. Rather than advancing along the line with slope $m = f(x_0, y_0)$ to the point with ordinate y_1^* obtained by the usual Euler method, we advance instead along the line through (x_0, y_0) with slope m_{ave}

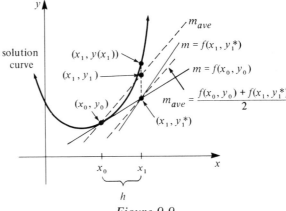

Figure 9.9

until we reach x_1. It seems plausible from inspection of the figure that y_1 is an improvement over y_1^*.

We might also say that the value of

$$y_1^* = y_0 + hf(x_0, y_0)$$

predicts a value of $y(x_1)$, whereas

$$y_1 = y_0 + h\frac{f(x_0, y_0) + f(x_1, y_1^*)}{2}$$

corrects this estimate.

EXAMPLE

Use the improved Euler formula to obtain the approximate value of $y(1.5)$ for the solution of

$$y' = 2xy$$

$$y(1) = 1.$$

Compare the results for $h = 0.1$ and $h = 0.05$.

Solution: For $n = 0$ and $h = 0.1$ we first compute

$$y_1^* = y_0 + (0.1)2x_0 y_0$$

$$= 1.2.$$

Then from (2)

$$y_1 = y_0 + (0.1)\frac{2x_0 y_0 + 2x_1 y_1^*}{2}$$

$$= 1 + (0.1)\frac{2(1)(1) + 2(1.1)(1.2)}{2}$$

$$= 1.232.$$

The comparative values of the calculations for $h = 0.1$ and $h = 0.05$ are given in Tables 9.5 and 9.6, respectively.

Table 9.5   Improved Euler's Method with $h = 0.1$

x_n	y_n	True Value	Error	% Rel Error
1.00	1.0000	1.0000	0.0000	0.00
1.10	1.2320	1.2337	0.0017	0.14
1.20	1.5479	1.5527	0.0048	0.31
1.30	1.9832	1.9937	0.0106	0.53
1.40	2.5908	2.6117	0.0209	0.80
1.50	3.4509	3.4904	0.0394	1.13

Table 9.6   Improved Euler's Method with $h = 0.05$

x_n	y_n	True Value	Error	% Rel Error
1.00	1.0000	1.0000	0.0000	0.00
1.05	1.1077	1.1079	0.0002	0.02
1.10	1.2332	1.2337	0.0004	0.04
1.15	1.3798	1.3806	0.0008	0.06
1.20	1.5514	1.5527	0.0013	0.08
1.25	1.7531	1.7551	0.0020	0.11
1.30	1.9909	1.9937	0.0029	0.14
1.35	2.2721	2.2762	0.0041	0.18
1.40	2.6060	2.6117	0.0057	0.22
1.45	3.0038	3.0117	0.0079	0.26
1.50	3.4795	3.4904	0.0108	0.31

A brief word of caution is in order here. We cannot compute all the values of y_n^* first and then substitute these values in the first formula of (2). In other words, we cannot use the data in Table 9.1 to help construct the values in Table 9.5. Why?

EXAMPLE   Use the improved Euler formula to obtain the approximate value of $y(0.5)$ for the solution of

$$y' = (x + y - 1)^2$$

$$y(0) = 2.$$

Solution: For $n = 0$ and $h = 0.1$, we have

$$y_1^* = y_0 + (0.1)(x_0 + y_0 - 1)^2$$

$$= 2.1$$

and so $$y_1 = y_0 + (0.1)\frac{(x_0 + y_0 - 1)^2 + (x_1 + y_1^* - 1)^2}{2}$$

$$= 2 + (0.1)\frac{1 + 1.44}{2}$$

$$= 2.122.$$

The remaining calculations are summarized in Tables 9.7 and 9.8 for $h = 0.1$ and $h = 0.05$, respectively.

Table 9.7 Improved Euler's Method with $h = 0.1$

x_n	y_n
0.00	2.0000
0.10	2.1220
0.20	2.3049
0.30	2.5858
0.40	3.0378
0.50	3.8254

Table 9.8 Improved Euler's Method with $h = 0.05$

x_n	y_n
0.00	2.0000
0.05	2.0553
0.10	2.1228
0.15	2.2056
0.20	2.3075
0.25	2.4342
0.30	2.5931
0.35	2.7953
0.40	3.0574
0.45	3.4057
0.50	3.8840

EXERCISES 9.2   *Answers to odd-numbered problems begin on page A-32.*

1. Solve the initial-value problem

$$y' = (x + y - 1)^2$$
$$y(0) = 2$$

in terms of elementary functions.

2. Let $y(x)$ be the solution of the initial-value problem given in Problem 1. Rounded to four decimal places, compute the exact values of $y(0.1)$, $y(0.2)$, $y(0.3)$, $y(0.4)$, and $y(0.5)$. Compare these values with the entries in Tables 9.3, 9.4, 9.7, and 9.8.

Given the initial-value problems in Problems 3–12. Use the Euler formula to obtain a four decimal approximation to the indicated value. First use **(a)** $h = 0.1$, and then **(b)** $h = 0.05$.

3. $y' = 2x - 3y + 1$, $y(1) = 5$; $y(1.5)$

4. $y' = 4x - 2y$, $y(0) = 2$; $y(0.5)$

5. $y' = 1 + y^2$, $y(0) = 0$; $y(0.5)$

★6. $y' = x^2 + y^2$, $y(0) = 1$; $y(0.5)$

7. $y' = e^{-y}$, $y(0) = 0$; $y(0.5)$

8. $y' = x + y^2$, $y(0) = 0$; $y(0.5)$

9. $y' = (x - y)^2$, $y(0) = 0.5$; $y(0.5)$

10. $y' = xy + \sqrt{y}$, $y(0) = 1$; $y(0.5)$

11. $y' = xy^2 - \dfrac{y}{x}$, $y(1) = 1$; $y(1.5)$

★**12.** $y' = y - y^2$, $\quad y(0) = 0.5$; $\quad y(0.5)$

13. As parts **(a)**–**(e)** of this problem, repeat the calculations of Problems 3, 5, 7, 9 and 11 using the improved Euler formula.

14. As parts **(a)**–**(e)** of this problem, repeat the calculations of Problems 4, 6, 8, 10, and 12 using the improved Euler formula.

15. Although it may not be obvious from the differential equation, its solution could "behave badly" near a point x at which we wish to approximate $y(x)$. Numerical procedures may then give widely differing results near this point. Let $y(x)$ be the solution of the initial-value problem

$$y' = x^2 + y^3, \qquad y(1) = 1.$$

Using the step size $h = 0.1$, compare the results obtained from the Euler formula with the results from the improved Euler formula in the approximation of $y(1.4)$.

EXAMPLE

The improved Euler formula can be used to obtain a sequence of approximations to $y(x_n)$ at a *fixed value* of x_n. If the basic Euler formula is denoted by

$$y_{n+1,1} = y_n + hf(x_n, y_n) \tag{3}$$

then we can define

$$y_{n+1,k+1} = y_n + h\frac{f(x_n, y_n) + f(x_{n+1}, y_{n+1,k})}{2} \tag{4}$$

for $n \geq 0$, $k \geq 1$. For $n = 0$ and $k = 1, 2, 3, \ldots$ equations (3) and (4) yield the sequence of values

$$y_{1,1}, \ y_{1,2}, \ y_{1,3}, \ y_{1,4}, \ \cdots$$

which are all approximations to $y(x)$ at $x = x_1$. For example, $y_{1,2}$ corresponds to the *original* improved Euler formula given in (2) and

$$y_{1,3} = y_0 + h\frac{f(x_0, y_0) + f(x_1, y_{1,2})}{2}.$$

It might be conjectured that since an average of two slopes (formula (2)) yields an improved approximation to $y(x)$ at a point, that an average including an average (formula (4)) *may* even give better results. See Problem 16.

★**16.** Consider the initial-value problem

$$y' = 2xy, \qquad y(1) = 1.$$

Convince yourself that "more" is not necessarily better by computing the values

$$y_{1,1}, \ y_{1,2}, \ y_{1,3}, \ y_{1,4}, \ y_{1,5}$$

with $h = 0.1$. By using the exact value of $y(1.1)$ (see Table 9.1), compute the percentage relative error at each step of the calculation.

Miscellaneous problems

17. Derive the basic Euler formula by integrating both sides of the equation $y' = f(x, y)$ on the integral $x_n \leq x \leq x_{n+1}$. Approximate the integral of the right side by replacing the function $f(x, y)$ by its value at the left end point of the interval of integration.

18. By following the procedure outlined in Problem 17, derive the improved Euler formula. [*Hint*: Replace the integrand of the right side by the average of its values at the end points of the interval of integration.]

9.3   The Three-Term Taylor Method

The numerical method that we shall consider in this section is more of theoretical interest than of practical importance since the results obtained using formula (5) below will not differ substantially from those obtained using the improved Euler method.

In the study of numerical solutions of differential equations, many computational algorithms can be derived from a Taylor series expansion. Recall from calculus that the form of this expansion about a point $x = a$ is

$$y(x) = y(a) + y'(a)\frac{(x-a)}{1!} + y''(a)\frac{(x-a)^2}{2!} + \cdots . \tag{1}$$

It is understood that the function $y(x)$ possesses derivatives of all orders and that the series (1) converges in some interval defined by $|x - a| < R$. Notice, in particular, that if we set $a = x_n$ and $x = x_n + h$, then (1) becomes

$$y(x_n + h) = y(x_n) + y'(x_n)h + y''(x_n)\frac{h^2}{2} + \cdots . \tag{2}$$

The Euler method revisited

Furthermore, let us now assume that the function $y(x)$ is a solution of the first-order differential equation

$$y' = f(x, y).$$

If we then truncate the series (2) after, say, two terms we obtain the approximation

$$y(x_n + h) \approx y(x_n) + y'(x_n)h$$

or $$y(x_n + h) \approx y(x_n) + f(x_n, y(x_n))h . \tag{3}$$

Observe that we can obtain the Euler formula

$$y_{n+1} = y_n + hf(x_n, y_n) \tag{4}$$

of the preceding section by formally replacing $y(x_n + h)$ and $y(x_n)$ in (3) by their approximations y_{n+1} and y_n, respectively. The approximation symbol $\approx$ is replaced by an equality since we are defining the left side of (4) by the numbers obtained from the right-hand member.

The Taylor method

By retaining three terms in the series (2), we can write

$$y(x_n + h) \approx y(x_n) + y'(x_n)h + y''(x_n)\frac{h^2}{2} .$$

After using the replacements noted above, it follows that

$$y_{n+1} = y_n + y_n'h + y_n''\frac{h^2}{2}.$$ (5)

The second derivative y'' can be obtained by differentiating $y' = f(x, y)$.

At this point let us re-examine the two initial-value problems of the preceding section.

EXAMPLE	Use the three-term Taylor formula to obtain the approximate value of $y(1.5)$ for the solution of

$$y' = 2xy$$

$$y(1) = 1.$$

Compare the results for $h = 0.1$ and $h = 0.05$.

Solution: Since $y' = 2xy$ it follows by the product rule that $y'' = 2xy' + 2y$. Thus, for example, when $h = 0.1$, $n = 0$, we can first calculate

$$y_0' = 2x_0y_0$$
$$= 2(1)(1)$$
$$= 2,$$

and then
$$y_0'' = 2x_0y_0' + 2y_0$$
$$= 2(1)(2) + 2(1).$$

Hence (5) becomes

$$y_1 = y_0 + y_0'(0.1) + y_0''\frac{(0.1)^2}{2}$$

$$= 1 + 2(0.1) + 6(0.005)$$

$$= 1.23.$$

The results of the iteration, along with the comparative exact values, are summarized in Tables 9.9 and 9.10.

Table 9.9   Three-Term Taylor Method with $h = 0.1$

x_n	y_n	True Value	Error	% Rel Error
1.00	1.0000	1.0000	0.0000	0.00
1.10	1.2300	1.2337	0.0037	0.30
1.20	1.5427	1.5527	0.0100	0.65
1.30	1.9728	1.9937	0.0210	1.05
1.40	2.5721	2.6117	0.0396	1.52
1.50	3.4188	3.4904	0.0715	2.05

Table 9.10   Three-Term Taylor Method with $h = 0.05$

x_n	y_n	True Value	Error	% Rel Error
1.00	1.0000	1.0000	0.0000	0.00
1.05	1.1075	1.1079	0.0004	0.04
1.10	1.2327	1.2337	0.0010	0.08
1.15	1.3788	1.3806	0.0018	0.13
1.20	1.5499	1.5527	0.0028	0.18
1.25	1.7509	1.7551	0.0041	0.23
1.30	1.9879	1.9937	0.0059	0.29
1.35	2.2681	2.2762	0.0081	0.36
1.40	2.6006	2.6117	0.0111	0.43
1.45	2.9967	3.0117	0.0150	0.50
1.50	3.4702	3.4904	0.0202	0.58

EXAMPLE

Use the three-term Taylor formula to obtain the approximate value of $y(0.5)$ for the solution of

$$y' = (x + y - 1)^2$$

$$y(0) = 2.$$

Solution: In this case we compute y'' by the power rule. We have

$$y'' = 2(x + y - 1)(1 + y').$$

The results are summarized in Tables 9.11 and 9.12 for $h = 0.1$ and $h = 0.05$, respectively.

Table 9.11   Three-term Taylor Method with $h = 0.1$

x_n	y_n
0.00	2.0000
0.10	2.1200
0.20	2.2992
0.30	2.5726
0.40	3.0077
0.50	3.7511

Table 9.12   Three-term Taylor Method with $h = 0.05$

x_n	y_n
0.00	2.0000
0.05	2.0550
0.10	2.1222
0.15	2.2045
0.20	2.3058
0.25	2.4315
0.30	2.5890
0.35	2.7889
0.40	3.0475
0.45	3.3898
0.50	3.8574

A comparison of the last two examples with the corresponding results obtained from the improved Euler method shows no startling dissimilarities. However, the next example is of some interest.

EXAMPLE

Compare the approximate values of $y(1.5)$ for
$$y' = x + y - 1$$
$$y(1) = 5$$
using the three-term Taylor method and the improved Euler method with $h = 0.1$.

Solution: In this case the differential equation is linear in y so that it is readily shown that the exact solution is
$$y = -x + 6e^{x-1}.$$
The results of the respective iterations are given in Table 9.13.

Table 9.13   Comparison of Numerical Methods with $h = 0.1$

x_n	Improved Euler	Three-Term Taylor	True Value
1.00	5.0000	5.0000	5.0000
1.10	5.5300	5.5300	5.5310
1.20	6.1262	6.1262	6.1284
1.30	6.7954	6.7954	6.7992
1.40	7.5454	7.5454	7.5510
1.50	8.3847	8.3847	8.3923

The fact that the values obtained in the preceding example are the same for both methods is no accident in this case. The Taylor method gives the same values as the improved Euler when $f(x,y)$ is linear in x and y (see Problem 12).

EXERCISES 9.3   *Answers to odd-numbered problems begin on page A-36.*

Given the initial-value problems in Problems 1–10. Use the three-term Taylor formula to obtain a four decimal approximation to the indicated value. First use **(a)** $h = 0.1$, and then **(b)** $h = 0.05$.

1. $y' = 2x - 3y + 1$, $y(1) = 5$; $y(1.5)$

2. $y' = 4x - 2y$, $y(0) = 2$; $y(0.5)$

3. $y' = 1 + y^2$, $y(0) = 0$; $y(0.5)$

★**4.** $y' = x^2 + y^2$, $y(0) = 1$; $y(0.5)$

5. $y' = e^{-y}$, $y(0) = 0$; $y(0.5)$

6. $y' = x + y^2$, $y(0) = 0$; $y(0.5)$

7. $y' = (x - y)^2$, $y(0) = 0.5$; $y(0.5)$

8. $y' = xy + \sqrt{y}$, $y(0) = 1$; $y(0.5)$

9. $y' = xy^2 - \dfrac{y}{x}$, $y(1) = 1$; $y(1.5)$

★**10.** $y' = y - y^2$, $y(0) = 0.5$; $y(0.5)$

11. Let $y(x)$ be the solution of the initial-value problem

$$y' = x^2 + y^3, \qquad y(1) = 1.$$

Use $h = 0.1$ and the three-term Taylor formula to obtain an approximation to $y(1.4)$. Compare your answer with the results obtained in Problem 15 of Section 9.2.

Miscellaneous problems

12. Consider the differential equation $y' = f(x, y)$, where f is linear in x and y. In this case prove that the improved Euler formula is the same as the three-term Taylor formula. [*Hint*: Recall from calculus that a Taylor series for a function g of two variables is

$$g(a + h, b + k) = g(a, b) + g_x(a, b)h + g_y(a, b)k$$

$$+ \frac{1}{2}(h^2 g_{xx} + 2hk g_{xy} + k^2 g_{yy})\Big|_{(a,b)}$$

$$+ \text{ terms involving higher order derivatives.}$$

Apply this result to $f(x_n + h, y_n + hf(x_n, y_n))$ in the improved Euler formula. Also use the fact that $y''(x) = \dfrac{d}{dx}y'(x) = f_x + f_y y'$.]

9.4   The Runge-Kutta Method

Probably one of the most popular as well as accurate numerical procedures used in obtaining approximate solutions to differential equations is the **fourth-order Runge-Kutta method**.* As the name suggests there are Runge-Kutta methods of different orders.

For the moment let us consider a **second-order** procedure. This consists of finding constants a, b, α, and β such that the formula

$$y_{n+1} = y_n + ak_1 + bk_2, \tag{1}$$

where

$$k_1 = hf(x_n, y_n)$$

$$k_2 = hf(x_n + \alpha h, y_n + \beta k_1) \tag{2}$$

agrees with a Taylor series expansion to as many terms as possible. The obvious purpose is to achieve the accuracy of the Taylor method without the

---

*Carl Runge (1856–1927) and Wilhelm Kutta (1867–1944), German applied mathematicians.

necessity of having to compute higher-order derivatives. Now it can be shown that whenever the constants satisfy

$$a + b = 1$$

$$b\alpha = \frac{1}{2}$$

$$b\beta = \frac{1}{2}$$

then (1) agrees with a Taylor expansion out to the h^2 or third term. It should be of interest to observe that when $a = 1/2$, $b = 1/2$, $\alpha = 1$, $\beta = 1$ then (1) reduces to the improved Euler method. Thus we can conclude that the three-term Taylor formula is essentially equivalent to the improved Euler formula. Also, the basic Euler method is a **first-order** Runge-Kutta procedure.

Notice, too, that the sum $ak_1 + bk_2$, $a + b = 1$, in equation (1) is simply a *weighted average* of k_1 and k_2. The numbers k_1 and k_2 are multiples of approximations to the slope at two different points.

The fourth-order Runge-Kutta formula

The **fourth-order** Runge-Kutta method consists of determining appropriate constants so that a formula such as

$$y_{n+1} = y_n + ak_1 + bk_2 + ck_3 + dk_4$$

agrees with a Taylor expansion out to h^4 or the fifth term. As in (2), the k_i are constant multiples of $f(x, y)$ evaluated at select points. The derivation of the actual method is tedious to say the least, so we state the results:

$$y_{n+1} = y_n + \frac{1}{6}(k_1 + 2k_2 + 2k_3 + k_4),$$
$$k_1 = hf(x_n, y_n)$$
$$k_2 = hf(x_n + \tfrac{1}{2}h, y_n + \tfrac{1}{2}k_1)$$
$$k_3 = hf(x_n + \tfrac{1}{2}h, y_n + \tfrac{1}{2}k_2)$$
$$k_4 = hf(x_n + h, y_n + k_3).$$

(3)

The reader is advised to look carefully at the formulas in (3); note that k_2 depends on k_1, k_3 depends on k_2, and so on. Also, k_2 and k_3 are approximations to the slope at the midpoint of the interval between x_n and $x_{n+1} = x_n + h$.

EXAMPLE

Use the Runge-Kutta method to obtain an approximation to $y(1.5)$ for the solution of

$$y' = 2xy$$

$$y(1) = 1.$$

Use $h = 0.1$.

Solution: For the sake of illustration let us compute the case when $n = 0$. From (3) we find

$$k_1 = (0.1)f(x_0, y_0)$$
$$= (0.1)2x_0y_0$$
$$= 0.2,$$

$$k_2 = (.01)f(x_0 + \tfrac{1}{2}(0.1), y_0 + \tfrac{1}{2}(0.2))$$
$$= (0.1)2(x_0 + \tfrac{1}{2}(0.1))(y_0 + \tfrac{1}{2}(0.2))$$
$$= 0.231,$$

$$k_3 = (0.1)f(x_0 + \tfrac{1}{2}(0.1), y_0 + \tfrac{1}{2}(0.231))$$
$$= (0.1)2(x_0 + \tfrac{1}{2}(0.1), y_0 + \tfrac{1}{2}(0.231))$$
$$= 0.234255,$$

$$k_4 = (0.1)f(x_0 + 0.1, y_0 + 0.234255)$$
$$= (0.1)2(x_0 + 0.1)(y_0 + 0.234255)$$
$$= 0.2715361,$$

and, therefore,

$$y_1 = y_0 + \tfrac{1}{6}(k_1 + 2k_2 + 2k_3 + k_4)$$
$$= 1 + \tfrac{1}{6}(0.2 + 0.231) + 0.234255 + 0.2715361)$$
$$= 1.23367435.$$

Rounded to the usual four decimal places we obtain

$$y_1 = 1.2337.$$

The accompanying table should convince the student why the Runge-Kutta method is so popular. Of course, there is no need to use any smaller step size.

Table 9.14   Runge–Kutta Method with $h = 0.1$

x_n	y_n	True Value	Error	% Rel Error
1.00	1.0000	1.0000	0.0000	0.00
1.10	1.2337	1.2337	0.0000	0.00
1.20	1.5527	1.5527	0.0000	0.00
1.30	1.9937	1.9937	0.0000	0.00
1.40	2.6116	2.6117	0.0001	0.00
1.50	3.4902	3.4904	0.0001	0.00

EXAMPLE   Use the Runge-Kutta method to compute an approximation to $y(0.5)$ for the solution to

$$y' = (x + y - 1)^2$$

$$y(0) = 2.$$

Solution: The results of the calculations for the case when $h = 0.1$ are given in Table 9.15.

Table 9.15   Runge–Kutta Method with $h = 0.1$

x_n	y_n
0.00	2.0000
0.10	2.1230
0.20	2.3085
0.30	2.5958
0.40	3.0649
0.50	3.9078

The reader might be interested in inspecting Tables 9.16 and 9.17 at this point. These tables compare the results obtained from the various formulas that we have examined applied to the two specific problems

$$y' = 2xy, \qquad y(1) = 1,$$

$$y' = (x + y - 1)^2, \qquad y(0) = 2,$$

that we have considered throughout the last three sections.

Table 9.16 $y' = 2xy, \qquad y(1) = 1$

x_n	Euler	Improved Euler	3-Term Taylor	Runge–Kutta	True Value
		Comparison of Numerical Methods with $h = 0.1$			
1.00	1.0000	1.0000	1.0000	1.0000	1.0000
1.10	1.2000	1.2320	1.2300	1.2337	1.2337
1.20	1.4640	1.5479	1.5427	1.5527	1.5527
1.30	1.8154	1.9832	1.9728	1.9937	1.9937
1.40	2.2874	2.5908	2.5721	2.6116	2.6117
1.50	2.9278	3.4509	3.4188	3.4902	3.4904

Table 9.16 continued

		Improved	3-Term	Runge–	True
x_n	Euler	Euler	Taylor	Kutta	Value

Comparison of Numerical Methods with $h = 0.05$

x_n	Euler	Improved Euler	3-Term Taylor	Runge–Kutta	True Value
1.00	1.0000	1.0000	1.0000	1.0000	1.0000
1.05	1.1000	1.1077	1.1075	1.1079	1.1079
1.10	1.2155	1.2332	1.2327	1.2337	1.2337
1.15	1.3492	1.3798	1.3788	1.3806	1.3806
1.20	1.5044	1.5514	1.5499	1.5527	1.5527
1.25	1.6849	1.7531	1.7509	1.7551	1.7551
1.30	1.8955	1.9909	1.9879	1.9937	1.9937
1.35	2.1419	2.2721	2.2681	2.2762	2.2762
1.40	2.4311	2.6060	2.6006	2.6117	2.6117
1.45	2.7714	3.0038	2.9967	3.0117	3.0117
1.50	3.1733	3.4795	3.4702	3.4903	3.4904

Table 9.17 $y' = (x + y - 1)^2$, $y(0) = 2$

Comparison of Numerical Methods with $h = 0.1$

x_n	Euler	Improved Euler	3-Term Taylor	Runge–Kutta	True Value
0.00	2.0000	2.0000	2.0000	2.0000	2.0000
0.10	2.1000	2.1220	2.1200	2.1230	2.1230
0.20	2.2440	2.3049	2.2992	2.3085	2.3085
0.30	2.4525	2.5858	2.5726	2.5958	2.5958
0.40	2.7596	3.0378	3.0077	3.0649	3.0650
0.50	3.2261	3.8254	3.7511	3.9078	3.9082
0.00	2.0000	2.0000	2.0000	2.0000	2.0000
0.05	2.0500	2.0553	2.0550	2.0554	2.0554
0.10	2.1105	2.1228	2.1222	2.1230	2.1230
0.15	2.1838	2.2056	2.2045	2.2061	2.2061
0.20	2.2727	2.3075	2.3058	2.3085	2.3085
0.25	2.3812	2.4342	2.4315	2.4358	2.4358
0.30	2.5142	2.5931	2.5890	2.5958	2.5958
0.35	2.6788	2.7953	2.7889	2.7998	2.7997
0.40	2.8845	3.0574	3.0475	3.0650	3.0650
0.45	3.1455	3.4057	3.3898	3.4189	3.4189
0.50	3.4823	3.8840	3.8574	3.9082	3.9082

EXERCISES 9.4   *Answers to odd-numbered problems begin on page A-38.*

Given the initial-value problems in Problems 1–10. Use the method of Runge-Kutta to obtain a four decimal approximation to the indicated value. Use $h = 0.1$.

1. $y' = 2x - 3y + 1$, $y(1) = 5$; $y(1.5)$

2. $y' = 4x - 2y$, $y(0) = 2$; $y(0.5)$

3. $y' = 1 + y^2$, $y(0) = 0$; $y(0.5)$

★4. $y' = x^2 + y^2$, $y(0) = 1$; $y(0.5)$

5. $y' = e^{-y}$, $y(0) = 0$; $y(0.5)$

6. $y' = x + y^2$, $y(0) = 0$; $y(0.5)$

7. $y' = (x - y)^2$, $y(0) = 0.5$; $y(0.5)$

8. $y' = xy + y$, $y(0) = 1$; $y(0.5)$

9. $y' = xy^2 - y/x$, $y(1) = 1$; $y(1.5)$

★10. $y' = y - y^2$, $y(0) = 0.5$; $y(0.5)$

11. Let $y(x)$ be the solution of the initial-value problem
$$y' = x^2 + y^3, \qquad y(1) = 1.$$
Determine whether the Runge-Kutta formula can be used to obtain an approximation for $y(1.4)$. Use $h = 0.1$.

Miscellaneous problems

12. Consider the differential equation $y' = f(x)$. In this case show that the fourth-order Runge-Kutta method reduces to Simpson's rule for the integral of $f(x)$ on the interval $x_n \leq x \leq x_{n+1}$.

[O]   9.5   Milne's Method, Second-Order Equations, Errors

There are many additional formulas which can be applied to obtain approximations to solutions of differential equations. Although it is not our intention to survey the vast field of numerical methods, one additional formula deserves mention. The **Milne method**, like the improved Euler formula, is a predictor-corrector method. By first using the *predictor*

$$y_{n+1}^* = y_{n-3} + \frac{4h}{3}(2y_n' - y_{n-1}' + 2y_{n-2}'), \tag{1}$$

where $n \geq 3$ and
$$y_n' = f(x_n, y_n)$$
$$y_{n-1}' = f(x_{n-1}, y_{n-1})$$
$$y_{n-2}' = f(x_{n-2}, y_{n-2})$$

we are then able to substitute the value of y_{n+1}^{*} into the *corrector*

$$y_{n+1} = y_{n-1} + \frac{h}{3}(y_{n+1}' + 4y_n' + y_{n-1}') \tag{2}$$

where $\qquad\qquad\qquad y_{n+1}' = f(x_{n+1}, y_{n+1}^{*}).$

Notice that formula (1) requires that we must know y_0, y_1, y_2, and y_3 in order to obtain y_4. Usually these last three values are computed by an accurate method such as the Runge-Kutta formula.

Since the Milne predictor-corrector formulas demand that we know more than just y_n to compute y_{n+1}, the procedure is called a **multistep** or **continuing** method. The Euler formulas, the three-term Taylor, and the Runge-Kutta formulas are examples of **single-step** or **starting** methods.

Higher-order equations

The numerical procedures that we have discussed in this chapter we applied only to the first-order equation $dy/dx = f(x, y)$ subject to an initial condition $y(x_0) = y_0$. To approximate a solution to, say, a second-order equation

$$\frac{d^2y}{dx^2} = f(x, y, y') \tag{3}$$

we first reduce the equation to a system of first-order equations. If we let $y' = u$, equation (3) becomes

$$y' = u$$
$$u' = f(x, y, u). \tag{4}$$

We now apply a particular method to *each* equation in the resulting system. For example, the basic Euler formulas would be

$$y_{n+1} = y_n + hu_n$$
$$u_{n+1} = u_n + hf(x_n, y_n, u_n). \tag{5}$$

EXAMPLE

Use the Euler method to obtain the approximate value of $y(0.2)$ where $y(x)$ is the solution of

$$y'' + xy' + y = 0$$
$$y(0) = 1 \qquad y'(0) = 2.$$

Solution: In terms of the substitution $y' = u$, the equation is equivalent to the system

$$y' = u$$
$$u' = -xu - y.$$

Thus from (5) we obtain

$$y_{n+1} = y_n + hu_n$$
$$u_{n+1} = u_n + h[-x_nu_n - y_n].$$

Using the step size $h = 0.2$, we find

$$y_1 = y_0 + (0.2)u_0$$
$$= 1 + (0.2)2$$
$$= 1.4,$$
$$u_1 = u_0 + (0.2)[-x_0 u_0 - y_0]$$
$$= 2 + (0.2)(-1)$$
$$= 1.8.$$

In other words, $y(0.2) \approx 1.4$ and $y'(0.2) \approx 1.8$.

Errors

In a serious and detailed study of numerical solutions of differential equations we would have to pay close attention to the various sources of errors. For some kinds of computation, accumulation of errors might reduce the accuracy of an approximation to the point of being useless.

By using only three terms of a Taylor series to approximate the value of a function, the method itself naturally will be a source of error. As we have seen, the Euler formula is essentially two terms of a Taylor series expansion; by advancing along a tangent line, we do not necessarily get to a point on or even near the solution curve. The errors inherent to these methods are known as **truncation errors**.*

Any calculator or computer can compute only to at most a finite number of decimal places. Suppose for the sake of illustration that we have a calculator that can display six digits while carrying eight digits internally. If we multiply two numbers, each having six decimals, then the product actually contains twelve decimal places. But the number that we see is rounded to six decimal places while the machine has stored a number rounded to eight decimal places. In one calculation such as this, the **round-off error** may not be deemed significant, but a problem could arise if many calculations are performed with rounded numbers. The effects of round-off can be minimized on a computer, provided it has double-precision capabilities.

When iterating a formula such as

$$y_{n+1} = y_n + hf(x_n, y_n)$$

we obtain a sequence of values

$$y_1, y_2, y_3, \ldots.$$

The value of y_1 is, of course, in error, and unfortunately, y_2 depends on y_1. Thus, y_2 must also be in error. In turn, y_3 inherits an error from y_2. The error resulting from the inheritance of errors in preceding calculations is known as **propagation error**. To make matters worse, formulas can be **unstable**. This

---

*This kind of error is also known as **formula error** or **discretization error**.

means that errors occuring in the early stages of calculation are not only propagated but are also *compounded* at each step of the iteration. The error may grow so fast so as to completely overwhelm the subsequent approximations. Under certain circumstances the corrector formula in Milne's method is unstable.

EXERCISES 9.5   *Answers to odd-numbered problems begin on page A-40.*

1. Obtain an approximation to $y(0.2)$ of the preceding example using the step size $h = 0.1$.

2. Generalize the improved Euler formula to a system such as

$$y' = u$$
$$u' = f(x, y, u).$$

3. Use the improved Euler method to approximate $y(0.2)$ where $y(x)$ is the solution of the initial-value problem given in the example of this section. Use $h = 0.2$ and $h = 0.1$.

★4. Generalize the Euler method to systems of the form

$$x' = f(x, y, t)$$
$$y' = g(x, y, t).$$

Use the results to approximate the values of $x(0.2)$, $y(0.2)$ where $x(t)$ and $y(t)$ are solutions of

$$x' = x + y$$
$$y' = x - y,$$
$$x(0) = 1, \qquad y(0) = 2.$$

Assume $h = 0.1$.

5. Use Milne's predictor-corrector method to approximate the value of $y(0.4)$ where $y(x)$ is solution of

$$y' = x + y - 1$$
$$y(0) = 1.$$

Obtain the values of y_1, y_2, and y_3 from the Runge-Kutta formula using $h = 0.1$.

★6. Consider the recurrence formula

$$y_{n+1} = k(1 - y_n)$$

where $n = 0, 1, 2, \ldots$, and k is a constant. Suppose that the initial value y_0 has an absolute error $\varepsilon = y_0 - y$, where y is the true value. Show that the formula is unstable for increasing n when $|k| > 1$ and stable when $|k| < 1$.

CHAPTER 9
SUMMARY

A solution of a differential equation may exist and yet we may not be able to determine it in terms of the familiar elementary functions. A way of convincing oneself that a first-order equation $y' = f(x, y)$ possesses a solution passing through a specific point (x_0, y_0) is to sketch the **direction field** associated with the equation. The equation $f(x, y) = c$ determines the **isoclines**, or curves of constant inclination. This means that every solution curve passing through a particular isocline does so with the same slope. The direction field is the totality of short line segments throughout two-dimensional space which have midpoints on the isoclines and possessing slope equal to the value of the parameter c. A carefully plotted sequence of these **lineal elements** can suggest the shape of a solution curve passing through the given point (x_0, y_0).

At best, a direction field can give only the crudest form of an approximation to a numerical value of the solution $y(x)$ of the initial-value problem when x is close to x_0.

To obtain the approximate values of $y(x)$, we used **Euler's formula**:

$$y_{n+1} = y_n + hf(x_n, y_n);$$

The **improved Euler formula**:

$$y_{n+1} = y_n + h\frac{f(x_n, y_n) + f(x_{n+1}, y_{n+1}^*)}{2}$$

where

$$y_{n+1}^* = y_n + hf(x_n, y_n);$$

The **three-term Taylor formula:**

$$y_{n+1} = y_n + y_n'h + y_n''\frac{h^2}{2};$$

The **Runge-Kutta formula**:

$$y_{n+1} = y_n + \tfrac{1}{6}(k_1 + 2k_2 + 2k_3 + k_4),$$

where

$$k_1 = hf(x_n, y_n)$$

$$k_2 = hf(x_n + \tfrac{1}{2}h, y_n + \tfrac{1}{2}k_1)$$

$$k_3 = hf(x_n + \tfrac{1}{2}h, y_n + \tfrac{1}{2}k_2)$$

$$k_4 = hf(x_n + h, y_n + k_3);$$

The **Milne formulas**:

$$y_{n+1}^* = y_{n-3} + \frac{4h}{3}(2y_n' - y_{n-1}' + 2y'_{n-2})$$

$$y_{n+1} = y_{n-1} + \frac{h}{3}(y_{n+1}' + 4y_n' + y_{n-1}'),$$

where

$$y_{n+1}' = f(x_{n+1}, y_{n+1}^*).$$

In each of the above formulas, the number h is the length of a uniform step.

In other words,

$$x_1 = x_0 + h, \; x_2 = x_1 + h = x_0 + 2h, \ldots, x_n = x_0 + nh.$$

**CHAPTER 9
SUMMARY**

Euler's method consists of approximating the solution curve by a sequence of straight lines. The improved Euler and Runge-Kutta methods use the idea of averaging slopes.

The first four methods are known as **single step** or **starting methods** while Milne's method is an example of a **multi-step** or a **continuing method**. To use the latter method, we must first compute y_1, y_2, and y_3 by some starting method such as the improved Euler or the Runge-Kutta formulas. Euler's method is not generally used if we desire accuracy to several decimal places. The improved Euler and Milne formulas are also particular examples of a class of approximating formulas known as **predictor-corrector** formulas. For example, when using the improved Euler formula, the value y_{n+1}^* obtained from the basic Euler formula is the predicted value which is then corrected through the new formula.

To obtain numerical approximations to higher order differential equations, we can reduce the differential equation to a system of first-order equations. We then apply a particular numerical technique to each equation of the system.

**CHAPTER 9
TEST**

Answers to odd-numbered problems begin on page A-40.

In Problems 1–2 sketch the direction field for the given differential equation. Indicate several possible solution curves.

1. $y\,dx - x\,dy = 0$ **2.** $y' = 2x - y$

In Problems 3–7 construct a table comparing the indicated values of $y(x)$ using the Euler, improved Euler, three-term Taylor, and Runge-Kutta methods. Compute to four rounded decimal places. Use $h = 0.1$ and $h = 0.05$.

3. $y' = 2 \ln xy$, $y(1) = 2$;

    $y(1.1)$, $y(1.2)$, $y(1.3)$, $y(1.4)$, $y(1.5)$

4. $y' = \sin x^2 + \cos y^2$, $y(0) = 0$;

    $y(0.1)$, $y(0.2)$, $y(0.3)$, $y(0.4)$, $y(0.5)$

5. $y' = \sqrt{x + y}$, $y(0.5) = 0.5$

    $y(0.6)$, $y(0.7)$, $y(0.8)$, $y(0.9)$, $y(1.0)$

6. $y' = xy + y^2$, $y(1) = 0$

    $y'(1.1)$, $y(1.2)$, $y(1.3)$, $y(1.4)$, $y(1.5)$

7. Use the Euler method to obtain the approximate value of $y(0.2)$ where $y(x)$ is the solution of the initial-value problem

$$y'' - (2x + 1)y = 0$$
$$y(0) = 3, \qquad y'(0) = 1.$$

First use one step with $h = 0.2$ and then repeat the calculations using $h = 0.1$.

8. Use Milne's predictor-corrector method to approximate the value of $y(0.4)$ where $y(x)$ is the solution of

$$y' = 4x - 2y$$
$$y(0) = 2.$$

Use the Runge-Kutta formula and $h = 0.1$ to obtain the values of y_1, y_2, and y_3.

Partial Differential Equations

Introduction

Throughout the preceding chapters, our attention has been focused on finding "general solutions" of ordinary differential equations. Also, we were primarily concerned with the theory and application of linear equations of order $n \leq 2$. In this chapter, we shall limit our consideration to a special kind of linear partial differential equation. However, we shall make no attempt to find, or even pursue, the concept of a general solution of such an equation. The emphasis will be on a specific procedure used in solving certain problems in the mathematical physics of temperature distributions and vibrations. These problems are described by relatively simple second-order partial differential equations, and hence the physical problems, depend on solving associated ordinary differential equations.

10.1 Orthogonal Functions

Terminology

In this and the next two sections, we set the stage for the material in Section 10.4. Fundamental to the entire discussion is the notion of **orthogonal functions**.

> **DEFINITION 10.1** Two functions f_1 and f_2 are said to be **orthogonal** on an interval $a \leq x \leq b$ if
>
> $$\int_a^b f_1(x)f_2(x) \, dx = 0. \qquad \square$$

EXAMPLE $f_1(x) = x^2$ and $f_2(x) = x^3$ are orthogonal on $-1 \leq x \leq 1$ since

$$\int_{-1}^{1} f_1(x)f_2(x) \, dx = \int_{-1}^{1} x^2 \cdot x^3 \, dx = \frac{1}{6}x^6 \Big|_{-1}^{1}$$

$$= \frac{1}{6}[1 - (-1)^6]$$

$$= 0.$$

DEFINITION 10.2   A set of real-valued functions

$$\phi_0(x), \ \phi_1(x), \ \phi_2(x), \ \ldots,$$

is said to be **orthogonal** on an interval $a \leq x \leq b$ if*

$$\int_a^b \phi_m(x)\phi_n(x) \, dx \quad \begin{cases} = 0, & m \neq n \\ \neq 0, & m = n. \end{cases} \qquad \square(1)$$

The positive number

$$\|\phi_n(x)\|^2 = \int_a^b \phi_n^2(x) \, dx \qquad (2)$$

is called the **square norm** and

$$\|\phi_n(x)\| = \sqrt{\int_a^b \phi_n^2(x) \, dx}$$

is the **norm** of the function $\phi_n(x)$. When $\|\phi_n(x)\| = 1$ for $n = 0, 1, 2, \ldots,$ the set $\{\phi_n(x)\}$ is said to be **orthonormal** on the interval.

EXAMPLE   Show that the set

$$1, \cos x, \cos 2x, \ldots,$$

is orthogonal on the interval $-\pi \leq x \leq \pi$.

Solution: If we make the identification $\phi_0(x) = 1$ and $\phi_n(x) = \cos nx$, we must then show $\int_{-\pi}^{\pi} \phi_0(x)\phi_n(x) \, dx = 0, n > 0$ and $\int_{-\pi}^{\pi} \phi_m(x)\phi_n(x) \, dx = 0$, $m > 0, n > 0, m \neq n$. We have in the first case

$$\int_{-\pi}^{\pi} \phi_0(x)\phi_n(x) \, dx = \int_{-\pi}^{\pi} \cos nx \, dx$$

$$= \frac{1}{n} \sin nx \Big|_{-\pi}^{\pi}$$

---

*The term *orthogonal* and condition (1) have no geometric significance.

$$= \frac{1}{n}\left[\sin n\pi - \sin (-n\pi) \right]$$

$$= 0, \qquad n > 0,$$

and in the second,

$$\int_{-\pi}^{\pi} \phi_m(x)\phi_n(x)\, dx = \int_{-\pi}^{\pi} \cos mx \cos nx\, dx$$

$$= \frac{1}{2}\int_{-\pi}^{\pi} \left[\cos(m + n)x + \cos(m - n)x \right] dx$$

$$= \frac{1}{2}\left[\frac{\sin(m + n)x}{m + n} + \frac{\sin(m - n)x}{m - n} \right]_{-\pi}^{\pi}$$

$$= 0, \qquad m \neq n.$$

EXAMPLE

Find the norms of each function in the orthogonal set given in the preceding example.

Solution: For $\phi_0(x) = 1$, we have from (2)

$$\|\phi_0(x)\|^2 = \int_{-\pi}^{\pi} dx$$

$$= 2\pi$$

so that $\|\phi_0(x)\| = \sqrt{2\pi}$. For $\phi_n(x) = \cos nx$, $n > 0$, it follows that

$$\|\phi_n(x)\|^2 = \int_{-\pi}^{\pi} \cos^2 nx\, dx$$

$$= \frac{1}{2}\int_{-\pi}^{\pi} [1 + \cos 2nx]\, dx$$

$$= \pi.$$

Thus for $n > 0$, $\|\phi_n(x)\| = \sqrt{\pi}$.

Any orthogonal set of functions $\{\phi_n(x)\}$, $n = 0, 1, 2, \ldots$, can be **normalized,** that is, made into an orthonormal set, by dividing each function by its norm.

EXAMPLE

It follows from the first two examples that the set

$$\frac{1}{\sqrt{2\pi}},\ \frac{\cos x}{\sqrt{\pi}},\ \frac{\cos 2x}{\sqrt{\pi}},\ \ldots,$$

is orthonormal on $-\pi \leq x \leq \pi$.

Generalized Fourier series

Suppose $\{\phi_n(x)\}$ is an infinite orthogonal set of functions on an interval $a \leq x \leq b$. We ask: If $y = f(x)$ is a function defined on the interval $a < x < b$, is it possible to determine a set of coefficients $c_n, n = 0, 1, 2, \ldots$, for which

$$f(x) = c_0\phi_0(x) + c_1\phi_1(x) + \cdots + c_n\phi_n(x) + \cdots? \tag{3}$$

Multiplying (3) by $\phi_m(x)$ and integrating over the interval gives

$$\int_a^b f(x)\phi_m(x)\, dx = c_0 \int_a^b \phi_0(x)\phi_m(x)\, dx + c_1 \int_a^b \phi_1(x)\phi_m(x)\, dx$$

$$+ \cdots + c_n \int_a^b \phi_n(x)\phi_m(x)\, dx + \cdots$$

By orthogonality, each term on the right-hand side of the last equation is zero except when $m = n$. In this case we have

$$\int_a^b f(x)\phi_n(x)\, dx = c_n \int_a^b \phi_n^2(x)\, dx.$$

It follows that the required coefficients are

$$c_n = \frac{\int_a^b f(x)\phi_n(x)\, dx}{\int_a^b \phi_n^2(x)\, dx}, \qquad n = 0, 1, 2, \ldots.$$

In other words,

$$f(x) = \sum_{n=0}^{\infty} c_n\phi_n(x) \tag{4}$$

where

$$c_n = \frac{\int_a^b f(x)\phi_n(x)\, dx}{\|\phi_n(x)\|^2}. \tag{5}$$

The series (4) with coefficients (5) is called a **generalized Fourier series.**

We note that the procedure outlined for determining the c_n was *formal*, that is, basic questions on whether a series expansion such as (4) is actually possible were ignored.

EXERCISES 10.1   *Answers to odd-numbered problems begin on Page A-42.*

In Problems 1–4 show that the given functions are orthogonal on the indicated interval.

1. $f_1(x) = x,\quad f_2(x) = x^2,\quad -2 \leq x \leq 2$

2. $f_1(x) = x^3,\quad f_2(x) = x^2 + 1,\quad -1 \leq x \leq 1$

3. $f_1(x) = e^x,\quad f_2(x) = xe^{-x} - e^{-x},\quad 0 \leq x \leq 2$

★4. $f_1(x) = \cos x,\quad f_2(x) = \sin^2 x,\ 0 \leq x \leq \pi$

In Problems 5–10 show that the given set of functions is orthogonal on the indicated interval. Find the norm of each function in the set.

5. $\sin x,\ \sin 3x,\ \sin 5x,\ \ldots,\quad 0 \leq x \leq \dfrac{\pi}{2}$

6. $\cos x,\ \cos 3x,\ \cos 5x,\ \ldots,\quad 0 \leq x \leq \dfrac{\pi}{2}$

7. $\{\sin nx\}, \quad n = 1, 2, 3, \ldots, \quad 0 \le x \le \pi$

8. $\left\{\sin \dfrac{n\pi}{p} x\right\}, \quad n = 1, 2, 3, \ldots, \quad 0 \le x \le p$

9. $\left\{1, \cos \dfrac{n\pi}{p} x\right\}, \quad n = 1, 2, 3, \ldots, \quad 0 \le x \le p$

★10. $\left\{1, \cos \dfrac{n\pi}{p} x, \sin \dfrac{m\pi}{p} x\right\}, \quad n = 1, 2, 3, \ldots,$
$\quad m = 1, 2, 3, \ldots, \quad -p \le x \le p$

A set of functions $\{\phi_n(x), n = 0, 1, 2, \ldots,$ is said to be **orthogonal with respect to a weight function** $w(x)$ on an interval $a \le x \le b$ if

$$\int_a^b w(x)\phi_m(x)\phi_n(x)\, dx \begin{cases} = 0, & m \ne n \\ \ne 0, & m = n. \end{cases}$$

EXAMPLE

The set $\qquad\qquad 1, \cos x, \cos 2x, \ldots,$

is orthogonal with respect to the constant weight function $w(x) = 1$ on the interval $-\pi \le x \le \pi$.

In Problems 11 and 12 verify by direct integration that the functions are orthogonal with respect to the indicated weight function on the given interval.

11. $H_0(x) = 1, \quad H_1(x) = 2x, \quad H_2(x) = 4x^2 - 2; \quad w(x) = e^{-x^2},$
$-\infty < x < \infty$

12. $L_0(x) = 1, \quad L_1(x) = -x + 1, \quad L_2(x) = \frac{1}{2}x^2 - 2x + 1; \quad w(x) = e^{-x},$
$0 \le x < \infty$

Miscellaneous problems

13. Let $\{\phi_n(x)\}$ be an orthogonal set of functions on $a \le x \le b$ such that $\phi_0(x) = 1$. Show that $\int_a^b \phi_n(x)\, dx = 0$ for $n = 1, 2, \ldots$.

14. Let $\{\phi_n(x)\}$ be an orthogonal set of functions on $a \le x \le b$ such that $\phi_0(x) = 1$ and $\phi_1(x) = x$. Show that $\int_a^b (\alpha x + \beta)\phi_n(x)\, dx = 0$ for $n = 2, 3, \ldots$, and any constants α and β.

15. Let $\{\phi_n(x)\}$ be an orthogonal set of functions on $a \le x \le b$. Show that $\|\phi_m(x) + \phi_n(x)\|^2 = \|\phi_m(x)\|^2 + \|\phi_n(x)\|^2$.

16. Let p, q, r, r' be continuous on $a \le x \le b$ and $p(x) > 0, r(x) > 0$ for every x in $a < x < b$, and let y_m and y_n be solutions of the boundary value problem

$$\frac{d}{dx}\left[r(x)\frac{dy}{dx}\right] + [q(x) + \lambda p(x)]y = 0 \tag{6}$$

$$\begin{aligned} \alpha_1 y(a) + \beta_1 y'(a) &= 0 \\ \alpha_2 y(b) + \beta_2 y'(b) &= 0 \end{aligned} \tag{7}$$

corresponding to distinct constants λ_m and λ_n. This is known as the **Sturm–Liouville problem.**

(a) Show that y_m and y_n are orthogonal on the interval $a \le x \le b$ with respect to the weight function $p(x)$. [*Hint:* Multiply the differential equation for y_m by y_n, and multiply the differential equation for y_n by y_m. Subtract and integrate the result over the interval $a \le x \le b$ and then use (7).]

(b) Show that the boundary conditions (7) are not needed for orthogonality of y_m and y_n when $r(a) = r(b) = 0$.

17. Consider Legendre's differential equation

$$(1 - x^2)y'' - 2xy' + n(n + 1)y = 0, \quad n = 0, 1, 2, \ldots,$$

Use the result from part (b) of Problem 16 to show that the Legendre polynomials $P_n(x)$ are orthogonal with respect to the weight function $p(x) = 1$ on $-1 \le x \le 1$.

★18. Let
$$f(x) = \begin{cases} 0, & -1 < x < 0 \\ 1, & 0 \le x < 1. \end{cases}$$

Use formula (5) and the Legendre polynomials given in (12) of Section 6.4 to find the first three coefficients in the expansion

$$f(x) = c_0 P_0(x) + c_1 P_1(x) + c_2 P_2(x) + \cdots.$$

19. Let
$$f(x) = \begin{cases} 0, & -1 < x < 0 \\ x, & 0 \le x < 1. \end{cases}$$

Find the first three coefficients in the expansion

$$f(x) = c_0 P_0(x) + c_1 P_1(x) + c_2 P_2(x) + \cdots.$$

20. From Problem 1 we know that $f_1(x) = x$ and $f_2(x) = x^2$ are orthogonal on $-2 \le x \le 2$. Find constants c_1 and c_2 such that $f_3(x) = x + c_1 x^2 + c_2 x^3$ is orthogonal to both f_1 and f_2 on the same interval.

10.2   Trigonometric Series

10.2.1   Fourier Series

The set of functions

$$1, \cos\frac{\pi}{p}x, \cos\frac{2\pi}{p}x, \ldots, \sin\frac{\pi}{p}x, \sin\frac{2\pi}{p}x, \sin\frac{3\pi}{p}x, \ldots, \qquad (1)$$

is orthogonal on the interval $-p \le x \le p$. (See Problem 10, Section 10.1.) Suppose f is a function defined on the interval $-p < x < p$ that can be expanded in the trigonometric series

$$f(x) = \frac{a_0}{2} + \sum_{n=1}^{\infty} \left(a_n \cos\frac{n\pi}{p}x + b_n \sin\frac{n\pi}{p}x \right). \qquad (2)$$

Then the coefficients $a_0, a_1, a_2, \ldots, b_1, b_2, \ldots,$ can be determined as follows.*

Integrating both sides of (2) from $-p$ to p gives

$$\int_{-p}^{p} f(x) \, dx = \frac{a_0}{2} \int_{-p}^{p} dx$$

$$+ \sum_{n=1}^{\infty} \left(a_n \int_{-p}^{p} \cos \frac{n\pi}{p} x \, dx + b_n \int_{-p}^{p} \sin \frac{n\pi}{p} x \, dx \right). \tag{3}$$

Since each function $\cos n\pi x/p$, $\sin n\pi x/p$, $n > 1$, is orthogonal to 1 on the interval the right side of (3) reduces to a single term and, consequently,

$$\int_{-p}^{p} f(x) \, dx = \frac{a_0}{2} \int_{-p}^{p} dx = \frac{a_0}{2} x \Big|_{-p}^{p} = p a_0.$$

Solving for a_0 yields

$$a_0 = \frac{1}{p} \int_{-p}^{p} f(x) \, dx. \tag{4}$$

Now multiply (2) by $\cos m\pi x/p$ integrate:

$$\int_{-p}^{p} f(x) \cos \frac{m\pi}{p} x \, dx = \frac{a_0}{2} \int_{-p}^{p} \cos \frac{m\pi}{p} x \, dx$$

$$+ \sum_{n=1}^{\infty} \left(a_n \int_{-p}^{p} \cos \frac{m\pi}{p} x \cos \frac{n\pi}{p} x \, dx \right.$$

$$\left. + b_n \int_{-p}^{p} \cos \frac{m\pi}{p} x \sin \frac{n\pi}{p} x \, dx \right). \tag{5}$$

Now

$$\int_{-p}^{p} \cos \frac{m\pi}{p} x \, dx = 0, \quad m > 0,$$

$$\int_{-p}^{p} \cos \frac{m\pi}{p} x \cos \frac{n\pi}{p} x \, dx \begin{cases} = 0, & m \neq n \\ = p, & m = n, \end{cases}$$

and

$$\int_{-p}^{p} \cos \frac{m\pi}{p} x \sin \frac{n\pi}{p} x \, dx = 0$$

so (5) reduces to

$$\int_{-p}^{p} f(x) \cos \frac{n\pi}{p} x \, dx = a_n p.$$

Therefore

$$a_n = \frac{1}{p} \int_{-p}^{p} f(x) \cos \frac{n\pi}{p} x \, dx. \tag{6}$$

---

*We have chosen to write the coefficient of 1 in the series (2) as $a_0/2$ rather than a_0. This is for convenience only; the formula for a_n will then reduce to a_0 when $n = 0$.

Finally, if we multiply (2) by sin $m\pi x/p$, integrate, and make use of the results

$$\int_{-p}^{p} \sin\frac{m\pi}{p}x \, dx = 0, \quad m > 0$$

$$\int_{-p}^{p} \sin\frac{m\pi}{p}x \cos\frac{n\pi}{p}x \, dx = 0$$

$$\int_{-p}^{p} \sin\frac{m\pi}{p}x \sin\frac{n\pi}{p}x \, dx \quad \begin{cases} = 0, & m \neq n \\ = p, & m = n \end{cases}$$

we find that
$$b_n = \frac{1}{p}\int_{-p}^{p} f(x)\sin\frac{n\pi}{p}x \, dx. \tag{7}$$

The trigonometric series (2) with coefficients a_0, a_n, and b_n defined (4), (6), and (7), respectively, is said to be the **Fourier series** of the function f.* The coefficients are sometimes referred to as the **Euler coefficients.**

As in the discussion of generalized Fourier series in the preceding section, the underlying assumption that f can be represented by such series (2) and the subsequent determination of the coefficients corresponding to this assumption was strictly formal. We assumed that f was integrable on the interval and that (2), as well as the series obtained by multiplying (2) by cos $m\pi x/p$, converged in such a manner as to permit term-by-term integration. Until (2) is shown to be convergent for a given function f, the equality sign is usually replaced by the symbol $\sim$. We summarize the above results.

> The **Fourier series** of a function f defined on the interval $-p < x < p$ is given by
>
> $$f(x) \sim \frac{a_0}{2} + \sum_{n=1}^{\infty}\left(a_n\cos\frac{n\pi}{p}x + b_n\sin\frac{n\pi}{p}x\right) \tag{8}$$
>
> where
> $$a_0 = \frac{1}{p}\int_{-p}^{p} f(x) \, dx \tag{9}$$
>
> $$a_n = \frac{1}{p}\int_{-p}^{p} f(x)\cos\frac{n\pi}{p}x \, dx \tag{10}$$
>
> $$b_n = \frac{1}{p}\int_{-p}^{p} f(x)\sin\frac{n\pi}{p}x \, dx. \tag{11}$$

*Named after the French mathematician, Joseph Fourier (1768–1830). Fourier used such series in his investigations into the theory of heat.

EXAMPLE

Expand

$$f(x) = \begin{cases} 0, & -\pi < x < 0 \\ \pi - x, & 0 < x < \pi \end{cases} \tag{12}$$

in a Fourier series.

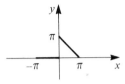

Figure 10.1

Solution: The graph of f is given in Figure 10.1.
With $p = \pi$, we have from (9) and (10) that

$$a_0 = \frac{1}{\pi} \int_{-\pi}^{\pi} f(x) \, dx = \frac{1}{\pi} \left[\int_{-\pi}^{0} 0 \, dx + \int_{0}^{\pi} (\pi - x) \, dx \right]$$

$$= \frac{1}{\pi} \left[\pi x - \frac{x^2}{2} \right]_{0}^{\pi}$$

$$= \frac{\pi}{2},$$

$$a_n = \frac{1}{\pi} \int_{-\pi}^{\pi} f(x) \cos nx \, dx = \frac{1}{\pi} \left[\int_{-\pi}^{0} 0 \, dx + \int_{0}^{\pi} (\pi - x) \cos nx \, dx \right]$$

$$= \frac{1}{\pi} \left[(\pi - x) \frac{\sin nx}{n} \Big|_{0}^{\pi} + \frac{1}{n} \int_{0}^{\pi} \sin nx \, dx \right]$$

$$= -\frac{1}{n\pi} \frac{\cos nx}{n} \Big|_{0}^{\pi}$$

$$= \frac{-\cos n\pi + 1}{n^2 \pi}$$

$$= \frac{1 - (-1)^n}{n^2 \pi}.$$

In like manner we find from (11) that

$$b_n = \frac{1}{\pi} \int_{0}^{\pi} (\pi - x) \sin nx \, dx$$

$$= \frac{1}{n}$$

Therefore $f(x) \sim \dfrac{\pi}{4} + \displaystyle\sum_{n=1}^{\infty} \left[\dfrac{1 - (-1)^n}{n^2 \pi} \cos nx + \dfrac{1}{n} \sin nx \right].$ \tag{13}

Note that a_n defined by (10) reduces to a_0 given by (9) when we set $n = 0$. But as the last example shows, this may not be the case *after* the integral for a_n is evaluated.

Convergence of a Fourier series

The following theorem gives sufficient conditions for convergence of a Fourier series to $f(x)$.

THEOREM 10.1 Let f and f' be piecewise continuous on $-p < x < p$, that is, let f and f' be continuous except at a finite number of points in the interval and have only finite discontinuities at these points. Then the Fourier series of f on the interval converges to $f(x)$ at a point of continuity. At a point of discontinuity, the Fourier series will converge to the average

$$\frac{f(x+) + f(x-)}{2}$$

where $f(x+)$ and $f(x-)$ denote the limit of f at x from the right and from the left, respectively.* $\qquad\square$

EXAMPLE

The function (12) given in the preceding example satisfies the conditions of Theorem 10.1. Thus for every x in $-\pi < x < \pi$, except at $x = 0$, the symbol $\sim$ in (13) can be replaced by an equality. At $x = 0$ the function is discontinuous and so the series (13) will converge to

$$\frac{f(0+) + f(0-)}{2} = \frac{\pi + 0}{2} = \frac{\pi}{2}.$$

Periodic extension

Observe that the functions in the basic set (1) have a common period $2p$. Hence the right side of (2) is periodic. We conclude that a Fourier series not only represents the function on the interval $-p < x < p$, but will also give the **periodic extension** of f outside this interval. We can now apply Theorem 10.1 to the periodic extension of f, or we may assume from the outset that the given function is periodic with period $2p$ (that is, $f(x + 2p) = f(x)$). When f is piecewise continuous and the right and left hand derivatives exist at $x = -p$ and $x = p$, respectively, then the series (8) will converge to the average $[f(p-) + f(-p+)]/2$ at these end points and to this value extended periodically to $\pm 3p$, $\pm 5p$, $\pm 7p$, and so on.

EXAMPLE

The Fourier series (13) converges to the periodic extension of (12) onto the entire x-axis. The solid dots given in Figure 10.2 represent the value

$$\frac{f(0+) + f(0-)}{2} = \frac{\pi}{2}$$

at 0, $\pm 2\pi$, $\pm 4\pi$, ... At $\pm \pi$, $\pm 3\pi$, $\pm 5\pi$, . . . , the series will converge to the value

$$\frac{f(\pi-) + f(-\pi+)}{2} = 0.$$

---

*That is, for x a point in the interval and $h > 0$,

$$f(x+) = \lim_{h\to 0} f(x + h), \qquad f(x-) = \lim_{h\to 0} f(x - h).$$

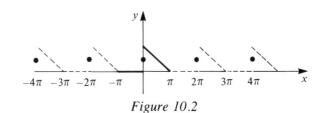

Figure 10.2

10.2.2 Cosine and Sine Series

Even and odd functions

The reader may recall that a function f is said to be **even** if

$$f(-x) = f(x).$$

Whereas, if

$$f(-x) = -f(x)$$

then f is said to be an **odd** function.

EXAMPLES

(a) $f(x) = x^2$ is even since

$$f(-x) = (-x)^2$$
$$= x^2$$
$$= f(x).$$

(b) $f(x) = x^3$ is odd since

$$f(-x) = (-x)^3$$
$$= -x^3$$
$$= -f(x)$$

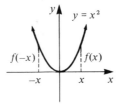

Figure 10.3

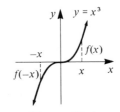

Figure 10.4

As illustrated in Figures 10.3 and 10.4, the graph of an even function is symmetric about the y-axis, and the graph of an odd function possesses symmetry about the origin.

EXAMPLE

Since $\cos(-x) = \cos x$ and $\sin(-x) = -\sin x$, the cosine and sine are even and odd functions, respectively.

Properties of even and odd functions

The proofs of the following properties are left as exercises.

  I. The product of two even functions is even.
  II. The product of two odd functions is even.
  III. The product of an even function with an odd function is odd.
  IV. If f is even, then $\int_{-a}^{a} f(x)\, dx = 2 \int_{0}^{a} f(x)\, dx$.
  V. If f is odd, then $\int_{-a}^{a} f(x)\, dx = 0$.

Cosine and sine series

If f is an even function on $-p < x < p$, then in view of the foregoing properties, the coefficients (9), (10), and (11) become

$$a_0 = \frac{1}{p} \int_{-p}^{p} f(x)\, dx = \frac{2}{p} \int_{0}^{p} f(x)\, dx$$

$$a_n = \frac{1}{p} \int_{-p}^{p} \underbrace{f(x) \cos \frac{n\pi}{p} x\, dx}_{\text{even}} = \frac{2}{p} \int_{0}^{p} f(x) \cos \frac{n\pi}{p} x\, dx$$

$$b_n = \frac{1}{p} \int_{-p}^{p} \underbrace{f(x) \sin \frac{n\pi}{p} x\, dx}_{\text{odd}} = 0.$$

Similarly, when f is odd on the interval $-p < x < p$,

$$a_n = 0, \ n = 0, 1, 2, \ldots , \qquad b_n = \frac{2}{p} \int_{0}^{p} f(x) \sin \frac{n\pi}{p} x\, dx.$$

We summarize the results:

The Fourier series of an even function on the interval $-p < x < p$ is the **cosine series**

$$f(x) \sim \frac{a_0}{2} + \sum_{n=1}^{\infty} a_n \cos \frac{n\pi}{p} x \qquad (14)$$

where
$$a_0 = \frac{2}{p} \int_{0}^{p} f(x)\, dx \qquad (15)$$

$$a_n = \frac{2}{p} \int_{0}^{p} f(x) \cos \frac{n\pi}{p} x\, dx. \qquad (16)$$

The Fourier series of an odd function on the interval $-p < x < p$ is the **sine series**

$$f(x) \sim \sum_{n=1}^{\infty} b_n \sin \frac{n\pi}{p} x \qquad (17)$$

where
$$b_n = \frac{2}{p} \int_{0}^{p} f(x) \sin \frac{n\pi}{p} x\, dx. \qquad (18)$$

EXAMPLE

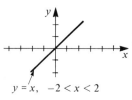

$y = x, \ -2 < x < 2$

Figure 10.5

Expand

$$f(x) = x, \qquad -2 < x < 2, \qquad (19)$$

in a Fourier series.

Solution: We expand f in a sine series since inspection of Figure 10.5 shows that the function is odd on the interval $-2 < x < 2$.

With the identification $2p = 4$, or $p = 2$, we can write (18) as

$$b_n = \int_0^2 x \ \sin \frac{n\pi}{2} x \ dx.$$

Integration by parts then yields

$$b_n = \frac{4(-1)^{n+1}}{n\pi}.$$

Therefore

$$f(x) \sim \frac{4}{\pi} \sum_{n=1}^{\infty} \frac{(-1)^{n+1}}{n} \sin \frac{n\pi}{2} x. \qquad (20)$$

EXAMPLE

The function (19) in the preceding example satisfies the conditions of Theorem 10.1. Hence the series (20) converges to the function on $-2 < x < 2$ and the periodic extension (of period 4) given in Figure 10.6.

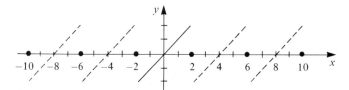

Figure 10.6

Half-range expansions

Throughout the preceding discussion it was understood that a function f was defined on an interval with the origin as midpoint, that is, $-p < x < p$. However, in many instances we are interested in representing a function that is defined only on $0 < x < L$ by a trigonometric series. This can be done in many different ways by supplying an arbitrary *definition* of the function to the interval $-L < x < 0$. For brevity we consider the two most important cases known as **half-range expansions.**

If $y = f(x)$ is defined on $0 < x < L$,

    (i) reflect the graph of the function about the y-axis onto $-L < x < 0$. The function is now even on $-L < x < L$. Use (14) with $p = L$. Or,

(ii) reflect the graph of the function through the origin onto $-L < x < 0$. The function is now odd on $-L < x < L$. Use (17) with $p = L$.

Note that the coefficients of the series (14) and (17) utilize only the definition of the function on $0 < x < p$ (that is, half of the interval $-p < x < p$). Hence in practice there is no actual need to make the reflections described in (i) and (ii); if f is defined on $0 < x < L$, we simply identify the half-period p with the length L of the interval. The coefficient formulas and the series themselves will effect either an even or an odd periodic extension (of period $2L$) of the original function.

EXAMPLE

$y = x^2, \quad 0 < x < L$

Figure 10.7

Expand $\qquad f(x) = x^2, \qquad 0 < x < L$

(a) in a cosine series, **(b)** in a sine series.

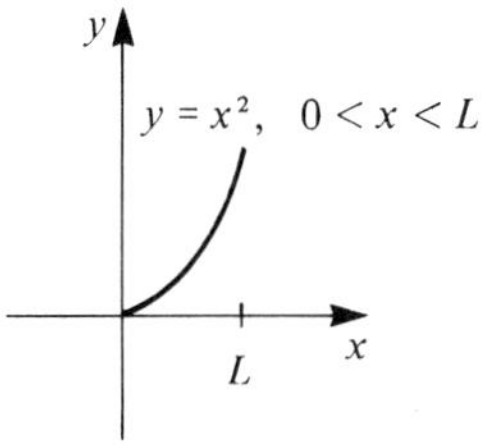

Solution: The graph of the function is given in Figure 10.7.

(a) We have

$$a_0 = \frac{2}{L} \int_0^L x^2 \, dx$$

$$= \tfrac{2}{3} L^2,$$

and, integrating by parts,

$$a_n = \frac{2}{L} \int_0^L x^2 \cos \frac{n\pi}{L} x \, dx$$

$$= \frac{2}{L} \left[\frac{Lx^2 \sin \dfrac{n\pi}{L} x}{n\pi} \Bigg|_0^L - \frac{2L}{n\pi} \int_0^L x \sin \frac{n\pi}{L} x \, dx \right]$$

$$= -\frac{4}{n\pi} \left[-\frac{Lx \cos \dfrac{n\pi}{L} x}{n\pi} \Bigg|_0^L + \frac{L}{n\pi} \int_0^L \cos \frac{n\pi}{L} x \, dx \right]$$

$$= \frac{4L^2 (-1)^n}{n^2 \pi^2}.$$

Thus $\qquad\qquad f(x) \sim \dfrac{L^2}{3} + \dfrac{4L^2}{\pi^2} \displaystyle\sum_{n=1}^{\infty} \dfrac{(-1)^n}{n^2} \cos \dfrac{n\pi}{L} x.$

(b) In this case

$$b_n = \frac{2}{L} \int_0^L x^2 \sin\frac{n\pi}{L}x \, dx.$$

After integrating by parts we find

$$b_n = \frac{2L^2(-1)^{n+1}}{n\pi} + \frac{4L^2}{n^3\pi^3}[(-1)^n - 1]$$

Thus

$$f(x) \sim \frac{2L^2}{\pi} \sum_{n=1}^{\infty} \left\{ \frac{(-1)^{n+1}}{n} + \frac{2}{n^3\pi^2}[(-1)^n - 1] \right\} \sin\frac{n\pi}{L}x.$$

We note that the series (a) and (b) of the foregoing example converge to the even periodic extension and the odd periodic extension of f given in Figures 10.8(a) and Figure 10.8(b), respectively.

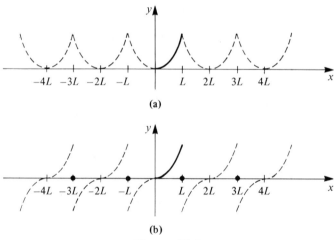

(a)

(b)

Figure 10.8

EXERCISES 10.2 *Answers to odd-numbered problems begin on page A-43.*

[10.2.1] In Problems 1–16 find the Fourier series of f on the given interval.

1. $f(x) = \begin{cases} 0, & -\pi < x < 0 \\ 1, & 0 \le x < \pi \end{cases}$

2. $f(x) = \begin{cases} -1, & -\pi < x < 0 \\ 2, & 0 \le x < \pi \end{cases}$

3. $f(x) = \begin{cases} 1, & -1 < x < 0 \\ x, & 0 \le x < 1 \end{cases}$

★4. $f(x) = \begin{cases} 0, & -1 < x < 0 \\ x, & 0 \le x < 1 \end{cases}$

5. $f(x) = \begin{cases} 0, & -\pi < x < 0 \\ x^2, & 0 \le x < \pi \end{cases}$

6. $f(x) = \begin{cases} \pi^2, & -\pi < x < 0 \\ \pi^2 - x^2, & 0 \le x < \pi \end{cases}$

7. $f(x) = x + \pi, \quad -\pi < x < \pi$

8. $f(x) = 3 - 2x, \quad -\pi < x < \pi$

EXAMPLE

The second-order equation

$$\frac{\partial^2 u}{\partial y^2} = 0 \qquad (1)$$

can be solved by integrating twice with respect to y

$$\frac{\partial u}{\partial y} = f(x),$$

$$u = yf(x) + g(x) \qquad (2)$$

where f and g are arbitrary functions.

EXAMPLE

Solve

$$\frac{\partial^2 u}{\partial x \partial y} + \frac{\partial u}{\partial y} = 1. \qquad (3)$$

Solution: If we let $v = \partial u / \partial y$ the equation becomes

$$\frac{\partial v}{\partial x} + v = 1.$$

By treating this latter equation as we would an ordinary linear first-order equation it is seen that an integrating factor is e^x. Therefore

$$\frac{\partial}{\partial x}(e^x v) = e^x$$

yields $v = 1 + F(y)e^{-x}.$

where F is arbitrary. Using the original substitution and integrating with respect to y then gives

$$u = y + f(y)e^{-x} + g(x)$$

where we have written $f(y) = \int F(y)\,dy$.

EXAMPLE

Solve $$\frac{\partial^2 u}{\partial x^2} - y^2 u = e^x. \qquad (5)$$

Solution: We solve the equation as we would be a nonhomogeneous linear second-order ordinary differential equation by first solving

$$\frac{\partial^2 u}{\partial x^2} - y^2 u = 0.$$

Treating y as a constant, it follows from Section 4.3 that

$$u_c = f(y)e^{xy} + g(y)e^{-xy.}$$

---

9. $f(x) = \begin{cases} 0, & -\pi < x < 0 \\ \sin x, & 0 \le x < \pi \end{cases}$ ★**10.** $f(x) = \begin{cases} 0, & -\pi/2 < x < 0 \\ \cos x, & 0 \le x < \pi/2 \end{cases}$

11. $f(x) = \begin{cases} 0, & -2 < x < -1 \\ -2, & -1 \le x < 0 \\ 1, & 0 \le x < 1 \\ 0, & 1 \le x < 2 \end{cases}$ **12.** $f(x) = \begin{cases} 0, & -2 < x < 0 \\ x, & 0 \le x < 1 \\ 1, & 1 \le x < 2 \end{cases}$

13. $f(x) = \begin{cases} 1, & -5 < x < 0 \\ 1 + x, & 0 \le x < 5 \end{cases}$ **14.** $f(x) = \begin{cases} 2 + x, & -2 < x < 0 \\ 2, & 0 \le x < 2 \end{cases}$

15. $f(x) = e^x, \quad -\pi < x < \pi$ **16.** $f(x) = \begin{cases} 0, & -\pi < x < 0 \\ e^x - 1, & 0 \le x < \pi \end{cases}$

17. (a) Use the result of Problem 5 to show

$$\frac{\pi^2}{6} = 1 + \frac{1}{2^2} + \frac{1}{3^2} + \frac{1}{4^2} + \cdots$$

and

$$\frac{\pi^2}{12} = 1 - \frac{1}{2^2} + \frac{1}{3^2} - \frac{1}{4^2} + \cdots.$$

(b) Find a series giving the numerical value of $\dfrac{\pi^2}{8}$.

★**18. (a)** Use the result of Problem 7 to show

$$\frac{\pi}{4} = 1 - \frac{1}{3} + \frac{1}{5} - \frac{1}{7} + \cdots.$$

(b) Use the result of Problem 9 to show

$$\frac{\pi}{4} = \frac{1}{2} + \frac{1}{1 \cdot 3} - \frac{1}{3 \cdot 5} + \frac{1}{5 \cdot 7} - \frac{1}{7 \cdot 9} + \cdots.$$

In Problems 19–22 expand the given function in a Fourier series (8).

EXAMPLE $$f(x) = x, \qquad 0 < x < 2\pi.$$

Solution: Although f is not defined on an interval $-p < x < p$, nonetheless, we can *define* f to be periodic of period 2π and use the fact that the value of an integral of a periodic function is the same over any interval of length equal to the period. Thus with $p = \pi$, the coefficients (9), (10), and (11) are

$$a_0 = \frac{1}{\pi} \int_0^{2\pi} x\,dx = 2\pi,$$

$$a_n = \frac{1}{\pi} \int_0^{2\pi} x \cos nx\,dx = 0,$$

$$b_n = \frac{1}{\pi} \int_0^{2\pi} x \sin nx\,dx = -\frac{2}{n}.$$

Figure 10.9

Hence

$$f(x) \sim \pi - 2 \sum_{n=1}^{\infty} \frac{1}{n} \sin nx.$$

Note that this is not a sine series since $a_0 \neq 0$. The series converges to the function whose graph is given in Figure 10.9.

---

19. $f(x) = x^2, \quad 0 < x < 2\pi$ **★20.** $f(x) = x, \quad 0 < x < \pi$

21. $f(x) = x + 1, \quad 0 < x < 1$ **22.** $f(x) = 2 - x, \quad 0 < x < 2$

[10.2.2]   In Problems 23–36 expand the given function in an appropriate cosine or sine series.

23. $f(x) = \begin{cases} -1, & -\pi < x < 0 \\ 1, & 0 \leq x < \pi \end{cases}$ **24.** $f(x) = \begin{cases} 1, & -2 < x < -1 \\ 0, & -1 < x < 1 \\ 1, & 1 < x < 2 \end{cases}$

25. $f(x) = |x|, \quad -\pi < x < \pi$ **26.** $f(x) = x, \quad -\pi < x < \pi$

27. $f(x) = x^2, \quad -1 < x < 1$ **28.** $f(x) = x|x|, \quad -1 < x < 1$

29. $f(x) = \pi^2 - x^2, \quad -\pi < x < \pi$ **30.** $f(x) = x^3, \quad -\pi < x < \pi$

31. $f(x) = \begin{cases} x - 1, & -\pi < x < 0 \\ x + 1, & 0 \leq x < \pi \end{cases}$ **★32.** $f(x) = \begin{cases} x + 1, & -1 < x < 0 \\ x - 1, & 0 \leq x < 1 \end{cases}$

33. $f(x) = \begin{cases} 1, & -2 < x < -1 \\ -x, & -1 \leq x < 0 \\ x, & 0 \leq x < 1 \\ 1, & 1 \leq x < 2 \end{cases}$ **34.** $f(x) = \begin{cases} -\pi, & -2\pi < x < \pi \\ x, & -\pi \leq x < \pi \\ \pi, & \pi \leq x < 2\pi \end{cases}$

35. $f(x) = |\sin x|, \quad -\pi < x < \pi$ **★36.** $f(x) = \cos x, \quad -\frac{\pi}{2} < x < \frac{\pi}{2}$

In Problems 37–46 find the half-range cosine and sine expansions of the given functions.

37. $f(x) = \begin{cases} 1, & 0 < x < \frac{1}{2} \\ 0, & \frac{1}{2} \leq x < 1 \end{cases}$ **★38.** $f(x) = \begin{cases} 0, & 0 < x < \frac{1}{2} \\ 1, & \frac{1}{2} \leq x < 1 \end{cases}$

---

39. $f(x) = \cos x, \quad 0 < x < \frac{\pi}{2}$ **40.** $f(x) = \sin x, \quad 0 < x <$

41. $f(x) = \begin{cases} x, & 0 < x < \frac{\pi}{2} \\ \pi - x, & \frac{\pi}{2} \leq x < \pi \end{cases}$ **42.** $f(x) = \begin{cases} 0, & 0 < x < \pi \\ x - \pi, & \pi \leq x \end{cases}$

43. $f(x) = \begin{cases} x, & 0 < x < 1 \\ 1, & 1 \leq x < 2 \end{cases}$ **44.** $f(x) = \begin{cases} 1, & 0 < x < 1 \\ 2 - x, & 1 \leq x < \end{cases}$

45. $f(x) = x^2 + x, \quad 0 < x < 1$ **46.** $f(x) = x(2 - x), \quad 0 < x <$

Miscellaneous problems

47. Prove Property I.   **48.** Prove Property II.

49. Prove Property III.   **50.** Prove Property IV.

51. Prove Property V.

52. Prove that any function f can be written as a sum of an even and an odd function. [*Hint:* Use the identity

$$f(x) = \frac{f(x) + f(-x)}{2} + \frac{f(x) - f(-x)}{2}.$$]

53. Find the Fourier series of

$$f(x) = \begin{cases} 0, & -\pi < x < 0 \\ x, & 0 \leq x < \pi \end{cases}$$

using the identity $f(x) = (|x| + x)/2, \; -\pi < x < \pi$, and the results of Problems 25 and 26. Observe that $|x|/2$ and $x/2$ are even and odd, respectively, on the interval. (See Problem 52.)

10.3   Partial Differential Equations

Linear equations

In this brief introduction to partial differential equations, we shall be interested in **linear** equations in two variables:

$$A(x, y)\frac{\partial^2 u}{\partial x^2} + B(x, y)\frac{\partial^2 u}{\partial x \partial y} + C(x, y)\frac{\partial^2 u}{\partial y^2} + D(x, y)\frac{\partial u}{\partial x}$$

$$+ E(x, y)\frac{\partial u}{\partial y} + F(x, y)u = G(x, y).$$

When $G(x, y) = 0$, the equation is said to be **homogeneous**, otherwise it is **nonhomogeneous.**

Solution by integration

Recall from calculus, integration of a partial derivative results in an arbitrary function rather than a constant. For example, the solution of $\frac{\partial u}{\partial x} = 0$ is $u = f(y)$ where f is a differentiable function.

To find a particular solution we use undetermined coefficients and assume
$$u_p = A(y)e^x.$$
Substituting this latter function into the given equation yields
$$Ae^x - y^2Ae^x = e^x.$$
and so $A(y) = 1/(1 - y^2)$. Hence a solution of the equation is
$$u = f(y)e^{xy} + g(y)e^{-xy} + \frac{e^x}{1 - y^2}. \tag{6}$$

Since (1), (3), and (5) are second-order and (2), (4), and (6) involve two arbitrary functions, each could be called a *general solution*. But we shall leave unanswered the question whether each solution yields every function satisfying the equation. Furthermore, it is not our intention in this section to focus upon those procedures leading to a general solution. Most partial differential equations cannot be solved as readily as the preceding three examples. However, in many applications involving linear partial differential equations it is sufficient to obtain particular solutions.

Separation of variables

For a homogeneous linear partial differential equation, it is sometimes possible to find particular solutions in the form of a product
$$u(x, y) = XY \tag{7}$$
where X is a function of x only, and Y is a function of y only. The use of the product (7), called the **method of separation of variables,** may enable us to reduce a partial differential equation to several *ordinary* differential equations. To this end we note
$$\frac{\partial u}{\partial x} = X'Y, \qquad \frac{\partial u}{\partial y} = XY'$$
and
$$\frac{\partial^2 u}{\partial x^2} = X''Y, \qquad \frac{\partial^2 u}{\partial y^2} = XY''$$
where the primes denote ordinary differentiation.

EXAMPLE

Find the product solutions of the equation
$$\frac{\partial^2 u}{\partial x^2} = 4\frac{\partial u}{\partial y}. \tag{8}$$

Solution: If $u = XY$, then (8) becomes
$$X''Y = 4XY'.$$
After dividing both sides by $4XY$, we have separated the variables:
$$\frac{X''}{4X} = \frac{Y'}{Y}.$$

Since the left-hand side of the last equation is independent of y and is identically equal to the right-hand side which is independent of x, we conclude that both sides must be a constant. (See Problem 37.) In practice it is convenient to write this real constant as either λ^2 or $-\lambda^2$. We distinguish the following cases.

CASE I   Using $\lambda^2 > 0$ the equalities

$$\frac{X''}{4X} = \frac{Y'}{Y} = \lambda^2$$

lead to $X'' - 4\lambda^2 X = 0$ and $Y' - \lambda^2 Y = 0.$

These latter equations have the solutions

$$X = c_1 \cosh 2\lambda x + c_2 \sinh 2\lambda x \quad \text{and} \quad Y = c_3 e^{\lambda^2 y}$$

respectively.* Thus a particular solution of (8) is

$$
\begin{aligned}
u &= XY \\
&= (c_1 \cosh 2\lambda x + c_2 \sinh 2\lambda x)(c_3 e^{\lambda^2 y}) \\
&= A_1 e^{\lambda^2 y} \cosh 2\lambda x + B_1 e^{\lambda^2 y} \sinh 2\lambda x,
\end{aligned}
\tag{9}
$$

where $A_1 = c_1 c_3$ and $B_1 = c_2 c_3$.   □

CASE II   Using $-\lambda^2 < 0$ the equalities

$$\frac{X''}{4X} = \frac{Y'}{Y} = -\lambda^2$$

give $X'' + 4\lambda^2 X = 0$ and $Y' + \lambda^2 Y = 0.$

Since the solutions of these equations are

$$X = c_4 \cos 2\lambda x + c_5 \sin 2\lambda x \quad \text{and} \quad Y = c_6 e^{-\lambda^2 y}$$

respectively, another solution of (8) is

$$u = A_2 e^{-\lambda^2 y} \cos 2\lambda x + B_2 e^{-\lambda^2 y} \sin 2\lambda x \tag{10}$$

where $A_2 = c_4 c_6$ and $B_2 = c_5 c_6$.   □

CASE III   If $\lambda^2 = 0$ it follows that

$$X'' = 0 \quad \text{and} \quad Y' = 0.$$

In this case $X = c_7 x + c_8$ and $Y = c_9$

so that $u = A_3 x + B_3$ (11)

where $A_3 = c_7 c_9$ and $B_3 = c_8 c_9$.   □

It is left as an exercise to verify that (9), (10), and (11) satisfy the given equation.

---

*Recall X can be written in the alternative form $X = c_1 e^{-2\lambda x} + c_2 e^{2\lambda x}$.

The superposition principle

The following theorem is analogous to Theorem 4.3 and is known as the **superposition principle.**

> **THEOREM 10.2** If $u_1, u_2, \ldots, u_k$ are solutions of a homogeneous linear partial differential equation, then the linear combination
>
> $$u = c_1 u_1 + c_2 u_2 + \cdots + c_k u_k,$$
>
> where the c_i, $i = 1, 2, \ldots, k$ are constants, is also a solution. $\square$

In the next section we shall make a formal assumption that whenever we have an infinite set

$$u_1, u_2, u_3, \ldots,$$

of solutions of a homogeneous linear equation that we can construct yet another solution u by forming the infinite series

$$u = \sum_{k=1}^{\infty} u_k.$$

EXERCISES 10.3   *Answers to odd-numbered problems begin on page A-44.*

In Problems 1–12 solve the given partial differential equation.

1. $\dfrac{\partial u}{\partial x} + y = 0$

2. $\dfrac{\partial u}{\partial y} = 2xy$

3. $\dfrac{\partial u}{\partial y} + 2u = e^y$

4. $x\dfrac{\partial u}{\partial x} + u = 4ye^{2x}$

5. $\dfrac{\partial^2 u}{\partial x^2} = 8xy^2 + 1$

★6. $\dfrac{\partial^2 u}{\partial y^2} + \sin xy = 0$

7. $\dfrac{\partial^2 u}{\partial x\, \partial y} = 1$

8. $\dfrac{\partial^2 u}{\partial x\, \partial y} = 2x + 4y$

9. $\dfrac{\partial^2 u}{\partial x\, \partial y} - \dfrac{\partial u}{\partial y} = 6xe^x$

10. $\dfrac{\partial^2 u}{\partial x\, \partial y} + 2y\dfrac{\partial u}{\partial x} = 4xy$

11. $\dfrac{\partial^2 u}{\partial y^2} - x^2 u = xe^{4y}$

12. $\dfrac{\partial^2 u}{\partial x^2} + y^2 u = y \sin 2x$

In Problems 13–28 determine whether the method of separation of variables is applicable to the given equation. If so, find the product solutions.

13. $\dfrac{\partial u}{\partial x} = \dfrac{\partial u}{\partial y}$

★14. $\dfrac{\partial u}{\partial x} + 3\dfrac{\partial u}{\partial y} = 0$

15. $\dfrac{\partial u}{\partial x} + \dfrac{\partial u}{\partial y} = u$

16. $\dfrac{\partial u}{\partial x} = \dfrac{\partial u}{\partial y} + u$

17. $x\dfrac{\partial u}{\partial x} = y\dfrac{\partial u}{\partial y}$

18. $y\dfrac{\partial u}{\partial x} + x\dfrac{\partial u}{\partial y} = 0$

19. $\dfrac{\partial^2 u}{\partial x^2} + \dfrac{\partial^2 u}{\partial x \, \partial y} + \dfrac{\partial^2 u}{\partial y^2} = 0$ **20.** $y\dfrac{\partial^2 u}{\partial x \, \partial y} + u = 0$

21. $k\dfrac{\partial^2 u}{\partial x^2} - u = \dfrac{\partial u}{\partial t}, \quad k > 0$ **22.** $k\dfrac{\partial^2 u}{\partial x^2} = \dfrac{\partial u}{\partial t}, \quad k > 0$

23. $a^2\dfrac{\partial^2 u}{\partial x^2} = \dfrac{\partial^2 u}{\partial t^2}$ **24.** $a^2\dfrac{\partial^2 u}{\partial x^2} = \dfrac{\partial^2 u}{\partial t^2} - 2k\dfrac{\partial u}{\partial t}, \quad k > 0$

25. $\dfrac{\partial^2 u}{\partial x^2} + \dfrac{\partial^2 u}{\partial y^2} = 0$ **★26.** $x^2\dfrac{\partial^2 u}{\partial x^2} + \dfrac{\partial^2 u}{\partial y^2} = 0$

27. $\dfrac{\partial^2 u}{\partial x^2} + \dfrac{\partial^2 u}{\partial y^2} = u$ **28.** $a^2\dfrac{\partial^2 u}{\partial x^2} - g = \dfrac{\partial^2 u}{\partial t^2},$

g a constant

In Problems 29 and 30 solve the given equation subject to the indicated conditions.

29. $\dfrac{\partial^2 u}{\partial x^2} = 6x; \quad u(0, y) = y, \quad u(1, y) = y^2 + 1$

★30. $y\dfrac{\partial^2 u}{\partial y^2} + \dfrac{\partial u}{\partial y} = 0; \quad u(x, 1) = x^2, \quad u(x, e) = 1$

In Problems 31–34 find product solutions satisfying the given equation and the indicated conditions.

EXAMPLE

$$k\dfrac{\partial^2 u}{\partial x^2} = \dfrac{\partial u}{\partial t}, \quad k > 0; \quad u(0, t) = 0, \quad u(L, t) = 0. \tag{12}$$

Solution: If $u = XT$, we can write the given equation as

$$\dfrac{X''}{X} = \dfrac{T'}{kT} = -\lambda^2 \tag{13}$$

which leads to

$$X'' + \lambda^2 X = 0 \tag{14}$$

$$T' + k\lambda^2 T = 0$$

and

$$X_1 = c_1 \cos \lambda x + c_2 \sin \lambda x \tag{15}$$

$$T = c_3 e^{-k\lambda^2 t}$$

respectively. Now since

$$u(0, t) = X(0)T(t) = 0$$

$$u(L, t) = X(L)T(t) = 0$$

we must have $X(0) = 0$ and $X(L) = 0$. These are boundary conditions for equation (14). Applying the first of these conditions in (15) immediately gives $c_1 = 0$. Therefore

$$X = c_2 \sin \lambda x.$$

The second boundary condition now implies

$$X(L) = c_2 \sin \lambda L = 0.$$

If $c_2 = 0$ then $X = 0$ so that $u = 0$. To obtain a *nonzero* solution u, we must have $c_2 \neq 0$ and so the last equation is satisfied when

$$\sin \lambda L = 0.$$

This implies that $\lambda L = n\pi$ or $\lambda = n\pi/L$, $n = 1, 2, 3, \ldots$.

Thus

$$u = (c_2 \sin \lambda x)(c_3 e^{-k\lambda^2 t})$$

$$= A_n e^{-k(n^2\pi^2/L^2)t} \sin \frac{n\pi}{L} x$$

satisfies the given equation and both side conditions. The coefficient $c_2 c_3$ is rewritten as A_n to emphasize the fact that a different solution is obtained for each n.* The reader should verify that using $\lambda^2 \geq 0$ in (13) does not lead to a solution of (12).

---

31. $k \dfrac{\partial^2 u}{\partial x^2} = \dfrac{\partial u}{\partial t}$, $\quad k > 0$; $\quad \left. \dfrac{\partial u}{\partial x} \right|_{x=0} = 0$, $\quad \left. \dfrac{\partial u}{\partial x} \right|_{x=5} = 0$

32. $a^2 \dfrac{\partial^2 u}{\partial x^2} = \dfrac{\partial^2 u}{\partial t^2}$; $\quad u(0, t) = 0$, $u(2, t) = 0$, $\quad \left. \dfrac{\partial u}{\partial t} \right|_{t=o} = 0$

33. $\dfrac{\partial^2 u}{\partial x^2} + \dfrac{\partial^2 u}{\partial y^2} = 0$; $\quad u(0, y) = 0$, $\quad u(x, 0) = 0$, $\quad u(x, 1) = 0$

★34. $\dfrac{\partial^2 u}{\partial x^2} + \dfrac{\partial^2 u}{\partial y^2} = 0$; $\quad \left. \dfrac{\partial u}{\partial x} \right|_{x=0} = 0$, $\quad \left. \dfrac{\partial u}{\partial x} \right|_{x=\pi} = 0$, $\quad u(x, 0) = 0$

Miscellaneous problems

35. Show that the equation

$$\frac{\partial u}{\partial t} = k \left(\frac{\partial^2 u}{\partial r^2} + \frac{1}{r} \frac{\partial u}{\partial r} \right), \qquad k > 0$$

possesses the product solution

$$u = e^{-k\lambda^2 t} \left(A_1 J_0(\lambda r) + B_2 J_0(\lambda r) \int \frac{dr}{r J_0^2(\lambda r)} \right).$$

36. Find the product solution of

$$\frac{\partial^2 u}{\partial t^2} = a^2 \left(\frac{\partial^2 u}{\partial r^2} + \frac{1}{r} \frac{\partial u}{\partial r} \right).$$

37. Prove that each side of the separated form of equation (8) is constant.

---

*Note that when $n = 0$, $\sin 0 = 0$ so that $u = 0$. Also, if n is a negative integer, say, $n = -k$, $k = 1, 2, \ldots$, we can use the trigonometric identity $\sin(-\theta) = -\sin\theta$ to rewrite $\sin(-k\pi x/L)$ as $-\sin k\pi x/L$. The factor of -1 can be absorbed in the arbitrary constant A_n. Thus, in this case, we need only consider the solutions obtained for the positive integers.

38. Verify that the products (9), (10), and (11) satisfy equation (8).

39. Consider the nonhomogeneous equation

$$\frac{\partial^2 u}{\partial x^2} + kx = \frac{\partial^2 u}{\partial t^2}, \qquad k > 0.$$

Find a function Ψ so that $v(x, t) = u(x, t) + \Psi(x)$ is a solution of the homogeneous equation

$$\frac{\partial^2 v}{\partial x^2} = \frac{\partial^2 v}{\partial t^2}.$$

40. (a) Show that the equation

$$a^2 \frac{\partial^2 u}{\partial x^2} = \frac{\partial^2 u}{\partial t^2} \tag{16}$$

can be put into the form $\partial^2 u / \partial \eta \partial \xi = 0$ by means of the substitutions $\xi = x + at$, $\eta = x - at$.
(b) Show that the solution of (16) is

$$u = f(x + at) + g(x - at)$$

where f and g are arbitrary functions of integration.

41. Use the result of part (b) of Problem 40 to solve (16) subject to

$$u(x, 0) = F(x), \qquad \left.\frac{\partial u}{\partial t}\right|_{t=0} = 0.$$

10.4 Boundary-Value Problems

Special equations   The following linear partial differential equations

$$k\frac{\partial^2 u}{\partial x^2} = \frac{\partial u}{\partial t}, \qquad k > 0, \tag{1}$$

$$a^2 \frac{\partial^2 u}{\partial x^2} = \frac{\partial^2 u}{\partial t^2}, \tag{2}$$

$$\frac{\partial^2 u}{\partial x^2} + \frac{\partial^2 u}{\partial y^2} = 0 \tag{3}$$

play an important role in many areas of physics and engineering. Equations (1) and (2) are known as the **one-dimensional heat equation** and the **one-dimensional wave equation,** respectively. "One-dimensional" refers to the fact that x denotes a spatial dimension whereas t usually represents time. Equation (3) is called **Laplace's equation.**

   We conclude this chapter by utilizing the method of separation of variables to solve several applied problems, each of which is described by one of the above equations along with certain side conditions. These side conditions consist of **boundary conditions:**

(a) u or $\partial u / \partial x$ specified at x = constant; u or $\partial u / \partial y$ specified at y = constant, and **initial conditions:**

(b) u at $t = 0$ for equation (1), or, u and $\partial u / \partial t$ at $t = 0$ for equation (2).

The collective mathematical description of such a problem is known as a **boundary-value problem.**

Equation (1) occurs in the theory of heat flow (that is, heat transferred by conduction) in a rod or a thin wire. The function $u(x, t)$ is temperature in the rod. Problems in mechanical vibrations often lead to the wave equation (2). For our purposes the solution $u(x, t)$ of (2) will represent the small displacements of an idealized vibrating string. Lastly, the solution $u(x, y)$ of Laplace's equation (3) can be interpreted as the steady-state (that is, time independent) temperature distribution in a thin flat plate. For a derivation of these equations in these three specific contexts, the reader is referred to any of the standard texts in engineering mathematics.*

Although we shall confine our attention to solving the problems described above, we note that the analysis of a wide variety of diverse phenomena yield equations (1), (2), or (3), or their generalizations involving a greater number of spatial variables. For example, (1) is sometimes called the **diffusion equation** since the diffusion of dissolved substances in solution is analogous to the flow of heat in a solid. The function $u(x, t)$ satisfying the partial differential equation in this case represents the concentration of the liquid. Similarly, equation (1) arises in the study of the flow of electricity in a long cable or a transmission line. In this setting, (1) is known as a **telegraph equation.** It can be shown that under certain assumptions the current and the voltage in the line are functions satisfying two equations identical in form with equation (1). The wave equation (2) also appears in the theory of high frequency transmission lines, fluid mechanics, acoustics, and elasticity. Laplace's equation (3) is encountered in engineering problems in static displacements of membranes, and most often, in problems dealing with potentials such as electrostatic, gravitational, and velocity potentials in fluid mechanics.

10.4.1   The Heat Equation

Consider a thin rod of length L with an initial temperature of $f(x)$ throughout and whose ends are held at a constant temperature of zero degrees for all time. See Figure 10.10.) If

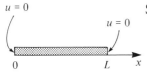

Figure 10.10

(a) the flow of heat takes place only in the x-direction,

(b) no heat escapes from the lateral surface of the rod,

(c) no heat is being generated in the rod,

*See C. R. Wylie, *Advanced Engineering Mathematics,* Fourth Edition, (N.Y.: McGraw-Hill Book Company, 1975).

(d) the rod is homogeneous, that is, its density per length is constant,

(e) its specific heat and thermal conductivity are constant,

then the temperature $u(x, t)$ in the rod is given by the solution of the boundary-value problem

$$k\frac{\partial^2 u}{\partial x^2} = \frac{\partial u}{\partial t}, \quad k > 0, \quad 0 < x < L, \quad t > 0, \tag{4}$$

$$u(0, t) = 0, \quad u(L, t) = 0, \quad t > 0, \tag{5}$$

$$u(x, 0) = f(x), \quad 0 < x < L. \tag{6}$$

The constant k is proportional to the thermal conductivity and is called the *diffusity*.

The solution

Using the product $u = XT$ and $-\lambda^2$ as a separation constant leads to

$$\frac{X''}{X} = \frac{T'}{T} = -\lambda^2$$

and
$$X'' + \lambda^2 X = 0, \quad X(0) = 0, \quad X(L) = 0, \tag{7}$$

$$T' + k\lambda^2 T = 0. \tag{8}$$

We have already obtained the solution of equation (4) subject to the boundary conditions (5) by solving (7). (See Exercise 10.3, page 460.) It was seen that (7) possessed a nonzero solution only if the value of the parameter λ took on the values

$$\lambda = \frac{n\pi}{L}, \quad n = 1, 2, 3, \ldots. \tag{9}$$

The corresponding solutions of (7) were then

$$X = c_1 \sin\frac{n\pi}{L}x, \quad n = 1, 2, 3, \ldots. \tag{10}$$

Eigenvalues and eigenfunctions

The values (9) for which (7) possess a nontrivial solution are known as **characteristic values,** or more commonly, as **eigenvalues.** The solutions (10) are called **characteristic functions** or **eigenfunctions.** We note that for a choice of λ, other than those given in (9), the only solution of (7) is the zero function $X \equiv 0$. In turn, this would imply that a function satisfying (4) and (5) is $u \equiv 0$. However, $u \equiv 0$ is *not* a solution of the original boundary-value problem when $f(x) \neq 0$. We naturally assume this last condition.

Since the solution of (8) is

$$T = c_3 e^{-k\lambda^2 t}$$

$$= c_3 e^{-k(n^2\pi^2/L^2)t}$$

the products
$$u_n = A_n e^{-k(n^2\pi^2/L^2)t} \sin\frac{n\pi}{L}x \tag{11}$$

satisfy the partial differential equation (4) and the boundary conditions (5) for each value of the positive integer n. For convenience we have replaced the constant $c_1 c_3$ by A_n. In order that the functions given in (11) satisfy the initial condition (6) we would have to choose the constant coefficients A_n in such a manner that

$$u(x, 0) = f(x) = A_n \sin\frac{n\pi}{L}x. \tag{12}$$

In general, we would not expect condition (12) to be satisfied for an arbitrary, but reasonable, choice of f. Therefore, we are forced to admit that (11) *is not a solution of the given problem.* However, by the superposition principle

$$u = \sum_{n=1}^{\infty} u_n$$

$$= \sum_{n=1}^{\infty} A_n e^{-k(n^2\pi^2/L^2)t} \sin\frac{n\pi}{L}x \tag{13}$$

must also, although formally, satisfy (4) and (5). Substituting $t = 0$ in (13) implies

$$u(x, 0) = f(x) = \sum_{n=1}^{\infty} A_n \sin\frac{n\pi}{L}x.$$

This last expression is recognized as the half-range expansion of f in a sine series. Thus, if we make the identification $A_n = b_n$, $n = 1, 2, 3, \ldots$, it follows from (18) of Section 10.2 that

$$A_n = \frac{2}{L} \int_0^L f(x) \sin\frac{n\pi}{L}x \, dx.$$

We conclude that the solution of the boundary-value problem described in (4), (5), and (6) is given by the infinite series

$$u(x, t) = \frac{2}{L} \sum_{n=1}^{\infty} \left(\int_0^L f(x) \sin\frac{n\pi}{L}x \, dx \right) e^{-k(n^2\pi^2/L^2)t} \sin\frac{n\pi}{L}x.$$

Insulated boundaries In the problem described in Section 10.4.1, the ends or boundaries of the rod could be **insulated.** At an insulated boundary the normal derivative of the temperature is zero. This fact follows from an empirical law that states the flux of heat across a surface (the time rate of flow of heat per unit area) is proportional to the value of the directional derivative of the temperature normal (perpendicular) to the surface.

EXAMPLE Give the boundary-value problem for the temperature u in a horizontal rod of length L if its ends are insulated and if its initial temperature throughout is given by $f(x)$, $0 < x < L$.

Solution: The ends of the rod are surfaces perpendicular to the x-axis and so $\partial u/\partial x = 0$ at both boundaries. Hence the temperature in the rod is given by the solution of

$$k\frac{\partial^2 u}{\partial x^2} = \frac{\partial u}{\partial t}, \qquad k > 0, \quad 0 < x < L, \quad t > 0,$$

$$\left.\frac{\partial u}{\partial x}\right|_{x=0} = 0, \qquad \left.\frac{\partial u}{\partial x}\right|_{x=L} = 0, \quad t > 0,$$

$$u(x, 0) = f(x), \qquad 0 < x < L.$$

The solution of the problem given in the foregoing example is left as an exercise. (See Problem 3.)

10.4.2   The Wave Equation

In the next example we consider the transverse vibrations of a string stretched between two points, say, $x = 0$ and $x = L$. As shown in Figure 10.11, the motion takes place in the xy-plane in such a manner that each point of the string moves in a direction perpendicular to the x-axis. If $u(x, t)$ denotes displacements of the string measured from the x-axis for $t > 0$, then u satisfies equation (2) under the following assumptions.

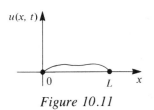

Figure 10.11

(a) The string is perfectly flexible.

(b) The string is homogeneous, that is, its mass per unit length is constant.

(c) The displacements u are small compared to the length of the string.

(d) The tension of the string is constant.

(e) The tension is large compared with the force of gravity.

(f) No other forces act on the string.

Thus a typical boundary-value problem is

$$a^2 \frac{\partial^2 u}{\partial x^2} = \frac{\partial^2 u}{\partial t^2}, \qquad 0 < x < L, \quad t > 0, \tag{14}$$

$$u(0, t) = 0, \qquad u(L, t) = 0, \quad t \geq 0, \tag{15}$$

$$u(x, 0) = f(x), \qquad \left.\frac{\partial u}{\partial t}\right|_{t=0} = g(x), \quad 0 < x < L. \tag{16}$$

The boundary conditions (15) simply state that the string is secured at the end points for all time. At $t = 0$, the functions f and g given in (16) specify the initial configuration and the initial velocity of each point of the string, respectively. It is implicit in this context that f is continuous and $f(0) = 0$, $f(L) = 0$.

The solution   Separating variables in (14) gives

$$\frac{X''}{X} = \frac{T''}{a^2 T} = -\lambda^2$$

so that

$$X'' + \lambda^2 X = 0$$

$$T'' + \lambda^2 a^2 T = 0$$

and therefore
$$X = c_1 \cos \lambda x + c_2 \sin \lambda x$$

$$T = c_3 \cos \lambda a t + c_4 \sin \lambda a t.^*$$

As before the boundary conditions (15) translate into $X(0) = 0$ and $X(L) = 0$. In turn we find

$$c_1 = 0 \quad \text{and} \quad c_2 \sin \lambda L = 0.$$

This last equation yields the eigenvalues $\lambda = n\pi/L$, $n = 1, 2, 3, \ldots$. The corresponding eigenfunctions are

$$X = c_2 \sin \frac{n\pi}{L} x, \quad n = 1, 2, 3, \ldots .$$

Thus solutions of equation (14) satisfying the boundary conditions (15) are

$$u_n = \left(A_n \cos \frac{n\pi a}{L} t + B_n \sin \frac{n\pi a}{L} t \right) \sin \frac{n\pi}{L} x,$$

and
$$u = \sum_{n=1}^{\infty} \left(A_n \cos \frac{n\pi a}{L} t + B_n \sin \frac{n\pi a}{L} t \right) \sin \frac{n\pi}{L} x. \tag{17}$$

Setting $t = 0$ in (17) gives

$$u(x, 0) = f(x) = \sum_{n=1}^{\infty} A_n \sin \frac{n\pi}{L} x$$

which is a half-range expansion for f in a sine series. As in the discussion of the heat equation, we can write $A_n = b_n$,

$$A_n = \frac{2}{L} \int_0^L f(x) \sin \frac{n\pi}{L} x \, dx. \tag{18}$$

To determine B_n we differentiate (17) with respect to t and then set $t = 0$:

$$\frac{\partial u}{\partial t} = \sum_{n=1}^{\infty} \left(-A_n \frac{n\pi a}{L} \sin \frac{n\pi a}{L} t + B_n \frac{n\pi a}{L} \cos \frac{n\pi a}{L} t \right) \sin \frac{n\pi}{L} x$$

$$\left. \frac{\partial u}{\partial t} \right|_{t=0} = g(x) = \sum_{n=1}^{\infty} \left(B_n \frac{n\pi a}{L} \right) \sin \frac{n\pi}{L} x.$$

In order that this last series to be the half-range sine expansion of g on the interval, the *total* coefficient $B_n n\pi a/L$ must be given by the form of (18) in Section 10.2. That is,

$$B_n \frac{n\pi a}{L} = \frac{2}{L} \int_0^L g(x) \sin \frac{n\pi}{L} x \, dx,$$

from which we obtain

$$B_n = \frac{2}{n\pi a} \int_0^L g(x) \sin \frac{n\pi}{L} x \, dx. \tag{19}$$

---

*You should convince yourself that a different choice of separation constant does not lead to a solution of the problem.

The formal solution of the problem consists of the series (17) with A_n and B_n defined by (18) and (19), respectively.

We note that when the string is released from *rest*, then $g(x) = 0$ for every x in $0 \leq x \leq L$ and, consequently, $B_n = 0$.

10.4.3   Laplace's Equation

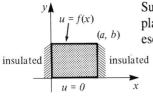

Figure 10.12

Suppose we wish to find the steady-state temperature $u(x, y)$ in a rectangular plate with boundary conditions indicated in Figure 10.12. When no heat escapes from the lateral faces of the plate, the problem is

$$\frac{\partial^2 u}{\partial x^2} + \frac{\partial^2 u}{\partial y^2} = 0, \qquad 0 < x < a, \quad 0 < y < b$$

$$\left.\frac{\partial u}{\partial x}\right|_{x=0} = 0, \qquad \left.\frac{\partial u}{\partial x}\right|_{x=a} = 0, \quad 0 < y < b$$

$$u(x, 0) = 0, \qquad 0 < x < a$$

$$u(x, b) = f(x), \qquad 0 < x < a.$$

The solution

Separation of variables leads to

$$\frac{X''}{X} = -\frac{Y''}{Y} = -\lambda^2,$$

$$X'' + \lambda^2 X = 0 \tag{20}$$

$$Y'' - \lambda^2 Y = 0, \tag{21}$$

$$X = c_1 \cos \lambda x + c_2 \sin \lambda x \tag{22}$$

$$Y = c_3 \cosh \lambda y + c_4 \sinh \lambda y,* \tag{23}$$

where the first three boundary conditions translate into $X'(0) = 0$, $X'(a) = 0$, and $Y(0) = 0$. Differentiating X and setting $x = 0$ implies $c_2 = 0$ and, therefore, $X = c_1 \cos \lambda x$. Differentiating again and then setting $x = a$ gives $-c_1 \lambda \sin \lambda a = 0$. This last condition is satisfied when $\lambda = 0$ or when $\lambda a = n\pi$, or $\lambda = n\pi/a$, $n = 1, 2, \ldots$. Observe that $\lambda = 0$ implies (20) is $X'' = 0$. The general solution of this equation is given by the linear function $X = c_1 + c_2 x$ and *not* by (22). In this case the boundary conditions $X'(0) = 0$, $X'(a) = 0$ demand that $X = C_1$. Unlike the previous two examples we are forced to conclude $\lambda = 0$ is an eigenvalue. Corresponding $\lambda = 0$ with $n = 0$, the eigenfunctions are

$$X = c_1, \quad n = 0 \quad \text{and} \quad X = c_1 \cos \frac{n\pi}{a} x, \quad n = 1, 2, \ldots.$$

---

*There are problems where a solution in terms of real exponential functions is more useful. See Problem 17.

Finally, the condition that $Y(0) = 0$ dictates that $c_3 = 0$ in (23) when $\lambda > 0$. However, when $\lambda = 0$ equation (21) becomes $Y'' = 0$ and thus the solution is given by $Y = c_3 + c_4 y$ rather than by (23). But $Y(0) = 0$ again implies $c_3 = 0$. Thus product solutions of the equation satisfying the first three boundary conditions are

$$A_0 y, \qquad n = 0 \ (\lambda = 0)$$

and
$$A_n \sinh\frac{n\pi}{a} y \, \cos\frac{n\pi}{a} x, \qquad n = 1, 2, \ldots .$$

The superposition principle yields another solution

$$u = A_0 y + \sum_{n=1}^{\infty} A_n \sinh\frac{n\pi}{a} y \, \cos\frac{n\pi}{a} x. \tag{24}$$

Substituting $y = b$ in (24) gives

$$u(x, b) = f(x) = A_0 b + \sum_{n=1}^{\infty} \left(A_n \sinh\frac{n\pi}{a} b \right) \cos\frac{n\pi}{a} x$$

which, in this case, is a half-range expansion of f in a cosine series. If we make the identifications $A_0 b = a_0/2$ and $A_n \sinh(n\pi b/a) = a_n$, $n = 1, 2, 3,$ $\ldots$, it follows from (15) and (16) of Section 10.2 that

$$2A_0 b = \frac{2}{a} \int_0^a f(x) \, dx$$

$$A_0 = \frac{1}{ab} \int_0^a f(x) \, dx, \tag{25}$$

and
$$A_n \sinh\frac{n\pi}{a} b = \frac{2}{a} \int_0^a f(x) \cos\frac{n\pi}{a} x \, dx$$

$$A_n = \frac{2}{a \, \sinh\dfrac{n\pi}{a} b} \int_0^a f(x) \cos\frac{n\pi}{a} x \, dx. \tag{26}$$

The formal solution of this problem consists of the series given in (24) where A_0 and A_n are defined by (25) and (26), respectively.

EXERCISES 10.4   *Answers to odd-numbered problems begin on page A-45.*

[10.4.1]   In Problems 1–4 solve the heat equation (1) subject to the given conditions. Assume a rod of length L.

1. $u(0, t) = 0, \quad u(L, t) = 0$

$$u(x, 0) = \begin{cases} 1, & 0 < x < \dfrac{L}{2} \\[2mm] 0, & \dfrac{L}{2} < x < L \end{cases}$$

2. $u(0, t) = 0, \quad u(L, t) = 0$

$$u(x, 0) = x(L - x)$$

3. $\left.\dfrac{\partial u}{\partial x}\right|_{x=0} = 0, \quad \left.\dfrac{\partial u}{\partial x}\right|_{x=L} = 0$

$u(x,\,0) = f(x)$

4. $\left.\dfrac{\partial u}{\partial x}\right|_{x=0} = 0, \quad \left.\dfrac{\partial u}{\partial x}\right|_{x=L} = 0$

$u(x,\,0) = \begin{cases} x, & 0 < x < \dfrac{L}{2} \\[2mm] 0, & \dfrac{L}{2} < x < L \end{cases}$

5. Solve the boundary-value problem

$$k\dfrac{\partial^2 u}{\partial x^2} = \dfrac{\partial u}{\partial t} + hu, \quad h > 0, \quad k > 0, \qquad 0 < x < L, \quad t > 0$$

$$\left.\dfrac{\partial u}{\partial x}\right|_{x=0} = 0, \quad \left.\dfrac{\partial u}{\partial x}\right|_{x=L} = 0$$

$$u(x,\,0) = f(x).$$

This problem describes the temperature $u(x,\,t)$ in the rod when heat is escaping from the lateral surface into a medium of zero degrees.

★**6.** Solve the boundary-value problem consisting of equation (1) and the conditions

$$u(0,\,t) = 100, \qquad u(L,\,t) = 100$$

$$u(x,\,0) = f(x), \qquad 0 < x < L.$$

[*Hint:* Try to find a solution of the form $v = u - 100$.]

[**10.4.2**] In Problems 7–13 solve the wave equation (2) subject to the given conditions. Assume a string stretched on the interval $0 \le x \le L$.

7. $u(0,\,t) = 0, \quad u(L,\,t) = 0$

$$u(x,\,0) = \begin{cases} \dfrac{2hx}{L}, & 0 < x < \dfrac{L}{2} \\[3mm] 2h\left(1 - \dfrac{x}{L}\right), & \dfrac{L}{2} \le x < L \end{cases}$$

$$\left.\dfrac{\partial u}{\partial t}\right|_{t=0} = 0$$

The constant h is positive but small compared to L. This is referred to as the "plucked string" problem.

8. $u(0,\,t) = 0, \quad u(L,\,t) = 0$

$u(x,\,0) = 0$

$\left.\dfrac{\partial u}{\partial t}\right|_{t=0} = x(L - x)$

9. $u(0,\,t) = 0, \quad u(L,\,t) = 0$

$u(x,\,0) = \frac{1}{4}x(L - x)$

$\left.\dfrac{\partial u}{\partial t}\right|_{t=0} = 0$

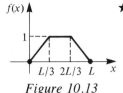

Figure 10.13

★**10.** $u(0, t) = 0$, $u(L, t) = 0$

$u(x, 0)$ as specified in Figure 10.13.

$$\left.\frac{\partial u}{\partial t}\right|_{t=0} = 0$$

11. $u(0, t) = 0$, $u(L, t) = 0$

$u(x, 0) = \frac{1}{6}x(L^2 - x^2)$

$$\left.\frac{\partial u}{\partial t}\right|_{t=0} = 0$$

12. $u(0, t) = 0$, $u(L, t) = 0$

$u(x, 0) = 0.01 \sin\dfrac{3\pi x}{L}$

$$\left.\frac{\partial u}{\partial t}\right|_{t=0} = 0$$

13. $u(0, t) = 0$, $u(L, t) = 0$

$u(x, 0) = 0$

$$\left.\frac{\partial u}{\partial t}\right|_{t=0} = \sin\frac{\pi x}{L}$$

14. Solve the boundary-value problem

$$a^2 \frac{\partial^2 u}{\partial x^2} = \frac{\partial^2 u}{\partial t^2}, \qquad 0 < x < L, \, t > 0$$

$$\left.\frac{\partial u}{\partial x}\right|_{x=0} = 0, \quad \left.\frac{\partial u}{\partial x}\right|_{x=L} = 0,$$

$$u(x, 0) = x,$$

$$\left.\frac{\partial u}{\partial t}\right|_{t=0} = 0.$$

This problem could describe the longitudinal displacements $u(x, t)$ of a vibrating elastic rod with free ends.*

[10.4.3]   In Problems 15–19 solve Laplace's equation (3) subject to the given conditions. Assume a flat plate such that $0 \leq x \leq a$ and $0 \leq y \leq b$.

15. $u(0, y) = 0$, $u(a, y) = 0$

$u(x, 0) = 0$, $u(x, b) = f(x)$

16. $u(0, y) = 0$, $u(a, y) = 0$

$$\left.\frac{\partial u}{\partial y}\right|_{y=0} = 0, \quad u(x, b) = f(x)$$

17. $u(0, y) = 0$, $u(a, y) = 0$

$u(x, 0) = f(x)$, $u(x, b) = 0$

★**18.** $\left.\dfrac{\partial u}{\partial x}\right|_{x=0} = 0, \quad \left.\dfrac{\partial u}{\partial x}\right|_{x=a} = 0$

$u(x, 0) = x$, $u(x, b) = 0$.

19. $u(0, y) = 0$, $u(a, y) = 0$

$u(x, 0) = f(x)$, $u(x, b) = g(x)$

20. Consider a semi-infinite plate extending in the positive y-direction bounded by $x = 0$, $x = a$ and $y = 0$. Find the steady-state temperature

---

*For a discussion of this theory see G. P. Tolstov, *Fourier Series*, Englewood Cliffs, N.J.: Prentice-Hall, 1962, pp. 275–82.

$u(x, y)$ if the boundary conditions are as indicated in Figure 10.14. [*Hint:* Use exponential functions in solving for Y and the intuitive condition that $u \to 0$ as $y \to \infty$].

Miscellaneous problems

21. Solve Problem 3 when $u(x, 0) = u_0, 0 < x < L$, where u_0 is a constant.

22. Consider the heat equation (1) subject to the conditions

$$u(0, t) = 0, \quad \frac{\partial u}{\partial x}\bigg|_{x=\pi} = 0$$

$$u(x, 0) = f(x), \quad 0 < x < \pi.$$

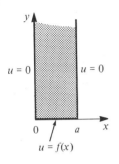

(a) Show that the eigenfunctions of the problem are $\left\{ \sin \dfrac{2n - 1}{2}x \right\}$, $n = 1, 2, 3, \ldots$.

(b) Show that the set of functions in part (a) is orthogonal on the interval $0 \le x \le \pi$.

(c) Use the concept of generalized Fourier series (Section 10.1) to solve the boundary-value problem.

23. Consider the boundary-value problem given in (14), (15) and (16) of this section. If $g(x) \equiv 0$ on $0 < x < L$ show that the solution of the problem can be written

$$u(x, t) = \tfrac{1}{2}[f(x + at) + f(x - at)].$$

[*Hint:* Use the identity

$$2 \sin \theta_1 \cos \theta_2 = \sin(\theta_1 + \theta_2) + \sin(\theta_1 - \theta_2).]$$

CHAPTER 10 SUMMARY

A set of functions $\{\phi_n(x)\}$, $n = 0, 1, 3, \ldots$, is said to be **orthogonal** on an interval $a \le x \le b$ if

$$\int_a^b \phi_m(x)\phi_n(x)\, dx \quad \begin{cases} = 0, & m \neq n, \\ \neq 0, & m = n. \end{cases}$$

A function f defined on the interval $-p < x < p$ can formally be expanded in terms of the trigonometric functions in the orthogonal set

$$1, \cos\frac{\pi}{p}x, \cos\frac{2\pi}{p}x, \ldots, \sin\frac{\pi}{p}x, \sin\frac{2\pi}{p}x, \ldots.$$

We say

$$f(x) \sim \frac{a_0}{2} + \sum_{n=1}^{\infty}\left(a_n \cos\frac{n\pi}{p}x + b_n \sin\frac{n\pi}{p}x \right)$$

where

$$a_0 = \frac{1}{p}\int_{-p}^{p} f(x)\, dx$$

**CHAPTER 10
SUMMARY**

$$a_n = \frac{1}{p} \int_{-p}^{p} f(x) \cos\frac{n\pi}{p}x \, dx$$

$$b_n = \frac{1}{p} \int_{-p}^{p} f(x) \sin\frac{n\pi}{p}x \, dx.$$

is the **Fourier series** corresponding to f. The coefficients are obtained using the concept of orthogonality. If f is an even function on the interval then $b_n = 0$, $n = 1, 2, 3, \ldots$, and

$$a_0 = \frac{2}{p} \int_{0}^{p} f(x) \, dx \tag{1}$$

and $$a_n = \frac{2}{p} \int_{0}^{p} f(x) \cos\frac{n\pi}{p}x \, dx. \tag{2}$$

Similarly, if f is odd on the interval then $a_n = 0$, $n = 0, 1, 2, 3, \ldots$,

and $$b_n = \frac{2}{p} \int_{0}^{p} f(x) \sin\frac{n\pi}{p}x \, dx. \tag{3}$$

When f is defined on an interval $0 < x < L$ it can be neither even nor odd. Nonetheless we can expand f in a cosine or a sine series. By defining $p = L$, we obtain a cosine series by using the coefficients (1) and (2) and a sine series by using (3). Such series are known as **half-range expansions.**

A particular solution of a linear partial differential equation in two variables can possibly be found by assuming a solution in the form of a product $u = XY$ where X is a function of x only, and Y is a function of y only. If applicable, this **method of separation of variables** leads to two ordinary differential equations.

A **boundary-value problem** consists of finding a function that satisfies a partial differential equation as well as side conditions consisting of perhaps both boundary conditions and initial conditions. We applied the method of separation of variables to obtain solutions of certain boundary-value problems involving the heat equation, the wave equation, and Laplace's equation. The procedure consisted of five basic steps.

(a) Separate the variables.

(b) Solve the separated ordinary differential equations and find the eigenvalues and eigenfunctions of the problem.

(c) Form the products u_n.

(d) Use the superposition principle to form an infinite series of the functions u_n.

(e) After using a boundary, or the initial condition(s), the coefficients in the series are obtained by making an appropriate identification with a half-range sine or cosine expansions.

**CHAPTER 10
TEST**

Answers to odd-numbered problems begin on page A-46.

Answer Problems 1–4 without referring back to the text. Fill in the blank or answer true/false.

1. To expand $f(x) = |x| + 1$, $-\pi < x < \pi$ in an appropriate Fourier series we would use a _____ series.

2. For a piecewise continuous function f defined on $-p < x < p$, a Fourier series will converge to _____ at a point of discontinuity.

3. Since $f(x) = x^2$, $0 < x < 2$, is not even it cannot be expanded in a Fourier cosine series. _____

4. Separation of variables will always yield particular solutions of a homogeneous linear partial differential equation. _____

5. Show that the set
$$\sin\frac{\pi}{2L}x, \ \sin\frac{3\pi}{2L}x, \ \sin\frac{5\pi}{2L}x, \ \ldots,$$
is orthogonal on the interval $0 \le x \le L$.

6. Find the norm of each function in Problem 1. Construct an orthonormal set.

7. Expand $f(x) = |x| - x$, $-1 < x < 1$, in a Fourier series.

8. Expand $f(x) = 2x^2 - 1$, $-1 < x < 1$, in a Fourier series.

9. Expand $f(x) = e^{-x}$, $0 < x < 1$, in a cosine series.

10. Expand the function given in Problem 9 in a sine series.

11. Solve: $\dfrac{\partial u}{\partial x} = xye^x + 1$

12. Solve: $\dfrac{\partial^2 u}{\partial x\,\partial y} = 8y^3\cos 6x$

13. Solve: $\dfrac{\partial u}{\partial x} + yu = 1$

14. Solve: $\dfrac{\partial^2 u}{\partial x^2} - y^2\dfrac{\partial u}{\partial x} = 0$

15. Use separation of variables to find product solutions of
$$\frac{\partial^2 u}{\partial x\,\partial y} = u.$$

16. Use separation of variables to find product solutions of
$$\frac{\partial^2 u}{\partial x^2} + \frac{\partial^2 u}{\partial y^2} + 2\frac{\partial u}{\partial x} + 2\frac{\partial u}{\partial y} = 0.$$
Is it possible to choose a constant of separation so that both X and Y are oscillatory functions?

**CHAPTER 10
TEST**

17. Find the steady-state temperature $u(x, y)$ in the square plate shown in Figure 10.15. The boundary conditions are as indicated.

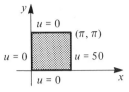

Figure 10.15

18. Find the steady-state temperature $u(x, y)$ in the semi-infinite plate of width π shown in Figure 10.16. The boundary conditions are as indicated.

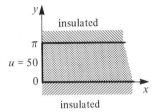

Figure 10.16

19. Solve Problem 18 if the boundaries $y = 0$ and $y = \pi$ are held at a constant zero degrees for all time.

Answers to Odd-Numbered Problems

Exercises 1.1, Page 9

1. linear, second-order

3. nonlinear, first-order

5. linear, fourth-order

7. nonlinear, second-order

9. linear, third-order

11. $2y' + y = 2(-\frac{1}{2}) e^{-x/2} + e^{-x/2} = 0$

13. $\dfrac{dy}{dx} - 2y - e^{3x} = (3e^{3x} + 20e^{2x})$
    $- 2(e^{3x} + 10e^{2x}) - e^{3x} = 0$

15. $y' - 25 - y^2 = 25 \sec^2 5x$
    $- 25(1 + \tan^2 5x)$
    $= 25 \sec^2 5x - 25 \sec^2 5x = 0$

17. $y' + y - \sin x = \frac{1}{2}\cos x + \frac{1}{2}\sin x$
    $- 10e^{-x} + \frac{1}{2}\sin x - \frac{1}{2}\cos x$
    $+ 10e^{-x} - \sin x = 0$

19. $yx^2 = -1$ implies $d(yx^2) = 0$
    or $\quad 2yx\, dx + x^2\, dy = 0$

21. $y - 2xy' - y(y')^2 = y - 2x\dfrac{c_1}{2y} - y\dfrac{c_1^2}{4y^2}$
    $= \dfrac{y^2 - (c_1 x + (c_1^2/4))}{y}$
    $= \dfrac{y^2 - y^2}{y} = 0$

23. $y' - \dfrac{1}{x}y - 1 = 1 + \ln x - \ln x - 1 = 0$

25. $\dfrac{d}{dt} \ln \dfrac{2 - X}{1 - X} = 1$
    $\left[\dfrac{-1}{2 - X} + \dfrac{1}{1 - X} \right] \dfrac{dX}{dt} = 1$
    simplifies to $\dfrac{dX}{dt} = (2 - X)(1 - X)$

27. The differential of $c_1 = xe^{y/x}/(x + y)^2$ is
    $\{(x + y)^2[xe^{y/x}(x\, dy - y\, dx/x^2) + e^{y/x}dx]$
    $- xe^{y/x}2(x + y)(dx + dy)\}/(x + y)^4 = 0$
    Multiplying by $-x^2(x + y)^3 e^{-y/x}$ and simplifying yields
    $(x^2 + y^2)\, dx + (x^2 - xy)\, dy = 0$

29. $y'' - 6y' + 13y$
    $= 5e^{3x}\cos 2x - 12e^{3x}\sin 2x$
    $+ 12e^{3x}\sin 2x - 18e^{3x}\cos 2x$
    $+ 13e^{3x}\cos 2x = 0$

31. $y'' = \cosh x + \sinh x = y$

33. $y'' + (y')^2 = -\dfrac{1}{(x + c_1)^2} + \dfrac{1}{(x + c_1)^2} = 0$

35. $x\dfrac{d^2y}{dx^2} + 2\dfrac{dy}{dx}$
    $= x(2c_2 x^{-3}) + 2(-c_2 x^{-2}) = 0$

37. $x^2 y'' - 3xy' + 4y$
    $= x^2(5 + 2\ln x) - 3x(3x + 2x\ln x)$
    $+ 4(x^2 + x^2 \ln x) = 9x^2 - 9x^2$
    $+ 6x^2 \ln x - 6x^2 \ln x = 0$

39. $y''' - 3y'' + 3y' - y$
    $= x^2 e^x + 6xe^x + 6e^x - 3x^2 e^x - 12xe^x$
    $- 6e^x + 3x^2 e^x + 6xe^x - x^2 e^x = 0$

41. For $x < 0$
    $xy' - 2y = x(-2x) - 2(-x^2) = 0$;
    for $x \geq 0$
    $xy' - 2y = x(2x) - 2(x^2) = 0$

43. $\left(\dfrac{dy}{dx}\right)^3 + 2x\dfrac{dy}{dx} - 2y - 1$
    $= t^3 - 3t^3 + 2t^3 + 1 - 1 = 0$

45. $y - xy' - (y')^2 + \ln y'$
    $= -t^2 - \ln t + 1 + 2t^2 - 1 - t^2 + \ln t = 0$

47. $y - xy' - (y')^2 = cx + c^2 - x(c) - c^2$
    $= 0$;
    $k = -1/4$

49. $y \equiv -1$

51. $m = 2 \quad$ and $\quad m = 3$

53. $m = \dfrac{1 \pm \sqrt{5}}{2}$

A-1

55. For $y = x^2$

$x^2y'' - 4xy' + 6y = x^2(2) - 4x(2x) + 6x^2$
$$= 8x^2 - 8x^2 = 0;$$

for $y = x^3$

$x^2y'' - 4xy' + 6y = x^2(6x) - 4x(3x^2) + 6x^3$
$$= 12x^3 - 12x^3 = 0;$$

yes; yes

57. (a) $y \equiv 0$; **(b)** no real solution;

(c) $y \equiv 1$ or $y \equiv -1$

Exercises 1.2, Page 26

1. $xy' = y - 2$ **3.** $y' + y = 0$

5. $2(x + 1)y' = y$

7. $(x^2 - y)y' = xy$

9. $y'' - y' = 0$ **11.** $y'' + \omega^2 y = 0$

13. $y'' - k^2 y = 0$ **15.** $y'' - 8y' + 16y = 0$

17. $xy'' + y' = 0$

19. $y''' - 6y'' + 11y' - 6y = 0$

21. $(1 + \cos\theta)\dfrac{dr}{d\theta} + r\sin\theta = 0$

23. $xy' - y = 0$ **25.** $2xyy' = y^2 - x^2$

27. $2xy' = y$ **29.** $yy'' + (y')^2 = 0$

31. $L\dfrac{di}{dt} + Ri = E(t)$

33. $\dfrac{dv}{dt} + \dfrac{k}{m}v^2 = g$

35. $\dfrac{dy}{dx} = -\dfrac{y}{\sqrt{s^2 - y^2}}$

37. $mx'' = -k\cos\theta$ $my'' = -mg - k\sin\theta$

$$= -k \cdot \dfrac{1}{v}\dfrac{dx}{dt} \qquad = -mg - k \cdot \dfrac{1}{v}\dfrac{dy}{dt}$$

$$= -|c|\dfrac{dx}{dt}, \qquad = -mg - |c|\dfrac{dy}{dt}$$

39. $\dfrac{dx}{dt} = r - kx,\ k > 0$

41. Using

$$\tan\phi = \dfrac{x}{y}, \quad \tan\left(\dfrac{\pi}{2} - \theta\right) = \dfrac{dy}{dx},$$

$$\tan\theta = \dfrac{dx}{dy}, \quad \text{and} \quad \tan\phi = \tan 2\theta$$

$$= \dfrac{2\tan\theta}{1 - \tan^2\theta} \quad \text{we obtain}$$

$$x\left(\dfrac{dx}{dy}\right)^2 + 2y\dfrac{dx}{dy} = x.$$

43. By combining Newton's second law of motion with his law of gravitation we obtain

$$m\dfrac{d^2y}{dt^2} = -k_1\dfrac{mM}{y^2}$$

where M is the mass of the earth and k_1 is a constant of proportionality. Dividing by m gives

$$\dfrac{d^2y}{dt^2} = -\dfrac{k}{y^2}$$

where $k = k_1 M$. The constant k is gR^2, where R is the radius of the earth. This follows from the fact that on the surface of the earth $y = R$ so that

$$k_1\dfrac{mM}{R^2} = mg$$

$$k_1 M = gR^2 \quad \text{or} \quad k = gR^2.$$

If $t = 0$ is the time at which burnout occurs then

$$y(0) = R + y_B$$

where y_B is the distance from the earth's surface to the rocket at the time of burnout, and

$$y'(0) = V_B$$

is the corresponding velocity at that time.

Chapter 1 Test, Page 31

1. (a) ordinary, first-order, nonlinear
(b) ordinary, third-order, nonlinear
(c) partial, second-order
(d) ordinary, second-order, linear

3. (a) $y = x^2$ **(b)** $y = \dfrac{x^2}{2}$

(c) $y \equiv 0$; $y = e^x$ **(d)** $y = 0$; $y = e^{5x}$

(e) $y \equiv 2$ **(f)** $y^2 = x$

5. $x < 0$ or $x > 1$

7. $(x - 2)y' = y - 1$

9. Water lost in time Δt = change in volume of water

$$\tfrac{1}{4}\sqrt{2gh}\,\Delta t = -\pi r^2\,\Delta h$$

where Δh is the change in height for a small change in time Δt, and r is the radius of A_1. Dividing by Δt gives

$$\frac{\Delta h}{\Delta t} = -\frac{\sqrt{2gh}}{4\pi r^2}.$$

Now by similar triangles it follows that $r = 2h/5$ at any time. Thus as $t\to 0$

$$\frac{dh}{dt} = -\frac{25\sqrt{2g}}{16\pi}h^{-3/2}.$$

Exercises 2.1, Page 36

1. Half planes defined by either $y > 0$ or $y < 0$

3. Half planes defined by either $x > 0$ or $x < 0$

5. The regions defined by either $y > 2$, $y < -2$, or $-2 < y < 2$

7. Any region not containing $(0,0)$

9. The entire xy-plane

11. $y \equiv 0$, $y = x^3$

13. There is some interval around $x = 0$ on which the unique solution is $y \equiv 0$.

15. $xy' - y = x(c) - cx = 0$ for every c; $y \equiv 0$, $y = x$; No, the given function is non-differentiable at $x = 0$.

17. Yes   **19.** No

Exercises 2.2, Page 44

1. $y = \dfrac{1}{2}\sin 2x + c$

3. $y = -\dfrac{1}{x} + c$

5. $y = x - \ln|x + 1| + c$

7. $y = cx^4$ **9.** $y^{-2} = 2x^{-1} + c$

11. $-3 + 3x\ln|x| = xy^3 + cx$

13. $-3e^{-2y} = 2e^{3x} + c$

15. $2 + y^2 = c(4 + x^2)$

17. $y^2 = x - \ln|x + 1| + c$

19. $\dfrac{x^3}{3}\ln x - \dfrac{1}{9}x^3 = \dfrac{y^2}{2} + 2y + \ln|y| + c$

21. $S = ce^{kr}$

23. $\dfrac{P}{1 - P} = ce^t$ or $P = \dfrac{ce^t}{1 + ce^t}$

25. $4\cos y = 2x + \sin 2x + c$

27. $-2\cos x + e^y + ye^{-y} + e^{-y} = c$

29. $(e^x + 1)^{-2} + 2(e^y + 1)^{-1} = c$

31. $\ln|N| = te^{t+2} - e^{t+2} - t + c$

33. $y - 5\ln|y + 3| = x - 5\ln|x + 4| + c$

or $\left(\dfrac{y + 3}{x + 4}\right)^5 = c_1 e^{y-x}$

35. $-\cot y = \cos x + c$

37. $y = \sin\left(\dfrac{x^2}{2} + c\right)$

39. $-y^{-1} = \tan^{-1}(e^x) + c$

41. $(1 + \cos x)(1 + e^y) = 4$

43. $\sqrt{y^2 + 1} = 2x^2 + \sqrt{2}$

45. $x = \tan(4y - 3\pi/4)$

47. $xy = e^{-(1 + 1/x)}$

49. **(a)** $y = 3\dfrac{1 - e^{6x}}{1 + e^{6x}}$,   **(b)** $y \equiv 3$,

(c) $y = 3\dfrac{2 - e^{6x-2}}{2 + e^{6x-2}}$

51. $y = -x - 1 + \tan(x + c)$

53. $2y - 2x + \sin 2(x + y) = c$

55. $4(y - 2x + 3) = (x + c)^2$

Exercises 2.3, Page 51

1. homogeneous of degree 2

3. homogeneous of degree 3

5. not homogeneous

7. homogeneous of degree 0

9. homogeneous of degree -2

11. $x \ln |x| + y = cx$

13. $(x - y) \ln |x - y| = y + c(x - y)$

15. $x + y \ln |x| = cy$

17. $\ln (x^2 + y^2) + 2 \tan^{-1}(y/x) = c$

19. $\ln |y| = 2\sqrt{x/y} + c$

21. $y^9 = c(x^3 + y^3)^2$

23. $(y/x)^2 = 2 \ln |x| + c$

25. $e^{2x/y} = 8 \ln |y| + c$

27. $x \cos \dfrac{y}{x} = c$ ·   29. $y + x = cx^2 e^{y/x}$

31. $y^3 + 3x^3 \ln |x| = 8x^3$

33. $y^2 = 4x(x + y)^2$

35. $\ln |x| = e^{y/x} - 1$

37. $4x \ln \left| \dfrac{y}{x} \right| + x \ln x + y - x = 0$

39. $3x^{3/2} \ln x + 3x^{1/2}y + 2y^{3/2} = 5x^{3/2}$

41. $(x + y) \ln |y| + x = 0$

43. $\ln |y| = -2\left(1 - \dfrac{x}{y}\right)^{1/2}$

45. $(y + 1)^2 + 2(y + 1)(x - 2)$
$\quad - (x - 2)^2 = c$

47. By homogeneity the equation can be written as

$$M\left(\frac{x}{y}, 1\right) dx + N\left(\frac{x}{y}, 1\right) dy = 0.$$

With $v = x/y$, it follows that

$M(v,1)(v\,dy + y\,dv) + N(v,1)\,dy = 0$

$[vM(v,1) + N(v,1)]\,dy + yM(v,1)\,dv = 0$

or $\dfrac{dy}{y} + \dfrac{M(v,1)\,dv}{vM(v,1) + N(v,1)} = 0.$

49. $\dfrac{dy}{dx} = -\dfrac{M(x,y)}{N(x,y)} = -\dfrac{y^n M\left(\dfrac{x}{y}, 1\right)}{y^n N\left(\dfrac{x}{y}, 1\right)}$

$\qquad = -\dfrac{M\left(\dfrac{x}{y}, 1\right)}{N\left(\dfrac{x}{y}, 1\right)} = G\left(\dfrac{x}{y}\right)$

Exercises 2.4, Page 59

1. $x^2 + 4x + \dfrac{3}{2}y^2 - y = c$

3. $\dfrac{5}{2}x^2 + 4xy - 2y^4 = c$

5. $x^2y^2 - 3x + 4y = c$

7. not exact, but is homogeneous

9. $xy^3 + y^2 \cos x - \dfrac{1}{2}x^2 = c$

11. not exact

13. $xy - 2xe^x + 2e^x - 2x^3 = c$

15. $x + y + xy - 3 \ln |xy| = c$

17. $x^3y^3 - \tan^{-1} 3x = c$

19. $-\ln |\cos x| + \cos x \sin y = c$

21. $y - 2x^2y - y^2 - x^4 = c$

23. $x^4y - 5x^3 - xy + y^3 = c$

25. $\dfrac{1}{3}x^3 + x^2y + xy^2 - y = \dfrac{4}{3}$

27. $4xy + x^2 - 5x + 3y^2 - y = 8$

29. $y^2 \sin x - x^3y - x^2 + y \ln y - y = 0$

31. $k = 10$

33. $M(x,y) = ye^{xy} + y^2 - \dfrac{y}{x^2} + h(x)$

35. $\dfrac{3}{2}x^2 + xy + 4y^2 = c_1$

37. $x^2y^2 + ye^x - y = c_1$

39. $\dfrac{\partial}{\partial x}[\mu N] = \mu \dfrac{\partial N}{\partial x} + N \dfrac{\partial \mu}{\partial x}$

$\quad \dfrac{\partial}{\partial y}[\mu M] = \mu \dfrac{\partial M}{\partial y} + N \dfrac{\partial \mu}{\partial y}$

Equating and rearranging gives

$N\dfrac{\partial \mu}{\partial x} - M\dfrac{\partial \mu}{\partial y} = \mu \dfrac{\partial M}{\partial y} - \mu \dfrac{\partial N}{\partial x}.$

41. $M(x, y) = 6xy^3$

$\quad N(x, y) = 4y^3 + 9x^2y^2$

$\quad \dfrac{\partial M}{\partial y} = 18xy^2 = \dfrac{\partial N}{\partial x}$

solution is $3x^2y^3 + y^4 = c$

43. $M(x,y) = -x^2y^2 \sin x + 2xy^2 \cos x$

$N(x,y) = 2x^2y \cos x$

$\dfrac{\partial M}{\partial y} = -2x^2y \sin x + 4xy \cos x = \dfrac{\partial N}{\partial x}$

solution is $x^2y^2 \cos x = c$

45. $M(x, y) = 2xy^2 + 3x^2$

$N(x, y) = 2x^2y$

$\dfrac{\partial M}{\partial y} = 4xy = \dfrac{\partial N}{\partial x}$

solution is $x^2y^2 + x^3$

47. The three resulting differential equations are equivalent to

$$d\left(\ln\frac{x}{y}\right) = 0, \; d\left(\frac{x}{y}\right) = 0, \; d\left(\tan^{-1}\frac{x}{y}\right) = 0.$$

Integrating and renaming the constants formally gives $y = cx$.

49. $\mu_1 = \dfrac{y^2}{x^4}, \quad \dfrac{y^3 dx - y^2x \, dy}{x^4} = 0$

$\mu_2 = \dfrac{x^2}{y^4}, \quad \dfrac{x^2y \, dx - x^3 dy}{y^4} = 0$

$\mu_3 = \dfrac{e^{x/y}}{y^2}, \quad \dfrac{ye^{x/y} dx - xe^{x/y} \, dy}{y^2} = 0$

51. A separable first-order differential equation can be written $h(y) \, dy - g(x) \, dx = 0$. Identifying $M(x,y) = -g(x)$ and $N(x,y) = h(y)$ it follows that $\dfrac{\partial M}{\partial y} = 0 = \dfrac{\partial N}{\partial x}$.

Exercises 2.5, Page 69

1. $y = ce^{4x}, \quad -\infty < x < \infty$

3. $y = \dfrac{1}{10} + ce^{-5x}, \quad -\infty < x < \infty$

5. $y = \dfrac{1}{4}e^{3x} + ce^{-x}, \quad -\infty < x < \infty$

7. $y = \dfrac{1}{3} + ce^{-x^3}, \quad -\infty < x < \infty$

9. $y = x^{-1}\ln x + cx^{-1}, \quad 0 < x < \infty$

11. $x = -\dfrac{4}{5}y^2 + cy^{-1/2}, \quad 0 < y < \infty$

13. $y = -\cos x + \dfrac{\sin x}{x} + \dfrac{c}{x}, \quad 0 < x < \infty$

15. $y = \dfrac{c}{e^x + 1}, \quad -\infty < x < \infty$

17. $y = \sin x + c \cos x, \quad -\pi/2 < x < \pi/2$

19. $y = \dfrac{1}{7}x^3 - \dfrac{1}{5}x + cx^{-4}, \quad 0 < x < \infty$

21. $y = \dfrac{1}{2x^2}e^x + \dfrac{c}{x^2}e^{-x}, \quad 0 < x < \infty$

23. $y = \sec x + c \csc x, \quad 0 < x < \pi/2$

25. $x = \dfrac{1}{2}e^y - \dfrac{1}{2y}e^y + \dfrac{1}{4y^2}e^y + \dfrac{c}{y^2}e^{-y},$

$0 < y < \infty$

27. $y = e^{-3x} + \dfrac{c}{x}e^{-3x}, \quad 0 < x < \infty$

29. $x = 2y^6 + cy^4, \quad 0 < y < \infty$

31. $y = e^{-x}\ln(e^x + e^{-x}) + ce^{-x}, \quad -\infty < x < \infty$

33. $x = \dfrac{1}{y} + \dfrac{c}{y}e^{-y^2}, \quad 0 < y < \infty$

35. $(\sec \theta + \tan \theta)r = \theta - \cos \theta + c,$

$-\pi/2 < \theta < \pi/2$

37. $y = \dfrac{5}{3}(x + 2)^{-1} + c(x + 2)^{-4},$

$-2 < x < \infty$

39. $y = 4 - 2e^{-5x}, \quad -\infty < x < \infty$

41. $i(t) = \dfrac{E}{R} + \left(i_0 - \dfrac{E}{R}\right)e^{-Rt/L}, \quad -\infty < t < \infty$

43. $y = \sin x \cos x - \cos x, \quad -\pi/2 < x < \pi/2$

45. $T(t) = 50 + 150 \, e^{kt}, \quad -\infty < t < \infty$

47. $(x + 1)y = x \ln x - x + 21, \quad 0 < x < \infty$

49. $y = \dfrac{2x}{x - 2}, \quad 2 < x < \infty$

51. $x = \dfrac{1}{2}y + \dfrac{8}{y}, \quad 0 < y < \infty$

53. $y = \begin{cases} \frac{1}{2}(1 - e^{-2}x), & 0 \le x \le 3, \\ \frac{1}{2}(e^6 - 1)e^{-2x}, & x > 3 \end{cases}$

55. $y = \begin{cases} \frac{1}{2} + \frac{3}{2}e^{-x^2}, & 0 \le x < 1, \\ (\frac{1}{2}e + \frac{3}{2})e^{-x^2}, & x \ge 1 \end{cases}$

A-5

Exercises 2.6, Page 72

1. $y^3 = 1 + cx^{-3}$

3. $y^{-3} = x + \frac{1}{3} + ce^{3x}$

5. $e^{x/y} = cx$

7. $y^{-3} = -\frac{9}{5}x^{-1} - \frac{49}{5}x^{-6}$

9. $x^{-1} = 2 - y^2 - e^{-y^2/2}$, the equation is Bernoulli in the variable x.

11. If $y = y_1 + u$, then $y' = y_1' + u'$ and so $dy/dx = P(x) + Q(x)y + R(x)y^2$ becomes

$$y_1' + u' = P + Q(y_1 + u) + R(y_1 + u)^2$$
$$y_1' + u' = P + Qy_1 + Ry_1^2 + Qu$$
$$+ 2y_1 Ru + Ru^2$$

Since y_1 is a solution of the Ricatti equation we obtain

$$u' - (Q + 2y_1 R)u = Ru^2$$

13. $y = 2 + \dfrac{1}{ce^{-3x} - 1/3}$

15. $y = \dfrac{2}{x} + \dfrac{1}{cx^{-3} - x/4}$

17. $y = -e^x + \dfrac{1}{ce^{-x} - 1}$

19. $y = -2 + \dfrac{1}{ce^{-x} - 1}$

21. If $y = cx + f(c)$ then $y' = c$ and $f(c) = f(y')$.
Therefore $y - xy' - f(y') = y - cx - f(c) = 0$.

23. $y = cx + 1 - \ln c$; $y = 2 + \ln x$

25. $y = cx - c^3$; $27y^2 = 4x^3$

27. $y = cx - e^c$; $y = x \ln x - x$

29. If $y = \dfrac{w'}{w}$ then $y' = \dfrac{ww'' - (w')^2}{w^2}$ and so $y' + y^2 - Q(x)y - P(x) = 0$ becomes

$$\dfrac{ww'' - (w')^2}{w^2} + \dfrac{(w')^2}{w^2} - Q\dfrac{w'}{w} - P = 0.$$

The last equation simplifies to
$$w'' - Qw' - Pw = 0.$$

Exercises 2.7, Page 79

1. $x^2 e^{2y} = 2x \ln x - 2x + c$

3. $e^{-x} = y \ln |y| + cy$

5. $-e^{-y/x^4} = x^2 + c$

7. $x^2 + y^2 = x - 1 + ce^{-x}$

9. $\ln(\tan y) = x + cx^{-1}$

11. $x^3 y^3 = 2x^3 - 9 \ln |x| + c$

13. $e^y = -e^{-x} \cos x + ce^{-x}$

15. $y^2 \ln x = ye^y - e^y + c$

17. $y = \ln |\cos(c_1 - x)| + c_2$

19. $y = -\dfrac{1}{c_1}(1 - c_1^2 x^2)^{1/2} + c_2$

21. The given equation is a Clairaut equation in $u = y'$. The solution is $y = c_1 x^2/2 + x + c_1^3 x + c_2$.

23. $y = c_1 + c_2 x^2$

25. $\dfrac{1}{3}y^3 - c_1 y = x + c_2$

27. $y = -\sqrt{1 - x^2}$

Exercises 2.8, Page 82

1. $y_1(x) = 1 - x$

$$y_2(x) = 1 - \dfrac{x}{1!} + \dfrac{x^2}{2!}$$

$$y_3(x) = 1 - \dfrac{x}{1!} + \dfrac{x^2}{2!} - \dfrac{x^3}{3!}$$

$$y_4(x) = 1 - \dfrac{x}{1!} + \dfrac{x^2}{2!} - \dfrac{x^3}{3!} + \dfrac{x^4}{4!};$$

$y_n(x) \to e^{-x}$ as $n \to \infty$

3. $y_1(x) = 1 + x^2$

$$y_2(x) = = 1 + \dfrac{x^2}{1!} + \dfrac{x^4}{2!}$$

$$y_3(x) = 1 + \dfrac{x^2}{1!} + \dfrac{x^4}{2!} + \dfrac{x^6}{3!}$$

$$y_4(x) = 1 + \dfrac{x^2}{1!} + \dfrac{x^4}{2!} + \dfrac{x^6}{3!} + \dfrac{x^8}{4!}$$

$y_n(x) \to e^{x^2}$ as $n \to \infty$

5. $y_1(x) = y_2(x) = y_3(x) = y_4(x) = 0$;
$y_n(x) \to 0$ as $n \to \infty$

7.(a) $y_1(x) = x$

$$y_2(x) = x + \frac{1}{3}x^3$$

$$y_3(x) = x + \frac{1}{3}x^3 + \frac{2}{15}x^5 + \frac{1}{63}x^7$$

(b) $y = \tan x$

(c) The Maclaurin series expansion of $\tan x$ is

$$x + \frac{1}{3}x^3 + \frac{2}{15}x^5 + \frac{17}{315}x^7 + \cdots, \quad |x| < \pi/2$$

Chapter 2 Test, Page 84

1. The regions defined by $x^2 + y^2 > 25$ and $x^2 + y^2 < 25$

3. false

5. (a) linear in x
   (b) homogeneous, exact, linear in y
   (c) Clairaut
   (d) Bernoulli in x
   (e) separable
   (f) separable, Ricatti
   (g) linear in x
   (h) homogeneous
   (i) Bernoulli
   (j) homogeneous, exact, Bernoulli
   (k) separable, homogeneous, exact, linear in x and in y
   (l) exact, linear in y
   (m) homogeneous
   (n) separable
   (o) Clairaut
   (p) Ricatti

7. $2y^2 \ln y - y^2 = 4xe^x - 4e^x - 1$

9. $2y^2 + x^2 = 9x^6$ **11.** $e^{xy} - 4y^3 = 5$

13. $y = \frac{1}{4} - 320(x^2 + 4)^{-4}$

15. $y = \dfrac{1}{x^4 - x^4 \ln |x|}$ **17.** $x^2 - \sin\dfrac{1}{y^2} = c$

19. $y_1(x) = 1 + x + \frac{1}{3}x^3$

$$y_2(x) = 1 + x + x^2 + \frac{2}{3}x^3 + \frac{1}{6}x^4 + \frac{2}{15}x^5$$

$$+ \frac{1}{63}x^7$$

1. $x^2 + y^2 = c_2^2$ **3.** $2y^2 + x^2 = c_2$

5. $2 \ln |y| = x^2 + y^2 + c_2$

7. $y^2 = 2x + c_2$ **9.** $2x^2 + 3y^2 = c_2$

11. $x^3 + y^3 = c_2$

13. $y^2 \ln |y| + x^2 = c_2 y^2$

15. $y^2 - x^2 = c_2 x$

17. $2y^2 = 2 \ln |x| + x^2 + c_2$

19. $y = \dfrac{1}{4} - \dfrac{1}{6}x^2 + c_2 x^{-4}$

21. $2y^3 = 3x^2 + c_2$

23. $2 \ln(\cosh y) + x^2 = c_2$

25. $y^{5/3} = x^{5/3} + c_2$

27. $y = 2 - x + 3e^{-x}$

29. The differential equation of the orthogonal family is $(x - y)\,dx + (x + y)\,dy = 0$. The verification follows by substituting $x = c_2 e^{-t} \cos t$ and $y = c_2 e^{-t} \sin t$ into the equation.

31. $r = c_2 \sin \theta$ **33.** $r^2 = c_2 \cos 2\theta$

35. $r = c_2 \csc \theta$

37. Let β be the angle of inclination, measured from the positive x-axis, of the tangent line to a member of the given family, and ϕ the angle of inclination of the tangent to a trajectory. At the point where the curves intersect, the angle between the tangents is α. From the following figures we conclude that there exist two possible cases, and that $\phi = \beta \pm \alpha$. Thus the slope of

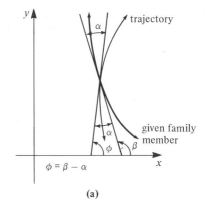

(a)

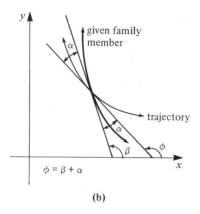

the tangent line to a trajectory is

$$\frac{dy}{dx} = \tan \phi = \tan (\beta \pm \alpha)$$

$$= \frac{\tan \beta \pm \tan \alpha}{1 \mp \tan \beta \tan \alpha}$$

$$= \frac{f(x,y) \pm \tan \alpha}{1 \mp f(x,y) \tan \alpha}.$$

39. $\mp \dfrac{2}{\sqrt{3}} \tan^{-1}(y/x) + \ln c_2(x^2 + y^2) = 0$

41. Since the given equation is quadratic in c_1 it follows from the quadratic formula that

$$2c_1 = -x \pm \sqrt{x^2 + y^2}.$$

Differentiating this last expression and solving for dy/dx gives

$$\frac{dy}{dx} = \frac{-x + \sqrt{x^2 + y^2}}{y}$$

and

$$\frac{dy}{dx} = \frac{-x - \sqrt{x^2 + y^2}}{y}.$$

These two equations correspond to choosing $c_1 > 0$ and $c_1 < 0$ in the given family, respectively. Forming the product of these derivatives yields

$$\left(\frac{dy}{dx}\right)_{(1)} \cdot \left(\frac{dy}{dx}\right)_{(2)} = \frac{x^2 - x^2 - y^2}{y^2} = -1.$$

This shows that the family is self-orthogonal.

43. The differential equation of the given family is $(x + yy')(xy' - y) = y'$. Replacing y' by $-1/y'$ results in exactly the same equation. This shows that the family is self-orthogonal.

1. 7.9 years;   10 years

3. $P_1 = P_0 e^{kt_1}$ and $P_2 = P_0{}^{kt_2}$ so that

$$\frac{P_1}{P_0} = e^{kt_1} \quad \text{and} \quad \frac{P_2^{t_1}}{P_0} = e^{kt_2}.$$

Thus

$$\left(\frac{P_1}{P_0}\right)^{t_2} = e^{kt_1 t_2} \quad \text{and} \quad \left(\frac{P_2}{P_0}\right)^{t_1} = e^{kt_2 t_1}.$$

The result follows by equating $e^{kt_1 t_2}$ with $e^{kt_2 t_1}$.

5. (a) $P(t) = P_0 e^{(k_1 - k_2)t}$
   (b) $k_1 > k_2$, births surpass deaths so population increases.
$k_1 = k_2$, a constant population since number of births equals the number of deaths.
$k_1 < k_2$, deaths surpass births so population decreases.

7. 136.5 hours

9. $I(15) = 0.00098 I_0$ or $I(15)$ is approximately 0.1% of I_0.

11. 15,600 years

13. $T(1) = 36.67$ degrees; approximately 3.06 minutes

15. $i(t) = \frac{3}{5} - \frac{3}{5} e^{-500t}$; $t \to \frac{3}{5}$ as $t \to \infty$.

17. $A(t) = 200 - 170 e^{-t/50}$

19. The differential equation for the number of pounds of salt $A(t)$ is

$$\frac{dA}{dt} + \frac{4}{100 + 2t} A = 3; \qquad A(30) = 64.38 \text{ lb}$$

21. $v(t) = \dfrac{mg}{k} + \left(v_0 - \dfrac{mg}{k}\right) e^{-kt/m}$;

$v \to \dfrac{mg}{k}$ as $t \to \infty$;

$s(t) = \dfrac{mg}{k} t - \dfrac{m}{k}\left(v_0 - \dfrac{mg}{k}\right) e^{-kt/m}$

$\qquad + \dfrac{m}{k}\left(v_0 - \dfrac{mg}{k}\right) + s_0$

Exercises 3.3, Page 111

1. 1834; 2000

3. 1,000,000; 52.9 months

5.(a) Separating variables gives

$$\frac{dP}{P(a - b \ln P)} = dt$$

so that

$$-\frac{1}{b} \ln|a - b \ln P| = t + c_1$$

$$a - b \ln P = c_2 e^{-bt} \qquad (e^{-bc_1} = c_2)$$

$$\ln P = \frac{a}{b} - ce^{-bt} \qquad \left(\frac{c_2}{b} = c\right)$$

$$P(t) = e^{a/b} \cdot e^{-ce^{-bt}}$$

(b) If $P(0) = P_0$ then

$$P_0 = e^{a/b} e^{-c} = e^{a/b - c}$$

and so

$$\ln P_0 = \frac{a}{b} - c$$

$$c = \frac{a}{b} - \ln P_0.$$

7. 29.3 grams; $X \to 60$ as $t \to \infty$; 0 grams of A and 30 grams of B.

9. For $\alpha \neq \beta$ the differential equation separates as

$$\frac{1}{\alpha - \beta}\left[-\frac{1}{\alpha - X} + \frac{1}{\beta - X}\right] dX = k \, dt.$$

It follows immediately that

$$\frac{1}{\alpha - \beta}[\ln|\alpha - X| - \ln|\beta - X|] = kt + c$$

or

$$\frac{1}{\alpha - \beta} \ln\left|\frac{\alpha - X}{\beta - X}\right| = kt + c.$$

For $\alpha = \beta$ the equation can be written as

$$(\alpha - X)^{-2} dX = k \, dt.$$

It follows that $(\alpha - X)^{-1} = kt + c$ or

$$X = \alpha - \frac{1}{kt + c}.$$

11. $v^2 = \dfrac{2gR^2}{y} + v_0^2 - 2gR.$

We note that as y increases v decreases. In particular, if $v_0^2 - 2gR < 0$ then there must be some

value of y for $v = 0$; the rocket stops and returns to earth under the influence of gravity. However, if $v_0^2 - 2gR \geq 0$ then $v > 0$ for all values of y. Hence we should have $v_0 \geq \sqrt{2gR}$. Using the values $R = 4000$ miles, $g = 32$ ft/sec², 1 ft = 1/5280 mi, 1 sec = 1/3600 hr, it follows that $v_0 \geq 25,067$ mi/hr.

13. Using the condition $y'(1) = 0$ we find

$$\frac{dy}{dx} = \frac{1}{2}[x^{v_1/v_2} - x^{-v_1/v_2}].$$

Now if $v_1 = v_2, y = \frac{1}{4}x^2 - \frac{1}{2}\ln x - \frac{1}{4}$; if $v_1 \neq v_2$ then

$$y = \frac{1}{2}\left[\frac{x^{1 + (v_1/v_2)}}{1 + \dfrac{v_1}{v_2}} - \frac{x^{1 - (v_1/v_2)}}{1 - \dfrac{v_1}{v_2}}\right] + \frac{v_1 v_2}{v_2^2 - v_1^2}$$

15. $2h^{1/2} = -\frac{1}{25}t + 2\sqrt{20};\quad t = 50\sqrt{20}$ sec

17. To evaluate the indefinite integral of the left side of

$$\frac{\sqrt{100 - y^2}}{y} dy = -dx$$

we use the substitution $y = 10 \cos \theta$. It follows that

$$x = 10 \ln\left(\frac{10 + \sqrt{100 - y^2}}{y}\right) - \sqrt{100 - y^2}.$$

19. Under the substitution $w = x^2$ the differential equation becomes

$$w = y\frac{dw}{dy} + \frac{1}{4}\left(\frac{dw}{dy}\right)^2$$

which is Clairaut's equation. The solution is

$$x^2 = cy + \frac{c^2}{4}.$$

If $2c_1 = c$, then we recognize

$$x^2 = 2c_1 y + c_1^2$$

as describing a family of parabolas.

21. $-\gamma \ln y + \delta y = \alpha \ln x - \beta x + c$

23.(a) The equation $2\dfrac{d^2\theta}{dt^2}\dfrac{d\theta}{dt} + 2\dfrac{g}{l}\sin \theta \dfrac{d\theta}{dt} =$

0 is the same as

$$\frac{d}{dt}\left(\frac{d\theta}{dt}\right)^2 + 2\frac{g}{l}\sin\theta \frac{d\theta}{dt} = 0.$$

Integrating this last equation with respect to t and using the initial conditions gives the result.

(b) From (a)

$$dt = \sqrt{\frac{l}{2g}}\frac{d\theta}{\sqrt{\cos\theta - \cos\theta_0}}.$$

Integrating this last equation gives the time for the pendulum to move from $\theta = \theta_0$ to $\theta = 0$,

$$t = \sqrt{\frac{l}{2g}}\int_0^{\theta_0}\frac{d\theta}{\sqrt{\cos\theta - \cos\theta_0}}.$$

The period is the total time T to go from $\theta = \theta_0$ to $\theta = -\theta_0$ and back again to $\theta = \theta_0$. This is

$$T = 4\sqrt{\frac{l}{2g}}\int_0^{\theta_0}\frac{d\theta}{\sqrt{\cos\theta - \cos\theta_0}}$$

$$= 2\sqrt{\frac{2l}{g}}\int_0^{\theta_0}\frac{d\theta}{\sqrt{\cos\theta - \cos\theta_0}}.$$

Chapter 3 Test, Page 116

1. $y^3 + 3/x = c_2$

3. $2(y - 2)^2 + (x - 1)^2 = c_2^2$

5. $P(45) = 8.99$ billion

7. $x(t) = \dfrac{\alpha c_1 e^{\alpha k_1 t}}{1 + c_1 e^{\alpha k_1 t}},$

$\quad y(t) = c_2(1 + c_1 e^{\alpha k_1 t})^{k_2/k_1}$

Exercises 4.1, Page 135

1. $y = \dfrac{1}{2}e^x - \dfrac{1}{2}e^{-x}$

3. $y = \dfrac{3}{5}e^{4x} + \dfrac{2}{5}e^{-x}$

5. $y = 3x - 4x\ln x$

7. $y \equiv 0, \quad y = x^2$

9. (a) $y = e^x\cos x - e^x\sin x$
   (b) no solution
   (c) $y = e^x\cos x + e^{-\pi/2}e^x\sin x$
   (d) $y = c_2 e^x\sin x$ where c_2 is arbitrary

11. $-\infty < x < 2$

13. $\lambda = n, n = 1, 2, 3, \ldots$

15. dependent   **17.** dependent

19. dependent   **21.** independent

23. $W(x^{1/2}, x^2) = \dfrac{3}{2}x^{3/2} \neq 0$ on $0 < x < \infty$

25. $W(\sin x, \csc x) = -2\cot x$.
    $W = 0$ only at $x = \pi/2$ in the interval.

27. $W(e^x, e^{-x}, e^{4x}) = -30e^{4x} \neq 0$

on $-\infty < x < \infty$

29. no

31.(a) $y'' - 2y^3 = \dfrac{2}{x^3} - 2\left(\dfrac{1}{x}\right)^3 = 0$

   (b) $y'' - 2y^3 = \dfrac{2c}{x^3} - 2\dfrac{c^3}{x^3} = \dfrac{2}{x^3}c(1 - c^2) \neq 0$

for $c \neq 0 \pm 1$

33. The functions satisfy the differential equation and are linearly independent on the interval since
$W(e^{-3x}, e^{4x}) = 7e^x \neq 0; \quad y = c_1 e^{-3x} + c_2 e^{4x}.$

35. The functions satisfy the differential equation and are linearly independent on the interval since
$W(e^x\cos 2x, e^x\sin 2x) = 2e^{2x} \neq 0;$
$y = c_1 e^x\cos 2x + c_2 e^x\sin 2x.$

37. The functions satisfy the differential equation and are linearly independent on the interval since
$W(x^3, x^4) = x^6 \neq 0; \quad y = c_1 x^3 + c_2 x^4.$

39. The functions satisfy the differential equation and are linearly independent on the interval since
$W(x, x^{-2}, x^{-2}\ln x) = 9x^{-6} \neq 0;$
$y = c_1 x + c_2 x^{-2} + c_3 x^{-2}\ln x.$

41. e^{2x} and e^{5x} form a fundamental set of solutions of the homogeneous equation; $6e^x$ is a particular solution of the nonhomogeneous equation.

43. e^{2x} and xe^{2x} form a fundamental set of solutions of the homogeneous equation; $x^2 e^{2x} + x - 2$ is a particular solution of the nonhomogeneous equation.

45.(a) The accompanying graphs show that y_1 and y_2 are not multiples of one another.

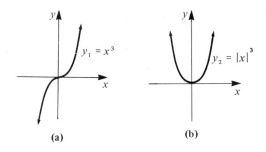

(a)   (b)

Also,

$$x^2y_1'' - 4xy_1' + 6y_1 = x^2(6x) - 4x(3x^2) + 6x^3$$
$$= 12x^3 - 12x^3 = 0.$$

For $x \geq 0$ the demonstration that y_2 is a solution of the equation is exactly as given above for y_1. For $x < 0$, $y_2 = -x^3$ and so

$$x^2y_2'' - 4xy_2' + 6y_2 = x^2(-6x)$$
$$- 4x(-3x^2) + 6(-x^3)$$
$$= -12x^3 + 12x^3 = 0.$$

(b) For $x \geq 0$

$$W(y_1, y_2) = \begin{vmatrix} x^3 & x^3 \\ 3x^2 & 3x^2 \end{vmatrix}$$
$$= 3x^5 - 3x^5 = 0.$$

For $x < 0$

$$W(y_1, y_2) = \begin{vmatrix} x^3 & -x^3 \\ 3x^2 & -3x^2 \end{vmatrix}$$
$$= -3x^5 + 3x^5 = 0.$$

Thus $W(y_1, y_2) = 0$ for every real value of x.
(c) No, $a_2(x) = x^2$ is zero at $x = 0$
(d) Since $Y_1 = y_1$ we need only show

$$x^2Y_2'' - 4xY_2' + 6Y_2 = x^2(2) - 4x(2x) + 6x^2$$
$$= 0$$
$$= 8x^2 - 8x^2 = 0$$

and $W(x^3, x^2) = -x^4$. Thus Y_1 and Y_2 are linearly independent solutions on the interval.
(e) $Y_1 = x^3$, $Y_2 = x^2$, or $y_2 = |x|^3$
(f) Neither; we form a general solution on an interval for which $a_2(x) \neq 0$ for every x in the interval. The linear combination

$$y = c_1Y_1 + c_2Y_2$$

would be a general solution of the equation on, say, $0 < x < \infty$.

47.(a) Since y_1 and y_2 are solutions of the given differential equation we have

$$a_2(x)y_1'' + a_1(x)y_1' + a_0(x)y_1 = 0$$
and $\quad a_2(x)y_2'' + a_1(x)y_2' + a_0(x)y_2 = 0.$

Now multiply the first equation by y_2 and the second by y_1 and subtract the first from the second:

$$a_2(x)[y_1y_2'' - y_2y_1'']$$
$$+ a_1(x)[y_1y_2' - y_2y_1'] = 0.$$

Now it is easily verified that

$$\frac{dW}{dx} = \frac{d}{dx}(y_1y_2' - y_2y_1')$$
$$= y_1y_2'' - y_2y_1''$$

and so it follows that

$$a_2(x)\frac{dW}{dx} + a_1(x)W = 0.$$

(b) Since this last equation is a linear first-order differential equation the integrating factor is

$$e^{\int (a_1(x)/a_2(x))\, dx}.$$

Therefore from

$$\frac{d}{dx}[e^{\int (a_1(x)/a_2(x))\, dx}W] = 0$$

we obtain $\quad W = ce^{-\int (a_1(x)/a_2(x))\, dx}$

(c) Substituting $x = x_0$ in the given result we find $c = W(x_0)$.

(d) Since an exponential function is never zero, when $W(x_0) \neq 0$ it follows from part (c) that $W \neq 0$. On the other hand, if $W(x_0) = 0$ we have immediately that $W = 0$.

49. From part (c) of Problem 47 we have

$$W(y_1, y_2) = W(y_1(x_0), y_2(x_0))e^{-\int_{x_0}^{x} dt/t}$$
$$= \begin{vmatrix} k_1 & k_3 \\ k_2 & k_4 \end{vmatrix} e^{-\ln (x/x_0)}$$
$$= (k_1k_4 - k_3k_2) \cdot \frac{x_0}{x}$$

Exercises 4.2, Page 144

1. $y_2 = e^{-4x}$ **3.** $y_2 = xe^{2x}$

5. $y_2 = \sin 4x$ **7.** $y_2 = \sinh x$

9. $y_2 = xe^{2x/3}$ **11.** $y_2 = x^4 \ln|x|$

13. $y_2 = 1$ **15.** $y_2 = x^2 + x + 2$

17. $y_2 = x \cos(\ln x)$ **19.** $y_2 = x$

21. $y_2 = x \ln x$ **23.** $y_2 = x^3$

25. $y_2 = x^2$ **27.** $y_2 = 3x + 2$

29. $y_2 = \frac{1}{2}[\tan x \sec x + \ln|\sec x + \tan x|]$

31. $y_2 = e^{2x}$, $y_p = -\frac{1}{2}$

33. $y_2 = e^{2x}$, $y_p = \frac{5}{2}e^{3x}$

Exercises 4.3, Page 152

1. $y = c_1 + c_2 e^{x/3}$ **3.** $y = c_1 e^{-4x} + c_2 e^{4x}$

5. $y = c_1 \cos 3x + c_2 \sin 3x$

7. $y = c_1 e^x + c_2 e^{2x}$

9. $y = c_1 e^{-4x} + c_2 x e^{-4x}$

11. $y = c_1 e^{(-3 + \sqrt{29})x/2} + c_2 e^{(-3 - \sqrt{29})x/2}$

13. $y = c_1 e^{2x/3} + c_2 e^{-x/4}$

15. $y = e^{2x}(c_1 \cos x + c_2 \sin x)$

17. $y = e^{-x/3}\left(c_1 \cos \dfrac{\sqrt{2}}{3}x + c_2 \sin \dfrac{\sqrt{2}}{3}x\right)$

19. $y = c_1 + c_2 e^{-x} + c_3 e^{5x}$

21. $y = c_1 e^x + e^{-x/2}\left(c_2 \cos \dfrac{\sqrt{3}}{2}x\right.$
$\left. + c_3 \sin \dfrac{\sqrt{3}}{2}x\right)$

23. $y = c_1 e^{-x} + c_2 e^{3x} + c_3 x e^{3x}$

25. $y = c_1 e^x + e^{-x}(c_2 \cos x + c_3 \sin x)$

27. $y = c_1 e^{-x} + c_2 x e^{-x} + c_3 x^2 e^{-x}$

29. $y = c_1 + c_2 x$
$+ e^{-x/2}\left(c_3 \cos \dfrac{\sqrt{3}}{2}x + c_4 \sin \dfrac{\sqrt{3}}{2}x\right)$

31. $y = c_1 \cos \dfrac{\sqrt{3}}{2}x + c_2 \sin \dfrac{\sqrt{3}}{2}x$
$+ c_3 x \cos \dfrac{\sqrt{3}}{2}x + c_4 x \sin \dfrac{\sqrt{3}}{2}x$

33. $y = c_1 + c_2 e^{-2x} + c_3 e^{2x}$
$+ c_4 \cos 2x + c_5 \sin 2x$

35. $y = c_1 e^x + c_2 x e^x + c_3 e^{-x}$
$+ c_4 x e^{-x} + c_5 e^{-5x}$

37. $\dfrac{d^3 y}{dx^3} + 6\dfrac{d^2 y}{dx^2} - 15\dfrac{dy}{dx} - 100y = 0$

39. $y = 2 \cos 4x - \frac{1}{2} \sin 4x$

41. $y = -\frac{3}{4}e^{-5x} + \frac{3}{4}e^{-x}$

43. $y = -e^{x/2} \cos x/2 + e^{x/2} \sin x/2$

45. $y \equiv 0$ **47.** $y = e^{2(x-1)} - e^{x-1}$

49. $y = \frac{5}{36} - \frac{5}{36}e^{-6x} + \frac{1}{6}x e^{-6x}$

51. $y = -\dfrac{1}{6}e^{2x} + \dfrac{1}{6}e^{-x} \cos \sqrt{3}x$
$- \dfrac{\sqrt{3}}{6}e^{-x} \sin \sqrt{3}x$

53. $y = 2 - 2e^x + 2xe^x - \frac{1}{2}x^2 e^x$

55. $y = e^{5x} - xe^{5x}$

57. $y = e^{-\sqrt{2}x/2}\left(c_1 \cos \dfrac{\sqrt{2}}{2}x + c_2 \sin \dfrac{\sqrt{2}}{2}x\right)$
$+ e^{\sqrt{2}x/2}\left(c_3 \cos \dfrac{\sqrt{2}}{2}x + c_4 \sin \dfrac{\sqrt{2}}{2}x\right)$

Exercises 4.4, Page 162

1. $(2D - 3)(2D + 3)$

3. $(D - 6)(D + 2)$

5. $D(D + 5)^2$

7. $(D - 1)(D - 2)(D + 5)$

9. $D(D + 2)(D^2 - 2D + 4)$

11. D^3 **13.** $D(D - 2)$

15. $D^3(D^2 + 16)$ **17.** $(D + 1)(D - 1)^3$

19. $D(D^2 - 2D + 5)$

21. $y = c_1 e^{-3x} + c_2 e^{3x} - 6$

23. $y = c_1 + c_2 e^{-x} + 3x$

25. $y = c_1 e^{-2x} + c_2 x e^{-2x} + \frac{1}{2}x + 1$

27. $y = c_1 + c_2 x + c_3 e^{-x} + \frac{2}{3}x^4 - \frac{8}{3}x^3 + 8x^2$

29. $y = c_1 e^{-3x} + c_2 e^{4x} + \frac{1}{7}x e^{4x}$

31. $y = c_1 e^{-x} + c_2 e^{3x} - e^x + 3$

33. $y = c_1 \cos 5x + c_2 \sin 5x + \frac{1}{4} \sin x$

35. $y = c_1 e^{-3x} + c_2 x e^{-3x} - \frac{1}{49} x e^{4x} + \frac{2}{343} e^{4x}$

37. $y = c_1 e^{-x} + c_2 e^x + \frac{1}{6} x^3 e^x - \frac{1}{4} x^2 e^x$
$\quad + \frac{1}{4} x e^x - 5$

39. $y = e^x(c_1 \cos 2x + c_2 \sin 2x) + \frac{1}{3} e^x \sin x$

41. $y = c_1 + c_2 x + c_3 e^x + c_4 x e^x$
$\quad + \frac{1}{2} x^2 e^x + \frac{1}{2} x^2$

43. $y = c_1 e^{x/2} + c_2 e^{-x/2} + c_3 \cos \dfrac{x}{2}$

$\quad + c_4 \sin \dfrac{x}{2} + \dfrac{1}{8} x e^{x/2}$

45. $y = c_1 e^x + c_2 x e^x + c_3 x^2 e^x + \frac{1}{6} x^3 e^x$
$\quad + x - 13$

47. $y = c_1 \cos 5x + c_2 \sin 5x - 2x \cos 5x$

49. $y = e^{-x/2}\left(c_1 \cos \dfrac{\sqrt{3}}{2} x + c_2 \sin \dfrac{\sqrt{3}}{2} x \right)$

$\quad + \sin x + 2 \cos x - x \cos x$

51. $y = c_1 + c_2 e^x + 2x + x e^x + \frac{1}{2} e^{-x}$

53. $y = \frac{5}{8} e^{-8x} + \frac{5}{8} e^{8x} - \frac{1}{4}$

55. $y = -\frac{41}{125} + \frac{41}{125} e^{5x} - \frac{1}{10} x^2 + \frac{9}{25} x$

57. $y = -\pi \cos x - \frac{11}{3} \sin x - \frac{8}{3} \cos 2x$
$\quad + 2x \cos x$

59. $y = 2 e^{2x} \cos 2x - \frac{3}{64} e^{2x} \sin 2x + \frac{1}{8} x^3$
$\quad + \frac{3}{16} x^2 + \frac{3}{32} x$

61. $y_p = A x e^x + B e^x \cos 2x + C e^x \sin 2x$
$\quad + E x e^x \cos 2x + F x e^x \sin 2x$

63. The operators do not commute.

Exercises 4.5, Page 169

1. $y = c_1 \cos x + c_2 \sin x + x \sin x$
$\quad + \cos x \ln |\cos x|; \quad -\pi/2 < x < \pi/2$

3. $y = c_1 \cos x + c_2 \sin x + \frac{1}{2} \sin x$
$\quad - \frac{1}{2} x \cos x$
$\quad = c_1 \cos x + c_3 \sin x - \frac{1}{2} x \cos x;$
$\quad -\infty < x < \infty$

5. $y = c_1 \cos x + c_2 \sin x + \frac{1}{2} - \frac{1}{6} \cos 2x;$
$\quad -\infty < x < \infty$

7. $y = c_1 e^x + c_2 e^{-x} + \frac{1}{4} x e^x - \frac{1}{4} x e^{-x}$
$\quad = c_1 e^x + c_2 e^{-x} + \frac{1}{2} x \sinh x;$
$\quad -\infty < x < \infty$

9. $y = c_1 e^{2x} + c_2 e^{-2x}$

$\quad + \dfrac{1}{4}\left(e^{2x} \ln |x| - e^{-2x} \displaystyle\int_{x_0}^{x} \dfrac{e^{4t}}{t} dt \right),$

$\quad x_0 > 0; \quad 0 < x < \infty$

11. $y = c_1 e^{-x} + c_2 e^{-2x} + (e^{-x} + e^{-2x})$
$\quad \times \ln(1 + e^x); \quad -\infty < x < \infty$

13. $y = c_1 e^{-2x} + c_2 e^{-x} - e^{-2x} \sin e^x;$
$\quad -\infty < x < \infty$

15. $y = c_1 e^x + c_2 x e^x - \frac{1}{2} e^x \ln(1 + x^2)$
$\quad + x e^x \tan^{-1} x; \quad -\infty < x < \infty$

17. $y = c_1 e^{-x} + c_2 x e^{-x} + \frac{1}{2} x^2 e^{-x} \ln x$
$\quad - \frac{3}{4} x^2 e^{-x}; \quad 0 < x < \infty$

19. $y = c_1 e^{x/2} + c_2 x e^{x/2} + \frac{8}{9} e^{-x} + x + 4;$
$\quad -\infty < x < \infty$

21. $y = \frac{3}{8} e^{-x} + \frac{5}{8} e^x + \frac{1}{4} x^2 e^x - \frac{1}{4} x e^x$

23. $y = \frac{4}{9} e^{-4x} + \frac{25}{36} e^{2x} - \frac{1}{4} e^{-2x} + \frac{1}{9} e^{-x}$

25. $y = c_1 x + c_2 x \ln x + \frac{2}{3} x (\ln x)^3$

27. $y = c_1 x^{-1/2} \cos x + c_2 x^{-1/2} \sin x + x^{-1/2}$

29. $y = c_1 + c_2 e^x + c_3 e^{-x} + \frac{1}{4} x^2 e^x - \frac{3}{4} x e^x$

31. $y_p = \frac{1}{8} e^{3x}$

Chapter 4 Test, Page 172

1. $y \equiv 0.$

3. false, the functions $f_1(x) \equiv 0$ and $f_2(x) = e^x$ are linearly dependent on $-\infty < x < \infty$ but f_2 is not a constant multiple of f_1

5. $-\infty < x < 0; \quad 0 < x < \infty$ **7.** false

9. $y_p = A + B x e^x$ **11.** $y_2 = \sin 2x$

13. $y = c_1 e^{(1 + \sqrt{3})x} + c_2 e^{(1-\sqrt{3})x}$

15. $y = c_1 + c_2 e^{-5x} + c_3 x e^{-5x}$

17. $y = c_1 e^{-x/3}$

$\quad + e^{-3x/2}\left[c_2 \cos \dfrac{\sqrt{7}}{2} x + c_3 \sin \dfrac{\sqrt{7}}{2} x \right]$

19. $y = e^{3x/2}\left[c_1 \cos \dfrac{\sqrt{11}}{2} x + c_2 \sin \dfrac{\sqrt{11}}{2} x \right]$

$\quad + \frac{4}{5} x^3 + \frac{36}{25} x^2 + \frac{46}{125} x - \frac{222}{625}$

21. $y = c_1 + c_2 e^{2x} + c_3 e^{3x} + \frac{1}{5} \sin x - \frac{1}{5} \cos x$
$\quad + \frac{4}{3} x$

23. $y = e^x[c_1 \cos x + c_2 \sin x]$
$- e^x \cos x \ln |\sec x + \tan x|$

25. $y = \frac{1}{2} \cos x + \frac{1}{2} \sin x + \frac{1}{2} \sec x$

Exercises 5.1, Page 182

1. A weight of 4 lb ($\frac{1}{8}$ slug), attached to a spring, is released from a point 3 units above the equilibrium position with an initial upward velocity of 2 ft/sec. The spring constant is 3 lb/ft.

3. $x(t) = 2\sqrt{2} \sin \left(5t - \dfrac{\pi}{4} \right)$

5. $x(t) = \sqrt{5} \sin (\sqrt{2}t + 3.6052)$

7. $x(t) = \dfrac{\sqrt{101}}{10} \sin (10t + 1.4711)$

9. 8 lb   **11.** $\sqrt{2}\pi/8$

13. $x(t) = -\frac{1}{4} \cos 4\sqrt{6}t$

15.(a) $x(\pi/12) = -1/4; \quad x(\pi/8) = -1/2;$
$x(\pi/6) = -1/4; \quad x(\pi/4) = 1/2;$
$x(9\pi/32) = \sqrt{2}/4$

(b) 4 ft/sec; downward

(c) $t = (2n + 1)\pi/16, \ n = 0, 1, 2, \ldots$

17.(a) the 20 kg mass

(b) the 20 kg mass; the 50 kg mass

(c) $t = n\pi, \ n = 0, 1, 2, \ldots$; at the equilibrium position; the 50 kg mass is moving upward whereas the 20 kg mass is moving upward when n is even and downward when n is odd.

19. $x(t) = \frac{1}{2} \cos 2t + \frac{3}{4} \sin 2t$
$= \dfrac{\sqrt{13}}{4} \sin (2t + 0.5880)$

21.(a) $x(t) = -\frac{2}{3} \cos 10t + \frac{1}{2} \sin 10t$
$= \frac{5}{6} \sin (10t - 0.927)$

(b) 5/6 ft; $\pi/5$

(c) 15 cycles

(d) 0.721 sec

(e) $(2n + 1)\pi/20 + 0.0927,$
$n = 0, 1, 2, \ldots$

(f) $x(3) = -0.597$ ft

(g) $x'(3) = -5.814$ ft/sec

(h) $x''(3) = 59.702$ ft/sec^2

(i) $\pm 8\frac{1}{3}$ ft/sec

(j) $0.1451 + n\pi/5; \quad 0.3545 + n\pi/5,$
$n = 0, 1, 2, \ldots$

(k) $0.3545 + n\pi/5, \quad n = 0, 1, 2, \ldots$

23. $x(t) = \frac{1}{8} \sin 16t$

25. Using $x(t) = c_1 \cos \omega t + c_2 \sin \omega t$, $x(0) = x_0$ and $x'(0) = v_0$ we find $c_1 = x_0$ and $c_2 = v_0/\omega$. The result follows from $A = \sqrt{c_1^2 + c_2^2}$.

27. $x(t) = 2\sqrt{2} \cos \left(5t + \dfrac{5\pi}{4} \right)$

29. At a maximum of $x(t)$, $x'(t) = 0$ and so $\cos (\omega t + \phi) = 0$ yields $\omega t_n + \phi = (2n + 1) \cdot \pi/2, \ n = 0, 1, 2, \ldots$. At two successive maxima, corresponding to t_k and t_{k+2}, we have $\omega t_k + \phi = (2k + 1)\pi/2$ and $\omega t_{k+2} + \phi = (2k + 5)\pi/2$. Subtracting gives $\omega(t_{k+2} - t_k) = 2\pi$ or $t_{k+2} - t_k = 2\pi/\omega$.

Exercises 5.2, Page 193

1. A 2-lb weight is attached to a spring whose constant is 1 lb/ft. The system is damped with a resisting force numerically equal to 2 times the instantaneous velocity. The weight starts from the equilibrium position with an upward velocity of 1.5 ft/sec.

3. $\frac{1}{4}$ sec; $\frac{1}{2}$ sec, $x(\frac{1}{2}) = e^{-2}$, that is, the weight is approximately 0.14 ft below the equilibrium position.

5.(a) $x(t) = \frac{4}{3}e^{-2t} - \frac{1}{3}e^{-8t}$
(b) $x(t) = -\frac{2}{3}e^{-2t} + \frac{5}{3}e^{-8t}$

7.(a) $x(t) = e^{-2t}[-\cos 4t - \frac{1}{2} \sin 4t]$

(b) $x(t) = \dfrac{\sqrt{5}}{2} e^{-2t} \sin (4t + 4.249)$

(c) $t = 1.294$ sec

9.(a) $\beta > 5/2$
(b) $\beta = 5/2$
(c) $0 < \beta < 5/2$

11. $x(t) = \frac{2}{7}e^{-7t} \sin 7t$

13. $v_0 > 2$ ft/sec

15. Suppose $\gamma = \sqrt{\omega^2 - \lambda^2}$ then the derivative of $x(t) = Ae^{-\lambda t} \sin (\gamma t + \phi)$ is

$$x'(t) = Ae^{-\lambda t}[\gamma \cos(\gamma t + \phi) - \lambda \sin(\gamma t + \phi)].$$

So $x'(t) = 0$ implies

$$\tan(\gamma t + \phi) = \frac{\gamma}{\lambda}$$

from which it follows that

$$t = \frac{1}{\gamma}\left[\tan^{-1}\frac{\gamma}{\lambda} + k\pi - \phi\right].$$

The difference between the t values between two successive maxima (or minima) is then

$$t_{k+2} - t_k = (k+2)\frac{\pi}{\gamma} - k\frac{\pi}{\gamma}$$

$$= \frac{2\pi}{\gamma}.$$

17. $t_{k+1}^* - t_k^* = \dfrac{(2k+3)\pi/2 - \phi}{\sqrt{\omega^2 - \lambda^2}}$

$$- \frac{(2k+1)\pi/2 - \phi}{\sqrt{\omega^2 - \lambda^2}}$$

$$= \frac{\pi}{\sqrt{\omega^2 - \lambda^2}}$$

19. Let the quasi period $2\pi/\sqrt{\omega^2 - \lambda^2}$ be denoted by T_q. From equation (15) we find

$$x_n/x_{n+2} = x(t)/x(t + T_q)$$

$$= \frac{e^{-\lambda t}\sin[\sqrt{\omega^2 - \lambda^2}\,t + \phi]}{e^{-\lambda(t+T_q)}\sin[\sqrt{\omega^2 - \lambda^2}(t + T_q) + \phi]}$$

$$= e^{\lambda T_q}$$

since

$$\sin[\sqrt{\omega^2 - \lambda^2}\,t + \phi]$$

$$= \sin[\sqrt{\omega^2 - \lambda^2}(t + T_q) + \phi].$$

Therefore

$$\ln(x_n/x_{n+2}) = \lambda T_q$$

$$= 2\pi\lambda/\sqrt{\omega^2 - \lambda^2}.$$

Exercises 5.3, Page 201

1. $x(t) = e^{-t/2}\left[-\dfrac{4}{3}\cos\dfrac{\sqrt{47}}{2}t\right.$

$$\left. - \frac{64}{3\sqrt{47}}\sin\frac{\sqrt{47}}{2}t\right] + \frac{10}{3}[\cos 3t + \sin 3t]$$

3. $x(t) = \frac{1}{4}e^{-4t} + te^{-4t} - \frac{1}{4}\cos 4t$

5. $x(t) = -\frac{1}{2}\cos 4t$
$\quad + \frac{9}{4}\sin 4t + \frac{1}{2}e^{-2t}\cos 4t - 2e^{-2t}\sin 4t$

7. $m\dfrac{d^2x}{dt^2} = -k(x - h) - \beta\dfrac{dx}{dt}$ or

$$\frac{d^2x}{dt^2} + 2\lambda\frac{dx}{dt} + \omega^2 x = \omega^2 h(t), \quad \text{where}$$

$$2\lambda = \frac{\beta}{m} \text{ and } \omega^2 = \frac{k}{m}.$$

9.(a) $x(t) = \frac{2}{3}\sin 4t - \frac{1}{3}\sin 8t$
(b) $t = n\pi/4, \quad n = 0, 1, 2, \ldots$
(c) $t = \pi/6 + n\pi/2, \quad n = 0, 1, 2, \ldots$
and $\quad t = \pi/3 + n\pi/2, \quad n = 0, 1, 2, \ldots$
(d) $\sqrt{3}/2$ cm, $\quad -\sqrt{3}/2$ cm
(e)

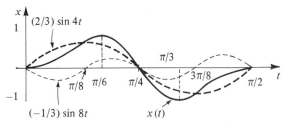

11.(a) $g'(\gamma) = 0$ implies $\gamma(\gamma^2 - \omega^2 + 2\lambda^2) = 0$ so that either $\gamma = 0$ or $\gamma = \sqrt{\omega^2 - 2\lambda^2}$. The first derivative test can be used to verify that $g(\gamma)$ is a maximum at the latter value.
(b) $g(\sqrt{\omega^2 - 2\lambda^2}) = F_0/2\lambda\sqrt{\omega^2 - \lambda^2}$

13. $x_p = -5\cos 2t + 5\sin 2t$

$$= 5\sqrt{2}\sin\left(2t - \frac{\pi}{4}\right)$$

15.(a) $x(t) = x_c + x_p$
$$= c_1\cos\omega t + c_2\sin\omega t$$
$$+ \frac{F_0}{\omega^2 - \gamma^2}\cos\gamma t$$

where the initial conditions imply that $c_1 = -F_0/(\omega^2 - \gamma^2)$ and $c_2 = 0$.
(b) By L'Hôpital's rule the given limit is the same as

$$\lim_{\gamma\to\omega}\frac{F_0(-t\sin\gamma t)}{-2\gamma} = \frac{F_0}{2\omega}t\sin\omega t.$$

17. $x(t) = -\cos 2t - \frac{1}{8} \sin 2t + \frac{3}{4}t \sin 2t$
$\qquad + \frac{5}{4}t \cos 2t$

19.(a) Recall that
$$\cos (u - v) = \cos u \cos v + \sin u \sin v$$
$$\cos (u + v) = \cos u \cos v - \sin u \sin v.$$
Subtracting gives
$$\sin u \sin v = \tfrac{1}{2}[\cos (u - v) - \cos (u + v)].$$
Setting $u = \frac{1}{2}(\gamma - \omega)t$ and $v = \frac{1}{2}(\gamma + \omega)t$ then gives
$$\sin \tfrac{1}{2}(\gamma - \omega)t \sin \tfrac{1}{2}(\gamma + \omega)t$$
$$= \tfrac{1}{2}[\cos \omega t - \cos \gamma t]$$
from which the result follows.

(b) For small ε, $\gamma \approx \omega$ so $\gamma + \omega \approx 2\gamma$ and therefore
$$\frac{-2F_0}{(\omega + \gamma)(\omega - \gamma)} \sin \frac{1}{2}(\gamma - \omega)t$$
$$\times \sin \frac{1}{2}(\gamma + \omega)t$$
$$\approx \frac{F_0}{2\gamma\varepsilon} \sin \varepsilon t \sin \frac{1}{2}(2\gamma)t.$$

(c) By L'Hôpital's rule the given limit is the same as
$$\lim_{\varepsilon \to 0} \frac{F_0 t \cos \varepsilon t \sin \gamma t}{2\gamma} = \frac{F_0}{2\gamma}t \sin \gamma t$$
$$= \frac{F_0}{2\omega}t \sin \omega t.$$

Exercises 5.4, Page 207

1. $\theta(t) = \frac{1}{2} \cos 4t + \frac{\sqrt{3}}{2} \sin 4t$;

1; $\pi/2$; $2/\pi$

3. Differentiate equation (1) with respect to t and use the fact that $dq/dt = i$. For example,
$$\frac{d^2q}{dt^2} = \frac{d}{dt}\left(\frac{dq}{dt}\right) = \frac{di}{dt}.$$

5. $i_p = \frac{1200k}{100 - k^2} \cos kt$

7. $q(t) = -\frac{1}{2}e^{-10t}(\cos 10t + \sin 10t) + \frac{3}{2}$; $\frac{3}{2}$ coulombs

Chapter 5 Test, Page 210

1. 8 ft $\qquad$ **3.** 5/4 m

5. false; there could be an impressed force driving the system

7. overdamped $\qquad$ **9.** 9/2 lb/ft

11. $x(t) = -\frac{2}{3}e^{-2t} + \frac{1}{3}e^{-4t}$

13. $0 < m \leq 2$ $\qquad$ **15.** $\gamma = 8\sqrt{3}/3$

17. $x(t) = e^{-4t}$
$$\times \left[\frac{26}{17} \cos 2\sqrt{2}t + \frac{28\sqrt{2}}{17} \sin 2\sqrt{2}t \right] + \frac{8}{17}e^{-t}$$

Exercises 6.1, Page 219

1. $y = c_1 x^{-1} + c_2 x^2$

3. $y = c_1 + c_2 \ln x$

5. $y = c_1 \cos (2 \ln x) + c_2 \sin (2 \ln x)$

7. $y = c_1 x^{(2-\sqrt{6})} + c_2 x^{(2 + \sqrt{6})}$

9. $y_1 = c_1 \cos (\frac{1}{5} \ln x) + c_2 \sin (\frac{1}{5} \ln x)$

11. $y = c_1 x^{-2} + c_2 x^{-2} \ln x$

13. $y = x[c_1 \cos (\ln x) + c_2 \sin (\ln x)]$

15. $y = x^{-1/2}$
$$\times \left[c_1 \cos \left(\frac{\sqrt{3}}{6} \ln x \right) + c_2 \sin \left(\frac{\sqrt{3}}{6} \ln x \right) \right]$$

17. $y = c_1 x^3 + c_2 \cos (\sqrt{2} \ln x)$
$\qquad + c_3 \sin (\sqrt{2} \ln x)$

19. $y = c_1 x^{-1} + c_2 x^2 + c_3 x^4$

21. $y = 2 - 2x^{-2}$

23. $y = \cos (\ln x) + 2 \sin (\ln x)$

25. $y = c_1 + c_2 \ln x + \frac{x^2}{4}$

27. $y = c_1 x^{-1/2} + c_2 x^{-1} + \frac{1}{15}x^2 - \frac{1}{6}x$

29. $y = c_1 x + c_2 x \ln x + x(\ln x)^2$

31. $y = c_1 x^{-1} + c_2 x^{-8} + \frac{1}{30}x^2$

33. $y = x^2[c_1 \cos (3 \ln x) + c_2 \sin (3 \ln x)]$
$\qquad + \frac{4}{13} + \frac{3}{10}x$

35. $y = c_1 x^2 + c_2 x^{-10} - \frac{1}{7}x^{-3}$

37. $y = c_1(x - 1)^{-1} + c_2(x - 1)^4$

39. $y = c_1 \cos (\ln (x + 2))$
$\qquad + c_2 \sin (\ln (x + 2))$

Exercises 6.2, Page 231

1. $y = ce^{-x}$; $\quad y = c_0 \sum\limits_{n=0}^{\infty} \dfrac{(-1)^n}{n!} x^n$

3. $y = ce^{x^3/3}$; $\quad y = c_0 \sum\limits_{n=0}^{\infty} \dfrac{1}{n!} \left(\dfrac{x^3}{3}\right)^n$

5. $y = c/(1-x)$; $\quad y = c_0 \sum\limits_{n=0}^{\infty} x^n$

7. $y = C_1 \cos x + C_2 \sin x$;

$y = c_0 \sum\limits_{n=0}^{\infty} \dfrac{(-1)^n}{(2n)!} x^{2n} + c_1 \sum\limits_{n=0}^{\infty} \dfrac{(-1)^n}{(2n+1)!} x^{2n+1}$

9. $y = C_1 + C_2 e^x$;

$y = c_0 + c_1 \sum\limits_{n=1}^{\infty} \dfrac{x^n}{n!} = c_0 - c_1 + c_1 \sum\limits_{n=0}^{\infty} \dfrac{x^n}{n!}$

$= c_0 - c_1 + c_1 e^x$

11. $y_1(x) = c_0 \left[1 + \dfrac{1}{3 \cdot 2} x^3 + \dfrac{1}{6 \cdot 5 \cdot 3 \cdot 2} x^6 \right.$

$\left. + \dfrac{1}{9 \cdot 8 \cdot 6 \cdot 5 \cdot 3 \cdot 2} x^9 + \cdots \right]$

$y_2(x) = c_1 \left[x + \dfrac{1}{4 \cdot 3} x^4 + \dfrac{1}{7 \cdot 6 \cdot 4 \cdot 3} x^7 \right.$

$\left. + \dfrac{1}{10 \cdot 9 \cdot 7 \cdot 6 \cdot 4 \cdot 3} x^{10} + \cdots \right]$

13. $y_1(x) = c_0 \left[1 - \dfrac{1}{2} x^2 \right.$

$\left. - \dfrac{3}{4!} x^4 - \dfrac{21}{6!} x^6 - \cdots \right]$

$y_2(x) = c_1 \left[x + \dfrac{1}{3!} x^3 \right.$

$\left. + \dfrac{5}{5!} x^5 + \dfrac{45}{7!} x^7 + \cdots \right]$

15. $y_1(x) = c_0 \left[1 - \dfrac{1}{3!} x^3 + \dfrac{4^2}{6!} x^6 \right.$

$\left. - \dfrac{7^2 \cdot 4^2}{9!} x^9 + \cdots \right]$

$y_2(x) = c_1 \left[x - \dfrac{2^2}{4!} x^4 + \dfrac{5^2 \cdot 2^2}{7!} x^7 \right.$

$\left. - \dfrac{8^2 \cdot 5^2 \cdot 2^2}{10!} x^{10} + \cdots \right]$

17. $y_1(x) = c_0$; $\quad y_2(x) = c_1 \sum\limits_{n=1}^{\infty} \dfrac{1}{n} x^n$

19. $y_1(x) = c_0 \sum\limits_{n=0}^{\infty} x^{2n}$; $\quad y_2(x) = c_1 \sum\limits_{n=0}^{\infty} x^{2n+1}$

21. $y_1(x) = c_0 \left[1 + \dfrac{1}{4} x^2 - \dfrac{7}{4 \cdot 4!} x^4 \right.$

$\left. + \dfrac{23 \cdot 7}{8 \cdot 6!} x^6 - \cdots \right]$

$y_2(x) = c_1 \left[x - \dfrac{1}{6} x^3 + \dfrac{14}{2 \cdot 5!} x^5 \right.$

$\left. - \dfrac{34 \cdot 14}{4 \cdot 7!} x^7 - \cdots \right]$

23. $y_1(x) = c_0 [1 + \tfrac{1}{2} x^2 + \tfrac{1}{6} x^3 + \tfrac{1}{6} x^4 + \cdots]$
$y_2(x) = c_1 [x + \tfrac{1}{2} x^2 + \tfrac{1}{2} x^3 + \tfrac{1}{4} x^4 + \cdots]$

25. $y(x) = -2[1 + \tfrac{1}{2!} x^2 + \tfrac{1}{3!} x^3 + \tfrac{1}{4!} x^4 + \cdots]$
$\qquad\qquad + 6x = 8x - 2e^x$

27. $y(x) = 3 - 12x^2 + 4x^4$

29. $y_1(x) = c_0 [1 - \tfrac{1}{6} x^3 + \tfrac{1}{120} x^5 + \cdots]$
$y_2(x) = c_1 [x - \tfrac{1}{12} x^4 + \tfrac{1}{180} x^6 + \cdots]$

31. $y_1(x) = c_0 [1 - \tfrac{1}{2} x^2 + \tfrac{1}{6} x^3 - \tfrac{1}{40} x^5 + \cdots]$
$y_2(x) = c_1 [x - \tfrac{1}{6} x^3 + \tfrac{1}{12} x^4 - \tfrac{1}{60} x^5 + \cdots]$

33. $y_1(x) = c_0 \left[1 + \dfrac{1}{3!} x^3 + \dfrac{4}{6!} x^6 \right.$

$\left. + \dfrac{7 \cdot 4}{9!} x^9 + \cdots \right]$

$+ c_1 \left[x + \dfrac{2}{4!} x^4 + \dfrac{5 \cdot 2}{7!} x^7 \right.$

$+ \dfrac{8 \cdot 5 \cdot 2}{10!} x^{10} + \cdots \Big]$

$+ \dfrac{1}{2!} x^2 + \dfrac{3}{5!} x^5 + \dfrac{6 \cdot 3}{8!} x^8$

$+ \dfrac{9 \cdot 6 \cdot 3}{11!} x^{11}$

$+ \cdots$

Exercises 6.3, Page 250

1. $x = 0$, irregular singular point

3. $x = -3$, regular singular point; $x = 3$, irregular singular point

5. $x = 0, 2i, -2i$, regular singular points

7. $x = -3, 2$, regular singular points

9. $x = 0$, irregular singular point; $x = -5, 5,$
2, regular singular points

11. $r_1 = \frac{3}{2}, r_2 = 0$;

$$y(x) = C_1 x^{3/2} \left[1 - \frac{2}{5}x + \frac{2^2}{7 \cdot 5 \cdot 2}x^2 \right.$$

$$\left. - \frac{2^3}{9 \cdot 7 \cdot 5 \cdot 3!}x^3 + \cdots \right]$$

$$+ C_2 \left[1 + 2x - 2x^2 \right.$$

$$\left. + \frac{2^3}{3 \cdot 3!}x^3 - \cdots \right]$$

13. $r_1 = \frac{7}{8}, r_2 = 0$;

$$y(x) = C_1 x^{7/8} \left[1 - \frac{2}{15}x + \frac{2^2}{23 \cdot 15 \cdot 2}x^2 \right.$$

$$\left. - \frac{2^3}{31 \cdot 23 \cdot 15 \cdot 3!}x^3 + \cdots \right]$$

$$+ C_2 \left[1 - 2x + \frac{2^2}{9 \cdot 2}x^2 \right.$$

$$\left. - \frac{2^3}{17 \cdot 9 \cdot 3!}x^3 + \cdots \right]$$

15. $r_1 = \frac{1}{3}, r_2 = 0$;

$$y(x) = C_1 x^{1/3} \left[1 + \frac{1}{3}x + \frac{1}{3^2 \cdot 2}x^2 \right.$$

$$\left. + \frac{1}{3^3 \cdot 3!}x^3 + \cdots \right]$$

$$+ C_2 \left[1 + \frac{1}{2}x + \frac{1}{5 \cdot 2}x^2 \right.$$

$$\left. + \frac{1}{8 \cdot 5 \cdot 2}x^3 + \cdots \right]$$

17. $r_1 = \frac{5}{2}, r_2 = 0$;

$$y(x) = C_1 x^{5/2} \left[1 + \frac{2 \cdot 2}{7}x + \frac{2^2 \cdot 3}{9 \cdot 7}x^2 \right.$$

$$\left. + \frac{2^3 \cdot 4}{11 \cdot 9 \cdot 7}x^3 + \cdots \right]$$

$$+ C_2 \left[1 + \frac{1}{3}x - \frac{1}{6}x^2 - \frac{1}{6}x^3 - \cdots \right]$$

19. $r_1 = \frac{2}{3}, r_2 = \frac{1}{3}$,

$$y(x) = C_1 x^{2/3} \left[1 - \frac{1}{2}x \right.$$

$$+ \frac{5}{28}x^2 - \frac{1}{21}x^3 + \cdots \right]$$

$$+ C_2 x^{1/3} \left[1 - \frac{1}{2}x + \frac{1}{5}x^2 \right.$$

$$\left. - \frac{7}{120}x^3 + \cdots \right]$$

21. $r_1 = 1, r_2 = -\frac{1}{2}$,

$$y(x) = C_1 x \left[1 + \frac{1}{5}x + \frac{1}{5 \cdot 7}x^2 \right.$$

$$\left. + \frac{1}{5 \cdot 7 \cdot 9}x^3 + \cdots \right]$$

$$+ C_2 x^{-1/2} \left[1 + \frac{1}{2}x + \frac{1}{2 \cdot 4}x^2 \right.$$

$$\left. + \frac{1}{2 \cdot 4 \cdot 6}x^3 + \cdots \right]$$

23. $r_1 = 0, r_2 = -1$;

$$y(x) = C_1 x^{-1} \sum_{n=0}^{\infty} \frac{1}{(2n)!}x^{2n}$$

$$+ C_2 x^{-1} \sum_{n=0}^{\infty} \frac{1}{(2n+1)!}x^{2n+1}$$

$$= \frac{1}{x}[C_1 \cosh x + C_2 \sinh x]$$

25. $r_1 = 4, r_2 = 0$;

$$y(x) = C_1 \left[1 + \frac{2}{3}x + \frac{1}{3}x^2 \right]$$

$$+ C_2 \sum_{n=0}^{\infty} (n+1)x^{n+4}$$

27. $r_1 = r_2 = 0$;

$$y(x) = C_1 y_1(x) + C_2 \left[y_1(x) \ln x + y_1(x) \right.$$

$$\times \left(-x + \frac{1}{4}x^2 - \frac{1}{3 \cdot 3!}x^3 \right.$$

$$\left. \left. + \frac{1}{4 \cdot 4!}x^4 - \cdots \right) \right],$$

where $y_1(x) = \sum_{n=0}^{\infty} \frac{1}{n!}x^n = e^x$

29. $r_1 = r_2 = 0$;

$$y(x) = C_1 y_1(x) + C_2[y_1(x) \ln x + y_1(x)$$

$$\times (2x + \frac{5}{4}x^2 + \frac{23}{27}x^3 + \cdots)],$$

where $y_1(x) = \sum_{n=0}^{\infty} \frac{(-1)^n}{(n!)^{2n}}x^n$

31. $r_1 = r_2 = 1$;

$y(x) = C_1 xe^{-x} + C_2 xe^{-x}$

$\times \left[\ln x + x + \dfrac{1}{4}x^2 + \dfrac{1}{3 \cdot 3!}x^3 + \cdots \right]$

33. $r_1 = 2$, $r_2 = 0$;

$y(x) = C_1 x^2 + C_2 \left[\dfrac{1}{2}x^2 \ln x - \dfrac{1}{2} + x \right.$

$\left. - \dfrac{1}{3!}x^3 + \cdots \right]$

35. The method of Frobenius yields only the trivial solution $y(x) \equiv 0$.

37. There is a regular singular point at ∞.

39. There is a regular singular point at ∞.

41. The assumption $y = \displaystyle\sum_{n=0}^{\infty} c_n x^{n+r}$ leads to $c_n[(n+r)(n+r+2) - 8] = 0$ for $n \geq 0$. For $n = 0$ and $c_0 \neq 0$ we have $r^2 + 2r - 8 = 0$ and so $r_1 = 2$, $r_2 = -4$. For these values we are forced to take $c_n = 0$ for $n > 0$. Hence a solution exists of the form $y = c_0 x^r$. It follows that the general solution on $0 < x < \infty$ is $y = C_1 x^2 + C_2 x^{-4}$.

Exercises 6.4, Page 259

1. $y = c_1 J_{1/3}(x) + c_2 J_{-1/3}(x)$

3. $y = c_1 J_{5/2}(x) + c_2 J_{-5/2}(x)$

5. $y = c_1 J_0(x) + c_2 J_0(x) \displaystyle\int \dfrac{dx}{x J_0^2(x)}$

7. If we let $t = \lambda x$, the differential equation becomes

$$t^2 y'' + ty' + (t^2 - \nu^2)y = 0.$$

Since the solution of the last equation is

$$y = c_1 J_\nu(t) + c_2 J_{-\nu}(t)$$

we find

$$y = c_1 J_\nu(\lambda x) + c_2 J_{-\nu}(\lambda x), \quad \nu \neq \text{integer}.$$

9. After using the change of variables, the differential equation becomes

$$x^2 v'' + xv' + (\lambda^2 x^2 - \tfrac{1}{4})v = 0.$$

Since the solution of the last equation is

$$v = c_1 J_{1/2}(\lambda x) + c_2 J_{-1/2}(\lambda x)$$

we find

$y = c_1 x^{-1/2} J_{1/2}(\lambda x) + c_2 x^{-1/2} J_{-1/2}(\lambda x)$.

11. After substituting into the differential equation we find

$xy'' + (1 + 2n)y' + xy$

$= x^{-n-1}[x^2 J_n'' + x J_n' + (x^2 - n^2)J_n]$

$= x^{-n-1} \cdot 0 = 0$.

13. From Problem 10 with $n = 1/2$ we find $y = x^{1/2} J_{1/2}(x)$; from Problem 11 with $n = -1/2$ we find $y = x^{1/2} J_{-1/2}(x)$.

15. From Problem 10 with $n = -1$ we find $y = x^{-1} J_{-1}(x)$; from Problem 11 with $n = 1$ we find $y = x^{-1} J_1(x)$ but since $J_{-1}(x) = -J_1(x)$ no new solution results.

17. From Problem 12 with $\lambda = 1$ and $\nu = \pm 3/2$ we find $y = \sqrt{x} J_{3/2}(x)$ and $y = \sqrt{x} J_{-3/2}(x)$.

19. Using the hint we can write

$xJ_\nu'(x) = -\nu \displaystyle\sum_{n=0}^{\infty} \dfrac{(-1)^n}{n! \, \Gamma(1+\nu+n)} \left(\dfrac{x}{2}\right)^{2n+\nu}$

$+ 2 \displaystyle\sum_{n=0}^{\infty} \dfrac{(-1)^n(n+\nu)}{n!(n+\nu)\Gamma(n+\nu)} \left(\dfrac{x}{2}\right)^{2n+\nu}$

$= -\nu \displaystyle\sum_{n=0}^{\infty} \dfrac{(-1)^n}{n! \, \Gamma(1+\nu+n)} \left(\dfrac{x}{2}\right)^{2n+\nu}$

$+ x \displaystyle\sum_{n=0}^{\infty} \dfrac{(-1)^n}{n! \, \Gamma(n+\nu)} \left(\dfrac{x}{2}\right)^{2n+\nu-1}$

$= -\nu J_\nu(x) + x J_{\nu-1}(x)$.

21. Subtracting the equations

$$xJ_\nu'(x) = \nu J_\nu(x) - xJ_{\nu+1}(x)$$
$$xJ_\nu'(x) = -\nu J_\nu(x) + xJ_{\nu-1}(x)$$

gives $2\nu J_\nu(x) = xJ_{\nu+1}(x) + xJ_{\nu-1}(x)$.

23. The result from the given example

$$xJ_\nu'(x) - \nu J_\nu(x) = -xJ_{\nu+1}(x)$$

is a linear first-order differential equation in $J_\nu(x)$. Dividing by x we find the integrating factor is $x^{-\nu}$. It follows from this theory that after multiplying both sides of the equation by the integrating factor we must have

$$\dfrac{d}{dx}[x^{-\nu} J_\nu(x)] = -x^{-\nu} J_{\nu+1}(x).$$

25. From Problem 22, $\dfrac{d}{dr}[rJ_1(r)] = rJ_0(r)$.

Therefore

$$\int_0^x rJ_0(r)\,dr = \int_0^x \frac{d}{dr}[rJ_1(r)]\,dr = \left. rJ_1(r)\right|_0^x$$

$$= xJ_1(x)$$

27. $J_{-1/2}(x) = \sqrt{\dfrac{2}{\pi x}}\ \cos x$

29. $J_{-3/2}(x) = \sqrt{\dfrac{2}{\pi x}}\left[-\sin x - \dfrac{\cos x}{x}\right]$

31. $J_{-5/2}(x) = \sqrt{\dfrac{2}{\pi x}}\left[\dfrac{3}{x}\sin x \right.$

$$\left. + \left(\frac{3}{x^2} - 1\right)\cos x\right]$$

33. $J_{-7/2}(x) =$

$$\sqrt{\frac{2}{\pi x}}\left[\left(1 - \frac{15}{x^2}\right)\sin x + \left(\frac{6}{x} - \frac{15}{x^3}\right)\cos x\right]$$

35. $y = c_1 I_\nu(x) + c_2 I_{-\nu}(x),\ \nu \neq$ integer

37. Since $1/\Gamma(1 - m + n) = 0$ when $n \leq m - 1$, m a positive integer,

$$J_{-m}(x) = \sum_{n=0}^{\infty} \frac{(-1)^n}{n!\,\Gamma(1 - m + n)}\left(\frac{x}{2}\right)^{2n-m}$$

$$= \sum_{n=m}^{\infty} \frac{(-1)^n}{n!\,\Gamma(1 - m + n)}\left(\frac{x}{2}\right)^{2n-m}$$

$$= \sum_{k=0}^{\infty} \frac{(-1)^{k+m}}{(k+m)!\,\Gamma(1 + k)}\left(\frac{x}{2}\right)^{2k+m}\ (n = k + m)$$

$$= (-1)^m \sum_{k=0}^{\infty} \frac{(-1)^k}{\Gamma(1 + k + m)k!}\left(\frac{x}{2}\right)^{2k+m}$$

$$= (-1)^m J_m(x)$$

39.(a) $P_5(x) = \frac{1}{8}(63x^5 - 70x^3 + 15x)$
$P_6(x) = \frac{1}{16}(231x^6 - 315x^4 + 105x^2 - 5)$
 (b) $y = P_5(x)$ satisfies $(1 - x^2)y'' - 2xy' + 30y = 0.$
$y = P_6(x)$ satisfies $(1 - x^2)y'' - 2xy' + 42y = 0.$

41. If $x = \cos\theta$ then $\dfrac{dy}{d\theta} = \dfrac{dy}{dx}\dfrac{dx}{d\theta} = -\sin\theta\dfrac{dy}{dx}$

and $\dfrac{d^2y}{d\theta^2} = \sin^2\theta\dfrac{d^2y}{dx^2} - \cos\theta\dfrac{dy}{dx}$. Now the orig-

inal equation can be written as

$$\frac{d^2y}{d\theta^2} + \frac{\cos\theta}{\sin\theta}\frac{dy}{d\theta} + n(n+1)y = 0$$

and so

$$\sin^2\theta\frac{d^2y}{dx^2} - 2\cos\theta\frac{dy}{dx} + n(n+1)y = 0.$$

since $x = \cos\theta$ and $\sin^2\theta = 1 - \cos^2\theta = 1 - x^2$ we obtain

$$(1 - x^2)\frac{d^2y}{dx^2} - 2x\frac{dy}{dx} + n(n+1)y = 0.$$

43. By the binomial theorem we have formally

$$(1 - 2xt + t^2)^{-1/2} =$$

$$1 + \frac{1}{2}(2xt - t^2) + \frac{1\cdot 3}{2^2 2!}$$

$$\times (2xt - t^2)^2 + \cdots.$$

Grouping by powers of t, we then find

$$(1 - 2xt + t^2)^{-1/2} = 1 \cdot t^0 + x\cdot t$$

$$+ \frac{1}{2}(3x^2 - 1)t^2 + \cdots$$

$$= P_0(x)t^0 + P_1(x)t + P_2(x)t^2 + \cdots.$$

45. For $k = 1$, $P_2(x) = \frac{1}{2}[3xP_1(x) - P_0(x)]$
$$= \frac{1}{2}(3x^2 - 1).$$

For $k = 2$, $P_3(x) = \frac{1}{3}[5xP_2(x) - 2P_1(x)]$
$$= \frac{1}{3}[5x\cdot\frac{1}{2}(3x^2 - 1) - 2x]$$
$$= \frac{1}{2}(5x^3 - 3x).$$

For $k = 3$, $P_4(x) = \frac{1}{4}[7xP_3(x) - 3P_2(x)]$
$$= \frac{1}{4}[7x\cdot\frac{1}{2}(5x^3 - 3x)$$
$$- \frac{3}{2}(3x^2 - 1)]$$
$$= \frac{1}{8}(35x^4 - 30x^2 + 3).$$

47. For $n = 0, 1, 2, 3$, the value of the integral is 2, 2/3, 2/5, and 2/7, respectively. In general

$$\int_{-1}^{1} P_n^2(x)\,dx = \frac{2}{2n + 1},\quad n = 0, 1, 2, \ldots$$

49. $c_0 = 0,\quad c_1 = 3/5,\quad c_2 = 0,\quad c_3 = 2/5$

51.(a) The equation becomes

$$r^2R''\Theta + 2rR'\Theta = -R\Theta'' - \frac{\cos\theta}{\sin\theta}R\Theta'.$$

By dividing both sides by $R\Theta$ we obtain

$$r^2\frac{R''}{R} + 2r\frac{R'}{R} = -\frac{\Theta''}{\Theta} - \frac{\cos\theta}{\sin\theta}\frac{\Theta'}{\Theta}$$

(b) If both sides of the equation are equal to $n(n+1)$, then the left side simplifies to

$$r^2R'' + 2rR' - n(n+1)R = 0$$

whereas the right side simplifies to

$$\sin\theta\,\Theta'' + \cos\theta\,\Theta' + n(n+1)\sin\theta\,\Theta = 0.$$

(c) The first equation in (b) is a Cauchy-Euler differential equation with general solution

$$R = c_1 r^{-n-1} + c_2 r^n.$$

By problem 41, a particular solution of the second equation in part (b) is

$$\Theta = P_n(\cos\theta).$$

Chapter 6 Test, Page 267

1. $y = c_1 x^{-1/3} + c_2 x^{1/2}$

3. $y(x) = c_1 x^2 + c_2 x^3 + x^4 - x^2 \ln x$

5. The singular points are $x = 0$, $x = -1 + \sqrt{3}i$, $x = -1 - \sqrt{3}i$. All other finite values of x, real or complex, are ordinary points.

7. RSP $x = 0$; ISP $x = 5$

9. RSP $x = -3$, $x = 3$; ISP $x = 0$

11. $|x| < \infty$

13. $y_1(x) = c_0\Big[1 - \frac{1}{3\cdot 2}x^3 + \frac{1}{6\cdot 5\cdot 3\cdot 2}x^6$

$$-\frac{1}{9\cdot 8\cdot 6\cdot 5\cdot 3\cdot 2}x^9 + \cdots\Big]$$

$y_2(x) = c_1\Big[x - \frac{1}{4\cdot 3}x^4 + \frac{1}{7\cdot 6\cdot 4\cdot 3}x^7$

$$-\frac{1}{10\cdot 9\cdot 7\cdot 6\cdot 4\cdot 3}x^{10} + \cdots\Big]$$

15. $y_1(x) = c_0[1 + \frac{3}{2}x^2 + \frac{1}{2}x^3 + \frac{5}{8}x^4 + \cdots]$
$y_2(x) = c_1[x + \frac{1}{2}x^3 + \frac{1}{4}x^4 + \cdots]$

17. $r_1 = 1$, $r_2 = -\frac{1}{2}$;

$y(x) = C_1 x\Big[1 + \frac{1}{5}x + \frac{1}{7\cdot 5\cdot 2}x^2$

$$+ \frac{1}{9\cdot 7\cdot 5\cdot 3\cdot 2}x^3 + \cdots\Big] +$$

$$C_2 x^{-1/2}\Big[1 - x - \frac{1}{2}x^2 - \frac{1}{3^2\cdot 2}x^3 - \cdots\Big]$$

19. $r_1 = 3$, $r_2 = 0$;
$y(x) = C_1 y_1(x)$

$$+ C_2\Big[-\frac{1}{36}y_1(x)\ln x + y_1(x)$$

$$\times\Big(-\frac{1}{3}\frac{1}{x^3} + \frac{1}{4}\frac{1}{x^2} + \frac{1}{16}\frac{1}{x} + \cdots\Big)\Big]$$

21. $r_1 = r_2 = 0$; $y(x) = C_1 e^x + C_2 e^x \ln x$

23. $y(x) = c_0\Big[1 - \frac{1}{2^2}x^2 + \frac{1}{2^4(1\cdot 2)^2}x^4$

$$-\frac{1}{2^6(1\cdot 2\cdot 3)^2}x^6 + \cdots\Big]$$

Chapter 6 Appendix, Page 269

1.(a) 24 **(b)** 720 **(c)** $4\sqrt{\pi}/3$
(d) $-8\sqrt{\pi}/15$

3. 0.297

5. $\Gamma(x) > \displaystyle\int_0^1 t^{x-1}e^{-t}\,dt > e^{-1}\int_0^1 t^{x-1}\,d$

$$= \frac{1}{xe}$$

for $x > 0$. As $x \to 0^+$, $1/x \to +\infty$.

Exercises 7.1, Page 283

1. $\dfrac{2}{s}e^{-s} - \dfrac{1}{s}$　　**3.** $\dfrac{1}{s^2} - \dfrac{1}{s^2}e^{-s}$

5. $\dfrac{e^7}{s-1}$　　**7.** $\dfrac{1}{(s-4)^2}$

9. $\dfrac{1}{s^2 + 2s + 2}$　　**11.** $\dfrac{s^2 - 1}{(s^2 + 1)^2}$

13. $\dfrac{48}{s^5}$　　**15.** $\dfrac{4}{s^2} - \dfrac{10}{s}$

17. $\dfrac{2}{s^3} + \dfrac{6}{s^2} - \dfrac{3}{s}$　　**19.** $\dfrac{6}{s^4} + \dfrac{6}{s^3} + \dfrac{3}{s^2} + \dfrac{1}{s}$

21. $\dfrac{1}{s} + \dfrac{1}{s-4}$　　**23.** $\dfrac{1}{s} + \dfrac{2}{s-2} + \dfrac{1}{s-4}$

25. $\dfrac{8}{s^3} - \dfrac{15}{s^2 + 9}$

27. Use $\sinh kt = \dfrac{e^{kt} - e^{-kt}}{2}$ to show that

$$\mathcal{L}\{\sinh kt\} = \dfrac{k}{s^2 - k^2}$$

29. $\dfrac{1}{2(s - 2)} - \dfrac{1}{2s}$ **31.** $\dfrac{2}{s^2 + 16}$

33. $\dfrac{1}{2}\left(\dfrac{s}{s^2 + 9} + \dfrac{s}{s^2 + 1}\right)$

35. $\dfrac{1}{2}\left(\dfrac{3}{s^2 + 9} - \dfrac{1}{s^2 + 1}\right)$

37. The result follows by letting $u = st$ in

$$\mathcal{L}\{t^\alpha\} = \int_0^\infty t^\alpha e^{-st}\, dt.$$

39. $\dfrac{\frac{1}{2}\Gamma(\frac{1}{2})}{s^{3/2}} = \dfrac{\sqrt{\pi}}{2s^{3/2}}$ **41.** $\frac{1}{2}t^2$

43. $1 + 3t + \frac{3}{2}t^2 + \frac{1}{6}t^3$ **45.** $t - 1 + e^{2t}$

47. $\frac{1}{4}e^{-t/4}$ **49.** $\cos\dfrac{t}{2}$ **51.** $\frac{1}{4}\sinh 4t$

53. $2 \cos 3t - 2 \sin 3t$

55. $\frac{1}{3} - \frac{1}{3}e^{-3t}$ **57.** $\frac{3}{4}e^{-3t} + \frac{1}{4}e^{t}$

59. $-\frac{1}{3}e^{-t} + \frac{8}{15}e^{2t} - \frac{1}{5}e^{-3t}$

61. $\frac{1}{4}t - \frac{1}{8}\sin 2t$

63. $\frac{1}{4}e^{-2t} + \frac{1}{4}\cos 2t + \frac{1}{4}\sin 2t$

65. $\dfrac{1}{s}$

67. On $0 \le t \le 1$, $e^{-st} \ge e^{-s}(s > 0)$. Therefore

$$\int_0^1 e^{-st}\dfrac{1}{t^2}\, dt \ge e^{-s}\int_0^1 \dfrac{1}{t^2}\, dt.$$

The latter integral diverges.

Exercises 7.2, Page 300

1. $\dfrac{1}{(s - 10)^2}$ **3.** $\dfrac{6}{(s + 2)^4}$

5. $\dfrac{3}{(s - 1)^2 + 9}$ **7.** $\dfrac{3}{(s - 5)^2 - 9}$

9. $\dfrac{1}{(s - 2)^2} + \dfrac{2}{(s - 3)^2} + \dfrac{1}{(s - 4)^2}$

11. $\dfrac{1}{2}\left[\dfrac{1}{(s + 1)} - \dfrac{s + 1}{(s + 1)^2 + 4}\right]$

13. $\frac{1}{2}t^2e^{-2t}$ **15.** $e^{3t} \sin t$

17. $e^{-2t} \cos t - 2e^{-2t} \sin t$

19. $e^{-t} - te^{-t}$

21. $5 - t - 5e^{-t} - 4te^{-t} - \frac{3}{2}t^2e^{-t}$

23. $\dfrac{e^{-s}}{s^2}$ **25.** $\dfrac{e^{-2s}}{s^2} + 2\dfrac{e^{-s}}{s}$

27. $\frac{1}{2}(t - 2)^2 \mathcal{U}(t - 2)$

29. $-\sin t\,\mathcal{U}(t - \pi)$

31. $\mathcal{U}(t - 1) - e^{-(t-1)}\mathcal{U}(t - 1)$

33. $\dfrac{s^2 - 4}{(s^2 + 4)^2}$ **35.** $\dfrac{6s^2 + 2}{(s^2 - 1)^3}$

37. $\dfrac{12s - 24}{[(s - 2)^2 + 36]^2}$ **39.** $\frac{1}{2}t \sin t$

41. $f(t) = t^2\mathcal{U}(t - 1)$
$= (t - 1)^2\mathcal{U}(t - 1)$
$\quad + 2(t - 1)\mathcal{U}(t - 1) + \mathcal{U}(t - 1);$

$$\mathcal{L}\{f(t)\} = \dfrac{e^{-s}}{s^3} + 2\dfrac{e^{-s}}{s^2} + \dfrac{e^{-s}}{s}$$

43. $f(t) = t - t\mathcal{U}(t - 2)$
$= t - (t - 2)\mathcal{U}(t - 2)$
$\quad - 2\mathcal{U}(t - 2);$

$$\mathcal{L}\{f(t)\} = \dfrac{1}{s^2} - \dfrac{e^{-2s}}{s^2} - 2\dfrac{e^{-2s}}{s}$$

45. $f(t) = \mathcal{U}(t - a) - \mathcal{U}(t - b);$

$$\mathcal{L}\{f(t)\} = \dfrac{e^{-as}}{s} - \dfrac{e^{-bs}}{s}$$

47.

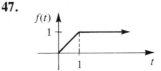

49. $\dfrac{e^{-t} - e^{3t}}{t}$ **51.** $\dfrac{\sin 2t}{t}$

53. Since $f'(t) = e^t$, $f(0) = 1$ it follows from Theorem 7.8 that $\mathcal{L}\{e^t\} = s\mathcal{L}\{e^t\} - 1$. Solving gives $\mathcal{L}\{e^t\} = 1/(s - 1)$.

55. $\dfrac{s+1}{s[(s+1)^2+1]}$ **57.** $\dfrac{1}{s^2(s-1)}$

59. $\dfrac{3s^2+1}{s^2(s^2+1)^2}$ **61.** $\dfrac{6}{s^5}$ **63.** $\dfrac{48}{s^8}$

65. $\dfrac{s-1}{(s+1)[(s-1)^2+1]}$

67. $1-e^{-t}$ **69.** $-\frac{1}{3}e^{-t}+\frac{1}{3}e^{2t}$

71. $\frac{1}{4}t\sin 2t$

73. $\dfrac{(1-e^{-as})^2}{s(1-e^{-2as})}=\dfrac{1-e^{-as}}{s(1+e^{-as})}$

75. $\dfrac{a}{s}\left(\dfrac{1}{bs}-\dfrac{1}{e^{bs}-1}\right)$ **77.** $\dfrac{\coth \pi s/2}{s^2+1}$

79. $\dfrac{1}{s^2+1}$

81. The result follows from letting $u=t-\tau$ in the first integral.

83. The result follows from $\cosh at = (e^{at}+e^{-at})/2$ and the first translation theorem.

85. The result follows from letting $u=at$, $a>0$.

87. $\ln\dfrac{s+1}{s-1}$

Exercises 7.3, Page 312

1. $y=-1+e^t$ **3.** $y=te^{-4t}+2e^{-4t}$

5. $y=\frac{4}{3}e^{-t}-\frac{1}{3}e^{-4t}$

7. $y=\frac{1}{9}t+\frac{2}{27}-\frac{2}{27}e^{3t}+\frac{10}{9}te^{3t}$

9. $y=\frac{1}{20}t^5e^{2t}$

11. $y=\cos t-\frac{1}{2}\sin t-\frac{1}{2}t\cos t$

13. $y=\frac{1}{2}-\frac{1}{2}e^t\cos t+\frac{1}{2}e^t\sin t$

15. $y=-\frac{8}{9}e^{-t/2}+\frac{1}{9}e^{-2t}+\frac{5}{18}e^t+\frac{1}{2}e^{-t}$

17. $y=\cos t$

19. $y=[5-5e^{-(t-1)}]\mathcal{U}(t-1)$

21. $y=-\frac{1}{4}+\frac{1}{2}t+\frac{1}{4}e^{-2t}-\frac{1}{4}\mathcal{U}(t-1)$
$\qquad -\frac{1}{2}(t-1)\mathcal{U}(t-1)$
$\qquad +\frac{1}{4}e^{-2(t-1)}\mathcal{U}(t-1)$

23. $y=\cos 2t-\frac{1}{6}\sin 2(t-2\pi)\mathcal{U}(t-2\pi)$
$\qquad +\frac{1}{3}\sin (t-2\pi)\mathcal{U}(t-2\pi)$

25. $y=\sin t+[1-\cos(t-\pi)]\mathcal{U}(t-\pi)$
$\qquad -[1-\cos(t-2\pi)]\mathcal{U}(t-2\pi)$

27. $y=\frac{1}{3}t^3+\frac{1}{2}ct^2$

29. $y=c\displaystyle\sum_{n=0}^{\infty}\dfrac{(-1)^nt^{2n}}{2^{2n}(n!)^2}$, the student may recognize this as the Bessel function $J_0(t)$.

31. $f(t)=\sin t$

33. $f(t)=-\frac{1}{8}e^{-t}+\frac{1}{8}e^t+\frac{3}{4}te^t+\frac{1}{4}t^2e^t$

35. $f(t)=e^{-t}$

37. $f(t)=\frac{3}{8}e^{2t}+\frac{1}{8}e^{-2t}+\frac{1}{2}\cos 2t+\frac{1}{4}\sin 2t$

39. $y=\sin t-\frac{1}{2}t\sin t$

41. $i(t)=20{,}000[te^{-100t}$
$\qquad -(t-1)e^{-100(t-1)}\mathcal{U}(t-1)]$

43. $i(t)=\dfrac{1}{101}e^{-10t}-\dfrac{1}{101}\cos t+\dfrac{10}{101}\sin t$

$\qquad -\dfrac{10}{101}e^{-10(t-3\pi/2)}\mathcal{U}\left(t-\dfrac{3\pi}{2}\right)$

$\qquad +\dfrac{10}{101}\cos\left(t-\dfrac{3\pi}{2}\right)\mathcal{U}\left(t-\dfrac{3\pi}{2}\right)$

$\qquad +\dfrac{1}{101}\sin\left(t-\dfrac{3\pi}{2}\right)\mathcal{U}\left(t-\dfrac{3\pi}{2}\right)$

45. $i(t)=\dfrac{t}{R}+\dfrac{L}{R^2}(e^{-Rt/L}-1)$

$\qquad +\dfrac{1}{R}\displaystyle\sum_{n=1}^{\infty}(e^{-R(t-n)/L}-1)\mathcal{U}(t-n)$;

For $0\le t<2$,

$i(t)=\begin{cases}\dfrac{t}{R}+\dfrac{L}{R^2}(e^{-Rt/L}-1),\ \ 0\le t<1\\[2mm]\dfrac{t}{R}+\dfrac{L}{R^2}(e^{-Rt/L}-1)\\[2mm]\qquad +\dfrac{1}{R}(e^{-R(t-1)/L}-1),\ \ 1\le t<2\end{cases}$

47. $q(t)=\frac{3}{5}e^{-10t}+6te^{-10t}-\frac{3}{5}\cos 10t$;
$\qquad i(t)=-60te^{-10t}+6\sin 10t$;
steady-state current is $6\sin 10t$

49. $x(t)=-\dfrac{3}{2}e^{-7t/2}\cos\dfrac{\sqrt{15}}{2}t$

$\qquad -\dfrac{7\sqrt{15}}{10}e^{-7t/2}\sin\dfrac{\sqrt{15}}{2}t$

Exercises 7.4, Page 319

1. $y = e^{3(t-2)}\mathcal{U}(t - 2)$

3. $y = \sin t + \sin t\,\mathcal{U}(t - 2\pi)$

5. $y = -\cos t\,\mathcal{U}\left(t - \dfrac{\pi}{2}\right) + \cos t\,\mathcal{U}\left(t - \dfrac{3\pi}{2}\right)$

7. $y = \frac{1}{2} - \frac{1}{2}e^{-2t} + [\frac{1}{2} - \frac{1}{2}e^{-2(t-1)}]\mathcal{U}(t - 1)$

9. $y = e^{-2(t-2\pi)}\sin t\,\mathcal{U}(t - 2\pi)$

11. $\mathcal{L}\{\delta_a(t - t_0)\} = \displaystyle\int_{t_0-a}^{t_0+a} \frac{1}{2a}e^{-st}\,dt$

$\qquad = \dfrac{1}{2a}\left(-\dfrac{e^{-st}}{s}\right)\Big|_{t_0-a}^{t_0+a}$

$\qquad = \dfrac{-1}{2sa}[e^{-s(t_0 + a)} - e^{-s(t_0-a)}]$

$\qquad = \dfrac{e^{-st_0}}{2sa}(e^{sa} - e^{-sa})$

13. $y = e^{-t}\cos t + e^{-(t-3\pi)}\sin t\,\mathcal{U}(t - 3\pi)$

15. $i(t) = \dfrac{1}{L}e^{-Rt/L};\quad$ no

Chapter 7 Test, Page 323

1. $\dfrac{1}{s^2} - \dfrac{2}{s^2}e^{-s}$ **3.** $\dfrac{1}{s + 7}$

5. $\dfrac{2}{s^2 + 4}$ **7.** $\dfrac{4s}{(s^2 + 4)^2}$

9. $\frac{1}{6}t^5$ **11.** $\frac{1}{2}t^2e^{5t}$

13. $e^{5t}\cos 2t + \frac{5}{2}e^{5t}\sin 2t$

15. $y = 5te^t + \frac{1}{2}t^2e^t$

17. $y = -\frac{2}{125} - \frac{2}{25}t - \frac{1}{5}t^2 + \frac{127}{125}e^{5t}$
$\qquad - [-\frac{37}{125} - \frac{12}{25}(t - 1) - \frac{1}{5}(t - 1)^2$
$\qquad + \frac{37}{125}e^{5(t-1)}]\mathcal{U}(t - 1)$

19. $y = 1 + t + \frac{1}{2}t^2$

Exercises 8.1, Page 331

1. $x = c_1e^t + c_2te^t$
$\qquad y = (c_1 - c_2)e^t + c_2te^t$

3. $x = c_1\cos t + c_2\sin t + t + 1$
$\qquad y = c_1\sin t - c_2\cos t + t - 1$

5. $x = \frac{1}{2}c_1\sin t + \frac{1}{2}c_2\cos t - 2c_3\sin\sqrt{6}t$
$\qquad -2c_4\cos\sqrt{6}t$
$\qquad y = c_1\sin t + c_2\cos t + c_3\sin\sqrt{6}t$
$\qquad +c_4\cos\sqrt{6}t$

7. $x = c_1e^{2t} + c_2e^{-2t} + c_3\sin 2t$
$\qquad + c_4\cos 2t + \frac{1}{5}e^t$
$\qquad y = c_1e^{2t} + c_2e^{-2t} - c_3\sin 2t$
$\qquad - c_4\cos 2t - \frac{1}{5}e^t$

11. $x = c_1e^t + c_2e^{-t/2}\cos\dfrac{\sqrt{3}}{2}t$
$\qquad + c_3e^{-t/2}\sin\dfrac{\sqrt{3}}{2}t$
$\qquad y = \left(-\dfrac{3}{2}c_2 - \dfrac{\sqrt{3}}{2}c_3\right)e^{-t/2}\cos\dfrac{\sqrt{3}}{2}t$
$\qquad + \left(\dfrac{\sqrt{3}}{2}c_2 - \dfrac{3}{2}c_3\right)e^{-t/2}\sin\dfrac{\sqrt{3}}{2}t$

13. $x = c_1e^{4t} + \frac{4}{3}e^t$
$\qquad y = -\frac{3}{4}c_1e^{4t} + c_2 + 5e^t$

15. $x = c_1 + c_2t + c_3e^t + c_4e^{-t} - \frac{1}{2}t^2$
$\qquad y = (c_1 - c_2 + 2) + (c_2 + 1)t + c_4e^{-t} - \frac{1}{2}t^2$

17. $x = c_1e^t + c_2e^{-t/2}\sin\dfrac{\sqrt{3}}{2}t$
$\qquad + c_3e^{-t/2}\cos\dfrac{\sqrt{3}}{2}t$
$\qquad y = c_1e^t + \left(-\dfrac{1}{2}c_2 - \dfrac{\sqrt{3}}{2}c_3\right)e^{-t/2}\sin\dfrac{\sqrt{3}}{2}t$
$\qquad + \left(\dfrac{\sqrt{3}}{2}c_2 - \dfrac{1}{2}c_3\right)e^{-t/2}\cos\dfrac{\sqrt{3}}{2}t$
$\qquad z = c_1e^t + \left(-\dfrac{1}{2}c_2 + \dfrac{\sqrt{3}}{2}c_3\right)e^{-t/2}\sin\dfrac{\sqrt{3}}{2}t$
$\qquad + \left(-\dfrac{\sqrt{3}}{2}c_2 - \dfrac{1}{2}c_3\right)e^{-t/2}\cos\dfrac{\sqrt{3}}{2}t$

19. $x = -6c_1e^{-t} - 3c_2e^{-2t} + 2c_3e^{3t}$
$\qquad y = c_1e^{-t} + c_2e^{-2t} + c_3e^{3t}$
$\qquad z = 5c_1e^{-t} + c_2e^{-2t} + c_3e^{3t}$

21. $x = -c_1e^{-t} + c_2 + \frac{1}{3}t^3 - 2t^2 + 5t$
$\qquad y = c_1e^{-t} + 2t^2 - 5t + 5$

23. $Dx - Dy = 0,\quad (D - 1)x - y = 0$

Exercises 8.2, Page 338

1. $x = -\frac{1}{3}e^{-2t} + \frac{1}{3}e^{t}$
$y = \frac{1}{3}e^{-2t} + \frac{2}{3}e^{t}$

3. $x = -\cos 3t - \frac{5}{3}\sin 3t$
$y = 2\cos 3t - \frac{7}{3}\sin 3t$

5. $x = -2e^{3t} + \frac{5}{2}e^{2t} - \frac{1}{2}$
$y = \frac{8}{3}e^{3t} - \frac{5}{2}e^{2t} - \frac{1}{6}$

7. $x = -\frac{1}{2}t - \frac{3}{4}\sqrt{2}\sin\sqrt{2}t$
$y = -\frac{1}{2}t + \frac{3}{4}\sqrt{2}\sin t$

9. $x = 8 + \dfrac{2}{3!}t^3 + \dfrac{1}{4!}t^4$

$y = -\dfrac{2}{3!}t^3 + \dfrac{1}{4!}t^4$

11. $x = \frac{1}{2}t^2 + t + 1 - e^{-t}$
$y = -\frac{1}{3} + \frac{1}{3}e^{-t} + \frac{1}{3}te^{-t}$

13. $x_1 = \dfrac{3}{5}\sin t + \dfrac{\sqrt{6}}{15}\sin\sqrt{6}t + \dfrac{2}{5}\cos t$

$\quad - \dfrac{2}{5}\cos\sqrt{6}t$

$x_2 = \dfrac{6}{5}\sin t - \dfrac{\sqrt{6}}{30}\sin\sqrt{6}t + \dfrac{4}{5}\cos t$

$\quad + \dfrac{1}{5}\cos\sqrt{6}t$

15. $i_2 = \frac{100}{9} - \frac{100}{9}e^{-900t}$
$i_3 = \frac{80}{9} - \frac{80}{9}e^{-900t}$
$i_1 = 20 - 20e^{-900t}$

17. $i_t(t) = \frac{6}{5} - \frac{6}{5}e^{-100t}\cos 100t$
$i_2(t) = \frac{6}{5} - \frac{6}{5}e^{-100t}\cos 100t$
$\quad - \frac{6}{5}e^{-100t}\sin 100t$

19. $i_2 = -\frac{20}{13}e^{-2t} + \frac{375}{1469}e^{-15t} + \frac{145}{113}\cos t$
$\quad + \frac{85}{113}\sin t$
$i_3 = \frac{30}{13}e^{-2t} + \frac{250}{1469}e^{-15t} - \frac{280}{113}\cos t$
$\quad + \frac{810}{113}\sin t$

21. $\theta_1(t) = \dfrac{1}{4}\cos\dfrac{2}{\sqrt{3}}t + \dfrac{3}{4}\cos 2t$

$\theta_2(t) = \dfrac{1}{2}\cos\dfrac{2}{\sqrt{3}}t - \dfrac{3}{2}\cos 2t$

Exercises 8.3, Page 346

1. $x_1' = x_2$
$x_2' = -4x_1 + 3x_2 + \sin t$

3. $x_1' = x_2$
$x_2' = x_3$
$x_3' = 10x_1 - 6x_2 + 3x_3 + t^2 + 1$

5. $x_1' = x_2$
$x_2' = x_3$
$x_3' = x_4$
$x_4' = -x_1 - 4x_2 + 2x_3 + t$

7. $x_1' = x_2$
$x_2' = x_3$
$x_3' = -x_1^2 + x_1x_3 + 2t - 1$

9. $\dfrac{d^2y}{dt^2} - 7\dfrac{dy}{dt} + 13y = 3t$

11. $Dx = t^2 + 5t - 2$
$Dy = -x + 5t - 2$

13. The system is degenerate.

15. $Dx = u$
$Dy = v$
$Du = w$
$Dv = 10t^2 - 4u - 3v$
$Dw = 4x + 4v - 3w$

17. $\dfrac{dx_1}{dt} = 6 - \dfrac{2}{25}x_1 + \dfrac{1}{50}x_2$

$\dfrac{dx_2}{dt} = \dfrac{2}{25}x_1 - \dfrac{2}{25}x_2$

19. $a_1a_4 - a_2a_3 = 0$ or
$\begin{vmatrix} a_1 & a_2 \\ a_3 & a_4 \end{vmatrix} = 0$

Exercises 8.4, Page 356

1.(a) $\begin{pmatrix} 2 & 11 \\ 2 & -1 \end{pmatrix}$ **(b)** $\begin{pmatrix} -6 & 1 \\ 14 & -19 \end{pmatrix}$

(c) $\begin{pmatrix} 2 & 28 \\ 12 & -12 \end{pmatrix}$

3.(a) $\begin{pmatrix} -11 & 6 \\ 17 & -22 \end{pmatrix}$ **(b)** $\begin{pmatrix} -32 & 27 \\ -4 & -1 \end{pmatrix}$

(c) $\begin{pmatrix} 19 & -18 \\ -30 & 31 \end{pmatrix}$ (d) $\begin{pmatrix} 19 & 6 \\ 3 & 22 \end{pmatrix}$

5.(a) $\begin{pmatrix} 9 & 24 \\ 3 & 8 \end{pmatrix}$ (b) $\begin{pmatrix} 3 & 8 \\ -6 & -16 \end{pmatrix}$

(c) $\begin{pmatrix} 0 & 0 \\ 0 & 0 \end{pmatrix}$ (d) $\begin{pmatrix} -4 & -5 \\ 8 & 10 \end{pmatrix}$

7.(a) 180 (b) $\begin{pmatrix} 4 & 8 & 10 \\ 8 & 16 & 20 \\ 10 & 20 & 25 \end{pmatrix}$ (c) $\begin{pmatrix} 6 \\ 12 \\ -5 \end{pmatrix}$

9.(a) $\begin{pmatrix} 7 & 38 \\ 10 & 75 \end{pmatrix}$ (b) $\begin{pmatrix} 7 & 38 \\ 10 & 75 \end{pmatrix}$

11. $\begin{pmatrix} -14 \\ 1 \end{pmatrix}$ 13. $\begin{pmatrix} -38 \\ -2 \end{pmatrix}$ 15. singular

17. nonsingular; $A^{-1} = \frac{1}{4}\begin{pmatrix} -5 & -8 \\ 3 & 4 \end{pmatrix}$

19. nonsingular; $A^{-1} = \frac{1}{2}\begin{pmatrix} 0 & -1 & 1 \\ 2 & 2 & -2 \\ -4 & -3 & 5 \end{pmatrix}$

21. nonsingular;

$A^{-1} = -\frac{1}{9}\begin{pmatrix} -2 & -2 & -1 \\ -13 & 5 & 7 \\ 8 & -1 & -5 \end{pmatrix}$

23. det $A(t) = 2e^{3t} \neq 0$ for every value of t;

$A^{-1}(t) = \frac{1}{2e^{3t}}\begin{pmatrix} 3e^{4t} & -e^{4t} \\ -4e^{-t} & 2e^{-t} \end{pmatrix}$

25. $\lambda = 7$ or $\lambda = -1$ 27. $\frac{dX}{dt} = \begin{pmatrix} -5e^{-t} \\ -2e^{-t} \\ 7e^{-t} \end{pmatrix}$

29. $\frac{dX}{dt} = 4\begin{pmatrix} 1 \\ -1 \end{pmatrix}e^{2t} - 12\begin{pmatrix} 2 \\ 1 \end{pmatrix}e^{-3t}$

31.(a) $\begin{pmatrix} 4e^{4t} & -\pi \sin \pi t \\ 2 & 6t \end{pmatrix}$

(b) $\begin{pmatrix} \frac{1}{4}e^8 - \frac{1}{4} & 0 \\ 4 & 6 \end{pmatrix}$

(c) $\begin{pmatrix} \frac{1}{4}e^{4t} - \frac{1}{4} & (1/\pi) \sin \pi t \\ t^2 & t^3 - t \end{pmatrix}$

33. $\frac{d}{dt}\begin{pmatrix} a_{11}(t) & a_{12}(t) \\ a_{21}(t) & a_{22}(t) \end{pmatrix}\begin{pmatrix} x_1(t) \\ x_2(t) \end{pmatrix}$

$= \frac{d}{dt}\begin{pmatrix} a_{11}(t)x_1(t) + a_{12}(t)x_2(t) \\ a_{21}(t)x_1(t) + a_{22}(t)x_2(t) \end{pmatrix}$

$= \begin{pmatrix} a_{11}(t)x_1'(t) + a_{11}'(t)x_1(t) + a_{12}(t)x_2'(t) \\ a_{21}(t)x_1'(t) + a_{21}'(t)x_1(t) + a_{22}(t)x_2'(t) \end{pmatrix}$

$\begin{matrix} + a_{12}'(t)x_2(t) \\ + a_{22}'(t)x_2(t) \end{matrix}$

$= \begin{pmatrix} a_{11}(t)x_1'(t) + a_{12}(t)x_2'(t) + a_{11}'(t)x_1(t) \\ a_{21}(t)x_1'(t) + a_{22}(t)x_2'(t) + a_{21}'(t)x_1(t) \end{pmatrix}$

$\begin{matrix} + a_{12}'(t)x_2(t) \\ + a_{22}'(t)x_2(t) \end{matrix}$

$= \begin{pmatrix} a_{11}(t) & a_{12}(t) \\ a_{21}(t) & a_{22}(t) \end{pmatrix}\begin{pmatrix} x_1'(t) \\ x_2'(t) \end{pmatrix}$

$+ \begin{pmatrix} a_{11}'(t) & a_{12}'(t) \\ a_{21}'(t) & a_{22}'(t) \end{pmatrix}\begin{pmatrix} x_1(t) \\ x_2(t) \end{pmatrix}$

$= A(t)X'(t) + A'(t)X(t)$

35. Since A^{-1} exists $AB = AC$ implies $A^{-1}(AB) = A^{-1}(AC)$,

$(A^{-1}A)B = (A^{-1}A)C$, $IB = IC$ or $B = C$.

37. No, since in general $AB \neq BA$.

Exercises 8.5, Page 372

1. $X' = \begin{pmatrix} 3 & -5 \\ 4 & 8 \end{pmatrix}X$, where $X = \begin{pmatrix} x \\ y \end{pmatrix}$

3. $X' = \begin{pmatrix} -3 & 4 & -9 \\ 6 & -1 & 0 \\ 10 & 4 & 3 \end{pmatrix}X$,

where $X = \begin{pmatrix} x \\ y \\ z \end{pmatrix}$

5. $X' = \begin{pmatrix} 1 & -1 & 1 \\ 2 & 1 & -1 \\ 1 & 1 & 1 \end{pmatrix}X + \begin{pmatrix} 0 \\ -3t^2 \\ t^2 \end{pmatrix} +$

$$\begin{pmatrix} t \\ 0 \\ -t \end{pmatrix} + \begin{pmatrix} -1 \\ 0 \\ 2 \end{pmatrix}, \text{ where } \mathbf{X} = \begin{pmatrix} x \\ y \\ z \end{pmatrix}$$

7. $\dfrac{dx}{dt} = 4x + 2y + e^t$

$\dfrac{dy}{dt} = -x + 3y - e^t$

9. $\dfrac{dx}{dt} = x - y + 2z + e^{-t} - 3t$

$\dfrac{dy}{dt} = 3x - 4y + z + 2e^{-t} + t$

$\dfrac{dz}{dt} = -2x + 5y + 6z + 2e^{-t} - t$

11. $\dfrac{d\mathbf{X}}{dt} = \begin{pmatrix} -5e^{-5t} \\ -10e^{-5t} \end{pmatrix}$

$\begin{pmatrix} 3 & -4 \\ 4 & -7 \end{pmatrix}\mathbf{X} = \begin{pmatrix} 3 - 8 \\ 4 - 14 \end{pmatrix}e^{-5t} = \begin{pmatrix} -5 \\ -10 \end{pmatrix}e^{-5t}$

$= \dfrac{d\mathbf{X}}{dt}$

13. $\mathbf{X}' = \begin{pmatrix} \frac{3}{2} \\ 3 \end{pmatrix}$

$\begin{pmatrix} -1 & \frac{1}{4} \\ 1 & -1 \end{pmatrix}\mathbf{X} = \begin{pmatrix} 1 + \frac{1}{2} \\ 1 + 2 \end{pmatrix}e^{-3t/2} = \begin{pmatrix} \frac{3}{2} \\ 3 \end{pmatrix}e^{-3t/2}$

$= \mathbf{X}'$

15. $\dfrac{d\mathbf{X}}{dt} = \begin{pmatrix} 0 \\ 0 \\ 0 \end{pmatrix}$

$\begin{pmatrix} 1 & 2 & 1 \\ 6 & -1 & 0 \\ -1 & -2 & -1 \end{pmatrix}\mathbf{X} = \begin{pmatrix} 1 + 12 - 13 \\ 6 - 6 \\ -1 - 12 + 13 \end{pmatrix}$

$= \begin{pmatrix} 0 \\ 0 \\ 0 \end{pmatrix} = \dfrac{d\mathbf{X}}{dt}$

17. Yes; $W(\mathbf{X}_1, \mathbf{X}_2) = -2e^{-8t} \neq 0$ implies $\mathbf{X}_1$ and $\mathbf{X}_2$ are linearly independent on $-\infty < t < \infty$.

19. No; $W(\mathbf{X}_1, \mathbf{X}_2, \mathbf{X}_3) =$

$$\begin{vmatrix} 1 + t & 1 & 3 + 2t \\ -2 + 2t & -2 & -6 + 4t \\ 4 + 2t & 4 & 12 + 4t \end{vmatrix} = 0 \text{ for every } t.$$

The solution vectors are linearly dependent on $-\infty < t < \infty$. Note that $\mathbf{X}_3 = 2\mathbf{X}_1 + \mathbf{X}_2$.

21. $\dfrac{d\mathbf{X}_p}{dt} = \begin{pmatrix} 2 \\ -1 \end{pmatrix}$

$\begin{pmatrix} 1 & 4 \\ 3 & 2 \end{pmatrix}\mathbf{X}_p + \begin{pmatrix} 2 \\ -4 \end{pmatrix}t - \begin{pmatrix} 7 \\ 18 \end{pmatrix}$

$= \begin{pmatrix} (2 - 4)t + 9 + 2t - 7 \\ (6 - 2)t + 17 - 4t - 18 \end{pmatrix}$

$= \begin{pmatrix} 2 \\ -1 \end{pmatrix} = \dfrac{d\mathbf{X}_p}{dt}$

23. $\mathbf{X}_p' = \begin{pmatrix} 2e^t + te^t \\ -te^t \end{pmatrix}$

$\begin{pmatrix} 2 & 1 \\ 3 & 4 \end{pmatrix}\mathbf{X}_p - \begin{pmatrix} 1 \\ 7 \end{pmatrix}e^t = \begin{pmatrix} 3e^t + te^t - e^t \\ 7e^t - te^t - 7e^t \end{pmatrix}$

$= \begin{pmatrix} 2e^t + te^t \\ -te^t \end{pmatrix} = \mathbf{X}_p'$

25. Let $\mathbf{X}_1 = \begin{pmatrix} 6 \\ -1 \\ -5 \end{pmatrix}e^{-t}$, $\mathbf{X}_2 = \begin{pmatrix} -3 \\ 1 \\ 1 \end{pmatrix}e^{-2t}$,

$\mathbf{X}_3 = \begin{pmatrix} 2 \\ 1 \\ 1 \end{pmatrix}e^{3t}$, and $\mathbf{A} = \begin{pmatrix} 0 & 6 & 0 \\ 1 & 0 & 1 \\ 1 & 1 & 0 \end{pmatrix}$. Then

$\mathbf{AX}_1 = \begin{pmatrix} -6 \\ 1 \\ 5 \end{pmatrix}e^{-t} = \mathbf{X}_1'$,

$\mathbf{AX}_2 = \begin{pmatrix} 6 \\ -2 \\ -2 \end{pmatrix}e^{-2t} = \mathbf{X}_2'$,

$\mathbf{AX}_3 = \begin{pmatrix} 6 \\ 3 \\ 3 \end{pmatrix}e^{3t} = \mathbf{X}_3'$ and

$W(\mathbf{X}_1, \mathbf{X}_2, \mathbf{X}_3) = \begin{vmatrix} 6e^{-t} & -3e^{-2t} & 2e^{3t} \\ -e^{-t} & e^{-2t} & e^{3t} \\ -5e^{-t} & e^{-2t} & e^{3t} \end{vmatrix}$

$= 20 \neq 0.$

Therefore $\mathbf{X}_1$, $\mathbf{X}_2$, $\mathbf{X}_3$ form a fundamental set of solutions of $\mathbf{X}' = \mathbf{A}\mathbf{X}$ on $-\infty < t < \infty$. By definition

$$\mathbf{X} = c_1\mathbf{X}_1 + c_2\mathbf{X}_2 + c_3\mathbf{X}_3$$

is the general solution.

27. $\boldsymbol{\Phi}(t) = \begin{pmatrix} e^{2t} & e^{7t} \\ -2e^{2t} & 3e^{7t} \end{pmatrix}$,

$\boldsymbol{\Phi}^{-1}(t) = \dfrac{1}{5e^{9t}}\begin{pmatrix} 3e^{7t} & -e^{7t} \\ 2e^{2t} & e^{2t} \end{pmatrix}$

29. $\boldsymbol{\Phi}(t) = \begin{pmatrix} -e^t & -te^t \\ 3e^t & 3te^t - e^t \end{pmatrix}$,

$\boldsymbol{\Phi}^{-1}(t) = \dfrac{1}{e^{2t}}\begin{pmatrix} 3te^t - e^t & te^t \\ -3e^t & -e^t \end{pmatrix}$

31. $\boldsymbol{\Psi}(t) = \begin{pmatrix} \frac{3}{5}e^{2t} + \frac{2}{5}e^{7t} & -\frac{1}{5}e^{2t} + \frac{1}{5}e^{7t} \\ -\frac{6}{5}e^{2t} + \frac{6}{5}e^{7t} & \frac{2}{5}e^{2t} + \frac{3}{5}e^{7t} \end{pmatrix}$

33. $\boldsymbol{\Psi}(t) = \begin{pmatrix} 3te^t + e^t & te^t \\ -9te^t & -3te^t + e^t \end{pmatrix}$

35. $\mathbf{X}(t_0) = \boldsymbol{\Phi}(t_0)\,\mathbf{C}$ implies $\mathbf{C} = \boldsymbol{\Phi}^{-1}(t_0)\,\mathbf{X}(t_0)$. Substituting in $\mathbf{X} = \boldsymbol{\Phi}(t)\,\mathbf{C}$ gives $\mathbf{X} = \boldsymbol{\Phi}(t) \times \boldsymbol{\Phi}^{-1}(t_0)\,\mathbf{X}_0$.

37. Comparing $\mathbf{X} = \boldsymbol{\Phi}(t)\,\boldsymbol{\Phi}^{-1}(t_0)\,\mathbf{X}_0$ and $\mathbf{X} = \boldsymbol{\Psi}(t)\,\mathbf{X}_0$ implies $[\boldsymbol{\Psi}(t) - \boldsymbol{\Phi}(t)\,\boldsymbol{\Phi}^{-1}(t_0)]\mathbf{X}_0 = \mathbf{0}$. Since this last equation is to hold for any $\mathbf{X}_0$ we conclude $\boldsymbol{\Psi}(t) = \boldsymbol{\Phi}(t)\,\boldsymbol{\Phi}^{-1}(t_0)$.

Exercises 8.6, Page 389

1. $\mathbf{X} = c_1\begin{pmatrix} 1 \\ 2 \end{pmatrix}e^{5t} + c_2\begin{pmatrix} 1 \\ -1 \end{pmatrix}e^{-t}$

3. $\mathbf{X} = c_1\begin{pmatrix} 2 \\ 1 \end{pmatrix}e^{-3t} + c_2\begin{pmatrix} 2 \\ 5 \end{pmatrix}e^{t}$

5. $\mathbf{X} = c_1\begin{pmatrix} 5 \\ 2 \end{pmatrix}e^{8t} + c_2\begin{pmatrix} 1 \\ 4 \end{pmatrix}e^{-10t}$

7. $\mathbf{X} = c_1\begin{pmatrix} 1 \\ 0 \\ 0 \end{pmatrix}e^{t} + c_2\begin{pmatrix} 2 \\ 3 \\ 1 \end{pmatrix}e^{2t} + c_3\begin{pmatrix} 1 \\ 0 \\ 2 \end{pmatrix}e^{-t}$

9. $\mathbf{X} = c_1\begin{pmatrix} -1 \\ 0 \\ 1 \end{pmatrix}e^{-t} + c_2\begin{pmatrix} 1 \\ 4 \\ 3 \end{pmatrix}e^{3t}$

$\qquad + c_3\begin{pmatrix} 1 \\ -1 \\ 3 \end{pmatrix}e^{-2t}$

11. $\mathbf{X} = 3\begin{pmatrix} 1 \\ 1 \end{pmatrix}e^{t/2} + 2\begin{pmatrix} 0 \\ 1 \end{pmatrix}e^{-t/2}$

13. $\mathbf{X} = c_1\begin{pmatrix} \cos t \\ 2\cos t + \sin t \end{pmatrix}e^{4t}$

$\qquad + c_2\begin{pmatrix} \sin t \\ 2\sin t - \cos t \end{pmatrix}e^{4t}$

15. $\mathbf{X} = c_1\begin{pmatrix} \cos t \\ -\cos t - \sin t \end{pmatrix}e^{4t}$

$\qquad + c_2\begin{pmatrix} -\sin t \\ \sin t - \cos t \end{pmatrix}e^{4t}$

17. $\mathbf{X} = c_1\begin{pmatrix} 5\cos 3t \\ 4\cos 3t + 3\sin 3t \end{pmatrix}$

$\qquad + c_2\begin{pmatrix} -5\sin 3t \\ -4\sin 3t + 3\cos 3t \end{pmatrix}$

19. $\mathbf{X} = c_1\begin{pmatrix} 1 \\ 0 \\ 0 \end{pmatrix} + c_2\begin{pmatrix} \cos t \\ -\cos t \\ -\sin t \end{pmatrix}$

$\qquad + c_3\begin{pmatrix} \sin t \\ -\sin t \\ \cos t \end{pmatrix}$

21. $\mathbf{X} = \begin{pmatrix} 28 \\ -5 \\ 25 \end{pmatrix}e^{2t}$

$\qquad + c_2\begin{pmatrix} 5\cos 3t \\ -4\cos 3t - 3\sin 3t \\ 0 \end{pmatrix}e^{-2t}$

$\qquad + c_3\begin{pmatrix} -5\sin 3t \\ 4\sin 3t - 3\cos 3t \\ 0 \end{pmatrix}e^{-2t}$

23. $X = -\begin{pmatrix} 25 \\ -7 \\ 6 \end{pmatrix} e^t - \begin{pmatrix} \cos 5t - 5\sin 5t \\ \cos 5t \\ \cos 5t \end{pmatrix}$

$\qquad -6 \begin{pmatrix} -5\cos 5t - \sin 5t \\ -\sin 5t \\ -\sin 5t \end{pmatrix}$

25. $X = c_1 \begin{pmatrix} 1 \\ 3 \end{pmatrix} + c_2 \left\{ \begin{pmatrix} 1 \\ 3 \end{pmatrix} t + \begin{pmatrix} \frac{1}{4} \\ -\frac{1}{4} \end{pmatrix} \right\}$

27. $X = c_1 \begin{pmatrix} 1 \\ 1 \end{pmatrix} e^{2t} + c_2 \left\{ \begin{pmatrix} 1 \\ 1 \end{pmatrix} t e^{2t} + \begin{pmatrix} -\frac{1}{3} \\ 0 \end{pmatrix} e^{2t} \right\}$

29. $X = c_1 \begin{pmatrix} 1 \\ 1 \\ 1 \end{pmatrix} e^t + c_2 \begin{pmatrix} 1 \\ 1 \\ 0 \end{pmatrix} e^{2t} + c_3 \begin{pmatrix} 1 \\ 0 \\ 1 \end{pmatrix} e^{2t}$

31. $X = c_1 \begin{pmatrix} -4 \\ -5 \\ 2 \end{pmatrix} + c_2 \begin{pmatrix} 2 \\ 0 \\ -1 \end{pmatrix} e^{5t}$

$\qquad + c_3 \left\{ \begin{pmatrix} 2 \\ 0 \\ -1 \end{pmatrix} t e^{5t} + \begin{pmatrix} -\frac{1}{2} \\ -\frac{1}{2} \\ -1 \end{pmatrix} e^{5t} \right\}$

33. $X = c_1 \begin{pmatrix} 0 \\ 1 \\ 1 \end{pmatrix} e^t + c_2 \left\{ \begin{pmatrix} 0 \\ 1 \\ 1 \end{pmatrix} t e^t + \begin{pmatrix} 0 \\ 1 \\ 0 \end{pmatrix} e^t \right\}$

$\qquad + c_3 \left\{ \begin{pmatrix} 0 \\ 1 \\ 1 \end{pmatrix} \frac{t^2}{2} e^t + \begin{pmatrix} 0 \\ 1 \\ 0 \end{pmatrix} t e^t + \begin{pmatrix} \frac{1}{2} \\ 0 \\ 0 \end{pmatrix} e^t \right\}$

35. $X = -7 \begin{pmatrix} 2 \\ 1 \end{pmatrix} e^{4t} + 13 \begin{pmatrix} 2t+1 \\ t+1 \end{pmatrix} e^{4t}$

37. $X = \begin{pmatrix} \frac{6}{5} e^{5t} - \frac{1}{5} e^{-5t} \\ \frac{2}{5} e^{5t} + \frac{3}{5} e^{-5t} \end{pmatrix}$

39. $X = c_1 t^2 \begin{pmatrix} 3 \\ 1 \end{pmatrix} + c_2 t^4 \begin{pmatrix} 1 \\ 1 \end{pmatrix}$

Exercises 8.7, Page 396

1. $X = c_1 \begin{pmatrix} -1 \\ 1 \end{pmatrix} e^{-t} + c_2 \begin{pmatrix} -3 \\ 1 \end{pmatrix} e^t + \begin{pmatrix} -1 \\ 3 \end{pmatrix}$

3. $X = c_1 \begin{pmatrix} 1 \\ -1 \end{pmatrix} e^{-2t} + c_2 \begin{pmatrix} 1 \\ 1 \end{pmatrix} e^{4t} + \begin{pmatrix} -\frac{1}{4} \\ \frac{3}{4} \end{pmatrix} t^2$

$\qquad + \begin{pmatrix} \frac{1}{4} \\ -\frac{1}{4} \end{pmatrix} t + \begin{pmatrix} -2 \\ \frac{3}{4} \end{pmatrix}$

5. $X = c_1 \begin{pmatrix} 1 \\ -8 \end{pmatrix} e^{-3t} + c_2 \begin{pmatrix} 1 \\ -1 \end{pmatrix} e^{4t}$

$\qquad + \begin{pmatrix} -\frac{13}{7} \\ \frac{13}{7} \end{pmatrix} t e^{4t} + \begin{pmatrix} 0 \\ \frac{1}{7} \end{pmatrix} e^{4t}$

7. $X = c_1 \begin{pmatrix} 1 \\ 0 \\ 0 \end{pmatrix} e^t + c_2 \begin{pmatrix} 1 \\ 1 \\ 0 \end{pmatrix} e^{2t} + c_3 \begin{pmatrix} 1 \\ 2 \\ 2 \end{pmatrix} e^{5t}$

$\qquad - \begin{pmatrix} \frac{3}{2} \\ \frac{7}{2} \\ 2 \end{pmatrix} e^{4t}$

9. $X = 13 \begin{pmatrix} 1 \\ -1 \end{pmatrix} e^t + 2 \begin{pmatrix} -4 \\ 6 \end{pmatrix} e^{2t} + \begin{pmatrix} -9 \\ 6 \end{pmatrix}$

Exercises 8.8, Page 400

1. $X = c_1 \begin{pmatrix} 1 \\ 1 \end{pmatrix} + c_2 \begin{pmatrix} 3 \\ 2 \end{pmatrix} e^t - \begin{pmatrix} 11 \\ 11 \end{pmatrix} t - \begin{pmatrix} 15 \\ 10 \end{pmatrix}$

3. $X = c_1 \begin{pmatrix} 2 \\ 1 \end{pmatrix} e^{t/2} + c_2 \begin{pmatrix} 10 \\ 3 \end{pmatrix} e^{3t/2}$

$\qquad - \begin{pmatrix} \frac{13}{2} \\ \frac{13}{4} \end{pmatrix} t e^{t/2} - \begin{pmatrix} \frac{15}{2} \\ \frac{9}{4} \end{pmatrix} e^{t/2}$

5. $X = c_1 \begin{pmatrix} 2 \\ 1 \end{pmatrix} e^t + c_2 \begin{pmatrix} 1 \\ 1 \end{pmatrix} e^{2t} + \begin{pmatrix} 3 \\ 3 \end{pmatrix} e^t$

$\qquad + \begin{pmatrix} 4 \\ 2 \end{pmatrix} t e^t$

7. $X = c_1 \begin{pmatrix} 4 \\ 1 \end{pmatrix} e^{3t} + c_2 \begin{pmatrix} -2 \\ 1 \end{pmatrix} e^{-3t} + \begin{pmatrix} -12 \\ 0 \end{pmatrix} t$

$\qquad - \begin{pmatrix} \frac{4}{3} \\ \frac{4}{3} \end{pmatrix}$

9. $\mathbf{X} = c_1 \begin{pmatrix} 1 \\ -1 \end{pmatrix} e^t + c_2 \begin{pmatrix} t \\ \frac{1}{2} - t \end{pmatrix} e^t$

$+ \begin{pmatrix} \frac{1}{2} \\ -2 \end{pmatrix} e^{-t}$

11. $\mathbf{X} = c_1 \begin{pmatrix} \cos t \\ \sin t \end{pmatrix} + c_2 \begin{pmatrix} \sin t \\ -\cos t \end{pmatrix}$

$+ \begin{pmatrix} \cos t \\ \sin t \end{pmatrix} t + \begin{pmatrix} -\sin t \\ \cos t \end{pmatrix} \ln |\cos t|$

13. $\mathbf{X} = c_1 \begin{pmatrix} \cos t \\ \sin t \end{pmatrix} e^t + c_2 \begin{pmatrix} \sin t \\ -\cos t \end{pmatrix} e^t$

$+ \begin{pmatrix} \cos t \\ \sin t \end{pmatrix} te^t$

15. $\mathbf{X} = c_1 \begin{pmatrix} \cos t \\ -\sin t \end{pmatrix} + c_2 \begin{pmatrix} \sin t \\ \cos t \end{pmatrix}$

$+ \begin{pmatrix} \cos t \\ -\sin t \end{pmatrix} t + \begin{pmatrix} -\sin t \\ \sin t \tan t \end{pmatrix}$

$- \begin{pmatrix} \sin t \\ \cos t \end{pmatrix} \ln |\cos t|$

17. $\mathbf{X} = c_1 \begin{pmatrix} 2 \sin t \\ \cos t \end{pmatrix} e^t + c_2 \begin{pmatrix} 2 \cos t \\ -\sin t \end{pmatrix} e^t$

$+ \begin{pmatrix} 3 \sin t \\ \frac{3}{2} \cos t \end{pmatrix} te^t$

$+ \begin{pmatrix} \cos t \\ -\frac{1}{2} \sin t \end{pmatrix} e^t \ln |\sin t|$

$+ \begin{pmatrix} 2 \cos t \\ -\sin t \end{pmatrix} e^t \ln |\cos t|$

19. $\mathbf{X} = c_1 \begin{pmatrix} 1 \\ -1 \\ 0 \end{pmatrix} + c_2 \begin{pmatrix} 1 \\ 1 \\ 0 \end{pmatrix} e^{2t}$

$+ c_3 \begin{pmatrix} 0 \\ 0 \\ 1 \end{pmatrix} e^{3t}$

$+ \begin{pmatrix} -\frac{1}{4} e^{2t} + \frac{1}{2} te^{2t} \\ -e^t + \frac{1}{4} e^{2t} + \frac{1}{2} te^{2t} \\ \frac{1}{2} t^2 e^{3t} \end{pmatrix}$

21. $\mathbf{X} = \begin{pmatrix} 2 \\ 2 \end{pmatrix} te^{2t} + \begin{pmatrix} -1 \\ 1 \end{pmatrix} e^{2t} + \begin{pmatrix} -2 \\ 2 \end{pmatrix} te^{4t}$

$+ \begin{pmatrix} 2 \\ 0 \end{pmatrix} e^{4t}$

23. $\mathbf{X} = \begin{pmatrix} -2 \\ 4 \end{pmatrix} e^{2t} + \begin{pmatrix} 7 \\ -9 \end{pmatrix} e^{7t} + \begin{pmatrix} 20 \\ 60 \end{pmatrix} te^{7t}$

Exercises 8.9, Page 403

1. $\begin{pmatrix} \cosh t & \sinh t \\ \sinh t & \cosh t \end{pmatrix}$

3. $\mathbf{X} = \begin{pmatrix} \cosh t & \sinh t \\ \sinh t & \cosh t \end{pmatrix} \begin{pmatrix} c_1 \\ c_2 \end{pmatrix}$

$= c_1 \begin{pmatrix} \cosh t \\ \sinh t \end{pmatrix} + c_2 \begin{pmatrix} \sinh t \\ \cosh t \end{pmatrix}$

5. $\mathbf{X} = c_1 \begin{pmatrix} \cosh t \\ \sinh t \end{pmatrix} + c_2 \begin{pmatrix} \sinh t \\ \cosh t \end{pmatrix} - \begin{pmatrix} 1 \\ 1 \end{pmatrix}$

7. $\mathbf{X} = c_1 \begin{pmatrix} 1 \\ 0 \end{pmatrix} e^t + c_2 \begin{pmatrix} 0 \\ 1 \end{pmatrix} e^{2t} + \begin{pmatrix} -t - 1 \\ \frac{1}{2} e^{4t} \end{pmatrix}$

9. $\mathbf{P} = \begin{pmatrix} 1 & 1 \\ 1 & 3 \end{pmatrix}, \quad \mathbf{P}^{-1} = \begin{pmatrix} \frac{3}{2} & -\frac{1}{2} \\ -\frac{1}{2} & \frac{1}{2} \end{pmatrix},$

$\mathbf{D} = \begin{pmatrix} 3 & 0 \\ 0 & 5 \end{pmatrix}$

$\mathbf{PDP}^{-1} = \begin{pmatrix} 2 & 1 \\ -3 & 6 \end{pmatrix}$

11. $e^{t\mathbf{A}} = e^{t\mathbf{PDP}^{-1}}$

$= \mathbf{I} + t\mathbf{PDP}^{-1} + \dfrac{t^2}{2!} (\mathbf{PDP}^{-1})^2 + \cdots$

$= \mathbf{PP}^{-1} + t\mathbf{PDP}^{-1} + \dfrac{t^2}{2!} \mathbf{PD}^2 \mathbf{P}^{-1} + \cdots$

$= \mathbf{P}[\mathbf{I} + t\mathbf{D} + \dfrac{t^2}{2!} \mathbf{D}^2 + \cdots]\mathbf{P}^{-1}$

$= \mathbf{P}e^{t\mathbf{D}}\mathbf{P}^{-1}$

13. $\mathbf{X} = \begin{pmatrix} \frac{3}{2} e^{3t} - \frac{1}{2} e^{5t} & -\frac{1}{2} e^{3t} + \frac{1}{2} e^{5t} \\ \frac{3}{2} e^{3t} - \frac{3}{2} e^{5t} & -\frac{1}{2} e^{3t} + \frac{3}{2} e^{5t} \end{pmatrix} \begin{pmatrix} c_1 \\ c_2 \end{pmatrix}$

Chapter 8 Test, Page 406

1. $x = -c_1 e^t - \frac{3}{2} c_2 e^{2t} + \frac{5}{2}$
$y = c_1 e^t + c_2 e^{2t} - 3$

3. $x = c_1 e^t + c_2 e^{5t} + t e^t$
$y = -c_1 e^t + 3c_2 e^{5t} - t e^t + 2e^t$

5. $x = -\frac{1}{4} + \frac{9}{8} e^{-2t} + \frac{1}{8} e^{2t}$
$y = t + \frac{9}{4} e^{-2t} - \frac{1}{4} e^{2t}$

7.(a) $\begin{pmatrix} t^3 + 3t^2 + 5t - 2 \\ -t^3 - t + 2 \\ 4t^3 + 12t^2 + 8t + 1 \end{pmatrix}$

(b) $\begin{pmatrix} 3t^2 + 6t + 5 \\ -3t^2 - 1 \\ 12t^2 + 24t + 8 \end{pmatrix}$

9. $Dx = u$
$Dy = v$
$Du = -2u + v - 2x - \ln t + 10t - 4$
$Dv = -u - x + 5t - 2$

11. $\mathbf{X} = c_1 \begin{pmatrix} 1 \\ -1 \end{pmatrix} e^t + c_2 \left\{ \begin{pmatrix} 1 \\ -1 \end{pmatrix} t e^t \right.$
$\left. + \begin{pmatrix} 0 \\ -1 \end{pmatrix} e^t \right\}$

13. $\mathbf{X} = c_1 \begin{pmatrix} \cos 2t \\ -\sin 2t \end{pmatrix} e^t + c_2 \begin{pmatrix} \sin 2t \\ \cos 2t \end{pmatrix} e^t$

15. $\mathbf{X} = c_1 \begin{pmatrix} -1 \\ 1 \\ 0 \end{pmatrix} + c_2 \begin{pmatrix} -1 \\ 0 \\ 1 \end{pmatrix} + c_3 \begin{pmatrix} 1 \\ 1 \\ 1 \end{pmatrix} e^{3t}$

17. $\mathbf{X} = c_1 \begin{pmatrix} -4 \\ 1 \end{pmatrix} + c_2 \begin{pmatrix} 2 \\ 1 \end{pmatrix} e^{6t} + \begin{pmatrix} 2 \\ 1 \end{pmatrix} t^2 e^{6t}$
$+ \begin{pmatrix} -\frac{2}{3} \\ \frac{1}{6} \end{pmatrix} t e^{6t} + \begin{pmatrix} \frac{1}{9} \\ -\frac{1}{36} \end{pmatrix} e^{6t}$

19. $\mathbf{X} = c_1 \begin{pmatrix} \cos t \\ \cos t - \sin t \end{pmatrix}$
$+ c_2 \begin{pmatrix} \sin t \\ \sin t + \cos t \end{pmatrix}$
$- \begin{pmatrix} 1 \\ 1 \end{pmatrix} + \begin{pmatrix} \sin t \\ \sin t + \cos t \end{pmatrix}$
$\times \ln |\csc t - \cot t|$

Exercises 9.1, Page 412

1. a family of vertical lines $x = c - 4$

3. a family of hyperbolas $x^2 - y^2 = c$

5. a family of circles $x^2 + (y + 1)^2 = c^2$
with center at $(0, -1)$

7. a family of hyperbolas $xy + y^2 = c$

9. a family of straight lines $y = c(x - 2) + 1$
passing through $(2, 1)$

11.

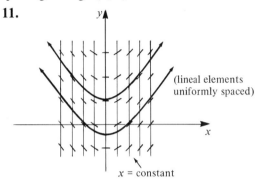

(lineal elements uniformly spaced)

$x = $ constant

13.

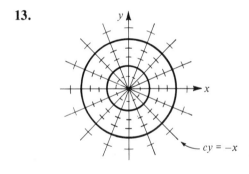

$cy = -x$

15.

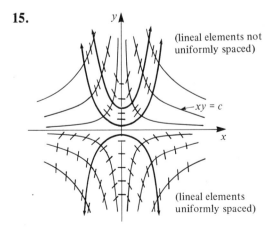

(lineal elements not uniformly spaced)

$xy = c$

(lineal elements uniformly spaced)

A-31

17.

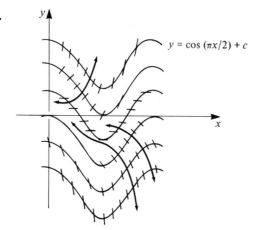

$y = \cos(\pi x/2) + c$

19. $y = \dfrac{\alpha - c\gamma}{c\delta - \beta}x$ **21.** $3x + 2y = -\dfrac{3}{2}$

23. $y = \pm\sqrt{2}x$ **25.** $y = 4x;\quad y = -x$

Exercises 9.2, Page 420

1. $y = 1 - x + \tan(x + \pi/4)$

3. **(a)**

x_n	y_n
1.00	5.0000
1.10	3.8000
1.20	2.9800
1.30	2.4260
1.40	2.0582
1.50	1.8207

(b)

x_n	y_n
1.00	5.0000
1.05	4.4000
1.10	3.8950
1.15	3.4707
1.20	3.1151
1.25	2.8179
1.30	2.5702
1.35	2.3647
1.40	2.1950
1.45	2.0557
1.50	1.9424

5. **(a)**

x_n	y_n
0.00	0.0000
0.10	0.1000
0.20	0.2010
0.30	0.3050
0.40	0.4143
0.50	0.5315

(b)

x_n	y_n
0.00	0.0000
0.05	0.0500
0.10	0.1001
0.15	0.1506
0.20	0.2018
0.25	0.2538
0.30	0.3070
0.35	0.3617
0.40	0.4183
0.45	0.4770
0.50	0.5384

7. **(a)**

x_n	y_n
0.00	0.0000
0.10	0.1000
0.20	0.1905
0.30	0.2731
0.40	0.3492
0.50	0.4198

(b)

x_n	y_n
0.00	0.0000
0.05	0.0500
0.10	0.0976
0.15	0.1429
0.20	0.1863
0.25	0.2278
0.30	0.2676
0.35	0.3058
0.40	0.3427
0.45	0.3782
0.50	0.4124

9. **(a)**

x_n	y_n
0.00	0.5000
0.10	0.5250
0.20	0.5431
0.30	0.5548
0.40	0.5613
0.50	0.5639

(b)

x_n	y_n
0.00	0.5000
0.05	0.5125
0.10	0.5232
0.15	0.5322
0.20	0.5395
0.25	0.5452
0.30	0.5496
0.35	0.5527
0.40	0.5547
0.45	0.5559
0.50	0.5565

11. **(a)**

x_n	y_n
1.00	1.0000
1.10	1.0000
1.20	1.0191
1.30	1.0588
1.40	1.1231
1.50	1.2194

(b)

x_n	y_n
1.00	1.0000
1.05	1.0000
1.10	1.0049
1.15	1.0147
1.20	1.0298
1.25	1.0506
1.30	1.0775
1.35	1.1115
1.40	1.1538
1.45	1.2057
1.50	1.2696

13. **(a)** $h = 0.1$

x_n	y_n
1.00	5.0000
1.10	3.9900
1.20	3.2545
1.30	2.7236
1.40	2.3451
1.50	2.0801

$h = 0.05$

x_n	y_n
1.00	5.0000
1.05	4.4475
1.10	3.9763
1.15	3.5751
1.20	3.2342
1.25	2.9452
1.30	2.7009
1.35	2.4952
1.40	2.3226
1.45	2.1786
1.50	2.0592

(b) $h = 0.1$

x_n	y_n
0.00	0.0000
0.10	0.1005
0.20	0.2030
0.30	0.3098
0.40	0.4234
0.50	0.5470

$h = 0.05$

x_n	y_n
0.00	0.0000
0.05	0.0501
0.10	0.1004
0.15	0.1512
0.20	0.2028
0.25	0.2554
0.30	0.3095
0.35	0.3652
0.40	0.4230
0.45	0.4832
0.50	0.5465

(c)

$h = 0.1$	x_n	y_n
	0.00	0.0000
	0.10	0.0952
	0.20	0.1822
	0.30	0.2622
	0.40	0.3363
	0.50	0.4053

$h = 0.05$	x_n	y_n
	0.00	0.0000
	0.05	0.0488
	0.10	0.0953
	0.15	0.1397
	0.20	0.1823
	0.25	0.2231
	0.30	0.2623
	0.35	0.3001
	0.40	0.3364
	0.45	0.3715
	0.50	0.4054

(d)

$h = 0.1$	x_n	y_n
	0.00	0.5000
	0.10	0.5215
	0.20	0.5362
	0.30	0.5449
	0.40	0.5490
	0.50	0.5503

$h = 0.05$	x_n	y_n
	0.00	0.5000
	0.05	0.5116
	0.10	0.5214
	0.15	0.5294
	0.20	0.5359
	0.25	0.5408
	0.30	0.5444
	0.35	0.5469
	0.40	0.5484
	0.45	0.5492
	0.50	0.5495

(e)

$h = 0.1$	x_n	y_n
	1.00	1.0000
	1.10	1.0095
	1.20	1.0404
	1.30	1.0967
	1.40	1.1866
	1.50	1.3260

$h = 0.05$	x_n	y_n
	1.00	1.0000
	1.05	1.0024
	1.10	1.0100
	1.15	1.0228
	1.20	1.0414
	1.25	1.0663
	1.30	1.0984
	1.35	1.1389
	1.40	1.1895
	1.45	1.2526
	1.50	1.3315

15.

x_n	Euler	Improved Euler
1.0	1.0000	1.0000
1.1	1.2000	1.2469
1.2	1.4938	1.6668
1.3	1.9711	2.6427
1.4	2.9060	8.7989

17. Using $y' = f(x, y)$ gives

$$\int_{x_n}^{x_{n+1}} y' \, dx = \int_{x_n}^{x_{n+1}} f(x, y) \, dx$$

$$y(x_{n+1}) - y(x_n) \approx f(x_n, y_n)(x_{n+1} - x_n)$$

$$= h f(x_n, y_n)$$

$$y(x_{n+1}) \approx y(x_n) + h f(x_n, y_n)$$

We write this as

$$y_{n+1} = y_n + h f(x_n, y_n).$$

Exercises 9.3, Page 425

1. (a)

x_n	y_n
1.00	5.0000
1.10	3.9900
1.20	3.2545
1.30	2.7236
1.40	2.3451
1.50	2.0801

(b)

x_n	y_n
1.00	5.0000
1.05	4.4475
1.10	3.9763
1.15	3.5751
1.20	3.2342
1.25	2.9452
1.30	2.7009
1.35	2.4952
1.40	2.3226
1.45	2.1786
1.50	2.0592

3. (a)

x_n	y_n
0.00	0.0000
0.10	0.1000
0.20	0.2020
0.30	0.3082
0.40	0.4211
0.50	0.5438

(b)

x_n	y_n
0.00	0.0000
0.05	0.0500
0.10	0.1003
0.15	0.1510
0.20	0.2025
0.25	0.2551
0.30	0.3090
0.35	0.3647
0.40	0.4223
0.45	0.4825
0.50	0.5456

5. (a)

x_n	y_n
0.00	0.0000
0.10	0.0950
0.20	0.1818
0.30	0.2617
0.40	0.3357
0.50	0.4046

(b)

x_n	y_n
0.00	0.0000
0.05	0.0488
0.10	0.0952
0.15	0.1397
0.20	0.1822
0.25	0.2230
0.30	0.2622
0.35	0.2999
0.40	0.3363
0.45	0.3714
0.50	0.4053

7. (a)

x_n	y_n
0.00	0.5000
0.10	0.5213
0.20	0.5355
0.30	0.5438
0.40	0.5475
0.50	0.5482

(b)

x_n	y_n
0.00	0.5000
0.05	0.5116
0.10	0.5213
0.15	0.5293
0.20	0.5357
0.25	0.5406
0.30	0.5441
0.35	0.5466
0.40	0.5480
0.45	0.5487
0.50	0.5490

9. (a)

x_n	y_n
1.00	1.0000
1.10	1.0100
1.20	1.0410
1.30	1.0969
1.40	1.1857
1.50	1.3226

(b)

x_n	y_n
1.00	1.0000
1.05	1.0025
1.10	1.0101
1.15	1.0229
1.20	1.0415
1.25	1.0663
1.30	1.0983
1.35	1.1387
1.40	1.1891
1.45	1.2518
1.50	1.3301

11.

x_n	Euler	Improved Euler	Three-Term Taylor
1.0	1.0000	1.0000	1.0000
1.1	1.2000	1.2469	1.2400
1.2	1.4938	1.6668	1.6345
1.3	1.9711	2.6427	2.4600
1.4	2.9060	8.7988	5.6353

Exercises 9.4, Page 431

1. (a)

x_n	y_n
1.00	5.0000
1.10	3.9724
1.20	3.2284
1.30	2.6945
1.40	2.3163
1.50	2.0533

(b)

x_n	y_n
1.00	5.0000
1.05	4.4452
1.10	3.9723
1.15	3.5700
1.20	3.2283
1.25	2.9389
1.30	2.6944
1.35	2.4886
1.40	2.3162
1.45	2.1724
1.50	2.0532

3. (a)

x_n	y_n
0.00	0.0000
0.10	0.1003
0.20	0.2027
0.30	0.3093
0.40	0.4228
0.50	0.5463

(b)

x_n	y_n
0.00	0.0000
0.05	0.0500
0.10	0.1003
0.15	0.1511
0.20	0.2027
0.25	0.2553
0.30	0.3093
0.35	0.3650
0.40	0.4228
0.45	0.4831
0.50	0.5463

5. **(a)**

x_n	y_n
0.00	0.0000
0.10	0.0953
0.20	0.1823
0.30	0.2624
0.40	0.3365
0.50	0.4055

(b)

x_n	y_n
0.00	0.0000
0.05	0.0488
0.10	0.0953
0.15	0.1398
0.20	0.1823
0.25	0.2231
0.30	0.2624
0.35	0.3001
0.40	0.3365
0.45	0.3716
0.50	0.4055

7. **(a)**

x_n	y_n
0.00	0.5000
0.10	0.5213
0.20	0.5358
0.30	0.5443
0.40	0.5482
0.50	0.5493

(b)

x_n	y_n
0.00	0.5000
0.05	0.5116
0.10	0.5213
0.15	0.5294
0.20	0.5358
0.25	0.5407
0.30	0.5443
0.35	0.5467
0.40	0.5482
0.45	0.5490
0.50	0.5493

9. **(a)**

x_n	y_n
1.00	1.0000
1.10	1.0101
1.20	1.0417
1.30	1.0989
1.40	1.1905
1.50	1.3333

(b)

x_n	y_n
1.00	1.0000
1.05	1.0025
1.10	1.0101
1.15	1.0230
1.20	1.0417
1.25	1.0667
1.30	1.0989
1.35	1.1396
1.40	1.1905
1.45	1.2539
1.50	1.3333

11.

x_n	y_n
1.00	1.0000
1.10	1.2511
1.20	1.6934
1.30	2.9425
1.40	903.0283

Exercises 9.5, Page 434

1. $y(0.2) \approx 1.39$

3. $h = 0.2$; $y(0.2) \approx 1.38$

$h = 0.1$:

x_n	y_n	u_n
0.0	1.0000	2.0000
0.1	1.1950	1.8808
0.2	1.3762	1.7254

5.

x_n	y_n
0.0	1.0000
0.1	1.0052
0.2	1.0214
0.3	1.0499
0.4	1.0918

Chapter 9 Test, Page 437

1. All isoclines $y = cx$ are solutions of the differential equation.

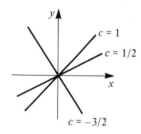

3.

	Comparison of Numerical Methods with $h = 0.1$			
x_n	Euler	Improved Euler	3-Term Taylor	Runge–Kutta
1.00	2.0000	2.0000	2.0000	2.0000
1.10	2.1386	2.1549	2.1556	2.1556
1.20	2.3097	2.3439	2.3453	2.3454
1.30	2.5136	2.5672	2.5694	2.5695
1.40	2.7504	2.8246	2.8277	2.8278
1.50	3.0201	3.1157	3.1198	3.1197

Comparison of Numerical Methods with $h = 0.05$				
x_n	Euler	Improved Euler	3-Term Taylor	Runge–Kutta
1.00	2.0000	2.0000	2.0000	2.0000
1.05	2.0693	2.0735	2.0735	2.0736
1.10	2.1469	2.1554	2.1556	2.1556
1.15	2.2329	2.2459	2.2462	2.2462
1.20	2.3272	2.3450	2.3454	2.3454
1.25	2.4299	2.4527	2.4532	2.4532
1.30	2.5410	2.5689	2.5695	2.5695
1.35	2.6604	2.6937	2.6944	2.6944
1.40	2.7883	2.8269	2.8278	2.8278
1.45	2.9245	2.9686	2.9696	2.9696
1.50	3.0690	3.1187	3.1198	3.1197

5.

Comparison of Numerical Methods with $h = 0.1$				
x_n	Euler	Improved Euler	3-Term Taylor	Runge–Kutta
0.50	0.5000	0.5000	0.5000	0.5000
0.60	0.6000	0.6048	0.6050	0.6049
0.70	0.7095	0.7191	0.7195	0.7194
0.80	0.8283	0.8427	0.8433	0.8431
0.90	0.9559	0.9752	0.9759	0.9757
1.00	1.0921	1.1163	1.1172	1.1169

Comparison of Numerical Methods with $h = 0.05$				
x_n	Euler	Improved Euler	3-Term Taylor	Runge–Kutta
0.50	0.5000	0.5000	0.5000	0.5000
0.55	0.5500	0.5512	0.5512	0.5512
0.60	0.6024	0.6049	0.6049	0.6049
0.65	0.6573	0.6609	0.6610	0.6610
0.70	0.7144	0.7193	0.7194	0.7194
0.75	0.7739	0.7800	0.7802	0.7801
0.80	0.8356	0.8430	0.8431	0.8431
0.85	0.8996	0.9082	0.9083	0.9083
0.90	0.9657	0.9755	0.9757	0.9757
0.95	1.0340	1.0451	1.0453	1.0452
1.00	1.1044	1.1168	1.1170	1.1169

7. $h = 0.2$: $y(0.2) \approx 3.2000$

$h = 0.1$:

x_n	y_n	u_n
0.0	3.0000	1.0000
0.1	3.1000	1.3000
0.2	3.2300	1.6720

Exercises 10.1, Page 441

1. $\displaystyle\int_{-2}^{2} x - x^2 \, dx = \dfrac{x^4}{4}\Big|_{-2}^{2} = 4 - 4 = 0$

3. $\displaystyle\int_{0}^{2} e^x(xe^{-x} - e^{-x}) \, dx = \int_{0}^{2} (x - 1) \, dx$

$$= \dfrac{x^2}{2} - x \Big|_{0}^{2}$$
$$= (2 - 2) - 0$$
$$= 0$$

5. $\displaystyle\int_{0}^{\pi/2} \sin(2m + 1)x \sin(2n + 1)x \, dx$

$= \dfrac{1}{2} \displaystyle\int_{0}^{\pi/2} [\cos 2(m - n)x$

$- \cos 2(m + n + 1)x] \, dx$

$= \dfrac{1}{4}\left[\dfrac{\sin 2(m - n)x}{m - n} - \dfrac{\sin 2(m + n + 1)x}{m + n + 1}\right]_{0}^{\pi/2}$

$= 0$, $m \neq n$; $\sqrt{\pi/2}$

7. $\displaystyle\int_{0}^{\pi} \sin mx \sin nx \, dx$

$= \dfrac{1}{2} \displaystyle\int_{0}^{\pi} [\cos(m - n)x - \cos(m + n)x] \, dx$

$= \dfrac{1}{2}\left[\dfrac{\sin(m - n)x}{m - n} - \dfrac{\sin(m + n)x}{m + n}\right]_{0}^{\pi}$

$= 0$, $m \neq n$; $\sqrt{\pi/2}$

9. $\displaystyle\int_{0}^{p} \cos\dfrac{n\pi}{p}x \, dx = \dfrac{p}{n\pi} \sin\dfrac{n\pi}{p}x \Big|_{0}^{p} = 0$,

$n \neq 0$;

$\displaystyle\int_{0}^{p} \cos\dfrac{m\pi}{p}x \cos\dfrac{n\pi}{p}x \, dx$

$= \dfrac{1}{2} \displaystyle\int_{0}^{p} \left[\cos\dfrac{(m - n)\pi}{p}x\right.$

$\left. + \cos\dfrac{(m + n)\pi}{p}x\right] dx$

$= \dfrac{p}{2\pi}\left[\dfrac{\sin\dfrac{(m - n)\pi}{p}x}{m - n} + \dfrac{\sin\dfrac{(m + n)\pi}{p}x}{m + n}\right]_{0}^{p}$

$= 0$, $m \neq n$; $\|1\| = \sqrt{p}$, $\left\|\cos\dfrac{n\pi}{p}x\right\|$

$= \sqrt{p/2}$

11. For example

$\displaystyle\int_{-\infty}^{\infty} e^{-x^2} H_0(x) H_1(x) \, dx = \int_{-\infty}^{\infty} e^{-x^2}(2x) \, dx$

$= -\displaystyle\int_{-\infty}^{0} e^{-x^2}(-2x \, dx) - \int_{0}^{\infty} e^{-x^2}(-2x \, dx)$

$= -e^{-x^2}\Big|_{-\infty}^{0} - e^{-x^2}\Big|_{0}^{\infty}$

$= -1 - (-1) = 0$.

The results $\displaystyle\int_{-\infty}^{\infty} e^{-x^2} H_0(x) H_2(x) \, dx = 0$

and $\displaystyle\int_{-\infty}^{\infty} e^{-x^2} H_1(x) H_2(x) \, dx = 0$

follow from integration by parts.

13. $\displaystyle\int_{a}^{b} \phi_n(x) \, dx = \int_{a}^{b} \phi_0(x)\phi_n(x) \, dx$

$= 0$ for $n = 1, 2, 3, \ldots$

15. $\|\phi_m(x) + \phi_n(x)\|^2$

$= \displaystyle\int_{a}^{b} [\phi_m(x) + \phi_n(x)]^2 \, dx$

$= \displaystyle\int_{a}^{b} \phi_m^2(x) \, dx + 2 \underbrace{\int_{a}^{b} \phi_m(x)\phi_n(x) \, dx}_{\text{zero by orthogonality}}$

$+ \displaystyle\int_{a}^{b} \phi_n^2(x) \, dx =$

$$= \int_a^b \phi_m^2(x)\, dx + \int_a^b \phi_n^2(x)\, dx$$

$$= \|\phi_m(x)\|^2 + \|\phi_n(x)\|^2$$

17. Legendre's equation can be written in the form

$$\frac{d}{dx}\left[(1 - x^2)\frac{dy}{dx}\right] + n(n + 1)y = 0.$$

With the identification $r(x) = 1 - x^2$, we see $r(-1) = r(1) = 0$. Since the weight function is $p(x) = 1$, $q(x) = 0$, $\lambda = n(n + 1)$ it follows from part (b) of Problem 16 that

$$\int_{-1}^1 P_m(x)P_n(x)\, dx = 0, \quad m \neq n.$$

19. $c_0 = 1/4$, $c_1 = 1/2$, $c_2 = 5/16$

Exercises 10.2, Page 452

1. $f(x) \sim \dfrac{1}{2} + \dfrac{1}{\pi}\displaystyle\sum_{n=1}^{\infty} \dfrac{1 - (-1)^n}{n} \sin nx$

3. $f(x) \sim \dfrac{3}{4} + \displaystyle\sum_{n=1}^{\infty}\left[\dfrac{(-1)^n - 1}{n^2\pi^2}\cos n\pi x\right.$

$$\left. - \frac{1}{n\pi}\sin n\pi x\right]$$

5. $f(x) \sim \dfrac{\pi^2}{6} + \displaystyle\sum_{n=1}^{\infty}\left\{\dfrac{2(-1)^n}{n^2}\cos nx\right.$

$$\left. + \left(\frac{(-1)^{n+1}\pi}{n} + \frac{2}{\pi n^3}[(-1)^n - 1]\right)\sin nx\right\}$$

7. $f(x) \sim \pi + 2\displaystyle\sum_{n=1}^{\infty}\dfrac{(-1)^{n+1}}{n}\sin nx$

9. $f(x) \sim \dfrac{1}{\pi} + \dfrac{1}{2}\sin x$

$$+ \frac{1}{\pi}\sum_{n=2}^{\infty}\frac{(-1)^n + 1}{1 - n^2}\cos nx$$

11. $f(x) \sim \dfrac{3}{4} + \dfrac{1}{\pi}\displaystyle\sum_{n=1}^{\infty}\left[-\dfrac{1}{n}\sin\dfrac{n\pi}{2}\cos\dfrac{n\pi}{2}x\right.$

$$\left. + \frac{3}{n}\left(1 - \cos\frac{n\pi}{2}\right)\sin\frac{n\pi}{2}x\right]$$

13. $f(x) \sim \dfrac{9}{4} + 5\displaystyle\sum_{n=1}^{\infty}\left\{\dfrac{[(-1)^n - 1]}{n^2\pi^2}\cos\dfrac{n\pi}{5}x\right.$

$$\left. + \frac{(-1)^{n+1}}{n\pi}\sin\frac{n\pi}{5}x\right\}$$

15. $f(x) \sim \dfrac{2\sinh\pi}{\pi}\left[\dfrac{1}{2} + \displaystyle\sum_{n=1}^{\infty}\dfrac{(-1)^n}{1 + n^2}(\cos nx\right.$

$$\left. - n\sin nx)\right]$$

17.(a) At the endpoint $x = \pi$ the series will converge to

$$\frac{f(\pi -) + f(-\pi +)}{2} = \frac{\pi^2}{2}.$$

Substituting $x = \pi$ into the series gives

$$\frac{\pi^2}{6} + 2\sum_{n=1}^{\infty}\frac{1}{n^2}.$$

Equating the two results then yields

$$\frac{\pi^2}{6} = \sum_{n=1}^{\infty}\frac{1}{n^2} = 1 + \frac{1}{2^2} + \frac{1}{3^2} + \frac{1}{4^2} + \cdots$$

Now at $x = 0$ the series converges to

$$f(0) = 0 = \frac{\pi^2}{6} + \sum_{n=1}^{\infty}\frac{2(-1)^n}{n^2}$$

This implies

$$\frac{\pi^2}{12} = \sum_{n=1}^{\infty}\frac{(-1)^{n+1}}{n^2} = 1 - \frac{1}{2^2} + \frac{1}{3^2} - \frac{1}{4^2} + \cdots$$

(b). The value of $\pi^2/8$ is obtained by adding the two series given in part (a) and dividing the result by 2:

$$\frac{\pi^2}{8} = 1 + \frac{1}{3^2} + \frac{1}{5^2} + \frac{1}{7^2} + \cdots$$

19. $f(x) \sim \dfrac{4\pi^2}{3} + 4\displaystyle\sum_{n=1}^{\infty}\left[\dfrac{1}{n^2}\cos nx\right.$

$$\left. - \frac{\pi}{n}\sin nx\right]$$

21. $f(x) \sim \dfrac{3}{2} - \dfrac{1}{\pi}\displaystyle\sum_{n=1}^{\infty}\dfrac{1}{n}\sin 2n\pi x$

23. $f(x) \sim \dfrac{2}{\pi}\displaystyle\sum_{n=1}^{\infty}\dfrac{1 - (-1)^n}{n}\sin nx$

25. $f(x) \sim \dfrac{\pi}{2} + \dfrac{2}{\pi}\displaystyle\sum_{n=1}^{\infty}\dfrac{(-1)^n - 1}{n^2}\cos nx$

27. $f(x) \sim \dfrac{1}{3} + \dfrac{4}{\pi^2} \displaystyle\sum_{n=1}^{\infty} \dfrac{(-1)^n}{n^2} \cos n\pi x$

29. $f(x) \sim \dfrac{2\pi^2}{3} + 4 \displaystyle\sum_{n=1}^{\infty} \dfrac{(-1)^{n+1}}{n^2} \cos nx$

31. $f(x) \sim \dfrac{2}{\pi} \displaystyle\sum_{n=1}^{\infty} \dfrac{1 - (-1)^n(1 + \pi)}{n} \sin nx$

33. $f(x) \sim \dfrac{3}{4} + \dfrac{4}{\pi^2} \displaystyle\sum_{n=1}^{\infty} \dfrac{\cos \dfrac{n\pi}{2} - 1}{n^2} \cos \dfrac{n\pi}{2} x$

35. $f(x) \sim \dfrac{2}{\pi} + \dfrac{2}{\pi} \displaystyle\sum_{n=2}^{\infty} \dfrac{1 + (-1)^n}{1 - n^2} \cos nx$

37. $f(x) \sim \dfrac{1}{2} + \dfrac{2}{\pi} \displaystyle\sum_{n=1}^{\infty} \dfrac{\sin \dfrac{n\pi}{2}}{n} \cos n\pi x$

$\qquad f(x) \sim \dfrac{2}{\pi} \displaystyle\sum_{n=1}^{\infty} \dfrac{1 - \cos \dfrac{n\pi}{2}}{n} \sin n\pi x$

39. $f(x) \sim \dfrac{2}{\pi} + \dfrac{4}{\pi} \displaystyle\sum_{n=1}^{\infty} \dfrac{(-1)^n}{1 - 4n^2} \cos 2nx$

$\qquad f(x) \sim \dfrac{8}{\pi} \displaystyle\sum_{n=1}^{\infty} \dfrac{n}{4n^2 - 1} \sin 2nx$

41. $f(x) \sim \dfrac{\pi}{4} + \dfrac{2}{\pi}$

$\qquad \times \displaystyle\sum_{n=1}^{\infty} \dfrac{2 \cos \dfrac{n\pi}{2} - (-1)^n - 1}{n^2} \cos nx$

$\qquad f(x) \sim \dfrac{4}{\pi} \displaystyle\sum_{n=1}^{\infty} \dfrac{\sin \dfrac{n\pi}{2}}{n^2} \sin nx$

43. $f(x) \sim \dfrac{3}{4} + \dfrac{4}{\pi^2} \displaystyle\sum_{n=1}^{\infty} \dfrac{\cos \dfrac{n\pi}{2} - 1}{n^2} \cos \dfrac{n\pi}{2} x$

$\qquad f(x) \sim \displaystyle\sum_{n=1}^{\infty} \left\{ \dfrac{4}{n^2\pi^2} \sin \dfrac{n\pi}{2} \right.$

$\qquad\qquad \left. - \dfrac{2}{n\pi}(-1)^n \right\} \sin \dfrac{n\pi}{2} x$

45. $f(x) \sim \dfrac{5}{6} + \dfrac{2}{\pi^2} \displaystyle\sum_{n=1}^{\infty} \dfrac{3(-1)^n - 1}{n^2} \cos n\pi x$

$f(x) \sim 4 \displaystyle\sum_{n=1}^{\infty} \left\{ \dfrac{(-1)^{n+1}}{n\pi} \right.$

$\qquad\qquad \left. + \dfrac{(-1)^n - 1}{n^3\pi^3} \right\} \sin n\pi x$

47. Let f and g be even functions. Define $h(x) = f(x)g(x)$. Then

$\qquad h(-x) = f(-x)g(-x) = f(x)g(x) = h(x).$

49. Let f be an even function and g be an odd function. Define $h(x) = f(x)g(x)$. Then

$\qquad h(-x) = f(-x)g(-x)$

$\qquad\qquad = f(x)[-g(x)] = -f(x)g(x)$

$\qquad\qquad = -h(x).$

51. Let f be an odd function. Then

$$\int_{-a}^{a} f(x)\, dx = \int_{-a}^{0} f(x)\, dx$$

$$+ \int_{0}^{a} f(x)\, dx = I_1 + I_2.$$

In I_1 let $-x = t$ and $-dx = dt$ so that

$$I_1 = \int_{a}^{0} f(-t)(-dt) = \int_{0}^{a} f(-t)\, dt$$

$$= -\int_{0}^{a} f(t)\, dt = -I_2.$$

Therefore $I_1 + I_2 = 0$.

53. Adding the results of Problems 25 and 26 and dividing by 2 gives:

$$\dfrac{\pi}{4} + \sum_{n=1}^{\infty} \left[\dfrac{(-1)^n - 1}{\pi n^2} \cos nx \right.$$

$$\left. + \dfrac{(-1)^{n+1}}{n} \sin nx \right]$$

Exercises 10.3, Page 459

1. $u = -xy + f(y)$

3. $u = \frac{1}{3}e^y + f(x)e^{-2y}$

5. $u = \frac{4}{3}x^3y^2 + \frac{1}{2}x^2 + xf(y) + g(y)$

7. $u = xy + f(x) + g(y)$

9. $u = 3x^2ye^x + f(y)e^x + g(x)$

11. $u = f(x)e^{xy} + g(x)e^{-xy} + \dfrac{xe^{4y}}{16 - x^2}$

13. The possible cases can be summarized in one form $u = c_1 e^{c_2(x + y)}$ where c_1 and c_2 are constants.

15. $u = c_1 e^{y + c_2(x - y)}$ **17.** $u = c_1(xy)^{c_2}$

19. Not separable

21. $u = e^{-t}[A_1 e^{k\lambda^2 t} \cosh \lambda x + B_1 e^{k\lambda^2 t} \sinh \lambda x]$

$u = e^{-t}[A_2 e^{-k\lambda^2 t} \cos \lambda x + B_2 e^{-k\lambda^2 t} \sin \lambda x]$

$u = (c_7 x + c_8)(-t + c_9)$

23. $u = (c_1 \cosh \lambda x + c_2 \sinh \lambda x)(c_3 \cosh at$
$\quad + c_4 \sinh at)$

$u = (c_5 \cos \lambda x + c_6 \sin \lambda x)(c_7 \cos at$
$\quad + c_8 \sin at)$

$u = (c_9 x + c_{10})(c_{11} t + c_{12})$

25. $u = (c_1 \cosh \lambda x + c_2 \sinh \lambda x)(c_3 \cos \lambda y$
$\quad + c_4 \sin \lambda y)$

$u = (c_5 \cos \lambda x + c_6 \sin \lambda x)(c_7 \cosh \lambda y$
$\quad + c_8 \sinh \lambda y)$

$u = (c_9 x + c_{10})(c_{11} y + c_{12})$

27. For $\lambda^2 > 0$ there are three possibilities:

$u = (c_1 \cosh \lambda x$
$\quad + c_2 \sinh \lambda x)(c_3 \cosh \sqrt{1 - \lambda^2} y$
$\quad + c_4 \sinh \sqrt{1 - \lambda^2} y), \lambda^2 < 1,$

$u = (c_1 \cosh \lambda x$
$\quad + c_2 \sinh \lambda x)(c_3 \cos \sqrt{\lambda^2 - 1} y$
$\quad + c_4 \sin \sqrt{\lambda^2 - 1} y), \lambda^2 > 1,$

$u = (c_1 \cosh x + c_2 \sinh x)$
$\quad \times (c_3 y + c_4), \lambda^2 = 1.$

The results for the case $-\lambda^2 < 0$ are similar. For $\lambda^2 = 0$ we have

$u = (c_9 x + c_{10})(c_{11} \cosh y + c_{12} \sinh y).$

29. $u = x^3 + x(y^2 - y) + y$

31. $u = A_n e^{-k(n^2 \pi^2/25)t} \cos \dfrac{n\pi}{5} x$, $n = 0, 1, 2, \ldots$

33. $u = A_n \sinh n\pi x \sin n\pi y$, $n = 1, 2, 3, \ldots$

35. Using $-\lambda^2$ as a separation constant we obtain

$$T' + k\lambda^2 T = 0$$
$$rR'' + R' + \lambda^2 rR = 0.$$

This last equation can be written as

$$r^2 R'' + rR' + \lambda^2 r^2 R = 0$$

which we recognize as Bessel's equation with $v = 0$. The solutions of the respective equations are as indicated in the problem.

37. Differentiate each side with respect to x:

$$\frac{d}{dx}\left[\frac{X''}{4X}\right] = \frac{d}{dx}\left[\frac{Y'}{Y}\right].$$

Since the right-hand side is zero we have

$$\frac{d}{dx}\left[\frac{X''}{4X}\right] = 0.$$

This implies $X''/4X$ is a constant. Similarly, if we differentiate both sides with respect to y we can show that Y'/Y is a constant.

39. $\Psi(x) = \dfrac{1}{6} kx^3 + k_1 x + k_2$, k_1 and k_2 arbitrary constants

41. $u = \dfrac{1}{2}[F(x + at) + F(x - at)]$

Exercises 10.4 Page 469

1. $u(x, t)$

$$= \frac{2}{\pi} \sum_{n=1}^{\infty} \left(\frac{-\cos \dfrac{n\pi}{2} + 1}{n}\right) e^{-k(n^2 \pi^2/L^2)t} \sin \frac{n\pi}{L} x$$

3. $u(x, t) = \dfrac{1}{L} \displaystyle\int_0^L f(x)\, dx$

$$+ \frac{2}{L} \sum_{n=1}^{\infty} \left(\int_0^L f(x) \cos \frac{n\pi}{L} x\, dx\right)$$
$$\times e^{-k(n^2 \pi^2/L^2)t} \cos \frac{n\pi}{L} x$$

5. $u(x, t) = e^{-ht}\left[\dfrac{1}{L} \displaystyle\int_0^L f(x)\, dx\right.$

$$+ \frac{2}{L} \sum_{n=1}^{\infty} \left(\int_0^L f(x) \cos \frac{n\pi}{L} x\, dx\right)$$
$$\left. \times e^{-k(n^2 \pi^2/L^2)t} \cos \frac{n\pi}{L} x\right]$$

7. $u(x, t) = \dfrac{8h}{\pi^2} \displaystyle\sum_{n=1}^{\infty} \frac{\sin \dfrac{n\pi}{2}}{n^2} \cos \frac{n\pi a}{L} t \sin \frac{n\pi}{L} x$

A-45

9. $u(x, t) = \dfrac{L^2}{\pi^3} \displaystyle\sum_{n=1}^{\infty} \dfrac{1 - (-1)^n}{n^3}$

$\times \cos \dfrac{n\pi a}{L} t \sin \dfrac{n\pi}{L} x$

11. $u(x, t) = \dfrac{2L^3}{\pi^3} \displaystyle\sum_{n=1}^{\infty} \dfrac{(-1)^{n+1}}{n^3}$

$\times \cos \dfrac{n\pi a}{L} t \sin \dfrac{n\pi}{L} x$

13. $u(x, t) = \dfrac{L}{\pi a} \sin \dfrac{\pi a}{L} t \sin \dfrac{\pi}{L} x$

15. $u(x, y) = \dfrac{2}{a} \displaystyle\sum_{n=1}^{\infty} \left(\dfrac{1}{\sinh \dfrac{n\pi}{a} b} \int_0^a \right.$

$\times f(x) \sin \dfrac{n\pi}{a} x \, dx \Bigg)$

$\times \sinh \dfrac{n\pi}{a} y \sin \dfrac{n\pi}{a} x$

17. $u(x, y) = \dfrac{2}{a} \displaystyle\sum_{n=1}^{\infty} \left(\dfrac{1}{\sinh \dfrac{n\pi}{a} b} \int_0^a \right.$

$\times f(x) \sin \dfrac{n\pi}{a} x \, dx \Bigg)$

$\times \sinh \dfrac{n\pi}{a} (b - y) \sin \dfrac{n\pi}{a} x$

19. $u(x, y)$

$= \displaystyle\sum_{n=1}^{\infty} \left(A_n \cosh \dfrac{n\pi}{a} y + B_n \sinh \dfrac{n\pi}{a} y \right) \sin \dfrac{n\pi}{a} x$

where $A_n = \dfrac{2}{a} \displaystyle\int_0^a f(x) \sin \dfrac{n\pi}{a} x \, dx$ and

$B_n = \dfrac{1}{\sinh \dfrac{n\pi}{a} b} \left(\dfrac{2}{a} \displaystyle\int_0^a g(x) \sin \dfrac{n\pi}{a} x \, dx \right.$

$\left. - A_n \cosh \dfrac{n\pi}{a} b \right)$

21. $u(x, t) = u_0$

23. $f(x) = \displaystyle\sum_{n=1}^{\infty} A_n \sin \dfrac{n\pi}{L} x$

so $f(x + at) = \displaystyle\sum_{n=1}^{\infty} A_n \sin \dfrac{n\pi}{L} (x + at)$

$f(x - at) = \displaystyle\sum_{n=1}^{\infty} A_n \sin \dfrac{n\pi}{L} (x - at)$

but $u(x, t) = \displaystyle\sum_{n=1}^{\infty} A_n \cos \dfrac{n\pi a}{L} t \sin \dfrac{n\pi}{L} x$

$= \dfrac{1}{2} \displaystyle\sum_{n=1}^{\infty} A_n \left[\sin \dfrac{n\pi}{L} (x + at) \right.$

$\left. + \sin \dfrac{n\pi}{L} (x - at) \right]$

$= \dfrac{1}{2} \left[\displaystyle\sum_{n=1}^{\infty} A_n \sin \dfrac{n\pi}{L} (x + at) \right.$

$\left. + \displaystyle\sum_{n=1}^{\infty} A_n \sin \dfrac{n\pi}{L} (x - at) \right]$

$= \dfrac{1}{2} \left[f(x + at) + f(x - at) \right]$

Chapter 10 Test, Page 474

1. cosine

3. false

5. $\displaystyle\int_0^L \sin \dfrac{(2m + 1)\pi}{2L} x \sin \dfrac{(2n + 1)\pi}{2L} x \, dx$

$= \dfrac{1}{2} \displaystyle\int_0^L \left[\cos \dfrac{(m - n)\pi}{L} x \right.$

$\left. - \cos \dfrac{(m + n + 1)\pi}{L} x \right] dx$

$= \dfrac{L}{2\pi} \left[\dfrac{\sin \dfrac{(m - n)\pi}{L} x}{m - n} \right.$

$\left. - \dfrac{\sin \dfrac{(m + n + 1)\pi}{L} x}{m + n + 1} \right]_0^L$

$= 0, \quad m \neq n.$

7. $f(x) \sim \dfrac{1}{2} + \dfrac{2}{\pi} \sum\limits_{n=1}^{\infty} \left\{ \dfrac{1}{n^2\pi} [(-1)^n - 1]\cos n\pi x \right.$

$\left. + \dfrac{2}{n}(-1)^n \sin n\pi x \right\}$

9. $f(x) \sim 1 - e^{-1}$

$+ 2 \sum\limits_{n=1}^{\infty} \dfrac{1 - (-1)^n e^{-1}}{1 + n^2\pi^2} \cos n\pi x$

11. $u = xye^x - ye^x + x + f(y)$

13. $u = \dfrac{1}{y} + f(y)e^{-xy}$

15. $u = c_1 e^{(c_2 x + y/c_2)}$

17. $u = \dfrac{100}{\pi} \sum\limits_{n=1}^{\infty} \dfrac{1 - (-1)^n}{n \sinh n\pi} \sinh nx \sin ny$

19. $u = \dfrac{100}{\pi} \sum\limits_{n=1}^{\infty} \dfrac{1 - (-1)^n}{n} e^{-nx} \sin ny$

Index

INDEX

Table I. Transforms of Some Basic Functions

$f(t)$	$\mathscr{L}\{f(t)\} = F(s)$
1. 1	$\dfrac{1}{s}$
2. $t^n,\ \ n = 1, 2, 3, \ldots$	$\dfrac{n!}{s^{n+1}}$
3. e^{at}	$\dfrac{1}{s - a}$
4. $\sin kt$	$\dfrac{k}{s^2 + k^2}$
5. $\cos kt$	$\dfrac{s}{s^2 + k^2}$
6. $\sinh kt$	$\dfrac{k}{s^2 - k^2}$
7. $\cosh kt$	$\dfrac{s}{s^2 - k^2}$

Table II. Operational Properties

8. $e^{at} f(t)$	$F(s - a)$
9. $f(t - a)\mathscr{U}(t - a),\ a > 0$	$e^{-as} F(s)$
10. $t^n f(t),\ \ n = 1, 2, 3, \ldots$	$(-1)^n \dfrac{d^n}{ds^n} F(s)$
11. $f^{(n)}(t),\ \ n = 1, 2, 3, \ldots$	$s^n F(s) - s^{n-1} f(0) - \cdots - f^{(n-1)}(0)$
12. $\displaystyle\int_0^t f(\tau)\,d\tau$	$\dfrac{F(s)}{s}$
13. $\displaystyle\int_0^t f(\tau) g(t - \tau)\,d\tau$	$F(s)G(s)$